PVP-Vol. 340

SEISMIC ENGINEERING

– 1996 –

presented at

THE 1996 ASME PRESSURE VESSELS AND PIPING CONFERENCE
MONTREAL, QUEBEC, CANADA
JULY 21–26, 1996

sponsored by

THE PRESSURE VESSELS AND PIPING DIVISION, ASME

edited by

M. A. SALEEM
AECL

M. L. AGGARWAL
ONTARIO HYDRO

THE AMERICAN SOCIETY OF MECHANICAL ENGINEERS
United Engineering Center / 345 East 47th Street / New York, N.Y. 10017

ISBN No. 0-7918-1787-3

Library of Congress Number 96-85630

FOREWORD

This volume will be the tenth annual publication of the Seismic Engineering Technical Committee of the ASME Pressure Vessels and Piping Division. In recognition of the ever increasing interest in this area of piping and pressure vessel design, the Seismic Engineering Technical Committee has organized 37 papers in eleven session encompassing various areas like high level dynamic response, seismic margins, seismic response control, damping and energy dissipation, soil structure interaction and developments in codes and standards. International contributions are recognized all through the volume.

M. A. Saleem
AECL, Mississauga, Ontario, Canada

M. C. Aggarwal
Ontario Hydro

CONTENTS

ADVANCED METHODS IN SEISMIC ENGINEERING

Introduction

L. H. Geraets
Tractebel Electricity & Gas International
Brussels, BELGIUM

and

J. P. Lafaille
Tractebel Energy Engineering
Brussels, BELGIUM

Increasing demand on designing for severe earthquake sites and qualification to more stringent requirements has resulted in continuing efforts to develope advanced methods in seismic engineering both in the area of analytical studies as well as experimental investigations. This chapter consists of seven papers that summarize recent developements in the area of seismic engineering technology.

Ni, Haliburton and Modarres present a new approach to evaluate the partial failure effect on current PRAs. The thermal-hydraulic analysis used in this approach is to identify partial operation effect of any PRA system function in a plant model. This methodology can be applied in the context of a full scope probability risk assessment to reduce core damage frequency.

Ni and Modarres also present an example of the application of Dynamic Master Logic Diagram with the reliability function diagram of a centrifugal pump.

G. Roussel and Ch. Cuvelliez present numerical verification of an analytical model predicting the modal crosscorrelation coefficient that takes into account the limited duration of the seismic excitation. In this study response calculations covering a range of damping values and a set of eigenfrequencies are presented, and the accuracy of the analytical model has been tested.

A detailed review of the site effects on seismic behavior of pipelines is presented by Jian-Wen Liang and Shaoping Sun in the third paper of this chapter. Liang and Sun note that pipelines are vulnerable to large ground movements, liquifaction, and irregularity of ground.

Yong Zhao presents nonlinear dynamic analysis of hydrodynamicly coupled stainless steel structures and discusses the application to design of free standing spent fuel racks submerged in water, contained in a storage pool on nuclear facility sites.

Many piping and equipment supports are unidirectional i.e. can take only tension or compression loading. Modelling of such supports usually requires expensive and time consuming time history analysis. Gurdal and Maxhum present an equivalent linear spectrum methodology for calculating the seismic response of a simple structural system with unidirectional supports.

Finally Basvaraju presets finite element evaluation of erosion/corrosion affected reducing elbow and demonstrates that FEA, though cumbersome, and time consuming is a valuable analytical tool in making critical decisions with regards to fitness for service assessment of components or replacements of border line situation cases.

The efforts of the authors and the reviewers, who contributed to bringing the state of the art technology and methods in seismic engineering, are appreciated.

PVP-Vol. 340, Seismic Engineering
ASME 1996

DEVELOPMENT OF PARTIAL FAILURE ANALYSIS METHOD IN PROBABILITY RISK ASSESSMENTS

Toni Ni
Halliburton NUS Corporation
910 Clopper Road, Gaithersburg, MD 20878

Mohammad Modarres
Center for Reliability Engineering
University of Maryland, College Park, MD 20742

Abstract

This paper presents a new approach to evaluate the partial failure effect on current PRAs. An integrated methodology of the thermal-hydraulic analysis and fuzzy logic simulation using the Dynamic Master Logic Diagram (DMLD) was developed. The thermal-hydraulic analysis used in this approach is to identify partial operation effect of any PRA system function in a plant model. The DMLD is used to simulate the system performance of the partial failure effect and inspect all minimal cut sets of system functions. This methodology can be applied in the context of a full scope probability risk assessment (PRA) to reduce core damage frequency. An example of this application of the approach is presented. The partial failure data used in the example is from a survey study of partial failure effects from the Nuclear Plant Reliability Data System (NPRDS).

Introduction

Event tree/fault tree methodology as classical modeling has been widely used in Probability Risk Assessments (PRAs) to identify the potential accident scenarios, to estimate the likelihood and consequences of accidents, and to improve plant/system safety designs and operations. This classical modeling is relative mature. However there is one of major problems of using the event tree/fault tree methodology [1]. The partial failure as degradation function of systems has been neglected since it has been believed that this effect is not significant. A system is considered to be unavailable if it is unable to operated at or above a required performance level that is usually called the success criteria, such as, one train must provide 100% of capacity of flow. No systematic approach in a PRA program has been pursued to assess the significance of system performance below 100% of capacity with a partial failure state of components.

Partial failure of a component is defined as " a failure that exhibits a partial or degraded loss of a component function" [2]. This failure mode is chosen when a component degrades to a level that, in effect, results in some loss of the ability to perform its required function although it may still be performing the function at a level below that required. Most of fluid components in nuclear power plant, such as pump, MOV, and check valve has a partial failure, that is, the component is operable, but unable to satisfy a success criterion from FSAR and Tech Spec. From a survey study, total partial failures caused by the functional degradation of all components contribute to approximately 74% of all the failure reported and completed failures only contribute approximately 26% of the total failure [3].

This paper attempts to develop a systematic procedure of partial failure analysis that combines the thermal-hydraulic analysis with RELAP5/MOD3 code and fuzzy logic simulation using the DMLD model. More specifically, partial failure analysis is developing technical bases to provide a less conservative assessment for current PRA. This method of partial failure analysis assumes that partial failures are a subset of all types of failures. As such, the probability of a partial failure can be expressed as a fraction of the probability of all types of failures. Thus, the probability associated with a specific level of system performance, in a degraded state, is easily requantified.

Methodology

The purpose of this paper is to develop an integrated approach for determining the impact of partial failures on the plant risk. The approach developed assesses the impact of partial failures on the accident sequences defined by a traditional PRA. It utilizes a plant model of thermal-hydraulic analysis to diagnostic or determine system partial operation effects based on the dominant accident sequences selected by a current PRA. Then a system DMLD model is developed to simulate the possibility of partial system operation. Finally, the physical and possibility results are integrated to requantify the dominant accident sequences with the partial failure data. The overall approach used is show in Figure 1.

Thermal-Hydraulic Model

A thermal-hydraulic code is used in the plant model to determine the sensitivity of the plant response to system partial operation. Dominant accident sequences are selected as inputs. The sequence outcomes are determined as a function of the performance level of the safety systems required. System performance levels (output flow from the system) below a success criteria or Tech Spec are tested to determine a critical system partial operation. For a sequence having significant effects of the partial system operation, this may require use of the possibility model to simulate all minimal cut sets with a partial failure state.

The Reactor Excursion and Loss-of-Coolant Analysis Program (RELAP) Version 5, Mod 3 code [4] and the Nuclear Plant Analyzer (NPA) [5] are used for the thermal-hydraulic analysis in this paper. The RELAP5/Mod 3 code is chosen specifically for its accuracy and its ability to handle cases in which the core is uncovered and approaches melting temperature. The NPA is an analysis tool that can be interactively linked to a simulation code, such as RELAP 5 to visually display data on a graphic workstation. In the NPA, experimental or calculated data are transformed through dynamic visual functions to color changes, arrows, dials, strip charts, moving graphic objects, levels, or other visual animation. Figure 2 is a typical NPA display.

The RELAP- NAP is used to evaluate plant response for each of the dominant sequences by changing different performance levels of systems from input data. The RELAP-NAP can provide both numerical and graphical results for a number of parameters including temperature, void fractions, and flow rates. These results are used to determine the critical partial operation for safety systems.

Possibility Model

If the thermal-hydraulic analysis indicates the partial failures have little impact on the sequence, then the sequence is retained in the PRA without any change. On the other hand, if the thermal-hydraulic analysis indicates large potential impact, then the sequence is requantified with partial failure analysis. This means the probability of the system operating at a given performance level. However, the probability associated with specific level of system performance, in a degraded condition, is not so easily quantified because both event tree and fault tree techniques are based on a binary logic.

Developing a partial failure evaluation model associated with the fault tree modeling could be easy adopted by the traditional PRAs. A new technique was developed that allows more accurate assessment of system unavailability as a function of performance level for the system with components that have degraded states. This technique is integrated with a DMLD model to simulate all minimal cut sets associated with a degraded state before the requalification of impacted systems.

Modarres and Hu applied fuzzy logic to evaluate system behavior through the DMLD model [6]. The DMLD is a possibility model that allows for an integration of dynamic fuzzy logic concept into the classical binary logic tree. In this concept, the physical behavior, such as partial failure and system structure are assumed to be represented by interaction among various combinations of component states. The degree of interaction is governed by the physics laws which describe this interaction. Accordingly, the logical relationships modeled in a fault tree is accompanied by a corresponding physical relationships by fuzzying such relationship. Therefore, the DMLD is used to decompose fault tree into more detail states. An example of this shown in Figure 3. In this example, assuming that full capacity of pump is 200 gpm and the current pump flow rate is 180 gpm, the Binary-logic-based GTST (Goal Tree Success Tree) will conclude full failure of the system. However the fuzzy-logic-based GTST will conclude 90% success for the system. It is obvious that the DMLD modeling represents a realist status of the system.

DMLD Modeling

A system diagram required by the classical fault tree modeling will be also used in DMLD modeling. The process involves constructing DMLDs for each component, combining these DMLDs including series and parallel structures to produce a system DMLD, and obtaining a simulation of a flow path system and performance with the partial failure state of any component. All component sets will be separated into two subsets, that is, partial failure and complete failure. These cut subsets will be simulated in the system DMLD modeling. Thus, the cut sets for a system performance level of 50% flow rate would be different from the cut sets from the same system at 80% flow rate because the fuzzy logic is used to replace the binary logic to approach three functional states of a system.

Fuzzy Operation

Different functional states of a component will be fuzzied to fuzzy sets, such as no failed, degraded, and fully failed states with a membership function. The fuzzy operations are defined in terms of arithmetic operations on these values regarded as "degrees of membership" to truth as follows [7]:

Logic	Fuzzy Value
$Z = X \cap Y \Rightarrow$	Z = min (X, Y, degree of membership)
$Z = X \cup Y \Rightarrow$	Z = max (X, Y, degree of membership)

After the minimal and maximum operation, fuzzy sets are transferred to numerical values by defuzzification equation as follows [8]:

$$y(x) = \frac{\sum_{i=1}^{m} w_i x y_i}{\sum_{i=1}^{m} w_i}$$

Where w_i is the degree of membership function of x_i to the ith fuzzy set, and y_i is the reference value of the ith fuzzy set.

Construction of the DMLD

The DMLD can be used to represent a single component or an entire system. A component DMLD can be used to associate input variables with the output that result from a fussy operation. The following component DMLD would represent the pump component shown in Figures 4 and 5. The input variables, "Flow In" and "Pump" are the flow rate into the pump suction, and the pump internal state respectively. The output variable "Flow Out" is the output flow rate. As showed in the figure 5, the flow into the pump suction is assumed to be full, partial, and none flow. Also the pump may not function, or it may be operating at a capacity of degraded or full flow.

If the input variable "Flow In" is replaced by another component, pump or valve, Figure 5 can represent a series path DMLD. Figure 6 shows that one of input variables has only the binary logic (0,1), such as a switch (on, off).

The DMLD constructs a system simulator form a series path DMLD, such as the one shown in Figure 5 and a parallel path DMLD shown in Figure 7. For example, if the both pumps, A and B in a parallel path provide unsatisfactory flow (such 50%) because of being degraded, a system still meet the total required flow. A minimal cut set PA *PB could produce four cut subsets, PA_C*PB_C, PA_C*PB_P, PA_P*PB_C, and PA_P*PB_P. The subscript c and p represent complete and partial failure.

An entire system DMLD with three functional states can be combined by several series or parallel path DMLDs. By starting at the system output that can be defined in terms of parameters that affect it directly, such as flow rate, then connecting all events that are non-component events, a system DMLD is produced that simulates the functional modes of the system. This connection is accomplished by integrating all DMLDs for the non-component events into the system DMLD. Each non-component event is replaced by the input events from and individual DMLD that has the non-component event as its out variable. From this system DMLD, all minimal cut subsets of the system can be examined to evaluate a special level of system output performance.

Application to a Sample Problem

The use of this technique for a system simulation with parallel paths will be demonstrated in the following example. The system shown in Figure 8 consisted of two pumps (PA and PB) that take suction from a common water source and pump fluid through the check valves (VA and VA) into a common header. Both pumps and check valves have partial failure states in this simple example. Figure 9 shows a DMLD to model such system with three functional states. The traditional minimal cut sets for Figure 8 are given quantified by a fault tree analysis.

Minimal Cut Sets =

$$PA*PB + PA*VB + VA*PB + VA*VB.$$

All above cut sets with a partial failure are separated into cut subsets with the 2^n rule. For example, a two-component cut set has 4 cut subsets and a three-component cut set has nine cut subsets. Minimal cut subsets for Figure 8 are listed as follows,

Minimal Cut Subsets =

$$PA_C *PB_C + PA_C *PB_P + PA_P*PB_C + PA_P *PB_P +$$
$$PA_C *VB_C + PA_C *VB_P + PA_P *VB_C + PA_P *VB_P +$$
$$VA_C *PB_C + VA_C *PB_P + VA_P *PB_C + VA_P *PB_P +$$
$$VA_C *VB_C + VA_C *VB_P + VA_P *VB_C + VA_P *VB_P .$$

Where the subscript c is a complete failure and p is a partial failure.

All cut subsets with partial failure will be tested in the system DMLD shown in Figure 9. Here the partial operation of pump and check valve is assumed 80% of component flow rate [9], that is, a full membership function of the fuzzy set for component degraded state is 80% of capacity flow. If the system thermal-hydraulic analysis shows that the system performance level of 85% of flow rate is successful, cut subsets of both partial failure will be reduced from total minimal cut subsets. If the system at 80% of flow rate below required Tech Spec provides a successful function, Minimal cut subsets will be further reduced as follows.

Minimal Cut Subsets (80% of system performance) =

$$PA_C *PB_C + PA_C *VB_C + VA_C *PB_C + VA_C *VB_C .$$

Obviously, unavailability of the system considering partial failure effects is less than traditional fault tree analysis. Since a sequence contains one or more than one system functions, This results decreasing of the total sequence frequencies.

Partial Failure Data

The objective of the data analysis is to obtain estimates of the probability of a component in a partial failure state. The data used in this analysis came from the Nuclear Plant Reliability Data System (NPRDS), a data base of the

Institute of Nuclear Power Operations (INPO) [2]. This data analysis consisted of computerized sorting of approximately 161,600 NPRDS component failure records from 35 different LWR components. 15 LWR safety and support systems most used in PRA analysis were selected, including systems used in both pressurized-water reactor (PWRs) and boiling-water reactor (BWRs). Two failure severity codes of NPRDS, Immediate Failure and Degraded Failure are classified into two functional failures: Complete Failure and Partial Failure.

Three NPRDS components, pump, valves, and motor/pneumatic-operated valves are selected based on for several reasons: (a) they tend to have high partial failure fractions, (b) they are presented in a majority of the system, and © they are used in PRA. The partial failure fractions were calculated by dividing total partial failure numbers for a particular component by the total failure numbers for that component within all operation or standby system. Table 1 exhibits different partial failure fractions for these three components. Other components with high partial fractions would be founded in the reference paper [3].

Using the same example in Figure 8, a comparison of the system unavailability with different performance levels are presented in Table 2. Assuming genetic failure rates of pump and check valve are 6.48e-4 and 1.2e-4 [10] in the mission time of 24 hours, partial failure fractions with different performance level can be used to re-evaluate system unavailability with the cut subsets.

Conclusion

The effects of partial failures can be quantified by the systematic integration approach to make a less conservative evaluation. Thermal-hydraulic analysis verifies the effects of the partial operation of system functions for dominant accident sequences. DMLD provides a possibility method for the evaluation of minimal cut sets of systems with partial failure state components. The application example has shown that partial failure caused by component degradation contributes to high percentage of unavailability of system functions at different performance levels. The methodology outlined in this paper can be adopted by current PRA program without fundamental changes of current PRA technology.

References

1. U.S. Nuclear Regulation Commission, "Severe Accident Risks: An Assessment for Five US Nuclear Power Plants", NUREG-1150, Washington, D.C. (1991)
2. NPRDS Reporting Guidance Manual, INPO (1994)
3. Ni T. & Modarres M., "Partial Failure effects of Safety Systems and Components in Light Water reactor",The 1995 Joint ASME/JSME Pressure Vessels and Piping Conference, Honolulu, Hawaii (1995)
4. Carlson K.E. et al., "RELAP5/MOD3 Code Manual", NRC, U.S.A (1993)
5. Snider, D.M. et al., "Nuclear Plant Analyzer Technical Report", EGG-EAST-9096 (1992)
6. Hu Y-S. & Modarres M., "An Introduction to Dynamic MPLD Model", Proceeding of the PSAM-II Conference, San Diego, CA (1994)
7. Zadeh, L. "Outline of a New Approach to the Analysis of Complex Systems", IEEE Trans. On Sys. Man and Cyb 3, (1973)
8. Takagi T. and Sugeno M., " Fuzzy Identification of Systems and Its Application to Modeling and Control", IEEE Trans. On Systems, Man, and Cybern., Vol. SMC-15(1) (1985)
9. Ni T., "Survey of Pump System Performance of Nuclear Power Industry", (1995)
10. U.S. Nuclear Regulation Commission, "The Reactor Safety Study", WASH-1400, Washington, D.C (1974)

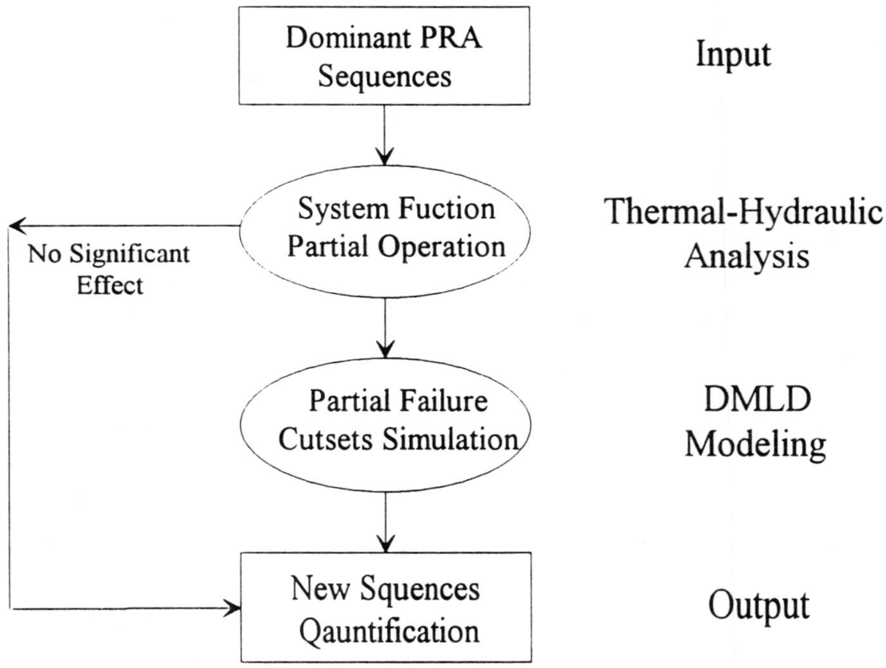

Figure 1. Methodology

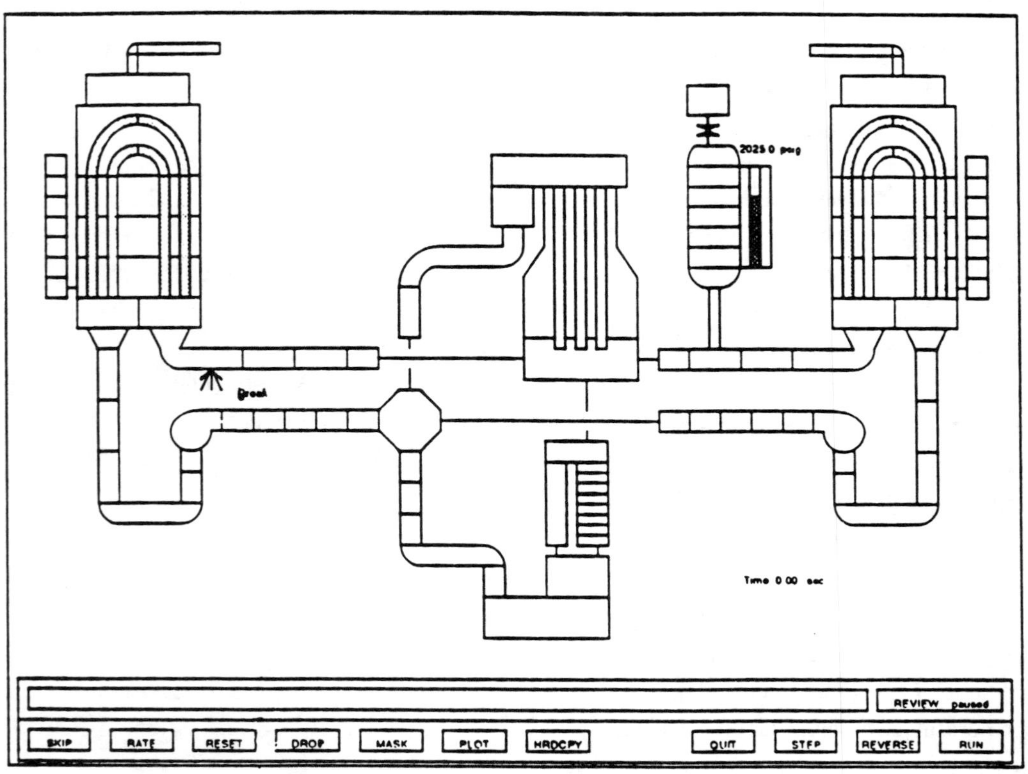

Figure 2. Typical NPA Display

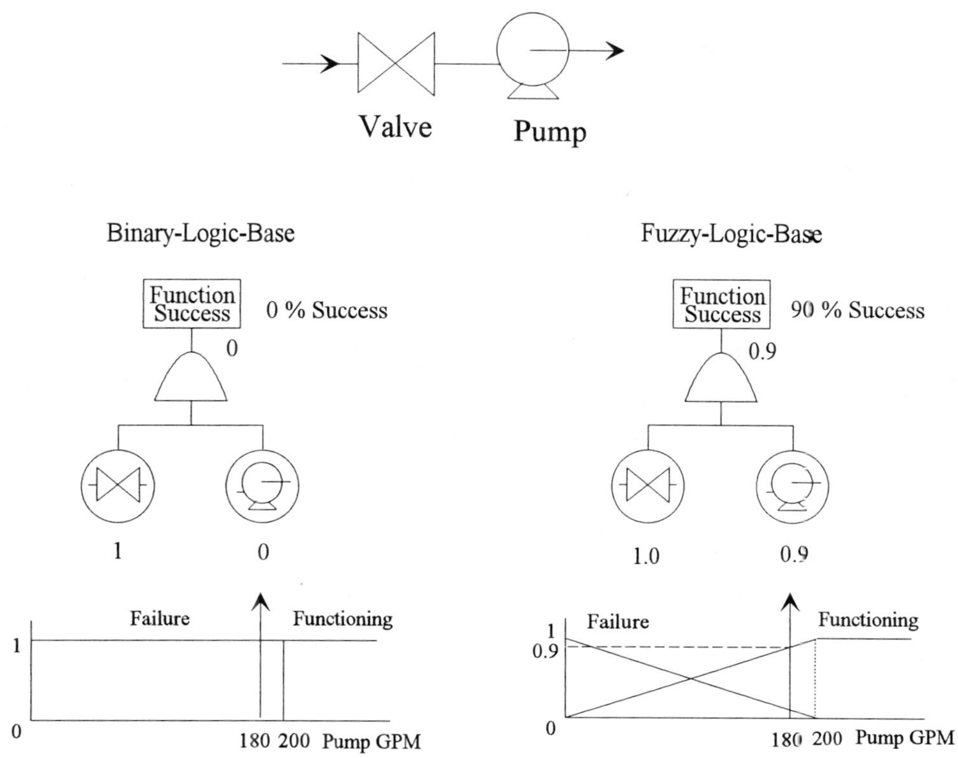

Figure 3. A Comparison Between Binary and Fuzzy Logic Base GTST

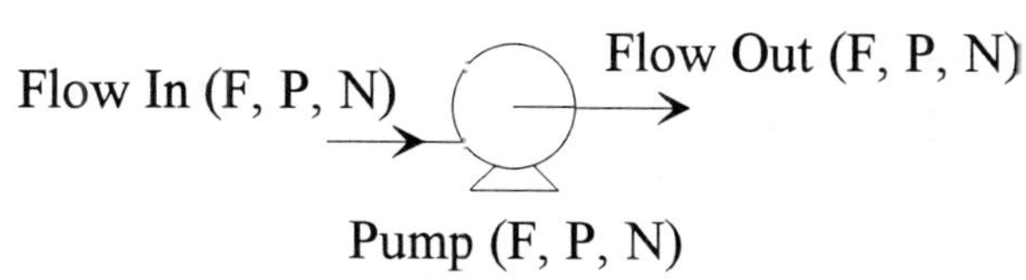

Figure 4. Simple Pump

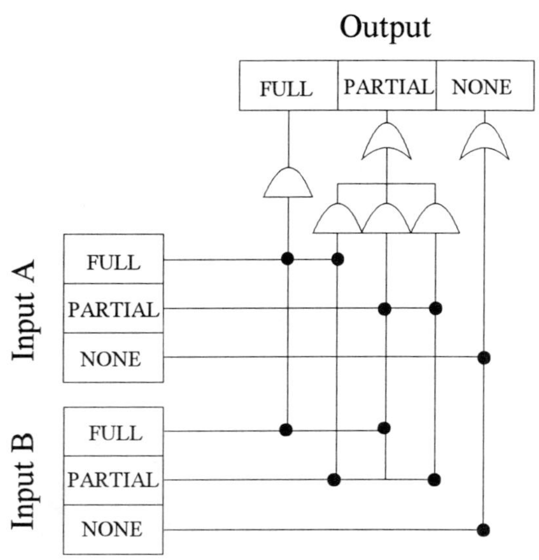

Figure 5. Series Path DMLD

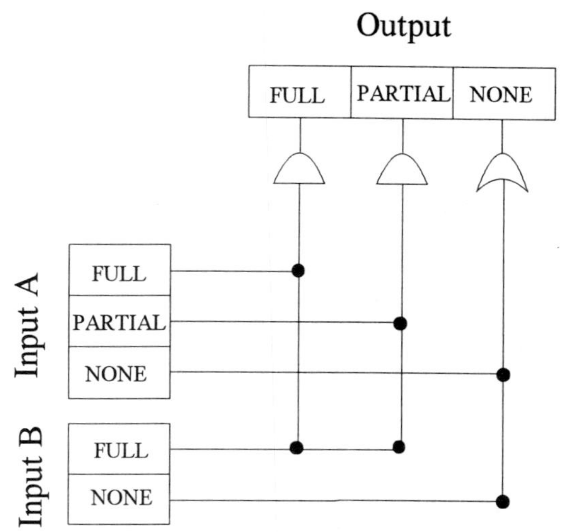

Figure 6. Series Path DMLD with One Binary Logic Component

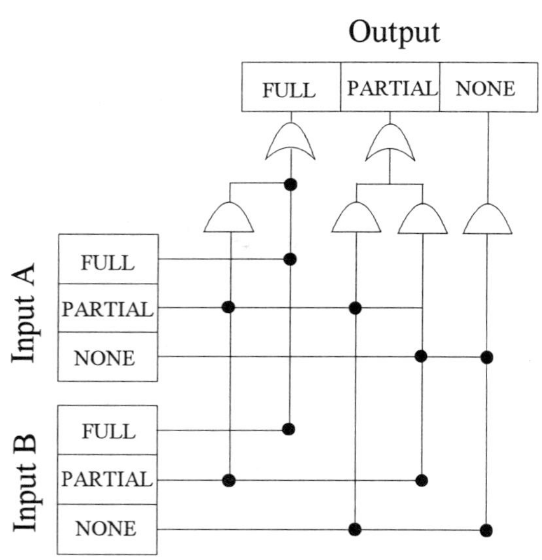

Figure 7. Parallel Path DMLD

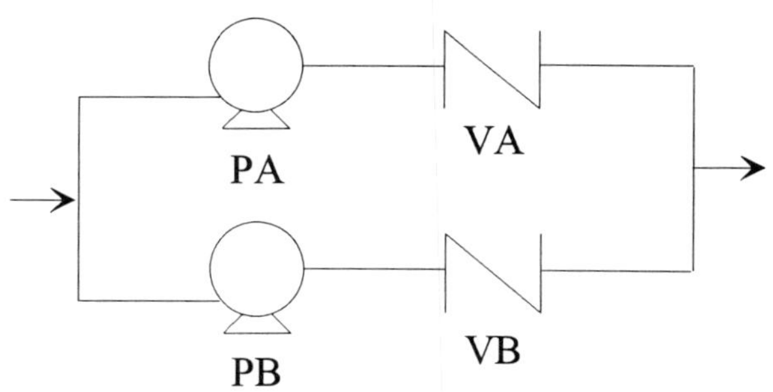

Figure 8. Parallel Pumping System

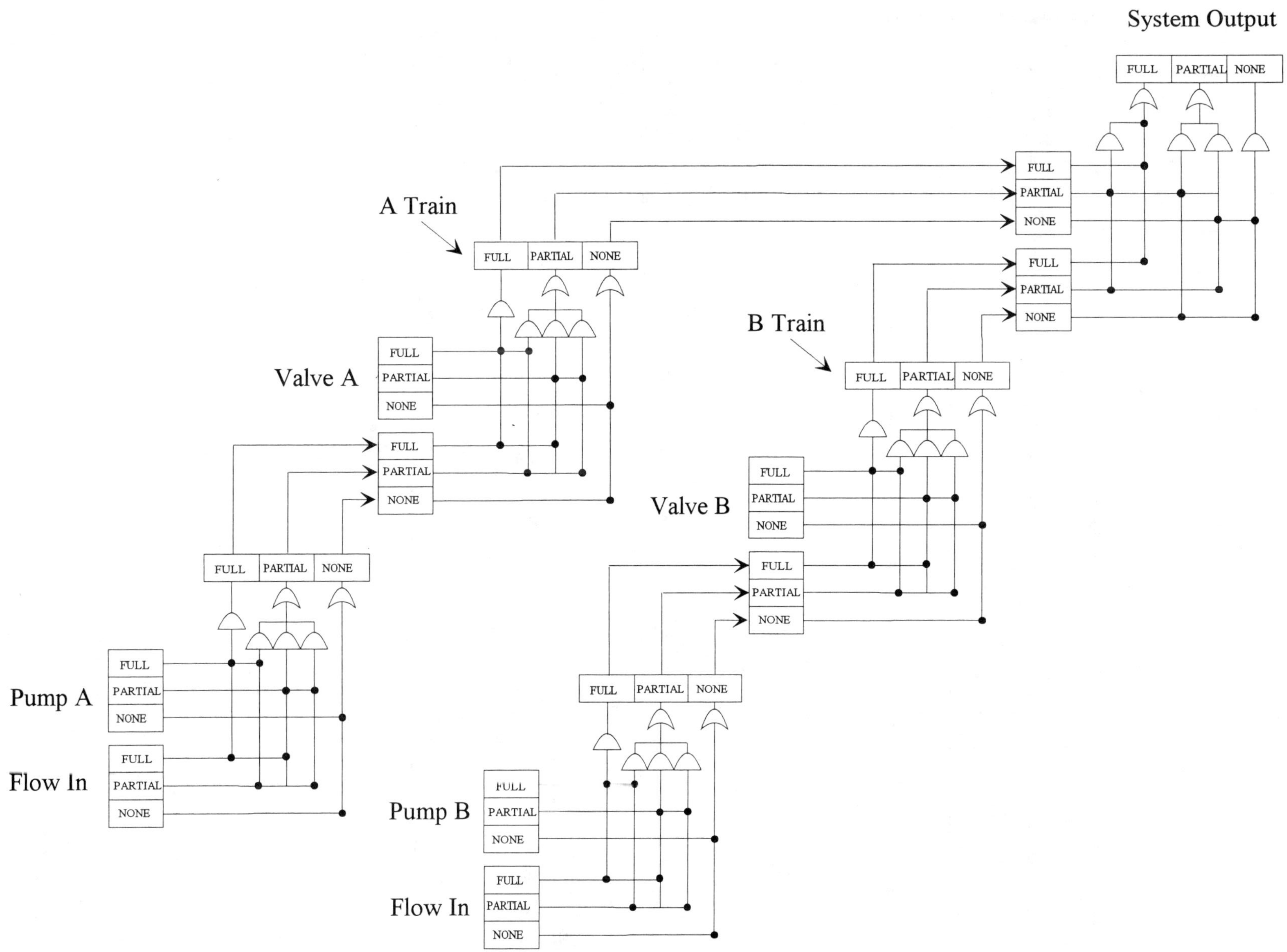

Figure 9. Parallel Pumping System DMLD

Table 1. Partial Failure Fraction

Components	Total Failure Number	Partial Failure Fraction in Operation Systems	Partial Failure Fraction in Standby Systems
Pumps	8117	90.8%	83.8%
Valves &Dampers	39310	86.9%	87.9%
Valve Operators	19457	55.4%	50.7%

Table 2. Comparison of System Unavailability with Differant Perforamnce Levels

Performance Level	100% of Flow	85% of Flow	80% of Flow
Function Unavailability	5.898E-7	1.471E-7	1.052E-8
Cut sets	PA*PB + PA*VB + VA*PB + VA*VB .	$PA_C*PB_C + PA_C*PB_P + PA_P*PB_C$ + $PA_C*VB_C + PA_C*VB_P + PA_P*VB_C$ + $VA_C*PB_C + VA_C*PB_P + VA_P*PB_C$ + $VA_C*VB_C + VA_C*VB_P + VA_P*VB_C$.	PA_C*PB_C + PA_C*VB_C + VA_C*PB_C + VA_C*VB_C .

PVP-Vol. 340, Seismic Engineering
ASME 1996

NUMERICAL VERIFICATION OF AN ANALYTICAL MODEL PREDICTING THE MODAL CROSSCORRELATION COEFFICIENT

G. Roussel and Ch. Cuvelliez
AIB-Vinçotte Nuclear
Brussels, Belgium

ABSTRACT

In the seismic analysis of linear systems using the Response Spectrum Method, the most probable maximum response is usually given by the "double sum equation". The contribution of the crosscorrelations to the response has been investigated by various authors. An analytical method was previously developed by one of the authors aiming at defining an expression of the crosscorrelation coefficient taking into account the limited duration of the seismic excitation. Numerical verification is here performed to test the accuracy of the proposed analytical method.

INTRODUCTION

The response spectrum method is extensively used in the seismic analysis and design of structures. The role of the modal crosscorrelations in calculating the response of a linear system to a seismic excitation has been studied in the past, mainly to define combination rules for modes with closely spaced frequencies.

In a previous study, G. Roussel (1993) developed an analytical model to study the effect of the earthquake duration on the modal crosscorrelation coefficient. The earthquake motion was assumed to be a nonstationary process obtained by modulating a white noise process by a deterministic time function.

Numerical verification was then performed to test the accuracy of this analytical method. The response history of a two-degree-of-freedom system was calculated for a variety of 18 earthquake records. Response calculations were performed for a set of damping ratios ξ_i and ξ_j ranging from 0.01 to 0.10 and from a set of eigenfrequencies f_i and f_j ranging from 0.1 to 20.0 Hz.

For each response history, the value of the modal crosscorrelation coefficient was then evaluated numerically and compared to the value given by the analytical model.

The results of this verification work are summarized hereafter. They show that the experimental modal crosscorrelation coeffcients are satisfactorily predicted by the analytical model.

MODAL COMBINATION

The equation of motion for an N-degree of freedom system can be transformed into a set of N independant equations using the eigenvalues and the eigenvectors of the undamped system This method for dynamic analysis is referred to as the response spectrum method.

The response history $R(t)$ is expressed as a linear combination of the response histories $R_i(t)$ for the various modes.

$$R(t) = \sum_i R_i(t) \tag{1}$$

Let us assume the system to be loaded by an earthquake ground acceleration with duration equal to t_d. The standard deviation of the response is defined as follows

$$\sigma^2 = \frac{1}{t_d}\int_0^{t_d} R^2(t)dt \qquad \sigma_i^2 = \frac{1}{t_d}\int_0^{t_d} R_i^2(t)dt \tag{2}$$

The standard deviation of the response $R(t)$ can then be written as

$$\sigma^2 = \frac{1}{t_d} \sum_i \sum_j \int_0^{t_d} R_i(t) R_j(t) dt \qquad (3)$$

Eq(3) can be then written as

$$\sigma^2 = \sum_i \sigma_i^2 + \sum_i \sum_{j \neq i} \varepsilon_{ij} \sigma_i \sigma_j \qquad (4)$$

in which ε_{ij} is the modal crosscorrelation coefficient and is defined by

$$\varepsilon_{ij} = \frac{1}{t_d} \frac{\int_0^{t_d} R_i(t) Rj(t) dt}{\sigma_i \sigma_j} \qquad (5)$$

If the earthquake is assumed to be a stationary ergodic process, L.E. Goodman *et al.* (1953) demonstrated that the maximum response of the system R_{max} can be written as

$$R_{max}^2 = \sum_i R_{i\,max}^2 + 2 \sum_i \sum_{j>i} \varepsilon_{ij} R_{i\,max} R_{j\,max} \qquad (6)$$

where $R_{i\,max}$ is the maximum response for mode i.

Eq(6) is known as the double sum equation.

If the modes are sufficiently separated the crosscorrelation coefficients can be considered as equal to zero and the double sum equation yields to the square root of the sum of the squares (SRSS) rule. L.E. Goodman *et al.* (1953) showed that the probable maximum value of response is equal to the SRSS of the modal responses.

One of the exceptions for the SRSS combination rule arises when the responses to be combined are from modes with closely spaced frequencies. An obvious case is when frequencies and dampings of the two modes are identical. In this situation the response histories of the two modes are in phase. The maximum values in the two modes occur simultaneously and they should be combined algebraically.

The effect of the close spacing of the mode is taken into account by using the modal crosscorrelation coefficient ε_{ij}.

Several authors gave expressions for ε_{ij}. Assuming earthquake ground motion to be a stationary white noise, Der Kiureghian (1981) obtained this expression for ε_{ij}.

$$\varepsilon_{ij} = \frac{\omega_{ij}^2 \sqrt{\xi_i \xi_j} \left(4\xi_{ij} + \Delta\xi_{ij} \Delta\omega_{ij}/\omega_{ij}\right)}{\Delta\Delta\omega_{ij}^2 + \omega_{ij}^2 \xi_{ij}^2} \qquad (7)$$

$$\omega_{ij} = \frac{\omega_i + \omega_j}{2} \qquad \Delta\omega_{ij} = \omega_i - \omega_j$$

$$\xi_{ij} = \frac{\xi_i + \xi_j}{2} \qquad \Delta\xi_{ij} = \xi_i - \xi_j$$

G. Roussel (1993) amended the Der Kiureghian's model and developed an analytical model to study the effect of the finite duration of the earthquake ground motion on the modal cross correlation coefficient. The theoretical model assumes a nonstationary stochastic process to represent the seismic ground acceleration. This is obtained by modulating a white noise process by the time modulating function $I(t)$ proposed by Shinozuka and Sato (1967).

$$I(t) = A\left(e^{-\alpha|t|} - e^{-\beta|t|}\right) \qquad (8)$$

in which $\beta > \alpha > 0$.

The analytical expression of the crosscorrelation coefficient ε_{ij} obtained with this model is given in Appendix A.

Numerical evaluation of the proposed model showed that, for frequencies greater than 1 Hz, the crosscorrelation coefficient could be accurately approximated by the coefficient given by the Der Kiureghian equation. For frequencies lower than 1 Hz the crosscorrelation values were found to differ significantly from the Der Kiureghian results when the ratio f_j/f_i is less than 1.5, particularly for low damping ratios ($\xi_i \cong 0.01; \xi_j \cong 0.01$).

The accuracy of the proposed model for determing the crosscorrelation coefficient ε_{ij} remained however to be tested by numerical verification.

NUMERICAL VERIFICATION

Numerical verification of the proposed model for determing the crosscorrelation coefficient ε_{ij} is performed by numerically evaluating the crosscorrelation coefficient for real earthquake records and by comparing the numerical values with the values predicted by the model.

As in the original analysis, the two-degrees-of-fredom system shown on Figure 1 is used for the numerical verification. The equation of motion is :

$$\mathbf{M}\ddot{\mathbf{X}} + \mathbf{C}\dot{\mathbf{X}} + \mathbf{K}\mathbf{X} = -\mathbf{M}\mathbf{X}_b \ddot{x}_g \qquad (9)$$

where $\mathbf{M}$, $\mathbf{C}$ and $\mathbf{K}$ are the mass, damping and stiffness matrices of the system.

$$\mathbf{M} = \begin{pmatrix} m & 0 \\ 0 & m \end{pmatrix}; \ \mathbf{K} = \begin{pmatrix} k(1+\varepsilon) & -k\varepsilon \\ -k\varepsilon & k(1+\varepsilon) \end{pmatrix}; \ \mathbf{C} = \begin{pmatrix} c(1+\gamma) & -\gamma c \\ -\gamma c & c(1+\gamma) \end{pmatrix}$$

$\mathbf{X}$ is the displacement vector, $\mathbf{X}_b$ is the static displacement vector when the support displaces by unity in the direction of the earthquake and $\ddot{x}_g$ is earthquake ground acceleration.

The matrix of the circular frequencies is :

$$\Omega = \sqrt{\frac{k}{m}} \begin{pmatrix} 1 & 0 \\ 0 & \sqrt{1+2\varepsilon} \end{pmatrix}$$

The matrix of the modal damping ratios is :

$$\xi = \frac{c}{2\sqrt{km}} \begin{pmatrix} 1 & 0 \\ 0 & \dfrac{1+2\gamma}{\sqrt{1+2\varepsilon}} \end{pmatrix}$$

For a given time history of $\ddot{x}_g$, Eq(9) can be solved by any numerical integration method. A fourth-order Runge-Kutta method was actually used to calculate the response histories $x_i(t)$ and $x_j(t)$.

Instead of calculating the crosscorrelation coefficient from the modal responses as given by Eq. (5), the coefficient ε_{ij} is calculated from the the response histories $x_i(t)$ and $x_j(t)$. The following equation is used which can be shown to be equivalent to the Eq. (5) for the specific case of a two-degrees-of-freedom system.

$$\varepsilon_{ij} = \frac{2\sigma_{x_i x_j}}{\sigma_{x_i}^2 + \sigma_{x_j}^2} \tag{10}$$

with

$$\sigma_{x_i}^2 = \frac{1}{t_d} \int_0^{t_d} x_i^2(t) dt \tag{11}$$

and

$$\sigma_{x_i x_j} = \frac{1}{t_d} \int_0^{t_d} x_i(t) x_j(t) dt \tag{12}$$

The time duration t_d of the ground motion considered in Eq. (11) and Eq. (12) is the actual duration of the earthquake record. Table 1 gives various ground motion data for the 18 earthquake records considered in the analysis.

The response histories $x_i(t)$ and $x_j(t)$ of the two-degree-of-freedom system and hence the crosscorrelation coefficient were calculated for a set of damping ratios ξ_i and ξ_j ranging from 0.01 to 0.10 and from a set of eigenfrequencies f_i and f_j ranging from 0.1 to 20.0 Hz.

The evaluation of the crosscorrelation coefficient ε_{ij} in accordance with the proposed theoretical model requires first to define the time modulating function $I(t)$ given by Eq. (8) for each of the earthquake records. The three parameters A, α and β are determined by requiring (i) the function to have its maximum equal to maximum ground acceleration, (ii) the maximum of the function to occur at the time of occurrence of the maximum ground acceleration, and (iii) the function to decay to 5 percent of its maximum value at the time when the recorded accelerogram reaches its last phase of constant low level amplitude.

The crosscorrelation coefficient is calculated for the considered sets of eigenfrequencies f_i and f_j and damping ratios ξ_i and ξ_j, using the formulae given in Appendix A.

The seismic excitation is assumed to initiate at time t_0 and to end at time $t_0 + t_S$, respectively at the times when the time modulating function reaches 5 percent of its maximum value.

RESULTS

As mentioned in the previous section, the accuracy of the proposed method is assessed by comparing the experimental and the calculated results. These results are also compared to the values obtained from the Der Kiureghian equation, Eq. (7).

Three frequency ranges are considered : 0.1-0.2 Hz, 1-2 Hz, and 10-20 Hz. Each frequency range is analyzed by fixing the value of the eigenfrequency f_i (0.1 Hz, 1 Hz and 10 Hz) and by varying the ratio f_j/f_i from 1.0 to 2.0.

For the first frequency range (0.1-0.2 Hz), numerical calculations are performed for three different sets of damping ratio value : $\xi_i = \xi_j = 0.01$; $\xi_i = 0.04$ and $\xi_j = 0.05$; $\xi_i = 0.08$ and $\xi_j = 0.10$.

For the 1-2 Hz and 10-20 Hz frequency ranges, numerical calculations are only performed for $\xi_i = \xi_j = 0.01$.

Preliminary calculations showed that for eigenfrequencies higher than 1.0 Hz, the proposed method and the Der Kiureghian equation gave results that are very close to each other and that also fit closely the experimental values. This justifies to concentrate the analysis on the first frequency range.

As a first illustration of the accuracy of the proposed method, two earthquake records are considered, El Centro SW and San Francisco Golden Gate Park WE. The analysis is made for the 0.1-0.2 Hz frequency range, assuming $\xi_i = \xi_j = 0.01$. Figures 2 to 7 summarize the results. Figures 2 and 5 show the recorded accelerograms enveloped by the adopted time modulating function $I(t)$, as given by Eq. (8). Figures 3 and 6 show a comparison between the experimental crosscorrelation values and those calculated in accordance with the proposed method. The figures also compares the experimental and calculated results to the crosscorrelation coefficients obtained from the Der Kiureghian equation. Figures 4 and 7 give the error in the combined response by using either the proposed method or the Der Kiureghian equation instead of numerically calculating the crosscorrelation coefficient from the recorded accelerogram. The error is shown for f_j/f_i ranging from 1.0 to 1.2.

The error in the combined response is defined as

$$\Delta R\,(\%) = \left(\sqrt{\frac{1+\varepsilon'_{ij}}{1+\varepsilon_{ij}^{exp}}} - 1 \right) \times 100 \tag{13}$$

where ε_{ij}^{exp} is the crosscorrelation coefficient numerically calculated from the recorded accelerogram and ε'_{ij} is the crosscorrelation coefficient calculated either by the proposed method or by the Der Kiureghian equation.

Indeed the value of the combined response is the most sensitive to any error in the value of the crosscorrelation coefficient when there are only two significant modes and when those are equal in amplitude. In this case the value of the combined response is given by

$$R^2 = 2R_i^2\left(1+\varepsilon_{ij}\right);\ R_i = R_j$$

and Eq. (13) gives the percentage change in the value of the combined response.

It can be seen from the examination of Figures 2 and 5 that the El Centro, Component SW accelerogram is satisfactorily enveloped by the time modulating function $I(t)$, which would not be apparently the case of the San Francisco, Golden Gate Park, WE component. However, Figures 3 and 6 as well as Figures 4 and 7 show that, for both ground motion accelerograms, the values of the crosscorrelation coefficient calculated by the proposed method are very close to those obtained numerically from the recorded accelerograms. Those figures also show that the proposed method gives better results than the Der Kiureghian equation in the 0.1-0.2 Hz frequency range.

The accuracy of the proposed method for determining the crosscorrelation coefficient in the frequency ranges 1.0-2.0 Hz and 10-20 Hz is illustrated by considering the El Centro SW earthquake record. The analysis is only performed for $\xi_i = \xi_j = 0.01$. Figures 8 and 9 show that the results of both the proposed method and the Der Kiureghian equation are quite identical and are very close to the experimental values.

The global assessment of the proposed method is achieved by calculating for each of the cases analyzed the error in the combined response ΔR when calculating the crosscorrelation coefficient with the proposed method. The results are summarized in Tables 2 and 3. The absolute value of the maximum error is only calculated for f_j/f_i ranging from 1.0 to 1.1 and for f_j/f_i ranging from 1.1 to 1.2. Those ranges are considered to be the ranges of concern since for higher value of f_j/f_i, the crosscorrelation terms do not contribute significantly to the response of the system.

The error in the combined response ΔR when calculating the crosscorrelation coefficient with the Der Kiureghian equation is also calculated. In order to assess how the determination of the crosscorrelation coefficient is improved by considering the limited duration of the earthquake motion, the errors in the combined response ΔR by using the proposed method or the Der Kiureghian equation are compared. Table 4 summarizes the results by giving the mean value of the maximum error ΔR for the eighteen considered accelerograms.

In the low frequency range (0.1-0.2 Hz), the proposed model gives crosscorrelation values which are very close to the numerical results, particularly for low values of the frequency ratio ($f_j/f_i < 1.1$) and low damping ratios.

Even the small negative values of the numerical results for higher values of f_j/f_i are predicted by the model.

In the medium (1.0-2.0 Hz) and high (10.0-20.0 Hz) frequency ranges, the proposed model give values of the crosscorrelation coefficient which can differ significantly from the numerical values, particularly for high values of the frequency ratio f_j/f_i where the crosscorrelation coefficient is very low and close to zero. However the error in the combined response ΔR, as defined by Eq. (13) remains lower than 25%.

In the low frequency range, the crosscorrelation values obtained from the Der Kiureghian equation differ significantly from those given by the proposed method. In the medium and high frequency ranges, the proposed model and the Der Kiureghian equation gives results very close to each other. This result shows the importance of the limited duration of the earthquake motion on the crosscorrelation value in the low frequency range (0.1-0.2 Hz). For those frequencies, the period of the eigen mode is of the same magnitude as the duration of the earthquake and the value of the crosscorrelation coefficient depends heavily on the transient response of the system.

CONCLUSIONS

The accuracy of an analytical model aiming at predicting the modal crosscorrelation coefficient when the limited duration of the earthquake motion is considered was tested by numerical experimentation. The numerical experimentation was performed by comparing the crosscorrelation values obtained from the proposed model with those calculated from the response of a two-degree-of-freedom system to real earthquake records. The model gives values in good agreement with the numerical values, particularly for low frequencies (0.1 -0.2 Hz). In this frequency range the proposed model improves significantly the Der Kiureghian model.

REFERENCES

Der Kiureghian A., 1981, "A Response Spectrum Method for Random Vibration Analyses in MDF Systems", *Earthquake Engineering and Structural Dynamics* - Vol.9, pp. 419-435

Goodman L.E., Rosenblueth E., Newmark N.M., 1953, "Aseismic Design of Elastic Structures Founded on Firm Ground", *Proceedings, ASCE,* pp.349-1, 349-27

Roussel G.,1993, "Effect of Finite Duration of the Earthquake Ground Motion on the Modal Crosscorrelation Coefficient", *Seismic Engineering - 1993*, PVP-Vol.256-1, pp.3-15, American Society of Mechanical Engineers, New York.

Shinozuka M., Sato Y., 1967, "Simulation of Non Stationary Random Process", *Proceedings, ASCE, J. Eng. Mech. Div.*, Vol.93

Table 1 : Various Ground Motion Data

No	Earthquake Record	Component	$\ddot{x}_g$ max. (m/sec²)	t_d (sec)
1	Imperial Valley, El Centro, May 18, 1940	UP	2.062	53.94
2	Imperial Valley, El Centro, May 18, 1940	SE	3.416	53.78
3	Imperial Valley, El Centro, May 18, 1940	SW	2.1	53.62
4	Friuli, Tarcento July 14, 1976	UP	-0.3305	10.60
5	Friuli, Tarcento July 14, 1976	WE	-0.4585	10.50
6	Friuli, Tarcento July 14, 1976	NS	-0.674	10.60
7	Friuli, San Rocco, Sept. 11, 1976	UP	0.1737	10.60
8	Friuli, San Rocco, Sept. 11, 1976	WE	0.6776	10.60
9	Friuli, San Rocco, Sept. 11, 1976	NS	-0.3859	10.60
10	Friuli, Tarcento May11, 1976	UP	0.1460	9.20
11	Friuli, Tarcento May11, 1976	WE	0.7675	9.20
12	Friuli, Tarcento May 11, 1976	NS	0.3909	9.20
13	Lytle Creek,Cedar Springs, Sept 12, 1970	UP	-0.9618	38.12
14	Lytle Creek,Cedar Springs, Sept 12, 1970	WE	-0.7948	38.13
15	Lytle Creek,Cedar Springs, Sept 12, 1970	NS	0.5586	38.13
16	San Francisco March 22, 1957	UP	-0.4855	39.72
17	San Francisco March 22, 1957	WE	1.1344	39.86
18	San Francisco March 22, 1957	NS	0.9653	39.86

Table 2 : Absolute value of the maximum error in the combined response ΔR (f_j/f_i ranging from 1.0 to 1.1)

Earthquake Nr	Maximum error (%) for f_j/f_i ranging from 1.0 to 1.1				
	f_i = 0.1Hz ξ_i = 0.01 ξ_j = 0.01	f_i = 0.1Hz ξ_i = 0.04 ξ_j = 0.05	f_i = 0.1Hz ξ_i = 0.08 ξ_j = 0.10	f_i = 1.0Hz ξ_i = 0.01 ξ_j = 0.01	f_i = 10.0Hz ξ_i = 0.01 ξ_j = 0.01
1	1.34	1.58	2.01	10.09	4.22
2	1.41	2.15	3.50	3.64	4.29
3	0.86	3.48	4.04	4.93	6.35
4	1.82	4.47	4.82	14.01	7.23
5	0.03	1.06	1.97	14.12	9.05
6	2.47	4.92	5.29	21.96	3.37
7	0.83	1.15	1.32	28.26	5.29
8	0.08	0.69	1.15	11.44	7.57
9	0.38	0.59	0.75	9.71	3.54
10	0.69	1.40	2.04	18.29	6.79
11	0.77	1.54	2.24	31.29	8.19
12	0.96	1.63	2.19	7.24	6.86
13	0.89	1.87	2.78	7.64	5.45
14	0.23	2.66	3.60	6.86	5.34
15	0.56	1.01	1.34	10.09	13.89
16	1.37	1.74	2.19	13.71	13.20
17	1.36	1.77	2.30	3.98	3.80
18	1.14	1.73	2.05	15.96	11.98

Table 3 : Absolute value of the maximum error in the combined response ΔR (f_j/f_i ranging from 1.1 to 1.2)

Earthquake Nr	Maximum error (%) for f_j/f_i ranging from 1.1 to 1.2				
	f_i = 0.1Hz ξ_i = 0.01 ξ_j = 0.01	f_i = 0.1Hz ξ_i = 0.04 ξ_j = 0.05	f_i = 0.1Hz ξ_i = 0.08 ξ_j = 0.10	f_i = 1.0Hz ξ_i = 0.01 ξ_j = 0.01	f_i = 10.0Hz ξ_i = 0.01 ξ_j = 0.01
1	6.06	2.07	1.88	4.46	1.69
2	1.41	4.30	4.92	9.39	4.16
3	6.09	4.54	4.31	2.60	6.76
4	2.30	9.73	9.51	13.34	8.03
5	0.34	1.61	2.74	14.05	6.12
6	3.33	10.64	10.30	11.92	1.12
7	0.83	1.14	1.31	23.07	1.04
8	0.89	0.63	1.12	15.70	3.03
9	0.41	0.59	0.70	13.88	3.11
10	2.31	1.32	2.03	17.58	6.07
11	1.44	2.32	3.13	31.14	3.64
12	0.95	1.63	2.21	6.82	4.01
13	5.16	2.95	3.39	8.77	5.00
14	3.72	3.58	4.13	7.35	2.56
15	6.51	7.85	5.78	6.83	7.11
16	10.21	4.93	3.33	12.83	13.11
17	7.67	4.78	3.62	2.05	3.74
18	1.13	1.47	1.05	15.70	5.39

Table 4 : Comparison of the error in the combined response ΔR, Der Kiureghian, Present work.

	Mean value of the absolute value of the maximum error (%) for f_j/f_i ranging from 1.0 to 1.1				
	$f_i = 0.1\,Hz$, $\xi_i = 0.01$, $\xi_j = 0.01$	$f_i = 0.1\,Hz$, $\xi_i = 0.04$, $\xi_j = 0.05$	$f_i = 0.1\,Hz$, $\xi_i = 0.08$, $\xi_j = 0.10$	$f_i = 1.0\,Hz$, $\xi_i = 0.01$, $\xi_j = 0.01$	$f_i = 10.0\,Hz$, $\xi_i = 0.01$, $\xi_j = 0.01$
Der Kiureghian	23.49	7.80	2.61	13.10	7.00
Present work	0.96	1.97	2.53	13.22	7.02

	Mean value of the absolute value of the maximum error (%) for f_j/f_i ranging from 1.1 to 1.2				
	$f_i = 0.1\,Hz$, $\xi_i = 0.01$, $\xi_j = 0.01$	$f_i = 0.1\,Hz$, $\xi_i = 0.04$, $\xi_j = 0.05$	$f_i = 0.1\,Hz$, $\xi_i = 0.08$, $\xi_j = 0.10$	$f_i = 1.0\,Hz$, $\xi_i = 0.01$, $\xi_j = 0.01$	$f_i = 10.0\,Hz$, $\xi_i = 0.01$, $\xi_j = 0.01$
Der Kiureghian	20.46	11.02	5.66	11.10	4.75
Present work	3.38	3.67	3.64	12.08	4.76

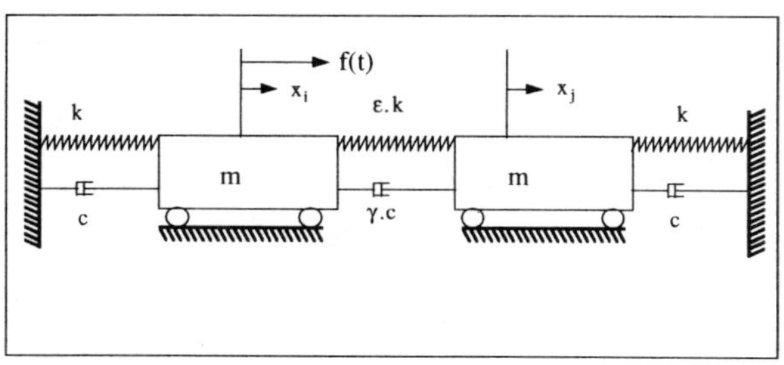

Figure 1: Two-Degrees of Freedom System Used in the Calculation of the Crosscorrelation Coefficient

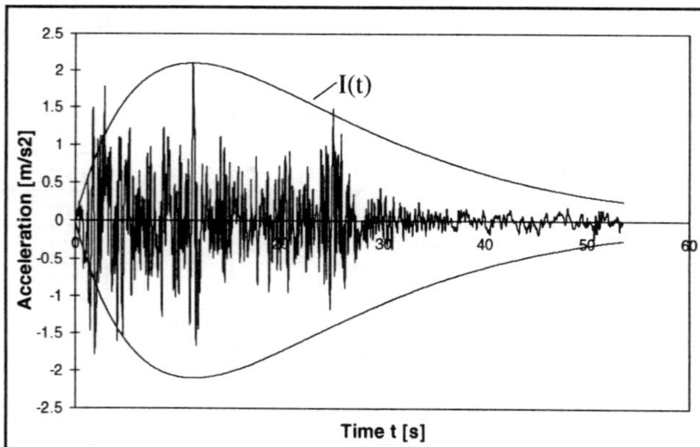

Figure 2: Ground Motion Time History, El Centro Record of May 18, 1940, Component SW

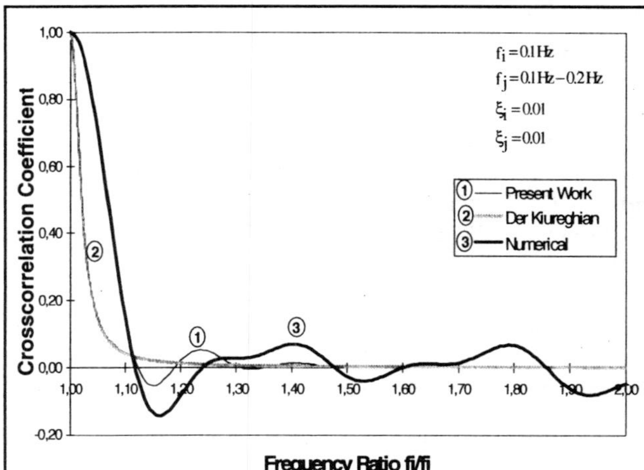

Figure 3: El Centro Earthquake of May 18, 1940, Component SW. Comparison of Crosscorrelation Coefficient, Numerical, Der Kiureghian, Present Work

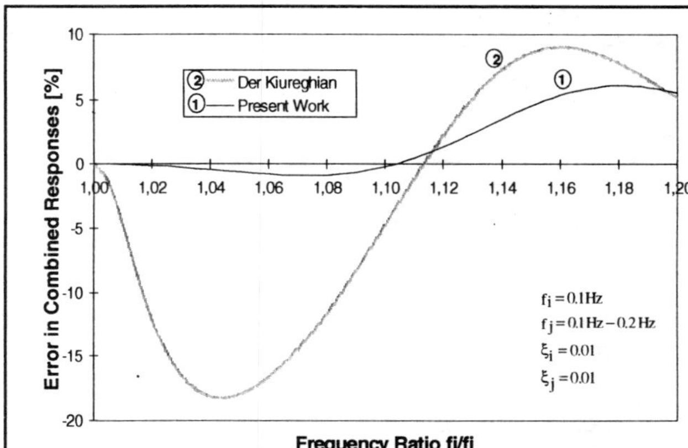

Figure 4: El Centro Earthquake of May 18, 1940, Component SW. Comparison of Error in the Combined Responses, Present Work, Der Kiureghian

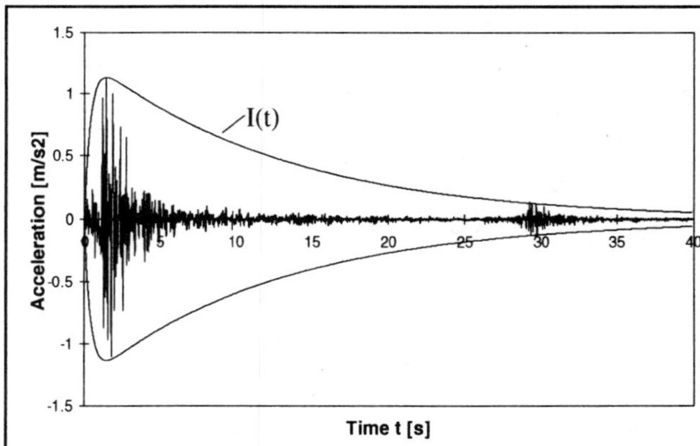

Figure 5: Ground Motion Time History, San Fransisco, Golden Gate Park Record of March 22, 1957, Component WE

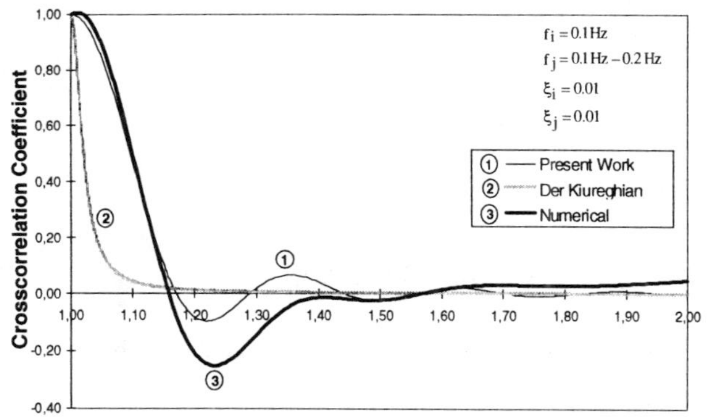

Figure 6: San Francisco, Earthquake of March 22, 1957, Component WE. Comparison of Crosscorrelation Coefficient, Numerical, Der Kiureghian,Present Work

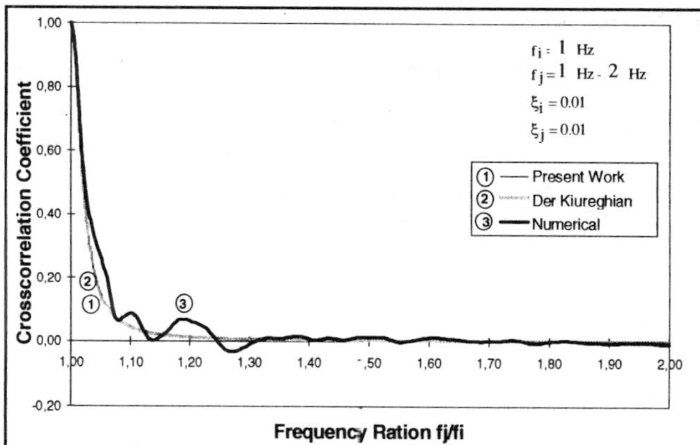

Figure 8: El Centro Earthquake of May 18,1940, Component SW. Comparison of Crosscorrelation Coefficient, Numerical, Der Kiureghian, Present Work

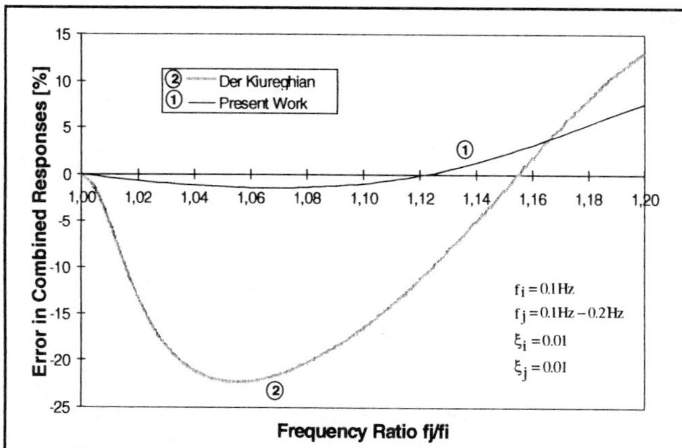

Figure 7: San Francisco, Earthquake of March 22,1957, Component WE. Comparison of Error in the Combined Responses, Present Work, Der Kiureghian

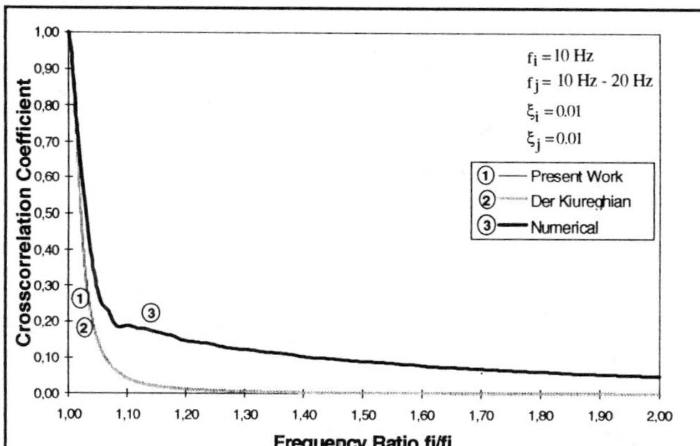

Figure 9: El Centro Earthquake of May 18, 1940, Component SW. Comparison of Crosscorrelation Coefficient, Numerical, Der Kiureghian, Present Work

APPENDIX A

In this Appendix, the cross correlation coefficient ε_{ij} is given. The strong motion part of the seismic excitation is assumed to iniate at time t_0 and to end at time t_0+t_S. The crosscorrelation coefficient is referred to as $\varepsilon_{ij}(t_0, t_0+t_S)$ and is given by

$$\varepsilon_{ij}(t_0,t_0+t_S)=\frac{B_{ij}(t_0,t_0+t_S)}{B_{ii}(t_0,t_0+t_S)+B_{jj}(t_0,t_0+t_S)}$$

with

$$B_{ij}(t_0,t_0+t_S)=B^*_{ij}(t_0+t_S)-B^*_{ij}(t_0)$$

$$B_{ii}(t_0,t_0+t_S)=B^*_{ii}(t_0+t_S)-B^*_{ii}(t_0)$$

$$B_{jj}(t_0,t_0+t_S)=B^*_{jj}(t_0+t_S)-B^*_{jj}(t_0)$$

$B^*_{ij}(t)$ can be expressed as

$$B^*_{ij}(t)=B^*_{ij,2\alpha}(t)+B^*_{ij,2\beta}(t)-\ 2B^*_{ij,\alpha+\beta}(t) \qquad \text{(A1)}$$

where $B^*_{ij,\delta}(t)$ is given, for $\delta=2\alpha,\ 2\beta$ and $\alpha+\beta$, by

$$B^*_{ij,\delta}(t)=8\frac{(\omega_i\xi_i+\omega_j\xi_j-\delta)}{C_{ij,\delta}^2+D_{ij,\delta}^2}\left\{-\frac{1}{\delta}e^{-\delta t}\right\}+8\frac{(\omega_i\xi_i+\omega_j\xi_j-\delta)}{C_{ij,\delta}^2+D_{ij,\delta}^2}\cdot\frac{1}{2}$$

$$\Bigg\{\left[\left(-1+\frac{\omega_{D,i}}{\omega_{D,j}}\right)-\left(\frac{C_{ij,\delta}}{D_{ij,\delta}}\right)\frac{\omega_i\xi_i+\omega_j\xi_j-\delta}{\omega_{D,j}}\right]\frac{e^{-(\omega_i\xi_i+\omega_j\xi_j)t}}{(\omega_i\xi_i+\omega_j\xi_j)^2+(\omega_{D,i}+\omega_{D,j})^2}\left[-(\omega_i\xi_i+\omega_j\xi_j)\cos((\omega_{D,i}+\omega_{D,j})t)+(\omega_{D,i}+\omega_{D,j})\sin((\omega_{D,i}+\omega_{D,j})t)\right]+$$

$$\left[\left(-1-\frac{\omega_{D,i}}{\omega_{D,j}}\right)+\left(\frac{C_{ij,\delta}}{D_{ij,\delta}}\right)\frac{\omega_i\xi_i+\omega_j\xi_j-\delta}{\omega_{D,j}}\right]\frac{e^{-(\omega_i\xi_i+\omega_j\xi_j)t}}{(\omega_i\xi_i+\omega_j\xi_j)^2+(\omega_{D,i}-\omega_{D,j})^2}\left[-(\omega_i\xi_i+\omega_j\xi_j)\cos((\omega_{D,i}-\omega_{D,j})t)+(\omega_{D,i}-\omega_{D,j})\sin((\omega_{D,i}-\omega_{D,j})t)\right]+$$

$$\left[\left(\frac{C_{ij,\delta}}{D_{ij,\delta}}\right)\left(1-\frac{\omega_{D,i}}{\omega_{D,j}}\right)-\frac{\omega_i\xi_i+\omega_j\xi_j-\delta}{\omega_{D,j}}\right]\frac{e^{-(\omega_i\xi_i+\omega_j\xi_j)t}}{(\omega_i\xi_i+\omega_j\xi_j)^2+(\omega_{D,i}+\omega_{D,j})^2}\left[-(\omega_i\xi_i+\omega_j\xi_j)\sin((\omega_{D,i}+\omega_{D,j})t)-(\omega_{D,i}+\omega_{D,j})\cos((\omega_{D,i}+\omega_{D,j})t)\right]+$$

$$\left[\left(\frac{C_{ij,\delta}}{D_{ij,\delta}}\right)\left(1+\frac{\omega_{D,i}}{\omega_{D,j}}\right)+\frac{\omega_i\xi_i+\omega_j\xi_j-\delta}{\omega_{D,j}}\right]\frac{e^{-(\omega_i\xi_i+\omega_j\xi_j)t}}{(\omega_i\xi_i+\omega_j\xi_j)^2+(\omega_{D,i}-\omega_{D,j})^2}\left[-(\omega_i\xi_i+\omega_j\xi_j)\sin((\omega_{D,i}-\omega_{D,j})t)-(\omega_{D,i}-\omega_{D,j})\cos((\omega_{D,i}-\omega_{D,j})t)\right]\Bigg\}$$

where

$$C_{ij,\delta}=\left(\omega_i^2-\omega_j^2\right)-2\,(\omega_i\xi_i-\delta)(\omega_i\xi_i-\omega_j\xi_j)-\delta^2$$

$$D_{ij,\delta}=\left(\omega_i\sqrt{1-\xi_i^2}\right)\left((\omega_i\xi_i-\omega_j\xi_j-\delta)\right)$$

$$\omega_{D,i}=\omega_i\sqrt{1-\xi_i^2}$$

$B^*_{ii}(t)$ and $B^*_{jj}(t)$ are obtained from Eq. (A1) by using the relationships:

$$B^*_{ii}(t)=\frac{1}{2}B^*_{ij}(t) \qquad \text{with } j=i$$

$$B^*_{jj}(t)=\frac{1}{2}B^*_{ij}(t) \qquad \text{with } i=j$$

The expression for $B^*_{ij,\delta}(t)$ can be more concisely written as follows

$$B^*_{ij,\delta}(t) = 8\frac{(\omega_i\xi_i + \omega_j\xi_j - \delta)}{C_{ij,\delta}^2 + D_{ij,\delta}^2}\left\{-\frac{1}{\delta}e^{-\delta t}\right\} + 8\frac{(\omega_i\xi_i + \omega_j\xi_j - \delta)}{C_{ij,\delta}^2 + D_{ij,\delta}^2}\cdot\frac{1}{2}$$

$$\left\{\sum_{+,-}\left[\left(-1 \pm \frac{\omega_{D,i}}{\omega_{D,j}}\right) \mp \left(\frac{C_{ij,\delta}}{D_{ij,\delta}}\right)\frac{\omega_i\xi_i + \omega_j\xi_j - \delta}{\omega_{D,j}}\right]\frac{e^{-(\omega_i\xi_i + \omega_j\xi_j)t}}{(\omega_i\xi_i + \omega_j\xi_j)^2 + (\omega_{D,i} \pm \omega_{D,j})^2}\left[-(\omega_i\xi_i + \omega_j\xi_j)\cos((\omega_{D,i} \pm \omega_{D,j})t) + (\omega_{D,i} \pm \omega_{D,j})\sin((\omega_{D,i} \pm \omega_{D,j})t)\right] + \right.$$

$$\left.\left[\left(\frac{C_{ij,\delta}}{D_{ij,\delta}}\right)\left(1 \mp \frac{\omega_{D,i}}{\omega_{D,j}}\right) \mp \frac{\omega_i\xi_i + \omega_j\xi_j - \delta}{\omega_{D,j}}\right]\frac{e^{-(\omega_i\xi_i + \omega_j\xi_j)t}}{(\omega_i\xi_i + \omega_j\xi_j)^2 + (\omega_{D,i} \pm \omega_{D,j})^2}\left[-(\omega_i\xi_i + \omega_j\xi_j)\sin((\omega_{D,i} \pm \omega_{D,j})t) \pm (\omega_{D,i} \pm \omega_{D,j})\cos((\omega_{D,i} \pm \omega_{D,j})t)\right]\right\}$$

or

$$B^*_{ij,\delta}(t) = 8\frac{(\omega_i\xi_i + \omega_j\xi_j - \delta)}{C_{ij,\delta}^2 + D_{ij,\delta}^2}\left\{-\frac{1}{\delta}e^{-\delta t}\right\} + 8\frac{(\omega_i\xi_i + \omega_j\xi_j - \delta)}{C_{ij,\delta}^2 + D_{ij,\delta}^2}\cdot\frac{1}{2}$$

$$\left\{\sum_{+,-}\sum_{n=0}^{1}\left[1 + \left(\frac{C_{ij,\delta}}{D_{ij,\delta}}\right) - \left(\frac{C_{ij,\delta}}{D_{ij,\delta}}\right)^n\right]\right.$$

$$\times\left[\left(\frac{C_{ij,\delta}}{D_{ij,\delta}}\right)^{(-1)^{n+1}}\left((-1)^{-n+1} \pm (-1)^n\frac{\omega_{D,i}}{\omega_{D,j}}\right) \mp \frac{\omega_i\xi_i + \omega_j\xi_j - \delta}{\omega_{D,j}}\right]\frac{e^{-(\omega_i\xi_i + \omega_j\xi_j)t}}{(\omega_i\xi_i + \omega_j\xi_j)^2 + (\omega_{D,i} \pm \omega_{D,j})^2}$$

$$\left.\times\left[-(\omega_i\xi_i + \omega_j\xi_j)\cos\left(\left(\frac{n\pi}{2}\right) + (-1)^n(\omega_{D,i} \mp \omega_{D,j})t\right) \pm (\omega_{D,i} \pm \omega_{D,j})\sin\left(\left(\frac{n\pi}{2}\right) + (-1)^n(\omega_{D,i} \pm \omega_{D,j})t\right)\right]\right\}$$

PVP-Vol. 340, Seismic Engineering
ASME 1996

SITE EFFECTS ON SEISMIC BEHAVIOR OF PIPELINES

Jian–Wen Liang
Department of Civil Engineering
Tianjin University
Tianjin 300072
China

Shaoping Sun
Beijing Municipal Engineering Research Institute
Beijing 100037
China

ABSTRACT

Post–earthquake investigations and earthquake observations have shown that the site conditions have spectacular effects on seismic behavior of pipelines. In this paper, an updated and detailed review of the site effects on seismic behavior of pipelines is presented, and research needs are pointed out.

INTRODUCTION

Buried gas, oil, water and sewer pipelines have been damaged heavily by earthquakes, such as Kanto 1923, Niigata 1964, San Fernando 1971, Haicheng 1975, Tangshan 1976, Miyagiken–Oki 1978, Loma Prieta 1989, Kobe 1995, and others. Because of the importance of lifelines to operation of cities, lifeline earthquake engineering has attracted the attention of researchers (Ariman and Muleski, 1981; Sun, 1992; Sun, 1994; Liang, 1995e; Liang, 1995f).

Buried pipelines are restrained by the surrounding soil, and the seismic behavior mainly depends on the soil conditions, which is quite different from that of the above ground structures whose damages are due to horizontal inertial force.

Many post–earthquake investigations (Sun, 1979; Kubo and Isoyama, 1980; Ariman and Muleski, 1981; Satake et al, 1982; Iwamoto, 1995) have indicated that site conditions have effects on seismic behavior of pipelines, and almost all earthquake damages to buried pipelines are found in irregular sites or due to large ground movements. Earthquake observations (Nishio, 1987) and theoretical studies (Nishio and Satake, 1983; Chen et al, 1980; Lee et al, 1980) have shown that a seismic wave cannot damage a pipeline which is buried in a homogeneous site unless it is near the epicenter where the amplitude of the seismic wave is very large. Therefore, it is a very important task to evaluate the site effects on seismic behavior of pipelines.

The aim of this paper is to review the effects of site conditions on seismic behavior of pipelines. For this purpose the observed damage, observed behavior, quantitative analysis, analytical study and experimental investigation are briefly discussed. It is hoped that this work could serve to stimulate discussion and interest on the problem.

QUANTITATIVE ANALYSIS OF DAMAGES

Site effects on seismic behavior of pipelines were firstly recognized by statistical analysis of earthquake damages of buried pipelines.

Okamoto (1973) pointed out that there were great differences in the local distribution between damage to wooden buildings and damage to water pipelines during 1923 Kanto earthquake. Damage to wooden buildings was shown to be concentrated in areas consisting of basins and alluvial plains where extremely soft and deep surface layers developed. In contrast, damage to pipelines was concentrated in the marginal areas of basins where the composition of the ground changes from alluvium to diluvium, creating complex geological conditions. He attributed the major cause of pipeline damage to uneven ground motion associated with the ground condition complexity.

Kubo et al (1975) and Katayama et al (1977) proposed a classification of ground conditions in a quantitative relation to the damage ratio of water pipelines during 1923 Kanto earthquake. They found that the seismic damages of pipelines were significantly affected by the degree of complexity of ground, and the damage ratio reached the highest in the transient zone from the loam layer to the soft alluvial layer, but not in the area of soft alluvial ground.

Ichihara and Yamada (1981) quantitatively analyzed the damages of water pipelines during 1923 Kanto earthquake, too, and they found that geological condition was the most important factor contributing to the damages except seismic intensity.

Sun (1979; 1983) pointed out that site condition was more

Table 1. Damage Ratio of Water Pipes in 1975 Haicheng Earthquake

Location	Seismic Intensity	Site	Pipeline Length	Number ofDamage	Damage Ratio
Anshan	7 °	II	537.40	3	0.006 / Km
Panshan	7 °	III	25.90	35	1.60 / Km
Yingkou	8 °	III	158.50	372	2.35 / Km
Yingkou	9 °	I ,II	26.10	26	1.00 / Km
Haicheng	9 °	III	21.35	216	10.00 / Km

Table 2. Damage Ratio of Water Pipes in 1976 Tangshan Earthquake

Location	Seismic Intensity	Site	Damage Ratio	Note
Tianjin	7–8 °	III	0.18 / km	
Tanggu	8 °	III	4.18 / km	The site soil worse than that in Tianjin
Hangu	9 °	III	10.00 / km	The site soil worse than that in Tanggu
Tangshan	10–11 °	II	4.00 / km	

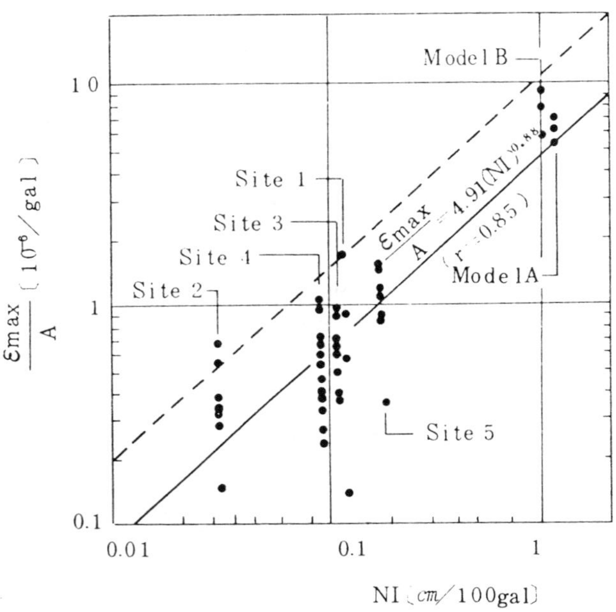

FIGURE 1. RELATIONSHIP BETWEEN *NI* AND MAXIMUM STRAIN IN A GROUND (Nishio, 1989b)

an important factor than seismic intensity after statistically analyzing the pipeline damages in 1975 Haicheng earthquake (Table 1) and 1976 Tangshan earthquake (Table 2).

Based on earthquake observations at five locations (Nishio et al, 1980; 1985; 1988; Tsukamoto et al, 1984) and model experiments (Nishio et al, 1983), Nishio and his colleagues established a close relationship between the nonuniformity of ground and seismic damage of pipelines (Nishio, 1987; 1989a; 1989b; 1992a; 1992b; 1994). He defined the degree of ground nonuniformity by "Nonuniformity Index" (NI). The relation between NI and maximum strain in a ground (of course in pipeline) is shown in Figure 1.

Kitaura and Miyajima (1988) presented statistical analysis of damage to pipelines induced by soil liquefaction, and noticed that remarkablely high damage ratios of pipelines were located at liquefied ground near the boundary between the liquefied and non–liquefied ground, and the damage was due to sharp change of ground characteristics at the boundary.

CLASSIFICATION OF SITE CONDITIONS

The site conditions were classified by different methods (Kubo et al, 1975; Sun, 1979; Nishio, 1987).

In this paper, site conditions are classified into irregular site and large ground movement, which is based on the damage mechanism of pipelines.

Large ground movement, or permanent ground deformation, can be caused by fault movement, subsidence, landslide, lateral spreading, etc. during or after earthquake. Pipeline damage is due to unability to resist the large ground movements, which may be regarded as a static problem.

Irregular site is the site with nonuniform topographies or subsurfaces, e.g., two–soil site, three–soil site, non–causative fault, cut–and–fill site, alluvial valley, riverbed, etc. Pipeline damage is due to amplification of seismic waves, which is a dynamic soil–pipeline interaction problem.

Liquefaction is a special phenomenon in which the stiffness of ground gradually reduces to zero during earthquake. Liquefaction can lead to large ground movement such as later spreading; and on the other hand, it can also be regarded as an irregular site.

LARGE GROUND MOVEMENT

OBSERVED DAMAGE

Large ground movements are the principal cause of pipeline damages in severe earthquakes.

For example, in 1971 San Fernando earthquake, there were 450 breaks in the underground pipelines in the area where severe damages occurred to residences, and this amounted to an average of one break per 1.5 miles of the pipelines in the area, and all but two of these breaks were at locations of sharp vertical or lateral dislocations or ruptures of ground. At locations where there was severe ground displacement but no ground ruptures the pipelines yielded but did not break. Another severe damage was sustained by 16–in transmission line. There were 52 separate breaks in an approximately 6–mile length of this transmission line. A number of sections were buckled when crossing faults (Ariman and Muleski, 1981). In addition, the earthquake triggered over 1000 landslides which caused severe damages to water and gas transmission lines (Ariman, 1984)

Similarly, the 1972 Managua earthquake caused ground displacement along four prominent strike–slip faults through the downtown area, and nearly all water mains crossing the faults ruptured (O'Rourke and Trautmann, 1980).

In 1976 Tangshan earthquake, all the pipelines crossing faults in Tangshan City were heavily damaged. Some riverbanks slided, and almost all pipelines in the slide area were seriously damaged. In Tangshan City, there were excavated hollow zones by mining, and most of the cast iron pipelines through these areas suffered serious damages. In Tianjin City, there were many backfill areas, pipelines through these areas were heavily damaged due to subsidence. (Sun, 1979).

In 1995 Kobe earthquake, many water pipeline damages were found in the place of subsidence and ground rupture (Iwamoto, 1995).

FIGURE 2. COMPRESSION FORCE BUCKLED A 16–INCH STEEL PIPE (Ariman, 1984)

FIGURE 3. BUCKLING FAILURE OF A WATER MAIN (Ariman, 1984)

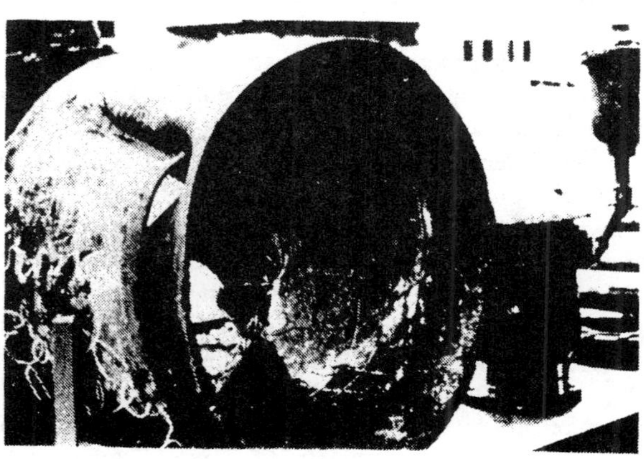

FIGURE 4. RUPTURE FAILURE OF A STEEL WATER MAIN (Ariman, 1984)

OBSERVED BEHAVIOR

O'Rourke and his colleagues (1980; 1984) examined the effects of large ground movements on buried pipelines based on ground displacement observations and records of pipeline damage during 1971 San Fernando earthquake. Some conclusions were obtained:

1. Pipelines were particularly vulnerable to local concentrations of permanent differential movements;
2. The most important parameters relating to ground movement included the distribution of displacement within the fault zone and the angle of pipeline / fault intersection;
3. Pipeline failure was also related closely to the pattern of ground displacement;
4. The weld quality could have a significant effect on the pipeline's ability to sustain permanent ground movement.

Ariman (1984) overviewed buckling and rupture failures of pipelines due to large ground movements in 1971 San Fernando earthquake. He found that shell–mode buckling failure was particularly dominent when crossing fault, and usually led rupture as well as crushing and breaking of the pipes, which could be clearly evidenced by Figure 2 to 4.

PIPELINE RESPONSE

FAULT MOVEMENT. As one of the most common large ground movements, fault movement effect on the response of pipelines has been paid attention by many researchers.

The earliest study was done by Newmark and Hall (1975) for the ability of a pipe to resist large fault displacement, and some conclusions were obtained on the basis of the numerical calculation:

1. The smaller the frictional force on the pipe, the greater the capability for surviving fault motion;
2. The lower the grade of steel, the more yielding could take place, and hense, the greater the capability;
3. Other things being equal, fault motion capability increased slightly with increasing thickness of pipe wall;

4. Pipeline should be placed in a relatively shallow trench and lifted bodily out of trench when fault motion occurred.

These conclusions were used for design of the Trans–Alaska pipelines, and this method later was extented by Kennedy et al (1977), Wang and Yeh (1985) and Wang and Wang (1995).

O'Rourke and Trautmann (1981) examined the various patterns of permanent differential ground movement caused by earthquakes, and a general method for analyzing the response of jointed pipelines to strike–slip displacement was developed to evaluate the influence of different couplings on pipeline performance.

Experimental study on the behavior of pipelines subjected to large ground deformation was conducted by Takada (1984). He found that joint rotation characteristics and joint arrangements were much more important factors than extraction for such as uplift and downward large deformation.

Ariman and Lee (1991) dealt with the tensile failures of buried pipelines subjected to abrupt fault movement, and pointed out that pipeline should be buried in a relatively shallow depth.

Performance of pipelines to strike–slip and reverse strike–slip fault was analyzed by Vougioukas and his colleagues (1991; 1992). The influence of parameters such as angle of pipe crossing the fault, geometric characteristics of pipe were presented and the ductility demands were calculated.

Buckling is a principal damage of pipeline crossing fault, which is due to compressive force by fault movement.

Kyriakides and his colleagues (Kyriakides et al, 1983; Yun and Kyriakides, 1990) identified and studied the beam and shell modes of buckling of buried pipelines due to large ground movements, and the studies showed that pipeline behaved basically as a beam and bended it self out of the ground in the beam–mode buckling, and the shell–mode buckling was a more localized failure due to combined axial compression and bending, and usually led to rupture of the pipe. The buckling failures were influenced by the pipe diameter, its radius / thickness ratio, the weight of its contents, the inelastic characteristics of the pipe material, the depth of the trench in which the pipe was buried, the flatness of the trench bottom, and the small geometric imperfections present in the pipe.

Lee and Ariman (1985) did some research on buckling of pipelines crossingan active fault and suggested that:

1. Pipeline should be in tension when crossing strike–slip fault;
2. Crossing angle with reverse–slip fault should be smaller to decrease compressive deformation;
3. Shallow burial depth of pipelines was recommended;
4. Thick and ductile steel pipes should be used.

Besides the above conclusions, backfill with sand instead of clay was also an important countmeasure (Desmond et al, 1995).

SUBSIDENCE. Subsidence usually appears at the boundary of a backfill area (Sun, 1979) and near revetment (Iwamoto, 1995).

Takada (1984) made an experiment on ductile iron pipelines subjected to large ground subsidence, and concluded that in the subsidence area the shorter the segmente and the more joints the pipelines, the more capability to resist the subsidence.

Three pattern of ground settlement were analyzed by Oishi (1985) and one pattern was also investigated experimentally. The most critical pattern was that the pipeline was fixed at one end. An effective countmeasure was proposed, and it was to lay a preventive rigid plate under the pipeline.

LANDSLIDE. The sliding of a relatively coherent soil mass ruptured three pipelines during 1994 Northridge earthquake, and the effect of sliding on buried pipelines was investigated numerically by O'Rourke and O'Rourke (1995).

LATERAL SPREADING. Lateral spreading is usually induced by liquefaction and has caused serious damages to pipelines (O'Rourke and Tawfik, 1983; Hamada et al, 1985; 1986).

LIQUEFACTION

As a geotechnical hazard, liquefaction was paid attention long ago, but the effect of liquefaction on pipelines had not been studied until the eighties.

Liquefaction caused heavily damages to buried pipelines during 1964 Niigata earthquake, 1975 Haicheng earthquake, 1976 Tangshan earthquake, 1983 Nihonkai–chubu earthquake, 1989 Loma Prita earthquake, and 1995 Kobe earthquake, etc. For example, during 1975 Haicheng earthquake, pipelines in backfill district of Haicheng City were heavily damaged by liquefaction, and in 1976 Tangshan earthquake, large areas of liquefaction in backfill district and alluvial valley damaged all the pipelines through the liquefaction areas (Sun, 1979). During 1989 Loma Prieta earthquake, there were about 120 failures or more to the water distribution pipelines in San Francisco. The failure rate, which is defined as the number of damage per length of pipeline, was about 0.06 / km. Most of them occurred in hydraulic fill district which was placed to created new land during 1912 to 1915 (Miyajima and Kitaura, 1991).

The statistical analysis of damages (Kitaura and Miyajima, 1988) showed that general failure modes of pipeline due to liquefaction coule be divided into damage to pipe body (breaking, bursting, cracking and thrusting) and damage to joint (breaking, pull–out and looseness), and markedly high damage ratios were obtained for the cast iron pipes.

The effect of liquefaction on lifelines can be classified into three types.

The first is the generation of buoyancy which leads to severe damage at boundary between pipelines and buildings, abutments, manholes, etc. Based on this point of view, some model experiments were carried out (Kitaura and Miyajima, 1983; Kuribayashi et al, 1985; Takada et al, 1987; Cai et al, 1992), and some countmeasures for uplift of the pipelines were proposed. The most effective countmeasure was to incerase the weight per unit volume and to apply extra weight.

The second is large ground movement such as lateral spreading (O'Rourke and Tawfik, 1983; Hamada et al, 1985; 1986).

The third is large dynamic response of ground during earthquake in which liquefaction zone can be regarded as a special irregular site, and pipeline failure will appear at the boundary between the liquefied and non–liquefied ground. Some model experiments (Katada and Hakuno, 1981; Kitaura and Miyajima, 1982; 1985; Nishio et al, 1987) were conducted on the dynamic behavior of pipelines in liquefaction process. Analytical models were proposed by many researchers (Wang, 1984; Takada and Tanabe, 1985; Yeh and Wang, 1985; Kitaura and Miyajima, 1987; Wang et al, 1990; Yeh and Wang, 1991; Wang et al, 1991; Miyajima and Kitaura, 1991; Wang and Zhang, 1992; Xiong et al, 1994).

Some anti–liquefaction countmeasures were proposed based on experiments and analytical investigations (Kitaura and Miyajima, 1984; Akiyoshi et al, 1992; Fuchida et al, 1995; Takada et al, 1995), they were compacting the surrounding ground, ground improvement by sand compacted piles, and backfill gravel layer to provide drainage.

FIGURE 5. BUCKLING FAILURE OF A PIPE CROSSING RIVERBED (Hou et al, 1992)

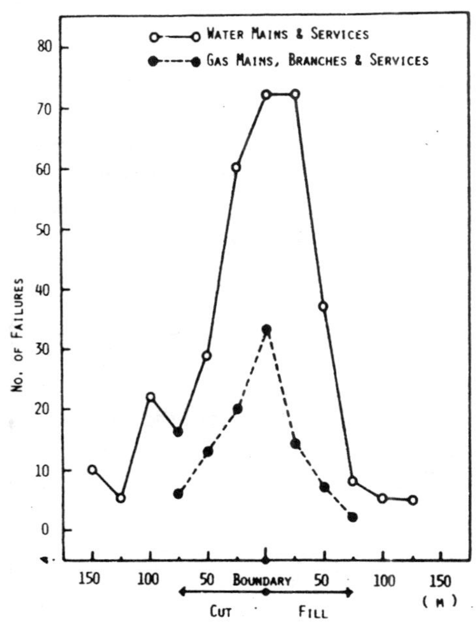

FIGURE 6. PIPELINE DAMAGES AT CUT–AND–FILL SITE (Kubo and Isoyama, 1980)

IRREGULAR SITE OBSERVED DAMAGE

Many post–earthquake investigations have indicated that when large ground movement and liquefaction area are excluded, almost all earthquake damages to buried pipelines are found in irregular sites.

For example, in 1923 Kanto earthquake, the seismic damages to water pipes in the region between firm ground at hillside and soft ground in downtown district were more severe than in the region of thick soft layer (Kubo, 1974).

Sun (1979; 1983) reported the effects of geography and terrain on damages of pipelines in 1975 Haicheng earthquake and 1976 Tangshan earthquake. He found that in location of backfill, river, canyon, and uneven surface, the pipelines suffered serious damages. In Haicheng earthquake, all pipelines along and across rivers were severely damaged. In Haicheng City, one pipeline through a riberbed and one pipeline through a canyon were broken. In Anshan City, one pipeline through a backfill basin was damaged, and both pipelines crossing Liao River were destroyed. In Tangshan earthquake, all six pipelines across Dou River sustained burst joints, and two pipelines across Hai River were damaged.

Also in Tangshan earthquake, two examples of buckling occurred to Qinhuangdao–Beijing oil pipelines through two riverbeds (Hou et al, 1992). Figure 5 showed one of the pipe failures.

In 1978 Miyagiken–Oki earthquake, approximately 350 damages (Figure 6) were found near the cut–and–fill boundary in the newly developed residential district (Kubo and Isoyama, 1980).

Iwamoto (1995) reported that damaged pipelines were mainly found in cutting and filling ground, near revetment (river, sea), banking part, etc.

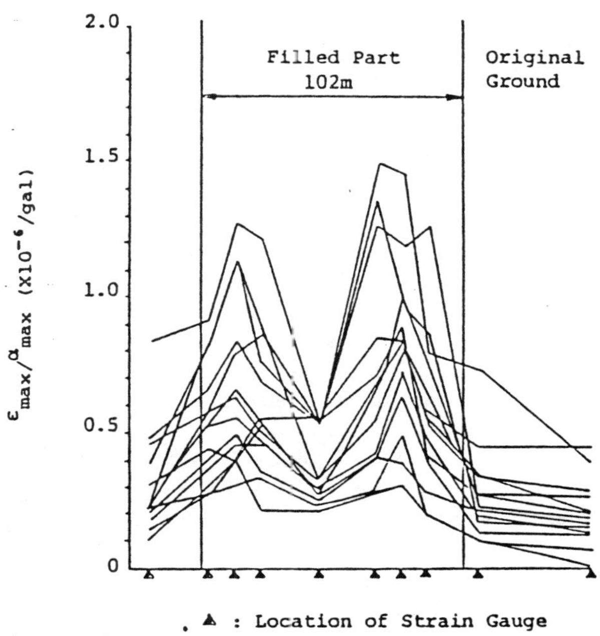

FIGURE 7. EARTHQUAKE OBSERVATION ON PIPELINE STRAIN AT ALLUVIAL VALLEY (Nishio et al, 1985)

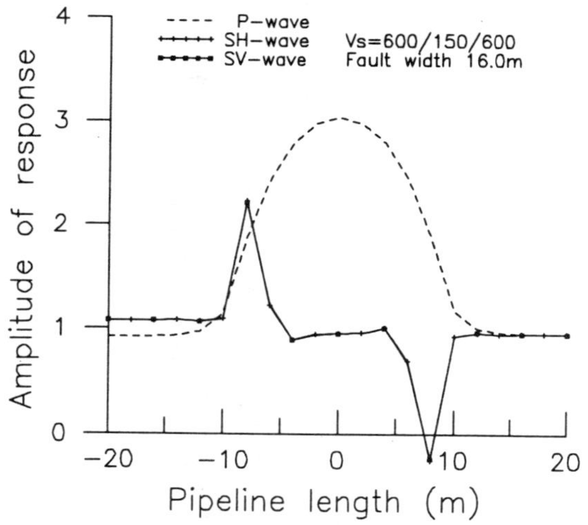

FIGURE 8. PIPELINE RESPONSE TO DIFFERENT WAVES (Liang, 1995b)

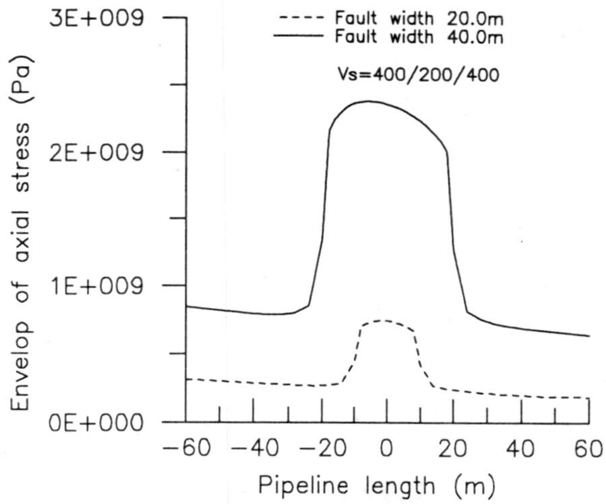

FIGURE 9. FAULT WIDTH EFFECT ON PIPELINE STRESS (Liang, 1995b)

OBSERVED BEHAVIOR

Nishio and his colleagues conducted earthquake observations on buried pipelines at five locations (Nishio et al, 1980; Tsukamoto et al, 1984; Nishio and Tsukamoto, 1985; Nishio et al, 1988), and found that nonuniformity of subsurface was the most essential cause of strain in the ground and subsequent strain in the pipeline through the ground; pipeline failure tended to concentrate to alluviums or river basins adjacent to plateaus or hilly regions. Figure 7 shows strain distribution of a pipeline crosses a valley filled with relatively less rigid soil observed during 15 earthquakes.

ANALYTICAL AND EXPERIMENTAL STUDY

The analytical study for the effect of irregular sites on seismic behavior of pipelines is a very important task (Liang, 1995e; Liang, 1995f), and have been conducted by many researchers.

TWO–SOIL SITE. Of the irregular sites, the simplest one is the site composed by two different soil.

The primary research was done by Goto et al (1973). They studied SH–wave propagation in sites with sharp changes of lateral stiffness and soil thickness, and relevant experiments were also carried out (Takada, 1974). They found that the difference of the vibrational characteristics of the sites was the reason why pipelines were damaged near the interfaces, and they also found that different directions of incident wave would lead to different responses.

Almost in the same time, Kubo (1974) noticed the problem. He found that in 1923 Kanto earthquake, the heaviest damage to water pipelines occurred in the transition region between firm soil to soft. He analyzed this problem by finite element model and pointed out that the maximum axial force in the pipe occurred near the border between the two soils.

Hindy and Novak (1979) analyzed the soil–pipeline interaction by calculating the amplitude of reflected and transmitted waves as well as their phase angles depending on the relative stiffness of the two media and the type of incident wave. The result indicated that the maximum stress in pipeline would appear in the case of P–wave propagating from firm soil to soft and S–wave from soft to firm.

Ohtsuki et al (1984) studied the effect of lateral inhomogeneity on seismic ground strain induced by Rayleigh wave and found that the maximum axial strain of pipeline would appear near the sloping interface.

This problem was also studied by other researchers (Wang and Cheng, 1979; Xie and He, 1988; Lin and Hu, 1991; Hall and O'Rourke, 1991; Liang and He, 1993; 1994a; 1994b).

THREE–SOIL SITE. Three–soil site is also a very common irregular site, e.g., deep canyon filled by soft deposit and non–causative fault. Since in one earthquake permanent movement only occurs to a few faults, and the others are still non–causative faults which should be treated as irregular sites. But, up to now, there is very little reference. (Liang, 1995f)

Liang (1995a, 1995b) noticed this problem and proposed an analytical model in which a half–space was composed by three horizontally adjacent soil media and a pipeline was laid across the interfaces at a right angle. He qualitatively studied seismic behavior of pipeline through the site. Figure 8 shows the amplification of pipeline response to P, SH and SV waves, and Figure 9 shows the effect of fault width on axial stress of a pipeline (length 120m). From the results we can know that the three soil site cannot be treated as two two–soil site, and maximum stress of pipeline will appear in the fault deposit (middle soil medium) and not at two interfaces. The damage position of pipeline will differ from that estimated by the theories of fault movement and two–soil site.

CUT–AND–FILL SITE. Cut–and–fill site, including alluvial valley, backfill site, etc. is another very common irregular site.

Nishio and his colleagues carried out earthquake observations (Nishio et al, 1980; Tsukamoto et al, 1984; Nishio and

Tsukamoto, 1985; Nishio et al, 1988) and two experimants (Nishio et al, 1983) on cut–and–fill site (alluvial valley), and they also made numerical simulation by finite element method. They demonstrated that the degree of nonuniformity of ground, especially the degree of the change of depth of surface layer beared a strong relationship with the occurrence of great strains in ground and pipe.

The same problem was also investigated by Satake et al (1982) using finite element method and Akiyoshi and Fuchida (1988) using boundary element method.

Liang (1995c) and Liang et al (1995) studied this problem by using theory of wave scattering and soil–pipeline interaction. The result showed that for a narrow valley the pipeline damage would appear at alluvium near two interfaces, and for a wider valley, the damage might also appear at the middle part of the alluvium, which has been evidenced by earthquake investigation (Figure 6).

RIVERBED. The earthquake damages of pipelines through riverbed in 1976 Tangshang earthquake have alreay been introduced above. Liang (1995d) investigated the damage of Qinhuangdao–Beijing pipeline by scattering of a river on seismic waves and the result indicated that damage would appear at the middle part of riverbed. The numerical result was in a very good agreement with the damages (Hou et al, 1992).

RANDOM GROUND CONDITION. The existing subsurface ground is necessarily irregular although the degree of irregularity may vary widely. Seismic waves are reflected and refracted by irregular site, and attenuate in propagation. From this point, loss of correlation of the random ground motion due to random ground conditions is to be expected and has to be counted when dealing with pipelines.

The spatial variation of seismic ground motion may be treated deterministically or stochastically. In the deterministic approach, a recorded time history at one point is used as the input, and the differential motion between two points is estimated by considering a delay in the arrival of the seismic wave between the points. In the stochastic approach, the ground motion at differential stations are modelled as a random process with a given power density, and the spatial variation is described by a correlation function.

Shinozuka and Kawakami (1977) firstly studied the effect of random variability of the soil property on the pipeline strain and a reasonable correlation was established between the strain and the damage statistics collected on the water pipelines under the 1923 Kanto earthquake.

In 1980, Hindy and Novak (1980) investigated the pipeline response theoretically, considering the ground motion as random and partialy correlated, and pointed out that the lack of correlation of seismic excitation could produce excessive stress in the pipe.

Zerva et al (1988) and Zerva (1993) examined the effect of spatial variation of ground motion on lifeline response, and pointed out that the assumption that perfectly correlated input motion is a safe approach in the seismic evaluation of pipelines was still a question, and pipeline response was sensitive to degree of the spatial variability of ground. The results showed that partially correlated motion could give high differential displacement at the pipeline joints whereas the perfectly correlated motion could not; and for a continuous pipeline, partially correlated motion would result in relatively high stress.

Lin et al (1994) demonstrated the importance of the ground motion spatial variations on the seismic response of buried pipelines. They found that pipeline response might be underestimated by neglecting the spatial variations and the coherence loss of strong ground motion, and the faster the apparent propagation velocity of waves, the more predominant the effects. They also pointed out that more recorded data from dense array observation were desirable for determining suitable coherency loss functions.

CONCLUSIONS

The effects of site conditions on seismic behavior of pipelines have been briefly discussed and some of the observed damages, observed behavior, quantitative analyses, analytical studies and experimental investigations are reviewed. Some conclusions are obtained:

1. Site condition has spectacular effect on seismic behavior of pipelines, and to some extent, has more important effect than seismic intensity;
2. Seismic waves cannot damage pipelines which are buried in a homogeneous site unless they are near the epicenter where the amplitude of seismic waves are very large;
3. Pipelines are very vulnerable to large ground movements, liquefaction, and irregularity of ground. When large ground movements are excluded, pipeline damages will appear at boundary between liquefied and non–liquefied soil, transition zone of irrgular site.
4. The effects of large ground movement such as sliding, lateral spreading, and surface rupture of ground on pipelines need further studies, and also the pipeline performance in liquefaction process and seismic behavior of pipelines in the more complex irregular site.
5. More post–earthquake investigations and earthquake observations are required to verify and establish the closer relationship between pipeline damage and irregularity of ground.
6. The influence of random ground condition on seismic behavior of pipelines is an important topic and more attention is needed on this subject.

ACKNOWLEDGEMENTS

It is a great honor for the first author to express his deep appreciation to Dr. N.Nishio, senior researcher of Tokyo Gas Co., Ltd.,for constant encouragement and advice. The author also gratefully acknowledges the financial support by the National Science Foundation of China under grant No.59308082.

REFERENCES

Akiyoshi, T., and Fuchida, K., 1988, "Seismic Response of Pipeline Systems Buried in Dipping Soil Layers," *Proceedings, 9th World Conference on Earthquake Engineering*, Tokyo–Kyoto, Vol.7, pp.67–72.

Akiyoshi, T., and Fuchida, K., 1992, "Stiffening Effect of Anti–Liquefaction Techniques for Buried Pipes," *Proceedings, 10th World Conference on Earthquake Engineering*, Madrid, pp.5417–5422.

Ariman, T., and Muleski, G.E., 1981, "A Review of Seismic Response of Buried Pipelines Under Seismic Excitations," *International Journal of Earthquake Engineering and Structural Dynamics*, Vol.9, pp.133–151.

Ariman, T., 1984, "Buckling and Rupture Failures of Pipelines due to Large Ground Deformations," *Proceedings, 8th World Conference on Earthquake Engineering*, San Francisco, Vol.7, pp.271–278.

Ariman, T., and Lee, B.J., 1991, "Tension / Bending Behavior of Buried Pipelines Under Large Deformation in Active Faults," *Proceedings, 3rd U.S. Conference on Lifeline Earthquake Engineering*, ASCE, Los Angeles, pp.226–233.

Cai, J., Liu, X., Hou, Z., and Chen, G., 1992, "Experimental Research on the Liquefaction Response of Oil Supply Pipeline," *Proceedings, Structural Technique of Pipeline Engineering*, Beijing, pp.284–291.

Chen, C.C., Ariman, T., and Lee, L.H.N., 1980, "Elastic Bucking Analysis of Buried Pipelines Under Seismic Loads," ASME Paper 80–C2 / PVP–76.

Desmond, T.P., Power, M.S., Taylor, C.L., and Lau, R.W., 1995, "Behavior of Large–Diameter Pipelines at Fault Crossing," *Proceedings, 4th U.S. Conference on Lifeline Earthquake Engineering*, ASCE, San Francisco, pp.296–303.

Fuchida, K., Akiyoshi, T., and Hyodo, T., 1995, "Behavior of Pipelines Buried in SCP–Improved Ground During Earthquake," *Proceedings, 4th U.S. Conference on Lifeline Earthquake Engineering*, ASCE, San Francisco, pp.104–111.

Hall, W.J., and O'Rourke, T.D., 1991, "Seismic Behavior and Vulnerability of Pipelines," *Proceedings, 3rd U.S. Conference on Lifeline Earthquake Engineering*, ASCE, Los Angeles, pp.761–773.

Hamada, M., Kubo, K., and Saito, K., 1985, "Large Ground Movement and Buried Pipe Failure by Soil Liquefaction During 1983 Nihonkai–chubu Earthquake," ASME,PVP–Vol.98–4, pp.11–18.

Hamada, M., Yasuda, S., Isoyama, R., and Emoto, K., 1986, "Observation of Permanent Ground Displacements Induced by Soil Liquefaction," *Proceedings of JSCE*, No.376, pp.211–220. (in Japanese)

Hindy, A., amd Novak, M., 1979, "Earthquake Response of Underground Pipelines," *International Journal of Earthquake Engineering and Structural Dynamics*, Vol.7, pp.451–476.

Hindy, A., and Novak, M., 1980, "Pipeline Response to Random Ground Motion," *Journal of Engineering Mechanics*, ASCE, Vol.106, pp.339–360.

Hou, Z., Li, P., and Yang, G., 1992, "Example Analysis of Seismic Damage on Buckling of Qinhuangdao–Beijing Oil Pipeline," *Proceedings, Structural Technique of Pipeline Engineering*, Beijing, pp.277–282.

Ichihara, M., and Yamada, K., 1981, "Relative Degree of Risk of Water–Supply Pipes in Nagoya City During Earthquake," *Proceedings of JSCE*, No.316, pp.51–63. (in Japanese)

Iwamoto, T., 1995, "Summary of Damaged Pipelines for Waterwork System," *Paper Presented at 4th U.S. Conference on Lifeline Earthquake Engineering*, ASCE, San Francisco.

Katada, T., and Hakono, M., 1981, "Experimental Analysis on Dynamic Behavior of Underground Structure in the Liquefaction Process," *Proceedings of JSCE*, No.306, pp.1–10. (in Japanese)

Katayama, T., Kubo, K., and Sato, N., 1977, "Quantitative Analysis of Seismic Damage to Buried Utility Pipelines," *Proceedings, 6th World Conference on Earthquake Engineering*, New Delhi, pp.3369–3374.

Kennedy, R.P., Chow, A.W., and Williamson, R.A., 1977, "Fault Movement Effects on Buried Oil Pipeline," *Journal of Transportation Engineering*, ASCE, Vol.103, pp.617–633.

Kitaura, M., and Miyajima, M., 1982, "Experimental Study on Strain Characteristics of Underground Pipe During Liquefaction," *Proceedings of JSCE*, No.323, pp.43–53. (in Japanese)

Kitaura, M., and Miyajima, M., 1983, "Dynamic Behavior of a Model Pipe Fixed at One End During Liquefaction," *Proceedings of JSCE*, No.336, pp.31–38.

Kitaura, M., and Miyajima, M., 1984, "Experiments on Damage Mitigation of Buried Pipelines due to Liquefaction," *Proceedings, 8th World Conference on Earthquake Engineering*, San Francisco, Vol.7, pp.239–245.

Kitaura, M., and Miyajima, M., 1985, "Dynamic Behavior of Buried Model Pipe During Incomplete Liquefaction," *Journal of Structural Engineering*, JSCE, Vol.31A, pp.421–426. (in Japanese)

Kitaura, M., and Miyajima, M., and Suzuki, H., 1987, "Response Analysis of Buried Pipelines Considering Rise of Ground Water Table in Liquefaction Processes," *Proceedings of JSCE*, No.380, pp.173–180.

Kitaura, M., and Miyajima, M., 1988, "Quantitative Evaluation of Damage to Buried Pipelines Induced by Soil Liquefaction," *Proceedings, 9th World Conference on Earthquake Engineering*, Tokyo–Kyoto, Vol.7, pp.11–16.

Kubo, K., 1974, "Behavior of Underground Waterpipes During an Earthquake," *Proceedings, 5th World Conference on Earthquake Engineering*, Rome, Vol.1, pp.569–578.

Kubo, K., and Isoyama, R., 1980, "Damage to Buried Utility Pipes in the Miyagiken–Oki Earthquake," *Proceedings, 7th World Conference on Earthquake Engineering*, Istanbul, Vol.8, pp.225–232.

Kubo, K., Katayama, T., and Sato, N., 1975, "Quantitative Analysis of Seismic Damage to Buried Pipelines," *Proceedings, 4th Japan Earthquake Engineering Symposium*, Tokyo, pp.655–662. (in Japanese)

Kuribayashi, E., Kawamura, M., Ieda, R., Aida, M., and Yui, Y., 1985, "An Experimental Behavior of Buried Pipes During Liquefaction of Saturated Sandy Soil," ASME, PVP–Vol.98–4, pp.19–24.

Kuribayashi, E., Jiang, T., Niiro, T., Nagasaka, H., Kuroiwa, S., and Nishioka, S., 1991, "Engineering Tactics on Lifelines Safety Against Earthquake," *Proceedings, 3rd U.S. Conference on Lifeline Earthquake Engineering*, ASCE, Los Angeles, pp.927–936.

Kyriakides, S., and Yun, H., and Yew, C.H., 1983, "Buckling of Buried Pipelines Due to Large Ground Movements," ASME PVP–Vol.77.

Lee, L.H.N., Ariman, T., and Chen,C.C., 1980, "On Buckling of Buried Pipelines Under Seismic Excitations," ASME

Paper 80–C2 / PVP–75.

Lee, B.J., and Ariman, T., 1985, "Bucking of Buried Pipelines Crossing an Active Fault," ASME, PVP–Vol.98–4, pp.25–33.

Liang, J.W., and He, Y.A., 1993, "Dynamic Response of Buried Pipelines Through Two Different Soils," *Proceedings, 1st National Conference on Structure–Media Interaction*, Luoyang, China, pp.848–852. (in Chinese)

Liang, J.W., and He, Y.A., 1994a, "3–D Seismic Response of Pipelines Through Nonhomogeneous Soil Media," *Journal of Tianjin University*, Vol.27, No.2, pp.200–204. (in Chinses)

Liang, J.W., and He, Y.A., 1994b, "3–D Seismic Responses of Underground Pipeline Systems in Nonhomogeneous Soil Media," *Proceedings, 5th U.S. National Conference on Earthquake Engineering*, Chicago, Vol.4, pp.451–458.

Liang, J.W., 1995a, "3–D Seismic Response of Pipelines Through Multiple Soil Media," *Proceedings, 1995 ASME Pressure Vessels and Piping Conference*, Honolulu.

Liang, J.W., 1995b, "3–D Seismic Response of Pipelines Laid Through Fault," *Proceedings, 4th U.S. Conference on Lifeline Earthquake Engineering*, ASCE, San Francisco, pp.200–207.

Liang, J.W., 1995c, "Dynamic Analysis of Pipelines Laid Through Alluvial Valley," *Transactions of Tianjin University*, Vol.1, No.1, pp.89–92.

Liang, J.W., 1995d, "Dymanic Response of Pipelines Laid Through Riverbed," *Proceedings, 1995 ASME Pressure Vessels and Piping Conference*, Honolulu.

Liang, J.W., 1995e, "A Review of Seismic Response and Buckling of Underground Pipelines," *Proceedings, International Symposium / Workshop on Progress of Structural Engineering*, National Science Foundation of China, Shanghai, pp.205–211.

Liang, J.W., 1995f, "A Review of the Earthquake Resistant Research of Buried Pipelines," *World Information on Earthquake Engineering*, Vol.44, pp.1–7. (in Chinese)

Liang, J.W., Jia, S., and Hou, Z., 1995, "Seismic Response of Pipelines Laid Through Alluvial Valley," *Proceedings, 4th U.S. Conference on Lifeline Earthquake Engineering*, ASCE, San Francisco, pp.328–335.

Lin, G., Wang, J., Li, B., and Shen, A., 1994, "Seismic Response of Buried Pipelines Based on Stochastic Model of Ground Motions," *Proceedings, 2nd China–Japan–US Trilateral Symposium on Lifeline Earthquake Engineering*, Xi'an, pp.105–111.

Lin, H., and Hu, Y., 1991, "Seismic Response Analysis of Pipelines in Inhomogeneous Site," *Proceedings, 3rd National Conference on Earthquake Engineering*, Harbin, China, pp.996–1001. (in Chinese)

Miyajima, M., amd Kitaura, M., 1991, "Performance of Pipelines During Soil Liquefaction," *Proceedings, 3rd U.S. Conference on Lifeline Earthquake Engineering*, ASCE, Los Angeles, pp.470–479.

Newmark, N.M., and Hall, W.J., 1975, "Pipeline Design to Resist Large Fault Displacement," *Proceedings, U.S. National Conference on Earthquake Engineering*, Oakland, pp.416–425.

Nishio, N., Ukaji, T., and Tsukamoto, K., 1980, "Experimental Studies and Observation of Pipeline Behavior During Earthquakes," ASME, PVP–Vol.43, pp.67–76.

Nishio, N., and Satake, M., 1983, "Characteristics of Deformation in a Buried Pipeline Under Sinusoidal Ground Motion," *Natural Disaster Science*, Vol.5, pp.53–68.

Nishio, N., Ukaji, T., Tsukamoto, K., and Ishita, O., 1983, "Model Experiments on the Behavior of Buried Pipelines During Earthquakes," ASME, PVP–Vol.77, pp.263–272.

Nishio, N., and Tsukamoto, K., 1985, "Seismic Behavior of a Buried Pipeline in a Non–Uniform Subsurface Layer," ASME, PVP–Vol.98–4, pp.119–124.

Nishio, N., 1987, "Mechanism of Seismic Strain in Buried Pipelines Based on Field Observations and Model Experiments," *Proceedings, 5th Canadian Conference on Earthquake Engineering*, Ottawa, pp.637–644.

Nishio, N., Tsukamoto, K., and Hamura, A., 1987, "Model Experiment on the Seismic Behavior of Buried Pipeline in Partially Liquefied Ground," *Proceedings of JSCE*, No.380, pp.449–458.

Nishio, N., Hamura, A., and Sase, T., 1988, "Earthquake Observation of a Buried Pipeline in a Non–Uniform Ground," *Proceedings, 9th World Conference on Earthquake Engineering*, Tokyo–Kyoto, Vol.7, pp.23–28.

Nishio, N., 1989a, "Quantitative Relationship Between Ground Conditions and Pipeline's Damage Susceptibility," *Proceedings, 3rd US–Japan Workshop on Earthquake Disaster Prevention for Lifeline Syatems*, Tsukuba Science City, pp.41–51.

Nishio, N., 1989b, "Studies on Seismic Behavior and Quantitative Prediction of Damage Ratio of Buried Pipelines," Ph.D. Dissertation, Tohoku University, Japan. (in Japanese)

Nishio, N., 1992a, "Seismic Damage Ratio Predicition for Buried Pipelines Considering Nonuniform Ground Displacement," *Proceeding of JSCE*, No.450, pp.63–74.

Nishio, N., 1992b, "Damage Ratio Predicition for Buried Pipelines Based on Deformability of Pipelines and Nonuniformity of Ground," ASME, PVP–Vol.227, pp.13–22

Nishio, N., 1994, "Damage Ratio Predicition for Buried Pipelines Based on the Deformability of Pipelines and Nonuniformity of Ground," *Journal of Pressure Vessel Technology*, ASME, Vol.116, pp.459–466.

Ohtsuki, A., Tazoh, T., and Shimizu, K., 1984, "Effect of Lateral Inhomogeneity on Seismic Waves and Ground Strains," *Proceedings of JSCE*, No.350, pp.291–300. (in Japanese)

Oishi, H., 1985, "Considerations on Behavior of Underground Pipelines Caused by Ground Settlement," *Proceedings of JSCE*, No.356, pp379–386. (in Japanese)

Okamoto, S., 1973, *Introduction to Earthquake Engineering*, The University of Tokyo Press, Tokyo.

O'Rourke, T.D., and Trautmann, C.H., 1980, "Buried Pipeline Response to Permanent Earthquake Ground Movements," ASME Paper 80–C2 / PVP–78.

O'Rourke, T.D., and Trautmann, C.H., 1981, "Earthquake Ground Rupture Effects on Jointed Pipe," *Proceedings, 2nd ASCE Speciality Conference on Lifeline Earthquake Engineering*, San Francisco, pp.65–79.

O'Rourke, T.D., and Tawfik, M.S., 1983, "Effects of Lateral Spreading on Buried Pipelines During the 1971 San Fernando

Earthquake," ASME, PVP–Vol.77, pp.124–132.

O'Rourke, T.D., and McCaffrey, M.A., 1984, "Buried Pipeline Response to Permanent Earthquake Ground Movements," *Proceedings, 8th World Conference on Earthquake Engineering*, San Francisco, Vol.7, pp.215–222.

O'Rourke, T.D., and O'Rourke, M.J., 1995, "Pipeline Response to Permanent Ground Deformation: A Benchmark Case," *Proceedings, 4th U.S. Conference on Lifeline Earthquake Engineering*, ASCE, San Francisco, pp.288–295.

Satake, M., Kishino, Y., and Asano, T., 1982, "Considerations on Earthquake Resistance of Pipeline Buried in Non–Uniform Ground," *Proceedings, 4th Japan Earthquake Engineering Symposium*, Tokyo, pp.1905–1912. (in Japanese)

Shinozuka, M., and Kawakami, H., 1977, "Underground Pipe Damages and Ground Characteristics," *Proceedings, ASCE Speciality Conference on Lifeline Earthquake Engineering*, Los Angeles, pp.239–307.

Sun, S., 1979, "Earthquake Damage to Pipelines," *Proceedings, 2nd U.S. National Conference on Earthquake Engineering*, Stanford, pp.61–67.

Sun, S., 1983, "Analysis on Seismic Damages to Buried Pipelines in Tangshan Earthquake," ASME, PVP–Vol.77, pp.365–367.

Sun, S., 1992, "The Recent State of Buried Pipeline Earthquake Engineering in China," ASME, PVP–Vol.227, pp.1–6. (Invited Paper)

Sun, S., 1994, "A Review of Buried Pipeline Earthquake Engineering in China," *Proceedings, 2nd China–Japan–US Trilateral Symposium on Lifeline Earthquake Engineering,*" Xi'an, pp.17–42. (Invited Paper)

Takada, S., 1974, "Vibrational Characteristics and Earthquake Response Analysis of Underground Tubular Structures," Ph.D. Dissertation, Kyoto University, Japan. (in Japanese)

Takada, S., 1984, "Model Analysis and Experimental Study on Mechnical Behavior of Buried Ductile Iron Pipelines Subjected to Large Ground Deformation," *Proceedings, 8th World Conference on Earthquake Engineering*, San Francisco, Vol.7, pp.255–262.

Takada, S., and Tanabe, K., 1985, "Dynamic Behaviors of Underground Structures subjected to Liquefaction," *Proceedings, 18th JSCE Earthquake Engineering Symposium*, pp.233–236. (in Japanese).

Takada, S., Tanabe, K., Yamayo, K., and Katagiri, S., 1987, "Liquefaction Analysis for Buried Pipelines," *Structures and Stochastic Methods*, Cakmak, A.S. Ed., Computational Mechanics Publications, pp.319–334.

Takada, S., Ogawa, Y., Shimizu, K., and Ueno, J., 1995, "Experiments on Seismic Behavior of Back–fill Gravel Layer as Liquefaction Countmeasure of Buried Gas Pipelines," *Proceedings, 4th U.S. Conference on Lifeline Earthquake Engineering*, ASCE, San Francisco, pp.232–239.

Tsukamoto, K., Nishio, N., Satake, M., and Asano, T., 1984, "Observation of Pipeline Behavior at Geographically Complex Site During Earthquake," *Proceedings, 8th World Conference on Earthquake Engineering*, San Francisco, Vol.7, pp.247–254.

Vougioukas, E.A., Theodossis, C., and Carydis, P.G., 1991, "Seismic Analysis of Buried Pipelines Subjected to Vertical Fault Movement," *Proceedings, 3rd U.S. Conference on Lifeline Earthquake Engineering*, ASCE, Los Angeles, pp.432–441.

Vougioukas, E.A., and Carydis, P.G., 1992, "Performance of Buried Pipelines Subjected to Fault Movement," *Proceedings, 10th World Conference on Earthquake Engineering*, Madrid, pp.5449–5452.

Wang, L.R.L., and Cheng, K.M., 1979, "Seismic Response Behavior of Buried Pipelines," *Journal of Pressure Vessel Technology*, ASME, Vol.101, pp.21–30.

Wang, L.R.L., 1984, "Parametric Investigation of Buried Pipelines Under Seismic Environments," *Proceedings, 8th World Conference on Earthquake Engineering*, San Francisco, Vol.7, pp.483–490.

Wang, L.R.L., and Yeh, Y.H., 1985, "A Refined Seismic Analysis and Design of Buried Pipelines for Faults Movement," *International Journal of Earthquake Engineering and Structural Dynamics*, Vol.13, pp.75–96.

Wang, L.R.L., Shim, J.S., Ishibashi, I., and Wang, Y., 1990, "Dynamic Responses of Buried Pipelines During a Liquefaction Process," *International Journal for Soil Dynamics and Earthquake Engineering*, Vol.9, No.1, pp.44–50.

Wang, L.R.L., Zhang, H., and Ishibashi, I., 1991, "Seismic Response of Buried Pipeline System in a Soil Liquefaction Environment," *Proceedings, 3rd U.S. Conference on Lifeline Earthquake Engineering*, ASCE, Los Angeles, pp.460–469.

Wang, L.R.L., and Zhang, H., 1992, "Buried Pipeline Systems in a Liquefaction Environment," *Proceedings, 10th World Conference on Earthquake Engineering*, Madrid, pp.5529–5534.

Xie, X., and He, Y.A., 1988, "Calculation of Earthquake responses to Embedded Pipeline Laid Through Different Media,", *Proceedings, 9th World Conference on Earthquake Engineering*, Tokyo–Kyoto, Vol.7, pp.49–53.

Xiong, J., Wang, T., Wang, G., and Lin, J., 1994, "Seismic Behavior of Buried Pipelines During Liquefaction," *Proceedings, 2nd China–Japan–US Trilateral Symposium on Lifeline Earthquake Engineering*, Xi'an, pp.185–192.

Yeh, Y.H., and Wang, L.R.L., 1985, "Combined Effects of Soil Liquefaction and Ground Displacement to Buried Pipeline," ASME, PVP–Vol.98–4, pp.43–52.

Yeh, Y.H., and Wang, C.M., 1991, "Inelastic Behaviors of Pipelines Buried Through Liquefaction Zone,"*Proceedings, 3rd U.S. Conference on Lifeline Earthquake Engineering*, ASCE, Los Angeles, pp.480–489.

Yun, H., and Kyriakides, S., 1990, "On the Beam ans Shell Modes of Buckling of Buried Pipelines," *International Journal for Soil Dynamics and Earthquake Engineering*, Vol.9, No.4, pp.179–193.

Zerva, A., Ang, A.H.–S., and Wen, Y.K., 1988, "Lifeline Response to Spatially Variable Ground Motions," *International Journal of Earthquake Engineering and Structural Dynamics*, Vol.16, pp.361–379.

Zerva. A., 1993, "Pipeline Response to Directionally and Spatially Correlated Seismic Ground Motions," *Journal of Pressure Vessel Technology*, ASME, Vol.115, pp.53–58.

PVP-Vol. 340, Seismic Engineering
ASME 1996

USING DYNAMIC MASTER LOGIC DIAGRAM FOR COMPONENT PARTIAL FAILURE ANALYSIS

Toni Ni
Halliburton NUS Corporation
910 Clopper Road, Gaithersburg, MD 20878

Mohammad Modarres
Center for Reliability Engineering
University of Maryland, College Park, MD 20742

Abstract

A methodology using the Dynamic Master Logic Diagram (DMLD) for the evaluation of component partial failure is presented. Since past PRAs have not focused on partial failure effects, the reliability of components are only based on the binary state assumption, i.e. defining a component as fully failed or functioning. This paper is to develop an approach to predict and estimate the component partial failure on the basis of the fuzzy state assumption. One example of the application of this methodology with the reliability function diagram of a centrifugal pump is presented.

Introduction

Classical reliability theory has been based upon binary state functions and probability theory. A binary structure function allows only two states: a perfect functioning or a complete failure. Since many real systems are more complex by nature, a degradation state must be allowed. By allowing the degradation state, we must extend the classical notion of reliability based upon the probability of failure or functioning of a component or system. Current research, however, showed that probability theory is not the only possible way of representing imprecision and uncertainty [1]. Possibility theory and fuzzy set theory can provide a useful alternative to the probabilistic approach of reliability.

Current PRA methods widely used for Individual Plant Examination (IPE) program are based on the binary state assumption [2]. A component is considered to be unavailable if it is unable to operated at a required performance level that is usually called the success criteria, such as, one pump must provide 200 gpm of flow rate during a normal operation. As shown in Figure 1, a classical binary state has only crisp state with sharp boundaries in a fluid system.

Suppose we consider a pump with approximate 90% of capacity flow, 180 gpm, this is considered as a failure or not? Zadeh [1] proposed to use membership functions that return a real value between 0 and 1. The membership function may be obtained by statistical analysis of appropriate data. Instead of fully failed or functioning, fuzzy logic makes use of its less precise, but more common expressions like, severe, moderate, and low. Figure 2 shows the membership function of the terms of the fuzzy logic, failure and functioning in a fluid system.

Most fluid components in a nuclear power plant, such as pump, MOV, check valve have a degradation state, that is, the component is operable but unable to satisfy the performance criterion. Total partial failures caused by functional degradation of components contribute to approximately 74% of all the failures reported [3]. Using fuzzy logic to represent imprecise knowledge and the uncertainty, three functional states of a component, i.e., failure, degradation, and functioning could be modeled by three fuzzy sets, showed in Figure 3.

Fuzzy set considers that three states change gradually from one to another with unclear boundary. In the case of Figure 3, flow capacity of a pump has degree of membership 0.85 for functioning and 0.15 for degradation for a given value of 180 gpm of flow. Fuzzy set indicates not only the flow value, but also the degree of the flow value. It can tell the difference between 100 gpm of flow rate (i.e., none of "functioning) and 180 gpm of flow rate (i.e. very high of "functioning") which cannot be presented by the binary state structure.

Expert Opinions for Fuzzy Set Model

The process of developing definition of functional states should be based on a systematic assessments of experienced individuals for the condition of a component at different functional performance levels. Since this process is based on subjective assessments of failure, different operators or experts might give different judgements regarding functional definitions.

Figure 4 shows a expert fuzzy set model of a pump performance with three states from a survey report [4]. The shape and location of these fuzzy sets is depend upon experts' opinion or judgment in relation to the performance of the pump operation. 0%-50% of pump flow capacity is considered a range of the failure fuzzy set; 20-100% of pump flow capacity with 80% as a full membership is considered a range of degraded fuzzy set; and 80%-100% pump flow capacity is considered a range of functioning fuzzy set.

Dynamic Master Logic Diagram Model

Moderres and Hu [5] developed a new diagram called Dynamic Master Logic Diagram (DMLD). This model can directly manipulate the linguistic terms to predict the partial failure risk and performance of a system, component or part with a failure mode. DMLD allows an analyst to evaluate the partial failure in a natural way. Appropriate predictions can be taken even though the information on which they are based is vague, ambiguous, qualitative, or imprecise.

DMLD is a logic-based diagram of combining logic, physiscal and linguistic modeling. This methodology was primarily developed from Goal Tree Success Tree and Master Plant Logic Diagram (GTST_MPLD). Fuzzy logic is used to replace binary state logic in GTST_MPLD modeling, as shown in Figure 5. Once GTST_MPLD is used to decompose a system or a component into locally subsystems or parts, DMLD is applied to model the physical behavior of a system or a component with mutilple states of the perfomance, such as, a partial failure state.

DMLD was further developed to a family of models except for representing a physical behavior. Users can come up with varieties of DMLDs and different numbers of nodes and layers, fuzzy sets , and transition logic. The structures in DMLDs can be group into several basic classes for rule-based relations, such as, static, uncertain output, etc. However these DMLDs should yield approximately similar results.

The overall DMLD procedure for making a fuzzy assessment is shown in Figure 6. The analysis uses linguistic variables as input to describe the functional states or severity. These inputs are then "fuzzified" using membership functions supplied by an application area expert to determine the degree of membership in each input. The resulting "fuzzy inputs" are evaluated using expert rule base and fuzzy logic operations to yield a real behavior of a component performance with an associated degree of membership. The "fuzzy conclusion" is then defuzified to give the prioritization level for the performance state.

Study Sample - A Centrifugal Pump

The centrifugal pump is a dynamic pump which is widely used in a nuclear power plant. A centrifugal pump uses centrifugal force to move liquid. Liquid enters the eye of the pump impeller and is rotated in a circular motion. Centrifugal force "pushes" the liquid outward from the center of the pump. When the move liquid hits the stationary casing its velocity slows because of friction. Since energy cannot be lost, the energy that was velocity must change o something else. Energy in the form of velocity changes to energy in the form of pressure and heat. The liquid moves through the pipe because liquid flows from an area of lower pressure.

A centrifugal pump consists of the totating parts and stationary parts. A hierarchical represention of a centrifugal pump is shown as Figure 7. The main parts of a centrifugal pump are the casing, impeller, shaft, bearing, and mechanical seal or packing. The function of these major parts are described in Table 1 for Failure Modes Effects Analysis of a centrifugal pump. Won [6] studied functional analysis of a centrifugal pump to identify failure modes. The primary failure mode of a pump, "no fluid flow" and one of supplementary failure modes, "external leakage" are considered two significant failure modes to effect a pump performance. Based on the FMEA, the reliability block diagram of a centrifugal pump with its major parts is developed shown in Figure 8.

DMLD Prediction for Pump Performance

DMLD for the pump equipment is shown in Figure 9. DMLD can logically and hierarchically display the physical interrelationship between five major parts functions of a pump and two major pump functions. The success paths are shown on the top. On the left side of the diagram, the pump parts function are displayed to support two major pump functions. Each notion presents the physical behavior and has a fuzzy logic instead of a binary logic. For example, assume that we find a noise on a pump bearing or a bearing was found dirt in a maintenance. This degraded state of the bearing, which may provide some unsatisfactory output, such as low flow, but the pump still appear functional normal. Using a fuzzy logic, the operator can predict the pump performance with a partial failure in DMLD.

Fuzzy logic allows the expert to use fuzzy ranges when specifying input parameters for partial failures and providing the rules for output uncertainty. The fuzzy sets can be defined the component or its part failure with different functional states. The resulting evaluation of the performance of a pump is given in qualitative simulation rather than numerical terms.

Fuzzification

When we express a failure mode, it is natural to use fuzzy expressions like "Fully Failed" or "Degraded". But such words provide little information about the severity quantitatively. In order to allow the users to have "more confidence", the range of the failure mode possibility is divided into several sections, each of which can be described by a fuzzy linguistic attribute for expressing the severity.

Severity class of the failure modes is used into the process of fuzzification. Severity is ranked according to the seriousness of failure mode effect associated with possible failure cause on the next higher level assembly or system. Usually, the effect degree of a failure mode could be described as the effects on the operation of pump. Since failure effect of each pump part is different, such as shaft is highly vibrated or bearing has excessive noise, severity class can provide an input of failure effects of parts with equivalent pump performance . The functional state variables express the possibility of failure effects through the linguistic severity classes Fuzzy expression is more preferable to allow the operators to express the severity as being severe on a range from minor to very severe, which, although imprecise, is more meaningful. Severity class of the failure mode for major parts of centrifugal pump is developed shown in Table 2.

The fuzzy sets representing three functional states of any part of a pump is shown in Figure 10. Observe that the linguistic terms describing the elements of each set are placed along an ordinal scale and trapezoidal membership functions showing the degree of membership of each term in the set are given. The membership functions tell the "degree" with which some input is an element of the set.

Observe that when an input approaches the boundary between two sets its membership value in the adjacent set increase.

To illustrate the fuzzification of the input parameters, Let us consider an evaluation of the pump partial failure. Assume that a pump has the degraded flow because of vibration and noise and operator thought the severity of bearing's wear out is high moderate and impeller wear is minor. As shown in Figure 10, from the intersection of slight and moderate of severity range, the fuzzy input are:

Pump Parts	Fuzzy Sets	Membership
Bearing	No Failed	0.12
	Degraded	0.88
Impeller	No Failed	0.75
	Degraded	0.25

The fuzzification process fuzzifies the input by determining the degree to which it is represented by each input set - i.e. the value of the membership function for each set corresponding to that input. This converts the inputs into their fuzzy representation which will execute the rules in the knowledge base that have the fuzzified inputs as part of their premise.

Rule Based Relations

The characterization of the fuzzy system is based on expert knowledge, usually in the form of "If-Then" rules, which are easily implemented by conditional statements in fuzzy logic. Fuzzy rules are more conveniently formulated in linguistic terms than in numerical terms. The proper choice of linguistic variables is fundamental in characterizing the operation of a fuzzy system. It has substantial effect on the accuracy of the system and a strong influence on the analysis: experience and engineering knowledge play an important role in selecting the appropriate rules.

DMLD nodes in Figure 9 can be described by a certain or uncertain relationship with fuzzy rules. There are seven basic structures of DMLD [7] designed to simulate a physical behavior with the rule base relations. These structures include static, uncertain output, uncertain input, scheduled, time-lagged, auto-correct, and feedback. Each DMLD model could be assembled from one or more than one structure.

In this paper we use a static structure with a certain relationship to form "If-Then" rules. A static DMLD represents a liner or nonlinear physical relation which is certain and not varied with time. With more information and experience of pump parts performance or time-depend behavior required, we can update the DMLD model combined with other DMLD structures to yield approximately similar results.

Let us using the same example as above shown in Figure 10 with membership value of 0.12 in the no failed fuzzy set and 0.88 in the degraded fuzzy set for a bearing, 0.75 in the no failed fuzzy set and 0.25 in the degraded fuzzy set for an impeller. Using a static DMLD, we can summarize the following rules for certain relationship of DMLD nodes (Figure 9) in Table 3.

The behavior of input can be also modeled by DMLD with more than three fuzzy sets of pump parts failure. As such, DMLD can predict more accurate outputs for all possible detail-transition failure modes.

Rule evaluation consists of the minimum and maximum operation [1]. In MIN interfacing, the output membership function is clipped off at a height corresponding to the rule premise's computed degree of truth (fuzzy logic AND). In MAX composition, the combined output fuzzy sets is constructed by taking the maximum over all of the fuzzy sets assigned the output variable by the inference rule (fuzzy logic OR). The result of rule evaluation is a set of fuzzy output values which reflect the effects of all rules whose truth value is greater than zero. The complete evaluation procedure involving the Rule- Base table (Table 3) for the given inputs is illustrated in Figure 11. In this example, we see that the fuzzy conclusion is contained in the three fuzzy states (Fully Failed, Degraded and No Failed) with memberships of 0.0, 0.75, and 0.15.

Defuzzification

The defuzzification process creates an assessment from the fuzzy conclusion set to predict how much output of a pump performance may have. Several techniques for defuzzifying the fuzzy conclusion have been developed [8]. Most analysts use the Weighted Mean of Maximum Method which represents the mean of all local actions whose membership functions reach the maximum. The output performance of the pump for this example using a DMLD model (diagram) can be represented as shown in Figure 11. From this result we can predict the performance in pump partial failure with a degree of 82.76% of flow capacity.

Conclusions

DMLD appears to be a useful fuzzy logic tool for prediction and estimation of the physical behavior or performance of a component or system with partial failures. More generally, it provides an effect tool for presenting and manipulating linguistic knowledge of failure modes and servility. Since linguistic failure modes are finally defuzzified to numerical values, DMLD modeling can be used to make a full-scale simulation for reliability analysis of a component or system.

References

1. Zedeh L.A., "Fuzzy sets as a Basis for a Theory of Possibility", Fuzzy Sets and Systems, 1, 3-28 (1978)
2. U.S. Nuclear Regulation Commission, "PRA Procedure Guide", NUREG/CR-2300 (1983)
3. Ni T. & Modarres M., "Partial Failure effects of Safety Systems and Components in Light Water reactor",The 1995 Joint ASME/JSME Pressure Vessels and Piping Conference, Honolulu, Hawaii (1995)
4. Ni T., "Survey of Pump Performance of Nuclear Power Plant", (1995)
5. Hu Y-S. & Modarres M., "An Introduction to Dynamic MPLD Model", Proceeding of the PSAM-II Conference, San Diego, CA (1994)
6. Won J-K., "Modeling and Diagnosis of Equipment Partial Failures in Process Plants", Ph.D. Thesis, University of Maryland (1995)
7. Hu Y-S., "Evaluating System Behavior through DMLD Modeling", Ph.D. Thesis, University of Maryland (1994)
8. Takagi T. & Sugeno, M. "Fuzzy Identification of Systems and its Applications to Modeling and Control", IEEE Trans. On Systems, Man, and Cybern., Vol. SMC-15(1) (1985)

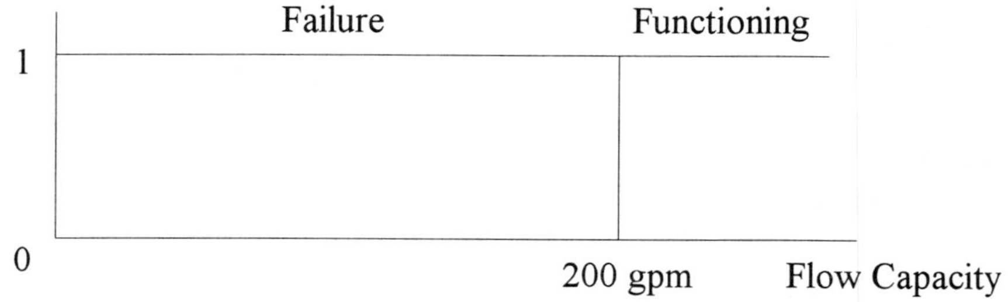

Figure 1. Binary Logic Function

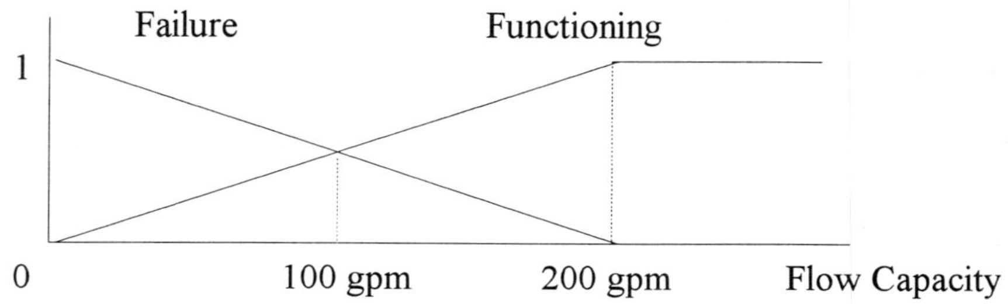

Figure 2. Fuzzy Logic Function

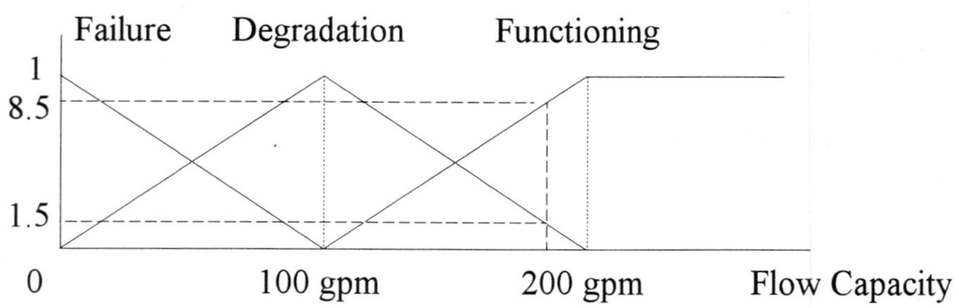

Figure 3. Three Fuzzy Sets of a Fluid Component

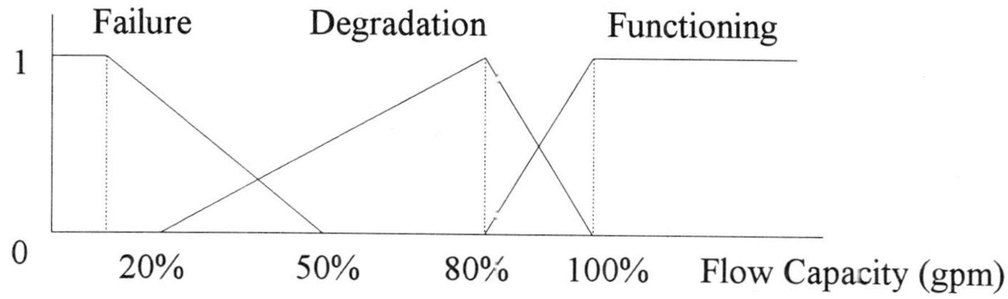

Figure 4. Expert Opinion for Three Fuzzy Sets of a Pump

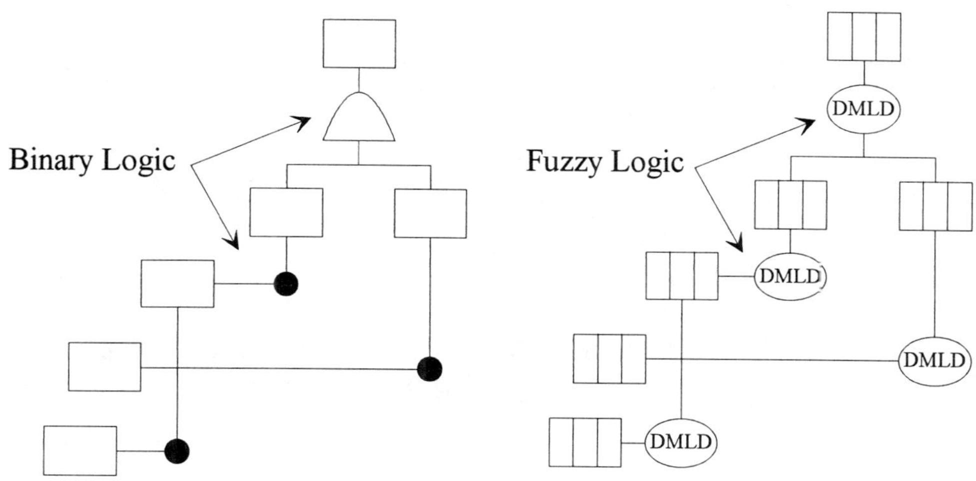

Figure 5. From Binary Logic to Fuzzy Logic Modeling

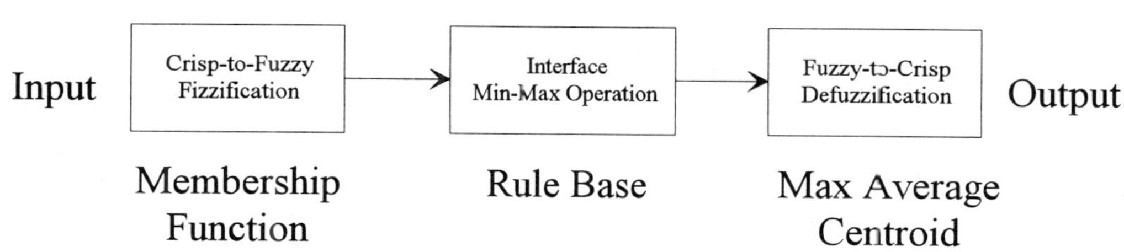

Figure 6. Tipical Fuzzy System Block Diagram

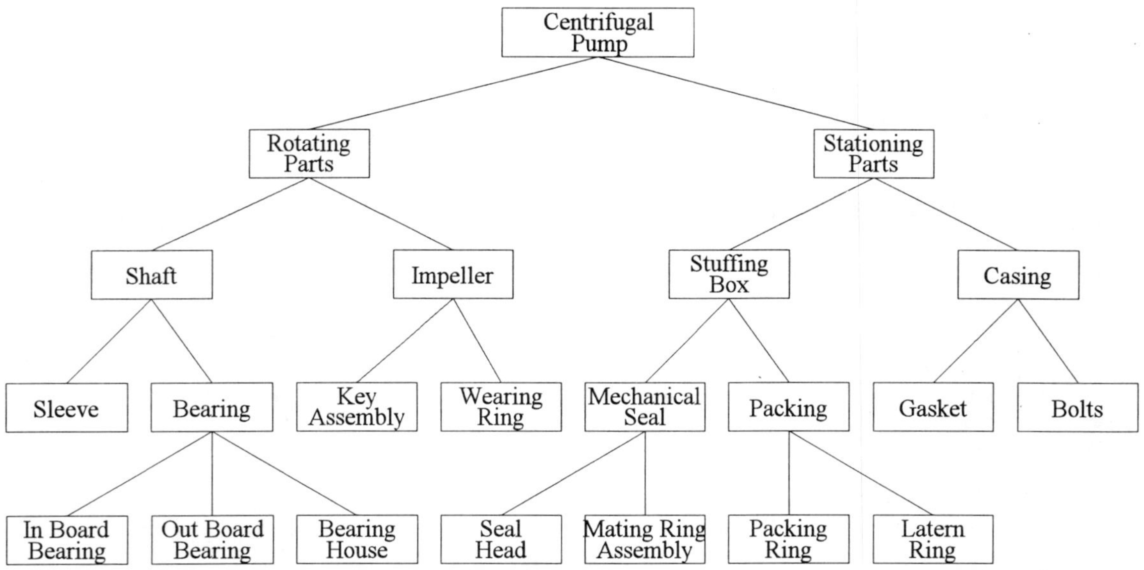

Figure 7. Hieratchical Model of a Centrifugal Pump Associated with Essential Parts

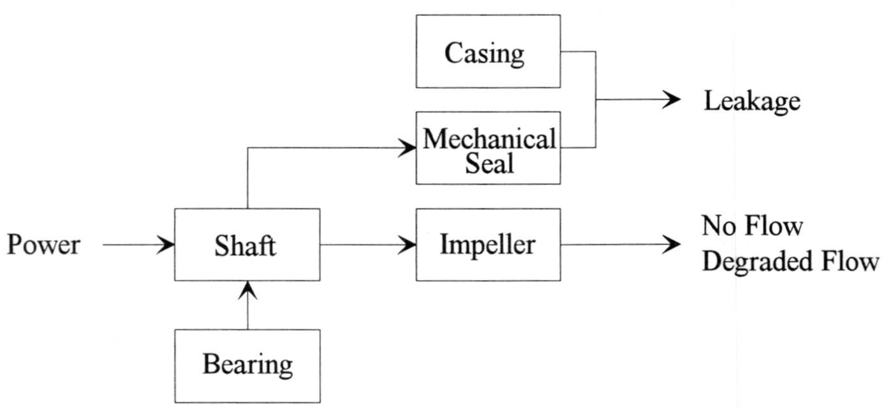

Figure 8. Simplified Reliability Block Diagram for a Centrifugal Pump Assocaited with Major Parts

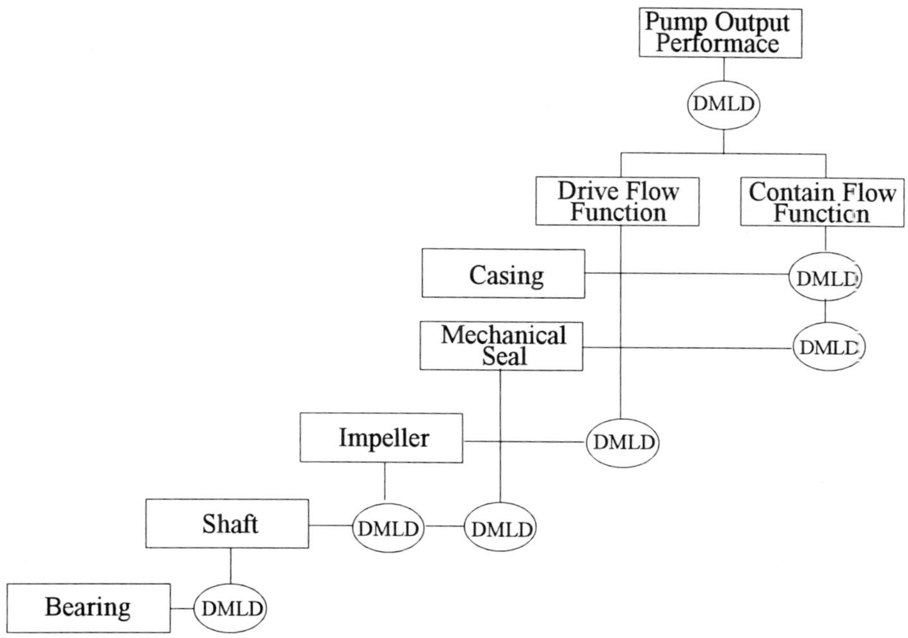

Figure 9. DMLD for a Centrifugal Pump Associated with Major Parts

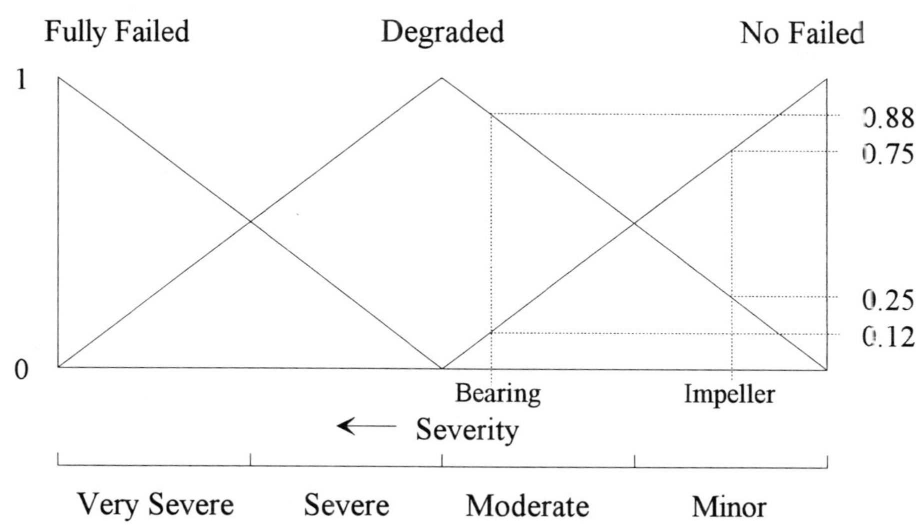

Figure 10. Linguistic Severity Input for Functional States of Pump Parts

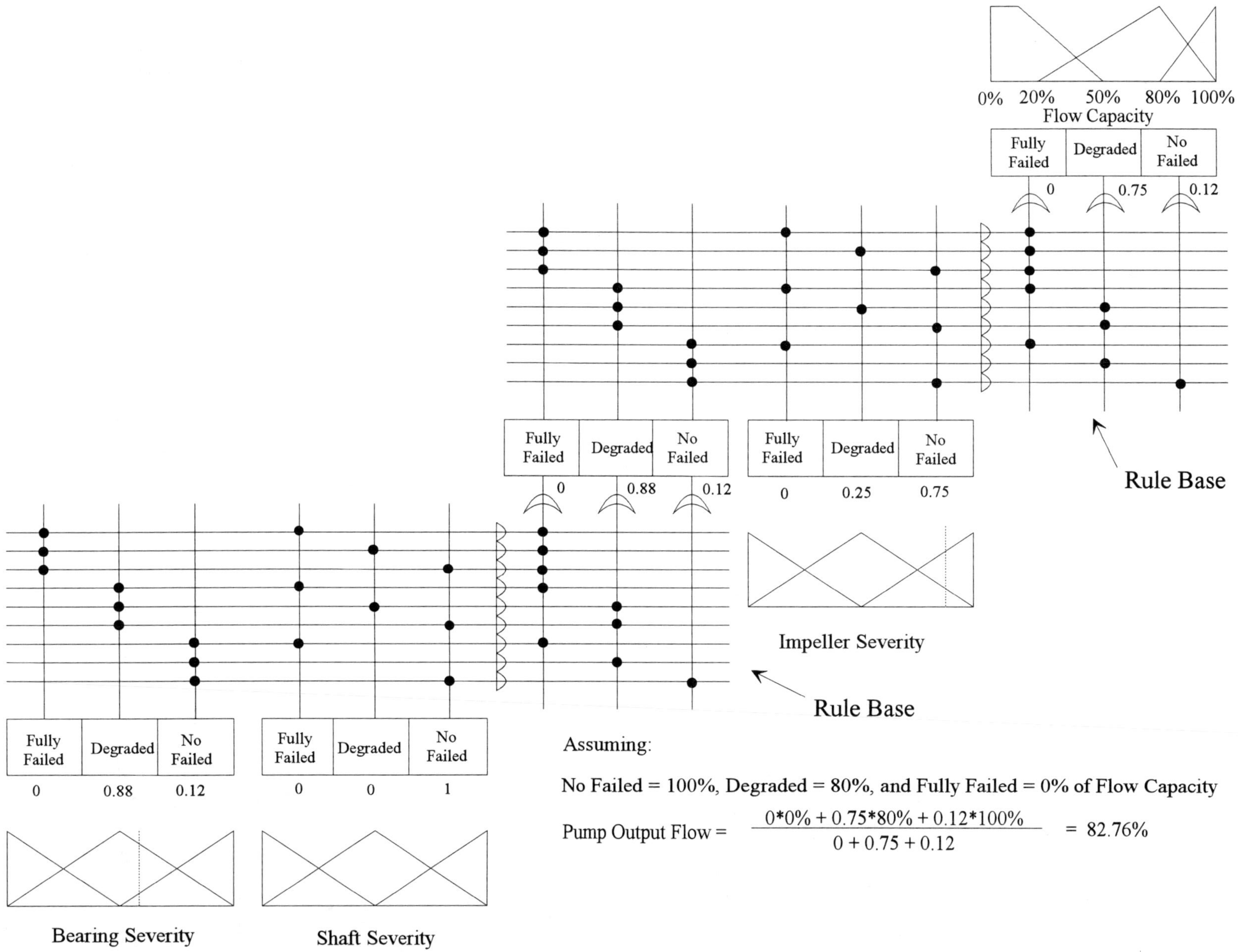

Assuming:

No Failed = 100%, Degraded = 80%, and Fully Failed = 0% of Flow Capacity

$$\text{Pump Output Flow} = \frac{0*0\% + 0.75*80\% + 0.12*100\%}{0 + 0.75 + 0.12} = 82.76\%$$

Figure 11. DMLD Prediction of Drive Flow Function of a Pump

Table 1. FMEA Analysis of a Centrifugal Pump

No	Major Parts	Function Mode	Failure Mode	Failure Cause	Effect	
					Local	System
1	Shaft	Transfer poser to impeller.	1. Fail to transfer power. 2. Reduced power transfer.	1. Shaft broken. 2. Shaft seizure. 1. Shaft bent. 2. Shaft running off center	1. Impeller stop. 1. Impeller wear. 2. Impeller partial broken. 3. Seal wear	No flow. Degraded flow. Leakage.
2	Impeller	Add energy to liquid.	1. Fail to rotate. 2. Degraded fluid acceleration.	1. Impeller broken. 2. Impeller clogged. 1. Impeller partial broken. 2. Impeller partial clogged. 3. Impeller wear.		No flow. Degraded flow.
3	Bearing	Support shaft and allow shaft to move smoothly by reducing friction.	1. Fail to support shaft. 2. Fail to maintain low friction.	1. Bearing broken. 2. Loss of lubrication. 1. Bearing wear. 2. Dirt getting into bearing.	1. Shaft fail to rotate. 2. Shaft running off center 3. Shaft partial blinding.	No flow. Degraded flow.
4	Casing	Collect liquid and convent velocity energy to pressure energy.	1. Leakage	1. Crack in casing. 2. Casing gasket failure. 3. Casing bolt inadequate tightening.		Degraded flow.
5	Mechanical Seal	Reduce or prevent leakage	1. Leakage	1. Seal/packing ring distortion. 2. Seal/packing wear. 3. No cooling liquid to lubricate. 4. Seal/packing improperly installed. 5. Incursion of hard solid particles into running clearances. 6. Shaft bent. 7. Shaft/sleeve wear.		Degraded flow

Table 2. Severity Evaluation of Pump Parts Failure

Rank	Description	Meaning	Equivalent Output Effected by Severity
1	Minor	Failure might have any real effect on a pump, operator probably will not even notice the failure.	90-100% of Flow Capacity
2	Moderate	Failure cause only a slight operator annoyance. Operator will probably only notice a slight deterioration of the performance of pump parts.	70-90% of Flow Capacity
3	Severe	Failure causes some dissatisfaction. Operator is made uncomfortable or is annoyed by the failure of pump parts.	50-70% of Flow Capacity
4	Very Severe	High degree of operator's high dissatisfaction due to the nature of the parts failure.	0-50% of Flow Capacity

Table 3. Rule Table for Pump Parts Failure

Parts States	Fully Failed	Degraded	No Failed
Fully Failed	FF	FF	FF
Degraded	FF	DF	DF
No Failed	FF	DF	NF

PVP-Vol. 340, Seismic Engineering
ASME 1996

NONLINEAR DYNAMIC ANALYSIS OF HYDRODYNAMICLY-COUPLED STAINLESS STEEL STRUCTURES

Yong Zhao
Engineering Mechanics Group
Battelle-Columbus
Columbus, Ohio

ABSTRACT

Spent nuclear fuel is usually stored temporarily on the site of nuclear power plants. The spent fuel storage racks are nuclear-safety-related stainless steel structures required to be analyzed for seismic loads. When the storage pool is subjected to three-dimensional (3-D) floor seismic excitations, rack modules, stored fuel bundles, adjacent racks and pool walls, and surrounding water are hydrodynamically coupled. Hydrodynamic coupling (HC) significantly affects the dynamic responses of the racks that are free-standing and submerged in water within the pool. A nonlinear time-history dynamic analysis is usually needed to describe the motion behavior of the racks that are both geometrically nonlinear and material nonlinear in nature. The nonlinerities include the friction resistance between the rack supporting legs and the pool floor, and various potential impacts of fuel-rack, rack-rack, and rack-pool wall. The HC induced should be included in the nonlinear dynamic analysis using the added-hydrodynamic-mass concept based on potential theory per the US Nuclear Regulatory Commission (USNRC) acceptance criteria. To this end, a finite element analysis constitutes a feasible and effective tool. However, most people perform somewhat simplified 1-D, or 2-D, or 3-D single rack and 2-D multiple rack analyses. These analyses are incomplete because a 3-D single rack model behaves quite differently from a 2-D mode. Furthermore, a 3-D whole pool multi-rack model behaves differently than a 3-D single rack model, especially when the strong HC effects are unsymmetrical. In this paper 3-D nonlinear dynamic time-history analyses were performed in a more quantitative manner using sophisticated finite element models developed for a single rack as well as all twelve racks in the whole-pool. Typical response results due to different HC effects are determined and discussed.

1. INTRODUCTION

Free standing spent fuel racks are submerged in water contained in a storage pool on nuclear facility sites. During a postulated strong motion earthquake, the water surrounding the racks is accelerated and the so-called HC effects are significantly induced between the water and the stored fuel assemblies, the racks, and the pool walls (Fritz, 1972; Dong, 1978). In the seismic response motion, the rack support legs may lift off from or slide on the pool floor steel liner due to inertial effects and the friction resistance to the motion of the rack legs from the pool floor liner. Such seismic response of the fluid-structure-coupling system may cause the highly nonlinear dynamic phenomena of impacts between the rack and the fuel assemblies that are free standing inside the rack cells, impacts of the rack with adjacent racks or the pool walls, and impacts of the rack support legs with the pool floor. The frictional motion of the free standing rack supports against the pool floor is also nonlinear in nature.

Due to HC effects that are usually unsymmetrical in practical problems, the seismic evaluation of the spent fuel storage racks is more complicated than for most other nuclear structural systems (Ashar and DeGrassi, 1989; DeGrassi, 1992). To this end, a 3-D nonlinear dynamic time-history analysis should be employed. The values of the Coulomb friction coefficient used to describe the friction resistance essentially govern the motion behavior of the free standing racks.

Most references on this subject cite analyses using 3-D single rack models with uniform gaps between racks and pool walls (Reed, 1979; Sturm, 1980; Durlofsky, 1981; Harstead, 1983) or nonuniform gaps (Scavuzzo, 1979; Soler and Singh, 1982, 1984; Pop, et al, 1990; Singh, et al, 1990; and Chang, 1994). The analysis results obtained are often used for the detailed design and construction of the rack modules. A few analyses using 2-D multiple rack models (Kabir, 1987; Stabel, 1993) or a 3-D whole-pool multiple-rack model (Singh and Soler, 1991) were performed to verify the dynamic motion phenomena found in the 3-D single rack analyses, because the multi-rack analysis is much more complicated in modeling and computationally expensive than the single rack efforts.

In this paper, the HC effects were determined using a 3-D single rack model and a 3-D whole-pool multiple-rack model developed herein and using the general purpose nonlinear finite element code SOLVIA (1992). Prescribed floor seismic motion time-histories specified in the three directions were applied simultaneously together with the rack deadweight. The results thus obtained for two typical different Coulomb friction coefficients (Ashar and DeGrassi, 1989; DeGrassi, 1992) were compared and discussed for the seismic design of racks.

2. 3-D SINGLE RACK MODEL

Figure 1 shows a typical 12 cell by 13 cell (12 x 13) free-standing spent fuel storage rack module referred to as N7, which is a welded honeycomb stainless steel structure. The location of the new rack N7 and its gap sizes around are shown in the pool layout in Figure 2, Table 1, and Figure 4. The spent fuel assemblies stored inside the rack cells are 17 x 17 standard fuel rods. Initial physical gaps of 0.0055 m and 0.0068 m exist between the fuel assemblies and rack cells for the new racks and existing racks, respectively. There are four and seven support legs for the new and existing racks, respectively.

As shown in Figure 3, the 3-D single rack stick finite element model developed using the SOLVIA (1992) code can effectively simulate the physical response phenomena of Rack N7. The model consists of 50 elements (including 28 nonlinear impact elements and 4 nonlinear contact-friction elements) with 112 degrees of freedom (DOF) and 13 1-D rigid links.

In the illustrative Figure 3, the general rack body and fuel bundle are represented separately by three lumped masses connected by two beam elements with the stiffness of the rack body and the fuel assembly cladding assuming all the fuel assemblies moving in phase. A 1-D rigid link vertically connecting the rack mass and the fuel bundle mass at the rack bottom is specified, assuming they remain in contact on a frictionless surface due to the relative large deadweight of the fuel assemblies moving freely inside the rack cells in the horizontal directions. The fuel-rack impacts at the three mass locations in the North, South, West, and East directions are simulated using twelve 1-D material nonlinear truss-gap impact elements and twelve 1-D rigid links, since only one directional impact is active at any instant when the impact happens. Four 3-D geometrical nonlinear contact-friction elements are defined to simulate the motion behavior of vertical contacting and horizontal frictional sliding at the support legs and pool floor liner interfaces with initially closed gaps. There are four impact elements simulating the potential vertical impact between the legs and the pool floor. Another group of twelve horizontal nonlinear impact elements are defined at the three mass locations between the rack and adjacent racks or pool walls, which will be active only when the corresponding rack displacement responses are larger than the physical gap dimensions in the same directions.

The HC effects can be rationally taken into account using the adding hydrodynamic mass concept based on potential theory (Fritz, 1972; Dong, 1978), which has been used by many investigators listed previously and accepted by the USNRC (1979, 1981). The hydrodynamic mass approach allows use of available structural analysis finite element codes and the stick finite element model described above. The hydrodynamic mass M_{Hx} for two submerged rectangular cylinders with non-uniform gaps filled with water as shown in Figure 4, due to relative motion in the X direction, was extended using the analytical, numerical, and experimental work by Fritz (1972), Dong (1978), Scavuzzo (1979), Pop (1990), and Singh (1990) as:

$$M_{Hx} = 2\rho h C^2 (C/3g_1 + C/3g_3 + B/g_2 + B/g_4) \qquad (2.1)$$

in which, ρ is water mass density, h is the cylinder height, C and B are the nominal dimensions, and g_1, g_2, g_3, and g_4 are the gap sizes. The HC between the two cylinder masses, Mass 1 representing a fuel bundle mass and Mass 2 a rack mass, is then described as the "added" forces b

$$\begin{vmatrix} F_{x1} \\ F_{x2} \end{vmatrix} = \begin{vmatrix} M_1 + M_2 + M_{Hx} & -(M_1 + M_{Hx}) \\ -(M_1 + M_{Hx}) & M_{Hx} \end{vmatrix} \begin{vmatrix} \ddot{X}_1 \\ \ddot{X}_2 \end{vmatrix} \qquad (2.2)$$

where, F_{x1} and F_{x2} are the added force acting on Mass 1 and Mass 2, respectively. M_1 and M_2 are the masses of water displaced by Mass 2 and enclosed by Mass 1 in the absence of Mass 2, respectively. X_1 and X_2 are the absolute accelerations of Mass 1 and Mass 2.

The mass matrix in Equation (2.2) for the added HC forces between the coupled Mass 1 and Mass 2 is easy to calculate and add to the structural analysis finite element model described previously when M_1, M_2, M_{Hx}, and M_{HZ} are computed separately in the two orthogonal horizontal motion directions. This approach can be used for the HC problem between the fuel bundle and the rack as well as between the rack and its adjacent racks or pool walls. An in-phase motion of the racks can be conservatively assumed (Ashar, 1989; DeGrassi, 1992; Pop, 1990).

From Equation (2.1), it is noted that the smaller the gap sizes, the larger the hydrodynamic masses. Thus, the masses are usually not equal to each other since the gap sizes are unsymmetrical in practice. The effect of the variation of the gap sizes on the hydrodynamic mass during a seismic event is not large until the gap on one side becomes very small (Singh, 1990), and therefore is usually not taken into account.

The vertical HC in the Y direction between the rack baseplate and the pool floor can also be considered similarly using the hydrodynamic mass M_{Hy} determined by (Fritz, 1972)

$$M_{Hy} = K(\pi \rho a^2 b/4) \qquad (2.3)$$

in which, $K = 0.487$, a and b are the dimensions of the baseplate, ρ is the water mass density. The M_{Hy} value calculated usually is less than 1% of the value of M_{Hx} or M_{HZ}.

3. 3-D WHOLE POOL MULTIPLE RACK MODEL

In engineering practice, since detailed nonlinear time-history analyses are usually conducted first for design purposes using 3-D single rack models, the major objective of a nonlinear dynamic time-history analysis using a 3-D whole pool multiple rack model is to obtain the information on the response behavior for all racks placed inside a pool and to verify whether the impact of rack-rack or rack-pool wall occurs. On the basis of the 3-D single rack model presented previously, a 3-D whole-pool multiple-rack model was developed as shown in Figure 5. The model includes all the nine new racks and three existing racks (Figure 2). In contrast to the 3-D single rack model in Figure 3, there are only two lumped rack and fuel mass pairs at the rack top and bottom in the 3-D whole pool model so as to reduce the total number of DOF and the associated expensive computing effort.

Each of the three existing rack stick models developed has seven rack support legs that are exactly modeled. The 3-D whole-pool multiple- rack model developed herein consists of a total of 407 elements (including 184 materially nonlinear impact elements and 57 geometrically nonlinear contact-friction elements), 1296 DOF, and 108 1-D rigid links. All the new and existing rack modules, surrounding water, and pool walls are hydrodynamically coupled in the whole-pool model.

4. 3-D NONLINEAR DYNAMIC ANALYSES

The friction effect on a pair of relatively moving and contacting solid bodies is a complicated phenomenon that depends on assumptions made and many internal and external factors. Nevertheless, the USNRC (1979, 1981) recommended and accepted the Coulomb static friction coefficient of 0.2 as a lower bound and 0.8 as a upper bound to describe the sliding friction resistance of the pool floor steel liner to the lateral motion of the rack leg bottom surfaces. This was determined statistically by Rabinowicz (1976) from his 199 stainless steel rack sliding test results including the effects of surface finishes and sliding speed. Therefore, constant static friction coefficients of 0.2 and 0.8 were assumed.

The damping effects due to the HC effects are small and can be neglected (Dong, 1978; USNRC, 1979, 1981). The damping considered is from the rack module structure only. Rayleigh damping (Clough and Penzien, 1975) may be used to conduct the nonlinear dynamic time-history analysis.

Prescribed uncorrelated 3-D floor seismic time histories of a safe shutdown earthquake (SSE) event with peak floor horizontal accelerations of 0.2g (Figure 6) were applied simultaneously in the three orthogonal directions plus the vertical deadweight of the rack modules. The vertical excitation peak values are equal to two thirds of the horizontal ones. The seismic excitation duration of 20 seconds was fully considered. Through a series of test runs a time step of 0.0015 seconds, plus a set of convergence control parameters properly defined was determined to achieve good numerical convergence. It is extremely important to get converged and stable results in the iterations of a nonlinear dynamic analysis of contact-impact and friction motion problems.

The general purpose nonlinear finite element code SOLVIA (1992) was employed. SOLVIA has been in commercial use about six years and was developed from the ADINA code which has been used widely and successfully in the nuclear industry for nonlinear problems for more than 15 years.

5. RESULTS AND DISCUSSIONS

Typical seismic response results for the 3-D single rack module N7 are shown in Figure 7 when the friction coefficient is equal to 0.8. The results indicate that the rack deflection is small but dominates the rack motion behavior. The rack maximum displacement of 4.32×10^{-4} m is found in the N-S direction at the instant of about 16 seconds. It is note that the smaller gaps, or larger added mass, and relative smaller structural stiffness (only 12 rather than 13 cells) exist in that direction. No tilting or uplift occurs due to no vertical displacements of the supports. The fuel bundle behaves like a rigid body inside the rack cell without uplift. However, it impacts the rack cell sometimes. The impact occurs at all the three elevations with the maximum value of 11,987 Newtons per cell at the rack top in the South direction. A more detailed plastic impact analysis of the fuel cladding tube may be necessary due to the strong fuel-rack impact behavior observed in this case. There is no rack-rack or rack-pool wall impact happening because the maximum rack displacements are always less than the physical gap dimensions presented earlier.

Figure 8 shows typical response results of the 3-D single rack module N7 in the case of the friction coefficient being equal to 0.2. The results demonstrate that sliding dominates the rack responses behavior. The maximum displacements at the rack top in the N-S and E-W directions are respectively about 5 times and 14 times those obtained when the friction coefficient is equal to 0.8. The largest rack displacement is 2.62×10^{-3} m in the E-W direction occurring at the instant of about 14 seconds when the displacement direction in the N-S direction is changed from the North (+X) to South (-X). It is noted that the largest displacement range at the rack top is still in the N-S direction due to the larger HC effects existing. Comparison of the displacement response time histories of the rack for the two different coefficients of friction demonstrates that the variation of friction coefficient significantly affects not only the value, but also the direction of the displacements which are loading history dependent. With the strong friction effects, the largest displacement in an absolute sense may happen in the direction with relatively weaker HC like this particular case when the difference of the HC effects in the two directions is not large, and when the excitations in the two directions have the same peak values but are statistically independent. From the results obtained it is found that there is no tilting or uplift occurring. Fuel-rack impact occurs occasionally, but only in the N-S direction with a maximum impact force of 2326 Newtons per cell. This is one fifth the value obtained when the coefficient of friction is equal to 0.8. This means the higher the coefficient of friction, the larger the fuel-rack impact force under the same prescribed excitations. No rack-rack or rack-pool wall impact is found.

The coefficient of friction of 0.2 was found to be the upper bound for the rack displacement responses in the 3-D single rack model analysis, therefore it was used for the 3-D whole-pool multiple-rack model analysis.

Typical response results for the whole-pool model (Figures 9 and 10) indicate that the rack motion is governed by sliding (Figure 9). No tilting or uplift and no rack-rack or rack-pool wall impact occur for all the 12 racks. Rack N4 exhibits the largest displacement of 0.01 m in the E-W direction (Figure 10). One explanation may be that Rack N4 is one of the largest (12 x 13) modules with larger inertia and hence larger displacement than other smaller modules. Another reason may be due to the smaller adjacent gaps (Table 1) and hence the larger HC effects on its motion than other 12 x 13 modules Further, for Rack N4 itself, a larger added mass was computed in the E-W direction than in the N-S direction. Even though Rack N5 has smaller gaps than Racks N4 and N7 (Table 1), it exhibits smaller displacement responses than Racks N4 and N7 because the adjacent racks are smaller and the HC effects are weaker. Fuel-rack impact happens only at a few instants with a maximum of 3990 Newtons per cell in Rack N4 in the East direction (Figure 10). Of the three existing racks, Rack E2, with the strongest HC effects, exhibits the largest displacements which are about 30 percent of Rack N4. No fuel-rack impact was found with the existing racks. This study indicates that the stronger the HC effects with a rack in a particular direction, the higher its dynamic responses in the same direction.

One of the important findings of the whole-pool multiple-rack analysis is that due to the HC effects, the displacements at the top of Rack N7 were found to be about 2.8 (in the N-S direction) to 3.2 (in the E-W direction) times those determined in the 3-D single rack model analyses with the same excitations and friction coefficient of 0.2. This means that the 3-D single rack model may under predict the responses by about a factor of three. This may be caused by the unsymmetrical HC effects from adjacent racks introduced in the whole pool model. A whole-pool multiple-rack model is recommended for a layout design of racks.

6. REFERENCES

Ashar, H. and DeGrassi, G., (1989) "Design and Analysis of Free-Standing Spent Fuel Racks in Nuclear Power Plants (An Overview)," SMiRT 10, K2, pp. 595-600.

Chang, S. J., (1994) "Seismic Analysis of Submerged Spent Fuel

Storage Structure," ASME PVP-Vol. 271, pp.73-82

Clough, R. W. and Penzien, J., (1975) "Dynamics of Structures," Second Edition, McGraw-Hill, Inc.

DeGrassi, G., (1992) "Review of the Technical Basis and Verification of Current Analysis Methods Used to Predict Seismic Response of Spent Fuel Storage Racks," NUREG/CR-5912, October.

Dong, R. G., (1978) "Effective Mass and Damping of Submerged Structures," UCRL-52342, Lawrence Livermore Laboratory, April

Durlofsky, H. and Sun, P. C., (1981) "Fuel/Module Impact Load in a Submerged Fuel Storage," SMiRT 6, August, J8/9

Fritz, R. J., (1972) "The Effects of Liquids on the Dynamic Motions of Immersed Solids," Journal of Engineering for Industry, Trans. ASME, pp.167-173, February

Harstead, G. A., et al, (1983) "Seismic Analysis of High Density Spent Fuel Storage Racks," ASME 83-NE-11

Kabir, A. F., et al, (1987) "Nonlinear Analysis of Spent-Fuel Racks for Consolidated Fuel Loading," SMiRT 9, August, Vol. K1, pp. 409-414.

Pop, J. Jr., Putman, S. F. and Singh, S., (1990) "Nonlinear Dynamic Analysis of Spent Fuel Storage Racks," ASME PVP-Vol. 191, pp. 113-118

Reed, J. W., Webster, F. A., and Sun, P. C., (1979) "Alternative Structural Systems for High Density Fuel Storage Racks in Existing Facilities," SMiRT 5, August, K4/4

Rabinowicz, E., (1976) "Friction Coefficients of water-Lubricated Stainless Steel for a Spent Fuel Rack Facility," Massachusetts Institute of technology, November

Scavuzzo, R. J., Stokey, W. F., and Ragke, E. F., (1979) "Dynamic Fluid-Structure Coupling of Rectangular Modules in Rectangular Pools," ASME PVP-39, June 25-29

Singh, K. P. and Soler, A. I., (1991) "Chin Shan Analyses Show Advantages of Whole Pool Multi-Rack Approach," Nuclear Engineering International, March, pp. 37-40

Singh, S., Putman, S. F. and Pop, J. Jr., (1990) "Structural Evaluation of Onsite Spent Fuel Storage: Recent Developments," Current Issues Related to Nuclear Power Plant Structures, Equipment and Piping," Proceedings of the third Symposium, Orlando, Florida, December, pp.V/4 1-18

Soler, A. I. and Singh, K. P., (1982) "Dynamic Coupling in a Closely Spaced Two-Body System Vibrating in a Liquid Medium: the Case of Fuel Racks," The 3rd Keswick International Conference-Vibration in Nuclear Plants, Keswick, U. K., May

Soler, A. I. and Singh, K. P., (1984) "Seismic Response of a Free Standing Fuel Rack Construction to 3-D Floor Motion," Nuclear Engineering and Design, Vol.80, pp.315-329

SOLVIA Engineering AB, (1992) "Users Manual for SOLVIA," Revision 90.2, April

Stabel, J., Ren, M. and Swelim, H., (1993) "Calculation of Seismic Loads on Fuel Storage Racks under Consideration of Fluid-Structure Interaction," SMiRT 12, August, K15/6, pp. 61-66

Sturm, A. J., Jr. and Song, C. S., (1980) "The Effects of Submergence on Structural Response in Confined Pools," Nuclear Engineering and Design, Vol. 60, pp. 287-296

USNRC, (1979) "OT Position for Review and Acceptance of Spent Fuel Storage and Handling Applications," January, modification to this document of April 1978.

USNRC, (1981), NUREG-0800, Standard Review Plan, Section 3.8.4, Appendix D, "Technical Position on Spent Fuel Pool Racks."

Table 1 Dimension of Racks and Adjacent Gaps 1 inch = 2.54 cm

Dimension (inch)	Rack Module I.D.											
	N1	N2	N3	N4	N5	N6	N7	N8	N9	E1	E2	E3
2B	110.59	110.59	110.59	109.34	109.34	109.34	109.59	109.59	109.59	111.50	111.50	111.50
2C	119.51	109.34	119.51	119.51	109.34	119.51	119.51	109.34	119.51	111.66	111.10	114.10
g_1	2.99	2.99	2.99	0.50	0.50	0.50	0.50	0.50	0.50	1.00	1.00	1.00
g_2	2.70	0.50	0.50	2.70	0.50	0.50	2.70	0.50	0.50	2.70	1.60	1.60
g_3	0.50	0.50	0.50	0.50	0.50	0.50	1.00	1.00	1.00	2.99	2.99	2.99
g_4	0.50	0.50	2.70	0.50	0.50	2.70	0.50	0.50	2.70	1.60	1.60	7.60

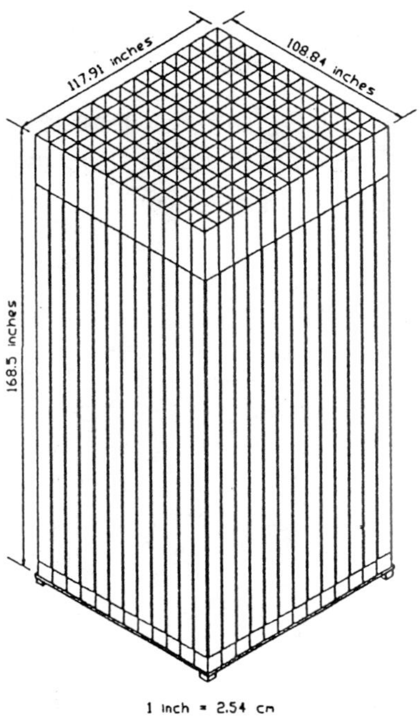

Figure 1 Typical 12x13 Cell Rack Module

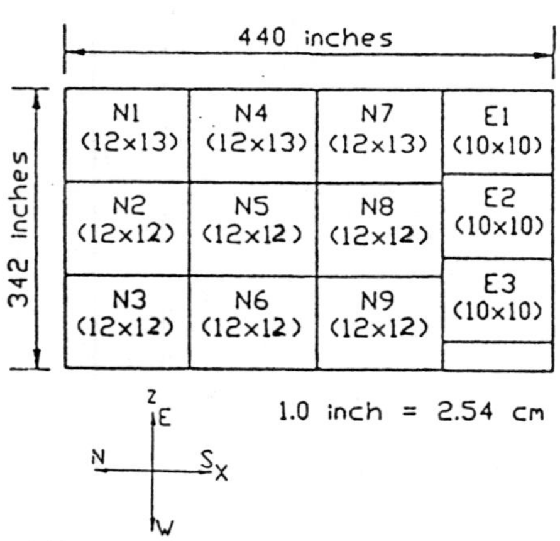

Figure 2 Rack Layout in a Storage Pool

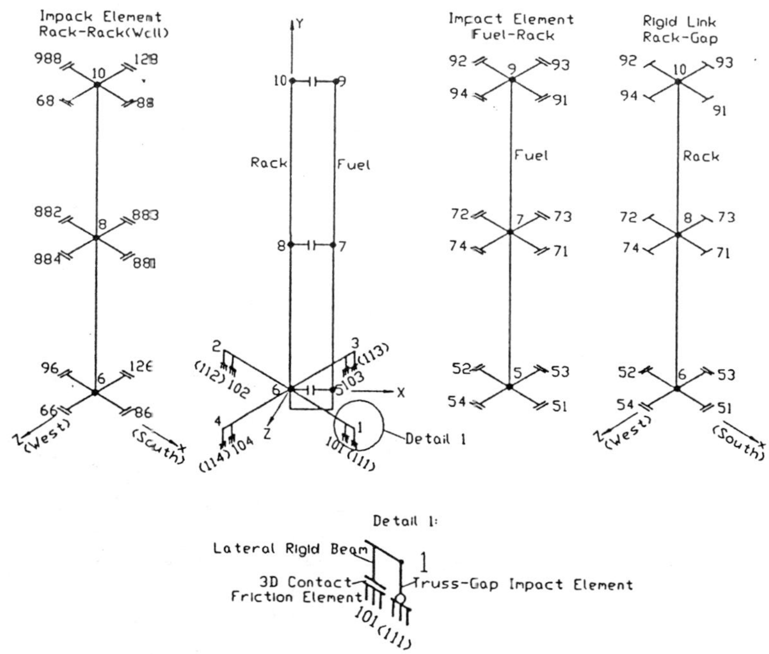

Figure 3 3-D Single Rack Stick Finite Element Model

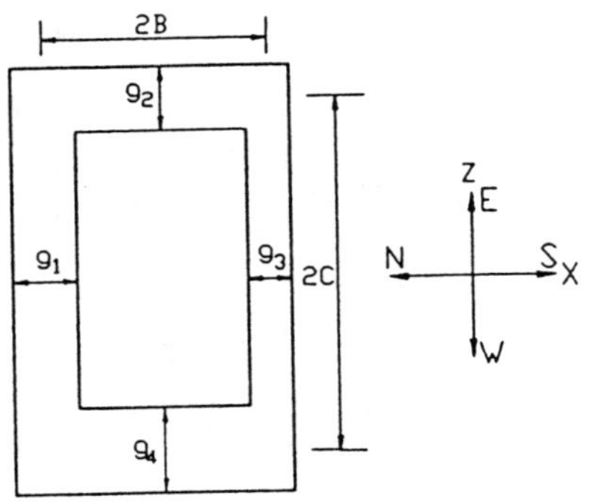

Figure 4 Cross Section Dimensions of Two Rectangular Cylinders Coupled by Fluid

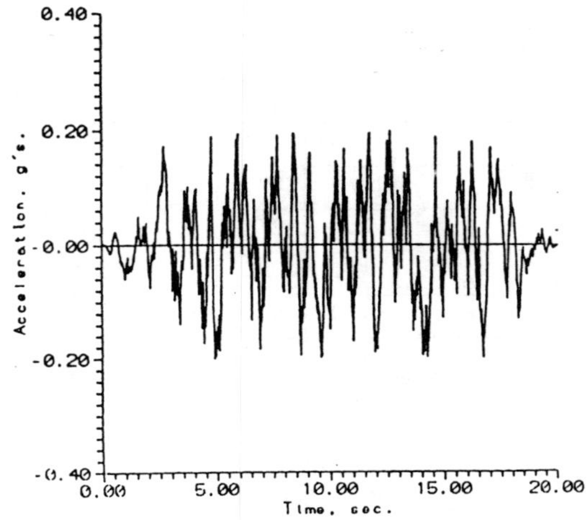

Figure 6 Floor Acceleration Response Time History in the N-S Direction

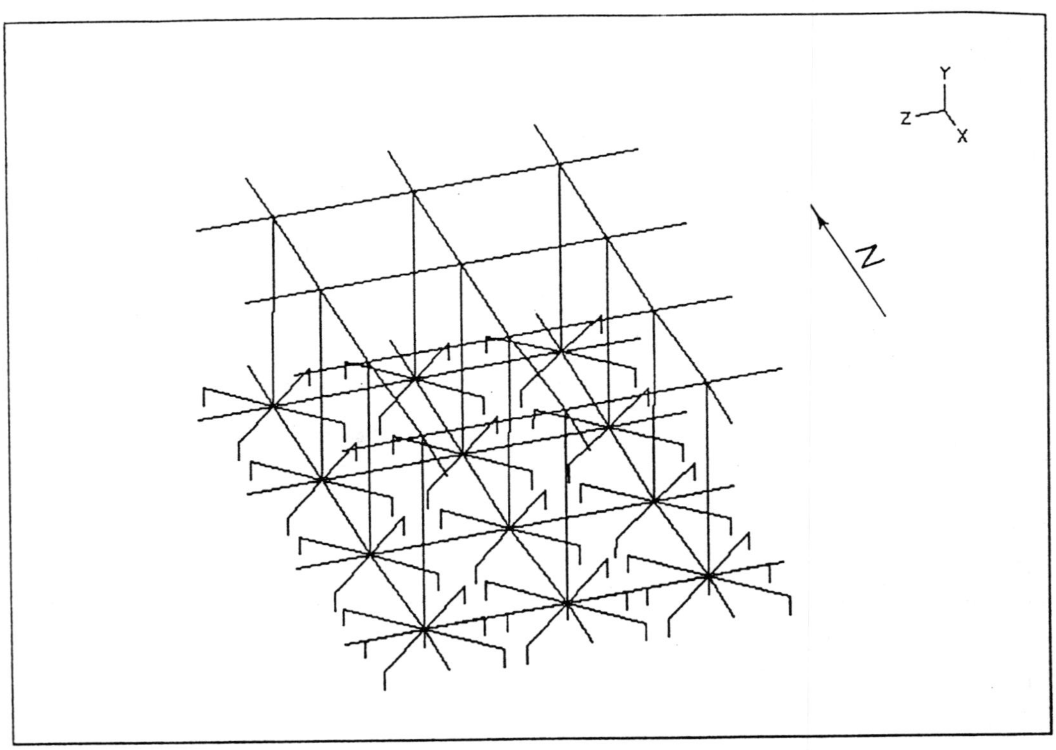

Figure 5 Whole Pool Multi-Rack Finite Element Model

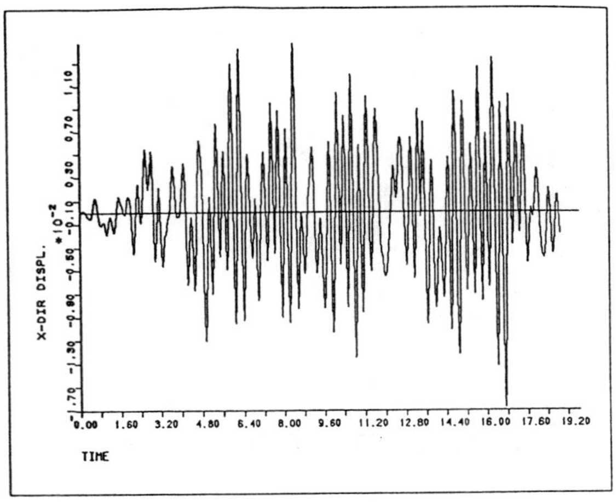

(7-a) Rack Top Displacement in the N-S Dir.

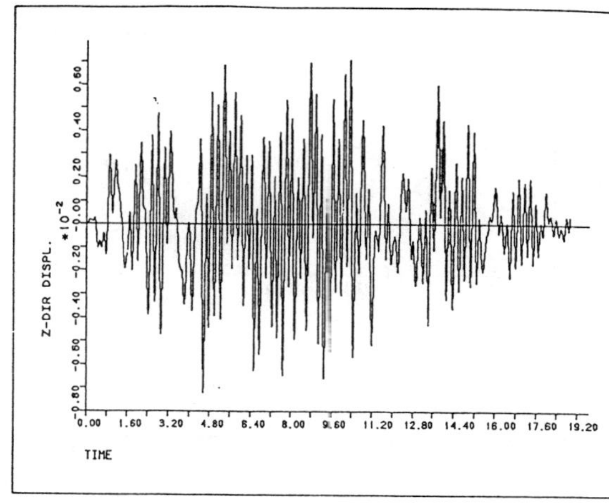

(7-b) Rack Top Displacement in the E-W Dir.

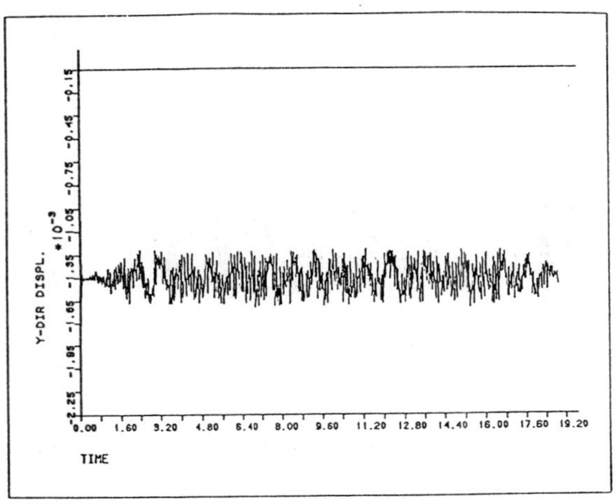

(7-c) Rack Top Displacement in the Vertical Dir.

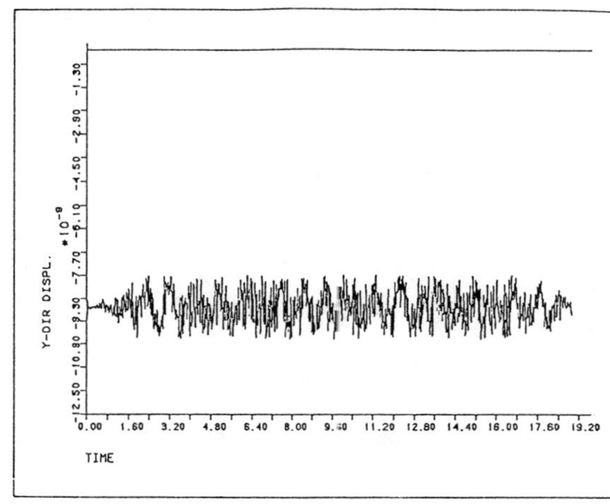

(7-d) Fuel Bottom Displacement in the Vertical Dir.

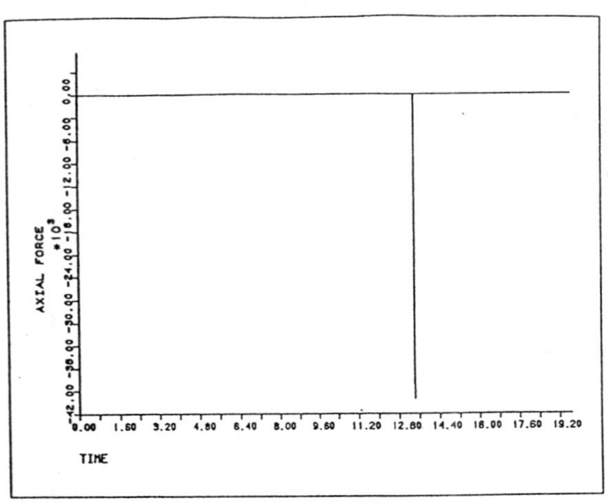

(7-e) Fuel-Rack Impact at Top in the South Dir.

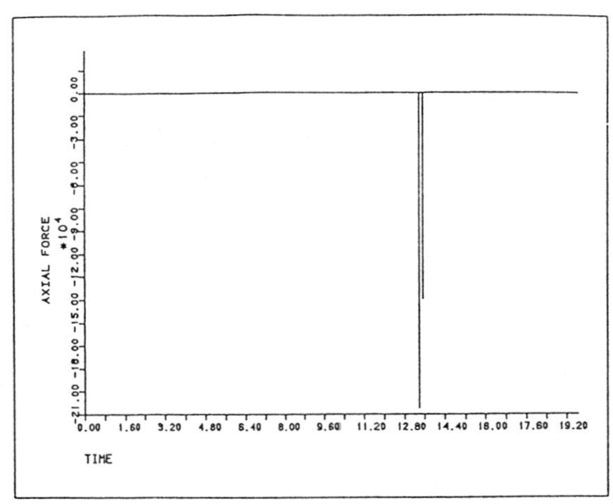

(7-f) Fuel-Rack Impact at Top in the North Dir.

Figure 7 Typical Responses of Rack N7 Using 3-D Single Rack Model with the Friction Coefficient Equal to 0.8
(1.0 in = 2.54 cm, 1.0 lbf = 4.448 N)

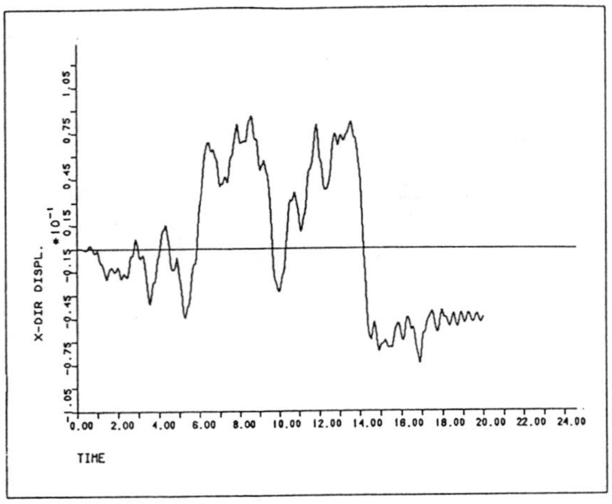

(8-a) Rack Top Displacement in the N-S Dir.

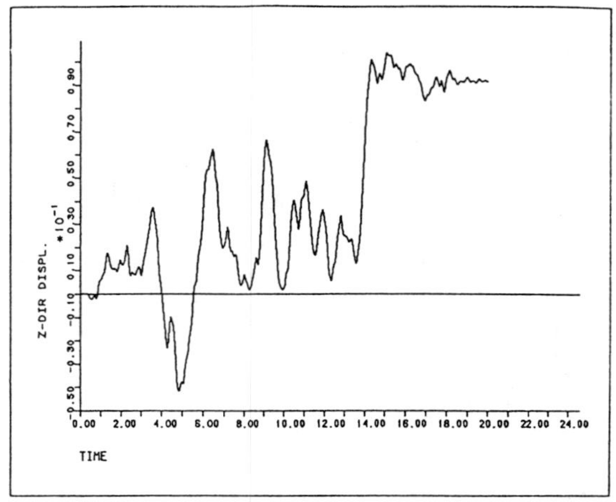

(8-b) Rack Top Displacement in the E-W Dir.

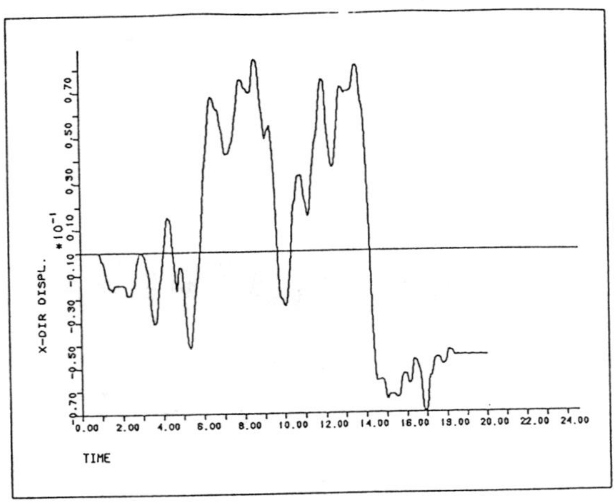

(8-c) Rack Bottom Displacement in the N-S Dir.

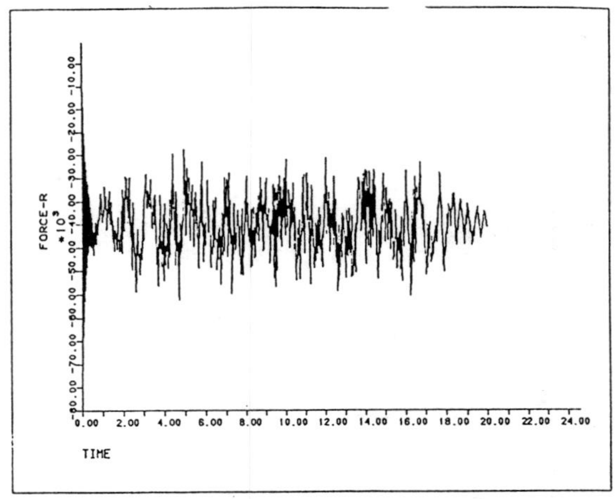

(8-d) Contact Force on the S-E Support Surface

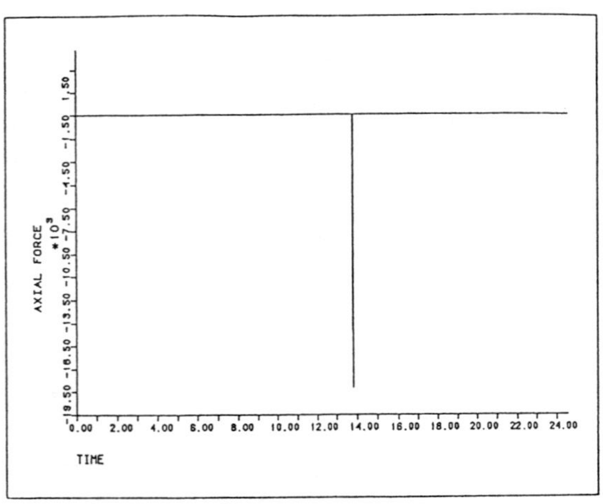

(8-e) Fuel-Rack Impact at Top in the South Dir.

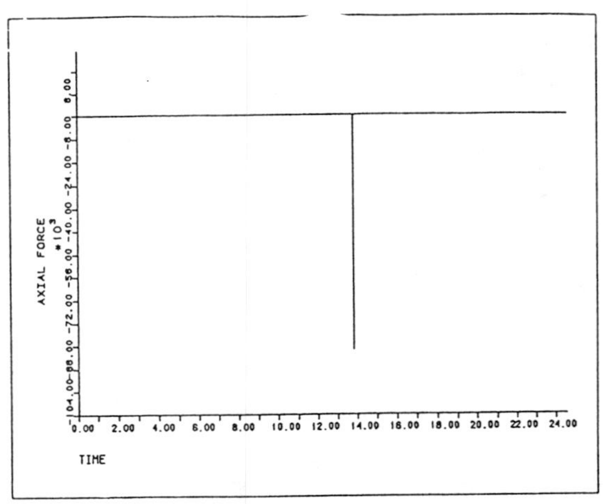

(8-f) Fuel-Rack Impact at Top in the North Dir.

Figure 8 Typical Responses of Rack N7 Using 3-D Single Rack Model with the Friction Coefficient Equal to 0.2
(1.0 in = 2.54 cm, 1.0 lbf = 4.448 N)

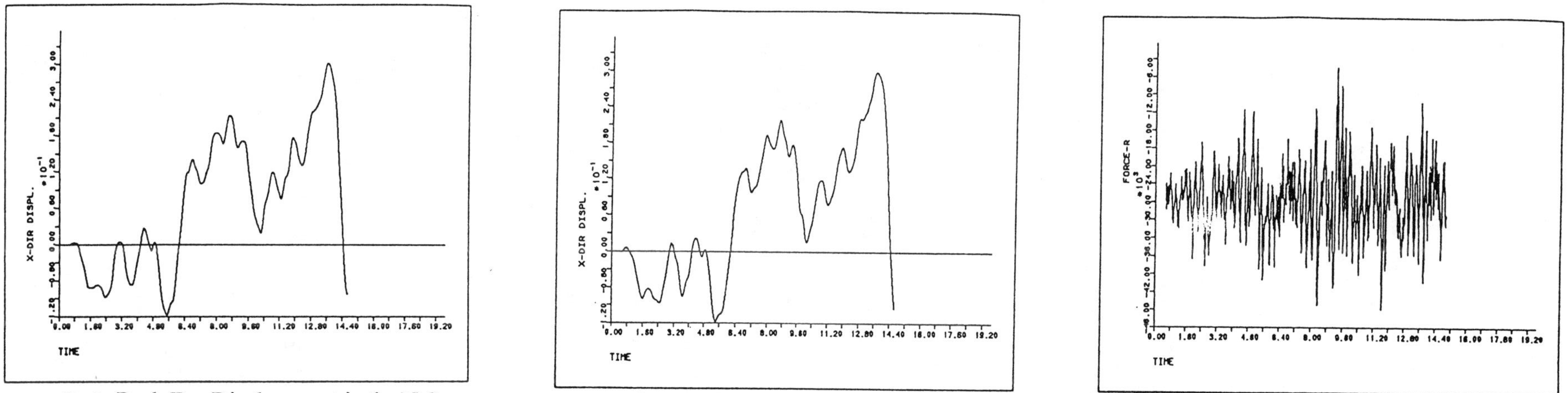

(9-a) Rack Top Displacement in the N-S (9-b) Rack Bottom Displacement in the N-S (9-c) Contact Force at Bottom of S-W Rack Support

Figure 9 Typical Responses of Rack N7 (1.0 in = 2.54 cm, 1.0 lbf = 4.448 N)

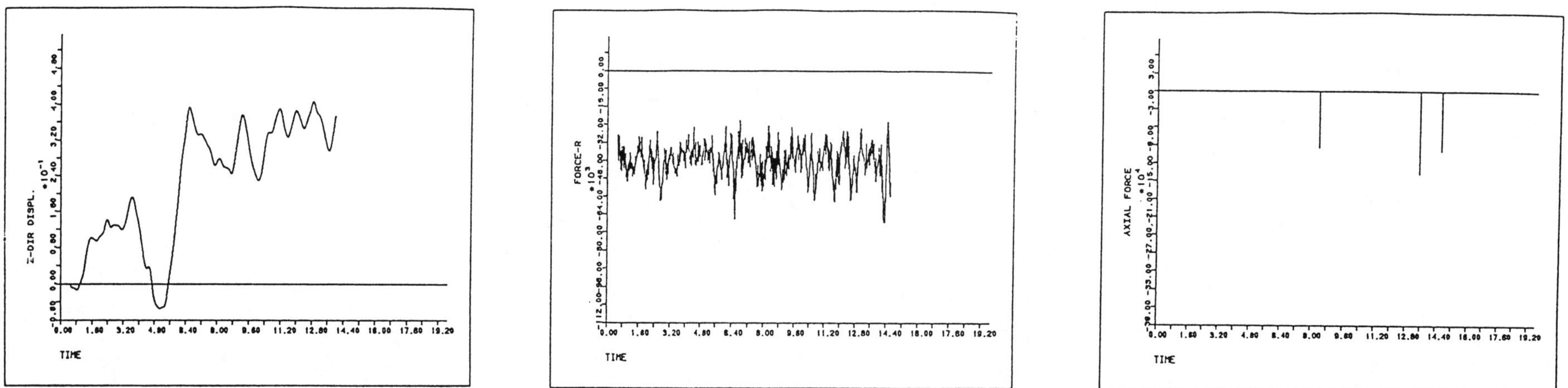

(10-a) Rack Top Displacement at in the E-W (10-b) Contact Force at Bottom of S-E Rack Support (10-c) Fuel-Rack Impact at Top in East Direction

Figure 10 Typical Responses of Rack N4 (1.0 in = 2.54 cm, 1.0 lbf = 4.448 N)

PVP-Vol. 340, Seismic Engineering
ASME 1996

EQUIVALENT LINEAR RESPONSE SPECTRUM METHODOLOGY FOR THE SEISMIC RESPONSE OF A SIMPLE STRUCTURAL SYSTEM WITH A UNI-DIRECTIONAL SUPPORT (TENSION ONLY OR COMPRESSION ONLY)

R.J. Gurdal
Integrated Nuclear Services
Framatome Technologies
Lynchburg, Virginia

W.D. Maxham
Integrated Nuclear Services
Framatome Technologies
Lynchburg, Virginia

ABSTRACT

Typically, structures with significant non-linear supports are analyzed using either time-history algorithms or response spectrum enveloping techniques, with multiple stiffness values to represent the non-linear supports. However, there are problems associated with these methods: (1) time-history analyses require considerable computer time, and, more importantly, the generated response files can become too large; (2) linear response spectrum analyses, when used to envelop the non-linear solution, are excessively conservative.

This paper presents an equivalent linear response spectrum methodology to analyze structural systems with a uni-directional support. The structure is free to move in one direction, and restricted with a support stiffness in the diametrically opposed direction. The mathematical formulas for this application are developed and presented.

The results using this methodology are compared with "actual" results from corresponding time-history non-linear structural analyses. These comparisons, performed for the one-mass structural system, show that the equivalent linear response spectrum method closely approximates the time-history results.

1.0 INTRODUCTION

Elastic non-linear supports of components, structures, or piping generally are analytically addressed for seismic excitation using non-linear time-history analysis algorithms or response spectrum methods, while accounting for multiple cases of support activity (a conservative envelop of multiple linear cases used to represent the total response). However, these analytical techniques have several drawbacks: 1) time-history analyses require considerable computer time and the generated response files can get extremely large resulting in handling, sorting, and storage problems, and 2) linear response spectrum analyses, when used to envelop the non-linear problem, get ultra-conservative very quickly. Further, the more non-linear supports present, the more complicated and conservative the envelop becomes.

The method presented herein uses an "equivalent" linear response spectrum method to analyze structural systems for seismic-type input with one-way bumper-type supports (no gap), e.g. a support that is active in compression but cannot take tension. The system and formulas given are for a simplified one-mass system with one linear spring and one non-linear, compression only, spring that has a zero gap when the system is at rest.

The results using this "equivalent" linear response spectrum methodology are compared with "actual" results from non-linear time-history structural analyses. These comparisons, performed for the one-mass structural system, show that the "equivalent" linear response spectrum method predicts the results of the non-linear ("actual") time-history analysis rather accurately, and, in fact, the "equivalent" method is slightly conservative.

We find that the maximum response of the one-mass system, in both the structure and the restraint, occurs for a sinusoidal forcing function with a forcing period equal to the average between the periods of the unrestrained and restrained structure. This is shown for two different damping values and for two different "restraint / structure" stiffness ratios. Based on these findings, the concept of an equivalent structural system is introduced. This system has two components. One element is an equivalent stiffness replacement for the bumper-type restraint. The other component is an energy-equality consideration that leads to deriving multiplication factors to be applied to the response spectrum results of the equivalent structural system to achieve final load distribution. The developed methodology is verified for harmonic input forcing functions and for an actual earthquake excitation. The results are discussed in Section 9.0, along with a brief outline of the initial impact problem and the generalization of the method for a multi-degree-of-freedom system.

It should finally be emphasized that this method is a "direct" method, in contrast to other equivalent linear methods which could require consecutive iterations to arrive at converging results.

2.0 NOMENCLATURE

subscript a — actual ("actual" is used in this paper for the results from the time-history analyses).

subscript E — Equivalent

M — Concentrated mass of the one-mass structural system

K_S — Stiffness of the structure

K_R — Stiffness of the Bumper-type restraint

f_{free} — Frequency of the unrestrained one-mass structural system

T_{free} — period of the unrestrained one-mass structural system.

f_{Restr} — Frequency of the restrained one-mass structural system

T_{restr} — period of the restrained one-mass structural system.

K_E — Equivalent restraint stiffness of the equivalent one-mass linear system.

λ — $\sqrt{(1 + \frac{K_R}{K_S})}$

δ_E — Maximum displacement of mass M in the equivalent one-mass linear system.

$\delta_{a(free)}$ — Maximum displacement of mass M in the free direction (bumper restraint not active)

$\delta_{a(Restr)}$ — Maximum displacement of mass M in the restrained direction (bumper restraint active)

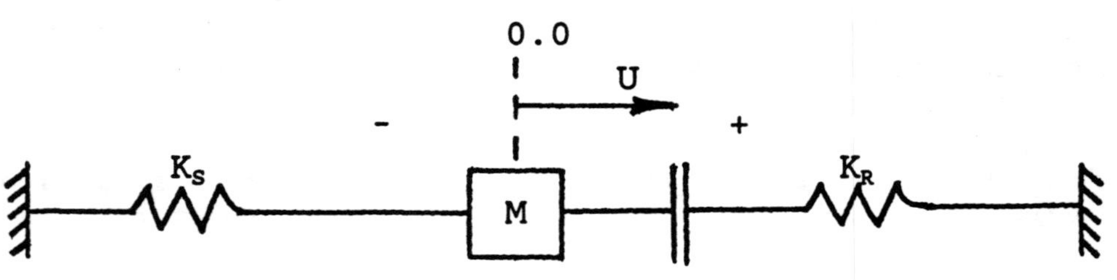

Figure 1: One-Mass System with Bumper-type Restraint
[K_R only active for U>0.0]

3.0 FORCING PERIOD EVALUATION

The evaluation performed in this section could be called "Forcing Frequency Evaluation". However, it seems more appropriate to refer to it as a "Forcing Period Evaluation" ,because, as explained below, the input parameter variation is a linear variation of the forcing period (where the forcing period is simply the inverse of the forcing frequency).

The one-mass structural system shown in Figure 1 is analyzed for a sinusoidal forcing acceleration time-history. The so-called free frequency of this one-mass structural system is equal to (References [1] and [2]):

$$f_{free} = \frac{1}{2\pi}\sqrt{\frac{K_S}{M}}$$

Also, the restrained frequency of this one-mass structural system is equal to:

$$f_{Restr} = \frac{1}{2\pi}\sqrt{\frac{(K_S + K_R)}{M}}$$

In each analysis documented later in this section (see Table 1), 11 dynamic non-linear computer runs (cases) are performed. The only difference between each one of these eleven computer runs is the period of the forcing function. These 11 periods ($T_{(1)}$, $T_{(2)}$, $T_{(3)}$,...$T_{(11)}$) are defined as follows:

$T_{(1)} = 1/f_{Restr} = T_{Restr}$ = period of restrained system,

$T_{(11)} = 1/f_{free} = T_{free}$ = period of free system.

The periods between the restrained and the free periods are evenly spaced via linear interpolation. The computer runs are performed for the four conditions shown in Table 1.

The results from analyses 1 through 4 are shown in Figures 2.1 through 2.4, respectively. From these figures, it can be seen that the maximum response in both the structure and the support occurs in case No. 6 in each analysis. This corresponds to period No. 6:

$$T_{(6)} = \frac{T_{(1)} + T_{(11)}}{2} = \frac{T_{Restr} + T_{free}}{2}$$

which is the one where the forcing function is the "Average Period" sinusoidal forcing function.

TABLE 1
Overview of Analyses for the Forcing Period Evaluation

	2% of critical damping	5% of critical damping
$K_R = K_S$	Analysis No. 1 (11 cases)	Analysis No. 2 (11 cases)
$K_R = 10 \times K_S$	Analysis No. 3 (11 cases)	Analysis No. 4 (11 cases)

K_R = Stiffness of the Bumper-type restraint

K_S = Stiffness of the structure

Figure 2.1 Analysis No.1

K(Restr.) = K(Structure) ; 2% damping

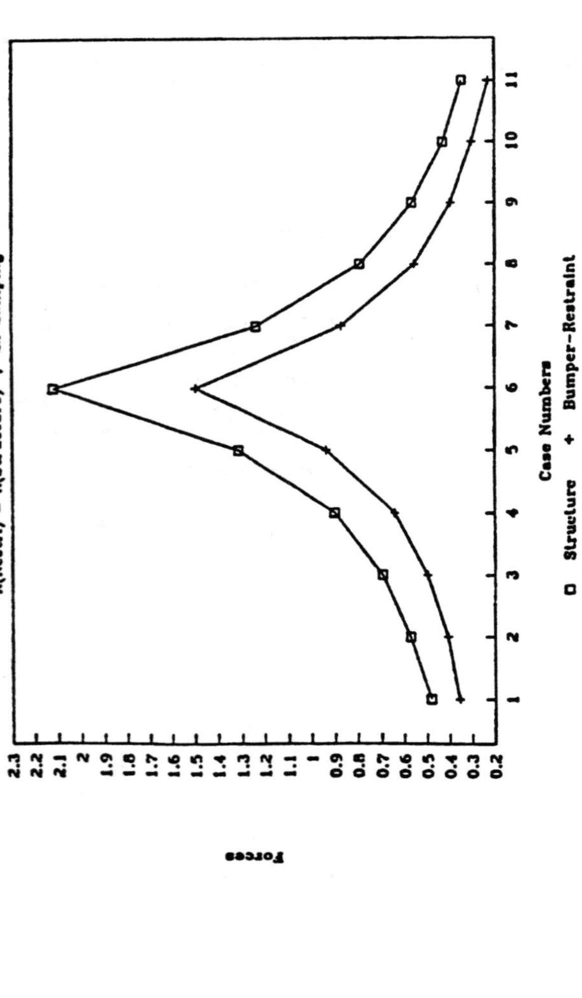

Figure 2.2 Analysis No.2

K(Restr.) = K(Structure) ; 5% damping

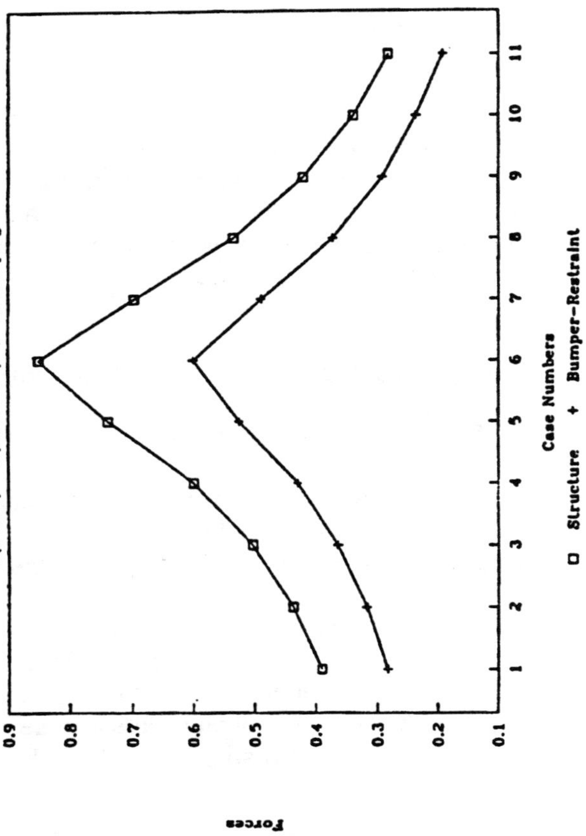

Figure 2.3 Analysis No.3

K(Restr.) = 10 * K(Str.) ; 2% damping

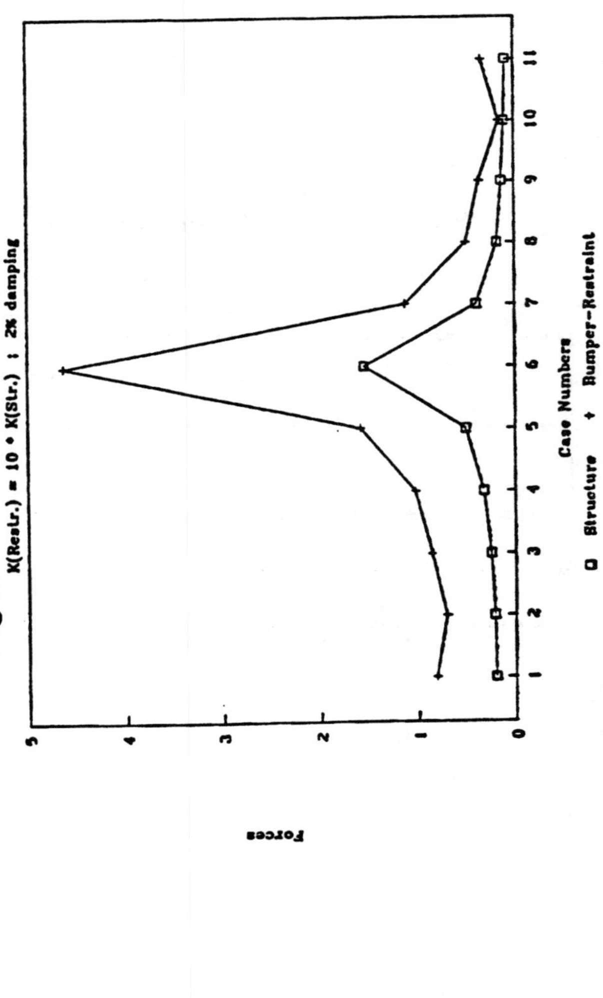

Figure 2.4 Analysis No.4

K(Restr.) = 10 * K(Str.) ; 5% damping

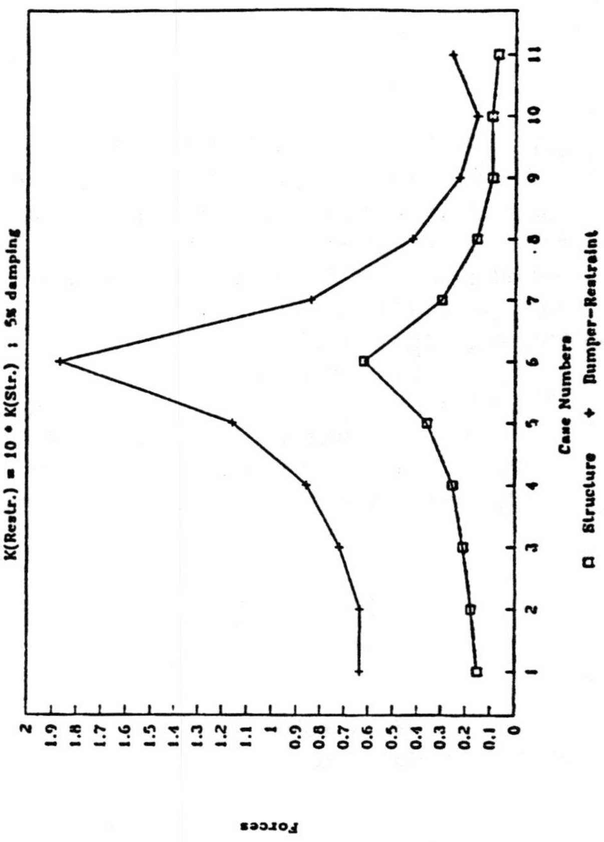

4.0 EQUIVALENT STRUCTURAL SYSTEM

As shown in the previous section, the forcing function with the average period leads to the highest response for both the structure and the restraint. Therefore, an Equivalent Structural System with an average period is built as shown in Figure 3 below. The calculation of the Equivalent Stiffness K_E is as follows:

$$\tfrac{1}{2}\ (T_{free} + T_{restr}) = T_E$$

$$\tfrac{1}{2}\ \left[2\pi\sqrt{\frac{M}{K_S}} + 2\pi\sqrt{\frac{M}{(K_S + K_R)}} \right]$$

$$= 2\pi\sqrt{\frac{M}{(K_S + K_E)}}$$

Introducing:

$$\lambda = \sqrt{(1 + \frac{K_R}{K_S})}$$

and solving for K_E yields:

$$K_E = K_S \frac{4\lambda^2}{(1+\lambda)^2} - 1 = K_S \frac{(3\lambda+1)\ (\lambda-1)}{(1+\lambda)^2}$$

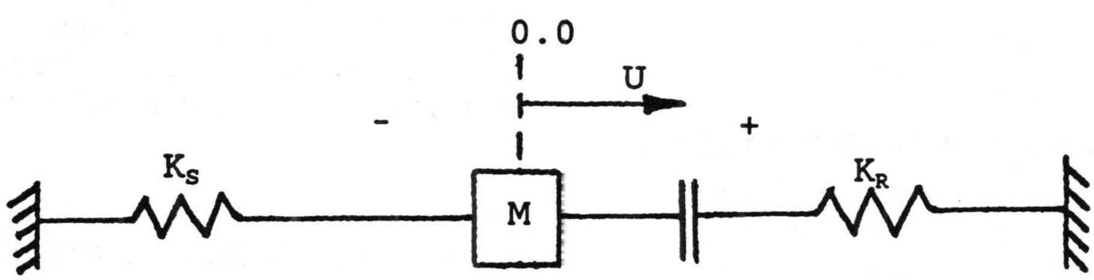

One-mass System with Bumper-type Restraint
[K_R only active for U>0.0]

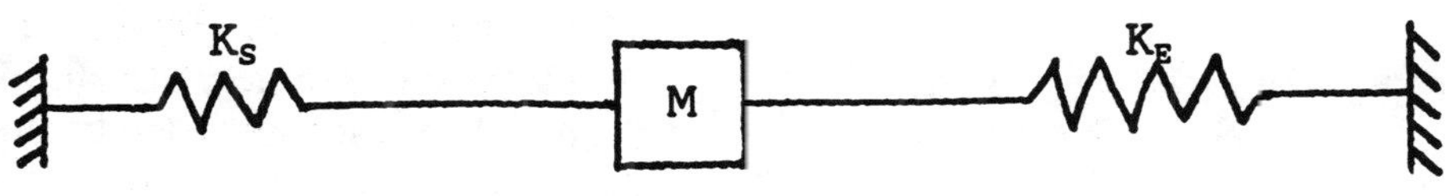

One-mass System with Average Period T_E

Figure 3: Overview of Average Period Equivalent Linear Methodology

5.0 ENERGY CONSIDERATIONS FOR CALCULATION OF MAXIMUM RESPONSES

The results from the Response Spectrum analyses on the Equivalent Structural System must be adjusted to produce the "actual" forces in both the structure and the restraint. These adjustments are based on the equality between the energy stored in the "actual" structure and that stored in the Equivalent structure.

a) Energy considerations in the free direction (mass M moving to the left of the zero position in Figure 3):

$$\tfrac{1}{2} K_S \cdot \delta^2_{a(free)} = \tfrac{1}{2} (K_S + K_E) \cdot \delta^2_E$$

Therefore: $\delta_{a(free)} = \dfrac{2 \cdot \lambda}{(1 + \lambda)} \cdot \delta_E$

b) Energy considerations in the restrained direction (mass M moving to the right of the zero position in Figure 3):

$$\tfrac{1}{2} \cdot (K_S + K_R) \cdot \delta^2_{a(Restr)} = \tfrac{1}{2} \cdot (K_S + K_E) \cdot \delta^2_E$$

Therefore: $\delta_{a(Restr)} = \dfrac{2}{(1 + \lambda)} \cdot \delta_E$

6.0 SUMMARY OF EQUATIONS (from Sections 4.0 and 5.0)

$$\lambda = \sqrt{1 + \frac{K_R}{K_S}}$$

Equivalent Stiffness $K_E = K_S \times \dfrac{(3\lambda + 1) \cdot (\lambda - 1)}{(1 + \lambda)^2}$

Required multiplication factors to obtain load distribution:

- for the structure:

$$\frac{2 \cdot \lambda}{(1 + \lambda)} \times F_E(\text{structure}) = F_a(\text{structure})$$

- for the restraint:

$$\frac{K_R}{K_E} \times \frac{2}{(1 + \lambda)} \times F_E(\text{restraint}) = F_a(\text{restraint})$$

7.0 PRACTICAL APPLICATIONS/TRIPLE SINE FORCING FUNCTION

Since an actual earthquake does not consist of a pure sine forcing function, a "triple" sine forcing function is built. This forcing function consists of the algebraic sum of three pure sine forcing functions of equal amplitude with the following three frequencies: frequency of the free structure, frequency of the restrained structure, and frequency corresponding to the average period between the free and the restrained structures.

The triple sine forcing function is applied as input to the one-mass structural system shown in Figure 1. The following two cases are analyzed:

Case 1: $K_R = K_S$
Case 2: $K_R = 10 \times K_S$.

All analyses are performed for 5% of the critical damping.

The maximum results from both the time-history non-linear analyses and the developed average period method are given in the upper half of Table 2. In this Table 2, the calculated results from the developed methodology are all based on the raw spectra from the time-histories (no peak-broadening is performed).

8.0 PRACTICAL APPLICATIONS / TAFT EARTHQUAKE

The triple sine forcing function in the analyses above is replaced with the actual Taft Earthquake acceleration time-history and response spectra. The same two cases are analyzed ($K_R = K_S$ and $K_R = 10 \times K_S$, both cases for 5% of the critical damping).

For Case 1 ($K_R = K_S$), Figure 4 shows both the Input Acceleration time-history from the Taft Earthquake and the resulting response in the bumper-type restraint. In the same Figure 4, the straight horizontal line indicates the calculated maximum response using the developed methodology (see the Equations in Section 6.0).

The results of both Case 1 ($K_R = K_S$) and Case 2 ($K_R = 10 \times K_S$) are shown in the lower half of Table 2. The trend seen in Table 2 holds true for other actual and synthetic acceleration time-histories.

Table 2 - Results from the Practical Applications

TRIPLE SINE FORCING FUNCTION	Time-history Non-linear Analysis (lbs.)	Average Period Method (lbs.)	Percentage Above Exact Solution
Case 1: Force in the Structure:	1.307	1.324	1%
Force in the Restraint:	0.924	0.936	1%
Case 2: Force in the Structure:	0.756	0.783	4%
Force in the Restraint:	2.374	2.359	1%
TAFT EARTHQUAKE	**Time-history Non-linear Analysis (lbs.)**	**Average Period Method (lbs.)**	**Percentage Above Exact Solution**
Case 1: Force in the Structure:	0.256	0.309	21%
Force in the Restraint:	0.219	0.219	0%
Case 2: Force in the Structure:	0.256	0.278	9%
Force in the Restraint:	0.881	0.839	-5%

Notes:
- For Case 1, $K_R = K_S$
- For Case 2, $K_R = 10 \times K_S$
- Damping: 5% of the critical damping

Figure 4 **Force in the Bumper Restraint/Comparison Between Time-History Response and Spectrum Response Using the Developed Methodology**

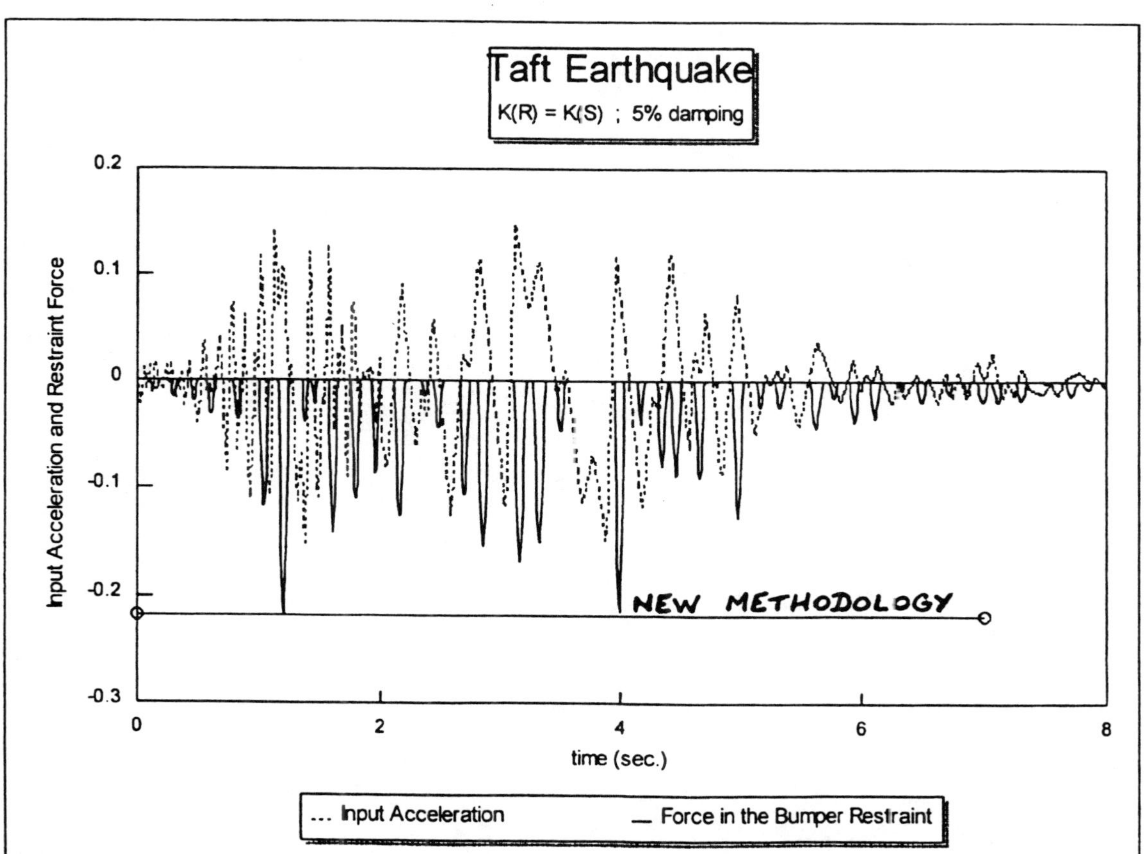

9.0 DISCUSSION

The results from the Practical Applications of Sections 7.0 and 8.0 are summarized in Table 2. These results show that, using the "equivalent" linear methodology, both the structural and restraint forces agree well with the maximum structural and restraint forces from the non-linear dynamic time-history analyses. Also, the equivalent linear methodology leads, in general, to results that are always slightly higher than the "actual" values (from the non-linear dynamic time-history analyses).

An additional methodology is developed to address the effect of impact in the initial cycles of vibration of the structure (in contrast to the methodology presented in this paper, which is for a resonance-type response). This alternative methodology could be required to predict the "actual" results from the non-linear dynamic analysis when one, or both, of the following conditions exist: 1) high damping (e.g. around 10%+), 2) forcing functions with a dominant frequency approximately equal to the main frequency of the free ("unrestrained") structural system. This alternative methodology is based on: a) the response build-up of the free structural system (Reference [2]), b) the calculation of the impact load through energy considerations, and c) the dynamic response of the restrained structure. This methodology, coupled with the Average Period Equivalent Linear method, completes the response for the one-way bumper-type system discussed in this paper.

The method presented in this technical paper is called the "Average Period Equivalent Linear" method. This method is a special case that fits one case of a large group of elastic non-linear systems. The method as described is equivalent to a modal solution and must be generalized and assimilated into the overall structural solution to obtain the dynamic response of the general multi-degree-of-freedom system.

10.0 REFERENCES

[1] Introduction to Structural Dynamics, John M. Biggs, McGraw - Hill Book Company, 1964.

[2] Dynamics of Structures, Ray W. Clough and Joseph Penzien, McGraw-Hill Book Company, 1975.

PVP-Vol. 340, Seismic Engineering
ASME 1996

FINITE ELEMENT EVALUATION OF EROSION/CORROSION AFFECTED REDUCING ELBOW

Chakrapani Basavaraju
Bechtel Corporation
Gaithersburg, Maryland

ABSTRACT

Erosion/corrosion is a primary source for wall thinning or degradation of carbon steel piping systems in service. A number of piping failures in the power industry have been attributed to erosion/corrosion. Piping elbow is one of such susceptible components for erosion/corrosion because of increased flow turbulence due to its geometry. In this paper, the acceptability of a 12"x8" reducing elbow in RHR service water pump discharge piping, which experienced significant degradation due to wall thinning in localized areas, was evaluated using finite element analysis methodology. Since the simplified methods showed very small margin and recommended replacement of the elbow, a detailed 3-D finite element model was built using shell elements and analyzed for internal pressure and moment loadings. The finite element analysis incorporated the U.T. measured wall thickness data at various spots that experienced wall thinning. The results showed that the elbow is acceptable as-is until the next fuel cycle. FEA, though cumbersome, and time consuming is a valuable analytical tool in making critical decisions with regard to component replacement of border line situation cases, eliminating some conservatism while not compromising the safety.

INTRODUCTION

Erosion/corrosion (E/C) of piping systems continues to be an area of concern for both the nuclear and fossil power industry [Walker, 1994]. As the age of the power plants is increasing, a significant proportion of the piping failures encountered are attributed to erosion/corrosion. Failures due to E/C are increasingly affecting the plant operations. As a result, most of the utilities established E/C programs to monitor the degradation of piping systems susceptible to erosion/corrosion. Significant work has been done in the area of predicting likely locations of wall thinning from E/C standpoint. Structural evaluation of E/C degraded components due to primary loads consideration was addressed in a report [EPRI, 1988] by the Electric Power Research Institute (EPRI).

Recent developments in the codes, such as appendices C & H, and code cases N-480 & N-494 are useful in evaluating flaws in ferritic components [Cloninger, 1994]. Acceptance standards for E/C affected piping items for continued service are provided in code case N-480 [ASME, 1990]. Technical, safety, and economic considerations used in developing long term corrosion management program for service water system was summarized in reference [Renwick, 1994].

BACKGROUND

A 12"x8" reducing 90° elbow connected to discharge side of RHR service water pump experienced wall thinning in localized spots due to E/C. Wall thickness measurements of locally thinned areas were taken utilizing ultrasonic examination technique (UT) and the sizes of the E/C affected localized spots (pits) were also identified. The wall thinning, though very localized, was very significant. The measured thickness in some localized areas is substantially smaller than 87.5% of nominal thickness (i.e. $t_{actual} << 0.875xt_{nom}$) and only marginally higher than minimum wall thickness requirement (t_{min}=0.1") from hoop stress considerations. In addition, the simplified methods using code case N-480 showed very little margins. Therefore, a detailed evaluation utilizing finite element analysis was considered to be appropriate in order to gain assurance concerning the safety for continued operation of the system with degraded elbow until the next outage. The purpose of the evaluation was to determine the acceptability of the fitting for the current as-is condition and projected condition for up to next six months of operation.

INPUT DATA

The material for the 12"x8" carbon steel elbow is A234 WPB. The small end of elbow is of 8.625" outside diameter (D_o) and 0.322" nominal wall thickness (t). The small end has 8"-300# welding neck flange of A105 material and is connected to the discharge nozzle of RHR service water pump. The large end of elbow is of 12.75" outside diameter and 0.375" nominal thickness. The radius of curvature (R_{elb}) of the elbow is 18". The large end of elbow is directly connected to 12" nominal diameter riser of the crossover piping of material A106 Gr B. The design pressure is 350 psig and the temperature is 350°F.

The wall thickness of a portion of the elbow was scanned using UT. The E/C affected area corresponds to the region located by $6.3° \leq \phi \leq 29°$ and $-50.4° \leq \Theta \leq 62.9°$ as shown in Figure 1. Here ϕ is the meridional angle of the elbow with $\phi=0°$ corresponding to the 8" end & $\phi=90°$ corresponding to the 12" end of the elbow, while Θ is the circumferential angle in any radial section with $\Theta=0°$ corresponding to the outer arc & $\Theta=180°$ corresponding to the inner arc of the elbow. This E/C affected portion covers an area which is approximately 9" long in the meridional direction & 16" long in the circumferential direction and is divided into 144 grids each of approximately 1"x1" size for UT scanning. There are a total of 11 isolated pits of size 1/8"x1/16" to 3/16"x1/8" located approximately in the middle region of 1"x1" grids. The wall thickness in these isolated pits ranged from 0.11" to 0.18". The field was requested to perform UT scanning and report thicknesses of less than 0.204", a base minimum value of current thickness for continued operation for next 6 months based on simplified calculations.

METHODOLOGY AND FINITE ELEMENT MODEL

The evaluation of E/C affected 12"x8" reducing elbow utilized finite element analysis (FEA) method employing ANSYS [1989] computer program. The finite element model (FEM) was built using ANSYS STIF63 elastic quadrilateral shell elements. Use of shell element is justified because the diameter to thickness ratio is greater than twenty (D_o/t = 12.750/0.375 = 34 for the large end and D_o/t = 8.625/0.322 = 26.8 for the small end of the elbow). The FEM includes the entire 12"x8" reducing 90° elbow, 8"-300# WNRF flange connected to the pump on the 8" end of elbow, and 4.5" length of 12" diameter straight pipe (riser portion) as attenuation length on 12" side of the elbow. The FEM is a full 3-D model (meridional angle ϕ =90° and circumferential angle Θ = 360° for the elbow). The FEM is very elaborate and consists of 4758 nodes (78 circumferential sections with 61 nodes in each circumferential section) which are conveniently generated in the cylindrical coordinate system with global origin at the center of curvature of the elbow. The radius of the curvature of the elbow is R_{elb} = 18" and is constant while the pipe diameter increases from 8" to 12". The nodes in the meridional direction are generated using toroidal coordinate system for the 12"x8" elbow. The FEM consists of a total of 4697 elements with 61 elements between any two radial sections. The finite element mesh is finer with an aspect ratio of one in the E/C affected area of pits and an aspect ratio of 2.2 in the vicinity. In the remote areas away from the E/C affected area, the element mesh has an aspect ratio of about three. A nodal plot of the FEM is shown in Figure 2.

The thickness of the shell elements for the E/C affected area is based on UT measured data for 144 (1"x1") grids. In the FEA, the size of the pit used for each of the 11 isolated pits is 3/16"x3/16" located in the middle portion of 1"x1" grid with the appropriate UT scanned thickness of the pit. Outside of the pit and within 1"x1" grid, the thickness used is 0.204". For grids where no pits were found, the smallest UT measured thickness out of 16 circumferential grids (1"x1") at that meridional direction (excluding those with pits if any) was used. These thicknesses ranged from 0.320" to 0.415". For portions of the elbow outside of the UT scanned region, nominal wall thickness was used.

ANALYSIS AND RESULTS

A linear elastic static analysis was performed for 4 separate unit loading cases, namely internal pressure and moment loadings from stress analysis of the piping system :(i) internal pressure of p=350 psig, (ii) M_x =58752 in-lb, (iii) M_y = 19668 in-lb, and (iv) M_z = 106656 in-lb using ANSYS computer program. The ANSYS program uses wavefront solution procedure for the system of simultaneous linear equations developed from the assembled finite elements.

$$\{F_{app}\} = [K].\{U\}$$

where $\{F_{app}\}$ = Applied Nodal Force Load Vector

[K] = Total Stiffness Matrix

{U} = Nodal Displacement Vector.

The FEA is based on the following material properties and boundary conditions.
Young's Modulus of Elasticity E = 27.9 x 10^6 psi, and Poisson's ratio μ = 0.3.
All 6 degrees of freedom for 61 boundary nodes on 8" flange side of elbow connected to RHR SW pump discharge nozzle are restrained for all 4 unit load cases considered in this analysis. In addition, the loadings are applied to the 61 boundary nodes on the 12" side of the FEM for the 4 unit loading cases as described below.

(i) Internal pressure loading case (p=350 psig)
Axial force F_y = p.[π (d^2)/4]/61 = 648.92 lb/node.
(ii) Moment M_x = 58752/61 = 963.15 in-lb/node.
(iii) Moment M_y = 19668/61 = 322.43 in-lb/node.
(iv) Moment M_z = 106656/61 = 1748.46 in-lb/node.

Two separate evaluations were made (a) one with UT scanned thicknesses for E/C affected area as described in the previous section for the as-is condition and (b) the second with reduced thicknesses for the UT scanned area to represent the future condition after 6 months based on an estimated wear rate described as follows.

$$t_p = t_a - (WR)(T)$$

where, t_p = Projected thickness of the E/C affected area after 6 months

t_a = UT measured thickness from current readings.

WR = Estimated wear rate = (t_1 - t_a)/(Time interval of operation between two inspections)
= 0.00148"/month (17.76 mils/year) was established for this evaluation.

t_1, t_a : Measured wall thicknesses at two intervals

T = Time in months = 6 months in this evaluation.

The results obtained from FEA for unit loading cases were scaled using factors for the appropriate magnitude of loadings and combined to obtain the various applicable combinations, namely Normal, Upset, Emergency & Faulted conditions. The review of the stress results indicates that the locations of maximum stresses are in the pits which experienced maximum E/C wall thinning. The maximum principal stresses obtained from FEA at inside (S1B, S2B, S3B), middle (S1M, S2M, S3M), outside (S1T, S2T, S3T) surface of elbow are reviewed and summarized for current as-is condition in Table 1 & for projected condition after 6 months in Table 2.

The criteria used to qualify the E/C affected 12"x8" reducing elbow is as follows.

$$\sigma_p + \sigma_{wt} \leq S_H$$

$$\sigma_p + \sigma_{wt} + \sigma_{th} \leq S_H + S_A$$

$$\sigma_p + \sigma_{wt} + \sigma_{obe} \leq 1.2\ S_H$$

$$\sigma_p + \sigma_{wt} + \sigma_{sse} \leq S_y$$

where, σ_p is pressure stress, and σ_{wt} , σ_{th} , σ_{obe} , & σ_{sse} are moment stresses from weight, thermal expansion, OBE seismic, & SSE seismic loadings respectively; and S_H, & S_A [=f($1.25S_C+0.25S_H$)] are allowable stresses for sustained & secondary stresses respectively; and S_y is yield stress.

The results show that the E/C affected areas are acceptable for as-is condition as well as for future condition after 6 months with a reasonable margin between computed and allowable stresses.

CONCLUSIONS

Based on a detailed finite element evaluation of 12"x8" reducing 90° elbow in service water system which experienced significant wall thinning in isolated areas, the following conclusions are drawn.

o Even though there has been significant wall thinning due to E/C in isolated and localized areas (pit size < 3/16"x1/8" & with wall thickness as low as 0.11"), the stress analysis of the elbow for internal pressure & moment loadings showed that the elbow is acceptable for both - the as-is and projected (up to 6 months) conditions.

o Detailed FEA approach, though cumbersome, is useful in making critical decisions regarding replacement of E/C affected fittings especially for borderline situation cases that showed very little margin by simplified methods without compromising the safety.

S.I. METRIC EQUIVALENTS

1 inch = 2.540000×10^{-2} m
1 lb_f = 4.448222 N
1 psi = 6.894757×10^{3} N/m^2

ACKNOWLEDGEMENTS

The author wishes to express his gratitude to Mr. G. L. Lushbaugh for his review and encouragement.

REFERENCES

Walker, S. M., "NDE for Erosion-Corrosion," PVP-Vol. 285, ASME 1994, pp. 13-16.

EPRI, Acceptance Criteria for Structural Evaluation of Erosion-Corrosion Thinning in Carbon Steel Piping, EPRI Report NP-5911SP, July 1988.

Cloninger, T. H., and Timmaraju, S., "Changing Priorities of Codes and Standards - Quality Engineering: A Utility Perspective," PVP-Vol. 285, ASME 1994, pp. 251-253.

ASME, Examination Requirements for Pipe Wall Thinning due to Single Phase Erosion and Corrosion, ASME Boiler and Pressure Vessel Code, Section XI, Division 1, Case No. N-480, 1990.

Renwick, B. L., and Wright, D. M., "Service Water System Management: Solving Problems at Reduced Costs," PVP-Vol. 285, ASME 1994, pp. 27-35.

ANSYS Engineering Analysis System User's Manual for ANSYS Revision 4.4, 1989, Houston, Pennsylvania.

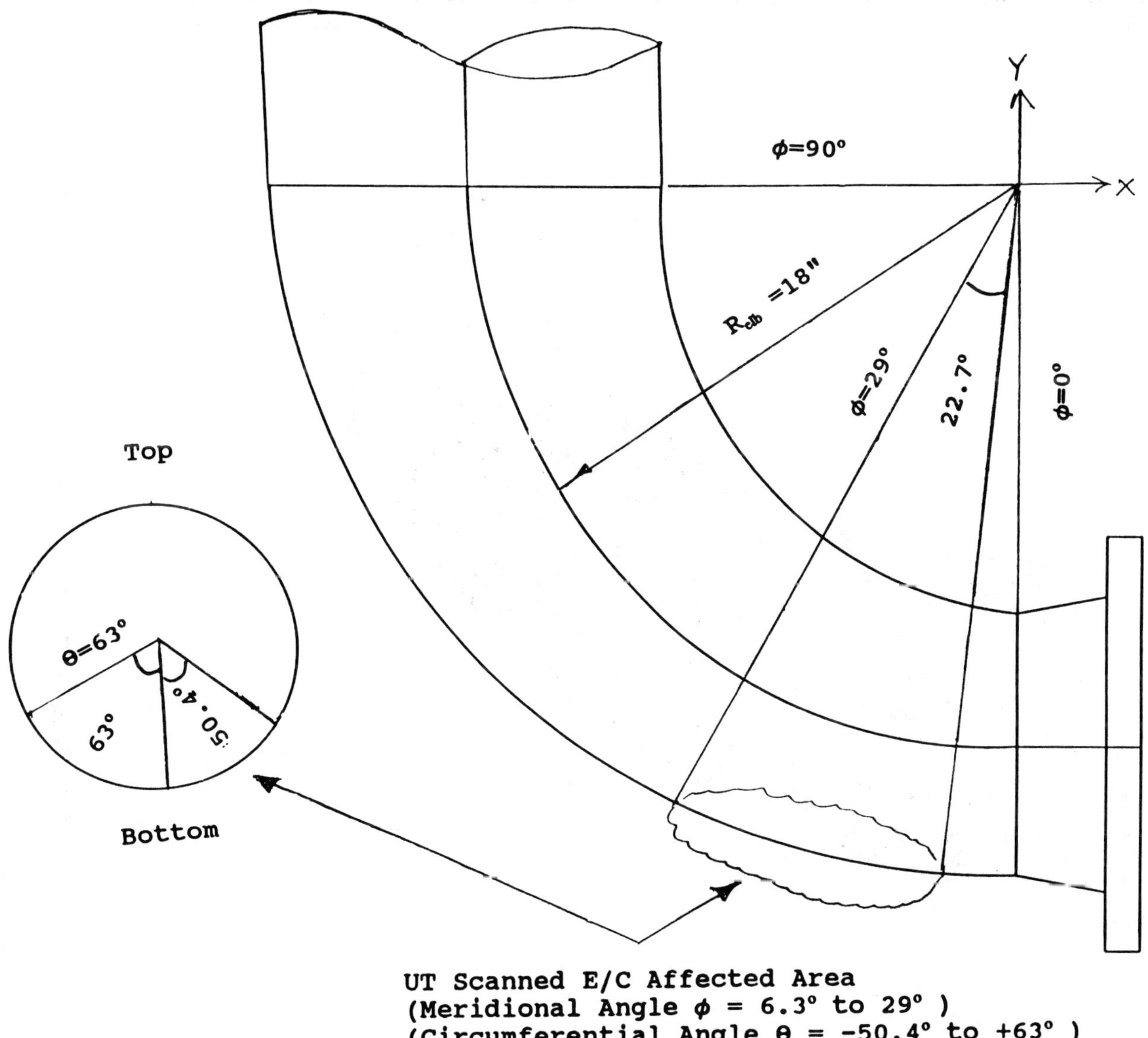

Fig. 1: 12"x8" Reducing Elbow

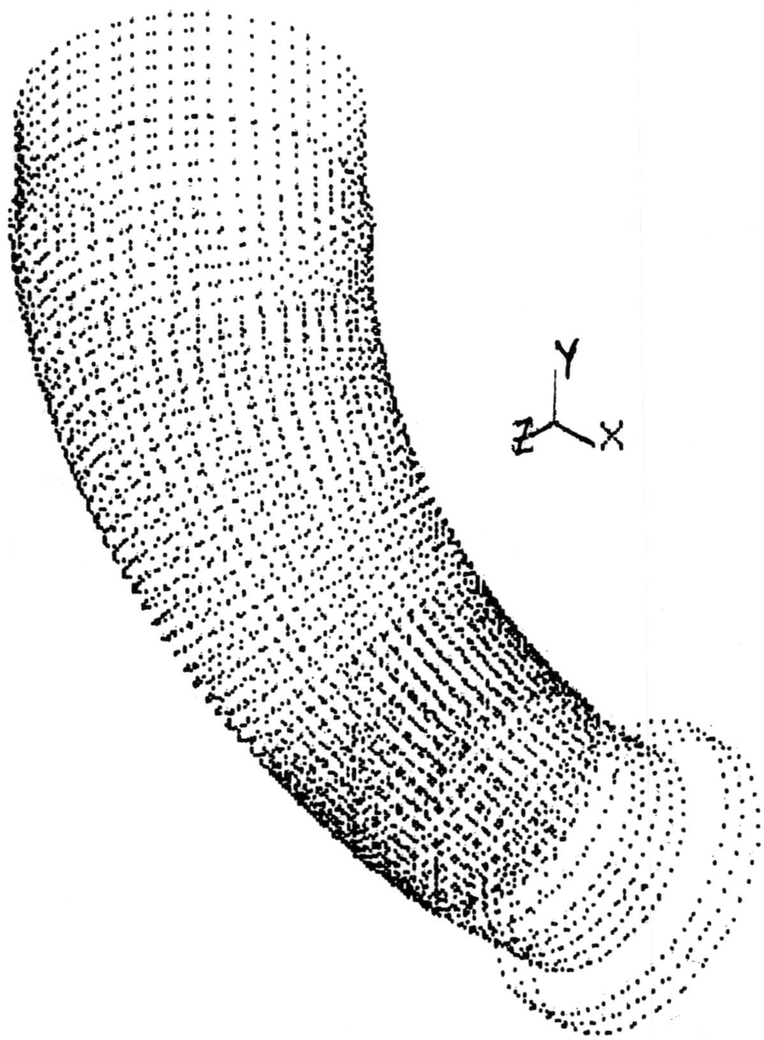

Fig. 2: FEM Nodal Plot of 12"x8" Reducing Elbow

Loadings	FEA Stress Results (psi)
Sustained (Pressure & Weight) p = 350 psig M_x = 336 in-lb M_y = 852 in-lb M_z = 120 in-lb	$\sigma_p + \sigma_{wt} = 8397 < S_H = 15000$
Sustained & Secondary (Pressure, Wt., & Therm. exp) p = 350 psig M_x = 4896 in-lb M_y = 4644 in-lb M_z = 2820 in-lb	$\sigma_p + \sigma_{wt} + \sigma_{th} = 9159 < S_H + S_A$ < 37500
Sustained: Normal & Occasional (Pressure, Wt., & OBE inertia) p = 350 psig M_x = 29544 in-lb M_y = 10260 in-lb M_z = 53388 in-lb	$\sigma_p + \sigma_{wt} + \sigma_{obe} = 13336 < 1.2\ S_H$ < 18000
Sustained: Normal & Occasional (Pressure, Wt., & SSE Inertia) p = 350 psig M_x = 58752 in-lb M_y = 19668 in-lb M_z = 106656 in-lb	$\sigma_p + \sigma_{wt} + \sigma_{sse} = 18261 < S_y$ < 30500

Table 1: Combined Stresses from FEA Results for Current (As-is) Condition

Loadings	FEA Stress Results (psi)
Sustained (Pressure & Weight) $p = 350$ psig $M_x = 336$ in-lb $M_y = 852$ in-lb $M_z = 120$ in-lb	$\sigma_p + \sigma_{wt} = 8771 < S_H = 15000$
Sustained & Secondary (Pressure, Wt., & Therm. exp) $p = 350$ psig $M_x = 4896$ in-lb $M_y = 4644$ in-lb $M_z = 2820$ in-lb	$\sigma_p + \sigma_{wt} + \sigma_{th} = 9580 < S_H + S_A$ < 37500
Sustained: Normal & Occasional (Pressure, Wt., & OBE inertia) $p = 350$ psig $M_x = 29544$ in-lb $M_y = 10260$ in-lb $M_z = 53388$ in-lb	$\sigma_p + \sigma_{wt} + \sigma_{obe} = 14179 < 1.2\ S_H$ < 18000
Sustained: Normal & Occasional (Pressure, Wt., & SSE Inertia) $p = 350$ psig $M_x = 58752$ in-lb $M_y = 19668$ in-lb $M_z = 106656$ in-lb	$\sigma_p + \sigma_{wt} + \sigma_{sse} = 19585 < S_y$ < 30500

Table 2: Combined Stresses from FEA Results for Projected (after 6 months) Condition

HIGH LEVEL DYNAMIC RESPONSE OF PIPING SYSTEMS

Introduction

G. C. Slagis
G. C. Slagis Associates
Walnut Creek, Ca, USA

Since the early 1970s, the seismic design of nuclear piping has been on linear elastic dynamic analysis methods, specifically the response spectra method. There have been a number of experimental studies on the dynamic response of piping at high levels of input loading. These studies indicate that ductile piping has inherent capability to withstand strong earthquake motions without failure. The EPRI/NRC Piping and Fitting Dynamic Reliability Program, initiated in 1985, tested piping components to extream levels of seismic input motions. Based on these tests, an increase to the Section III Level D stress limits for piping for seismic loads was proposed and incorporated into the Code in 1994. There have been regulatory concerns about the seismic margins at these higher stress limits -is there adequate seismic margin? The Section III code committee is presently reviewing the adequacy of the revised code rules. The five papers in this chapter all deal with the behavior of piping during high level dynamic response from five different perspectives.

The Kennedy, Chokshi, Chen paper addresses the issue of required seismic capacity of a piping component. A proposed methodology for determining appropriate seismic margins is developed. The approach is based on current seismic PRA (probabilistic risk assessment) evaluation methods. The evaluation starts with the plant seismic margin in terms of the High-Confidence-Low-Probability-of-Failure (HCLPF) seismic capacity. It is suggested that if the probability of failure of a piping system is limited to about 0.1 % of the plant HCLPF seismic capacity level, piping will not control the plant HCLPF capacity. Then, current seismic PRA methods can be maintained. The required seismic capacity of a piping component, as demonstrated by test, is determined using this methodology.

Lazzeri summarizes various one degree of freedom studies on the impact of pipe yielding (ductility) on pipe response. In particular, the effect of yielding on a pipe with a typical seismic floor response spectrum is evaluated. Limited ductility causes reduction in the response and flattening of the peaks of the spectrum. Lazzeri concludes that the peaks of the spectrum may be ignored, and the design of the fossil piping may be based on the zero period acceleration. Several piping systems not designed for lateral loads are analyzed for ZPA loading. Problems were found only for very flexible systems.

One of the experimental studies on pipe response was the work done by Beany at the Berkeley Nuclear Laboratories. The test data were summarized by Slagis in the 1995 Seismic Engineering volume (PVP-Vol.312). In the Slagis and Litton paper, nonlinear finite element dynamic analyses of the response of a simple pipe span to controlled displacement sinusoidal vibration are documented. Analysis results were in good agreement with the Beany test data, and demonstrate the attenuation of dynamic response by material yielding. The authors conclude that analytical simulation will be useful for assessing seismic margins.

When pressurized piping is subjected to high levels of dynamic loading, the cyclic bending moments acting together with pressure cause hoop ratcheting. The amount of incremental plastic strain per cycle demonstrated by test have been difficult to predict by analysis. Conventional inelastic analysis methods typically overpredict the ratchet strain. Hassan, Zhu & Matzen have incorporated an improved cyclic plasticity model into an ANSYS subroutine. The cyclic plasticity model is based on the Armstrong-Frederick kinematic hardening rule in conjunction with the Drucker-Palgen plastic modulus equation. Analytical studies show significant improvement in the prediction of incremental ratchet strain.

A record of the 1995 Forum on "Appropriate Criteria and Methods for Seismic Design of Nuclear Piping" is provided by Slagis in the last paper. The Forum is a discussion session at the PVP conference organized by the Seismic Engineering Subcommittee. The 1995 Forum was the fourth annual discussion session. The topic was the earthquake experience data base. Under discussion was the question of whether the data base demonstrates that seismic inertia loads will not cause failure in ductile piping systems.

The efforts of the authors and reviewers in making these contributions to the understanding of seismic response of piping are deeply appreciated.

PVP-Vol. 340, Seismic Engineering
ASME 1996

TOWARDS APPROPRIATE SEISMIC MARGINS IN NUCLEAR PLANT PIPING[1]

Robert P. Kennedy[2]
Structural Mechanics Consulting, Inc.
Yorba Linda, California

Nilesh C. Chokshi[2]
U.S. Nuclear Regulatory Commission
Washington, D.C

W. Paul Chen[2]
Energy Technology Engineering Center
Canoga Park, California

ABSTRACT

Some results of ongoing research being conducted for the U.S. Nuclear Regulatory Commission (NRC) at the Energy Technology Engineering Center (ETEC) are reported. These results include the development of a methodology for establishing and estimating appropriate seismic margins in nuclear plant piping.

INTRODUCTION

The 1994 Addenda to the ASME Code revised the Code, Section III, seismic design criteria for nuclear plant piping. In part, the allowable stress for the Code Service Level D load combination which includes the Safe Shutdown Earthquake (SSE) was increased by 50%. These revised criteria are the culmination of a number of initiatives in response to reviews by the NRC and the Pressure Vessel Research Council (PVRC) in the early 1980's. On the issue of overall design margins in piping systems, the reviews recommended in part that test programs for verifying piping seismic design margins and identifying failure modes be supported.

Subsequently, the Electric Power Research Institute (EPRI)/US NRC Piping and Fitting Dynamic Reliability (PFDR) Program (Guzy, 1988; EPRI, 1994) was implemented. This program included dynamic seismic, hydrodynamic and water hammer testing and analyses of piping components and systems. Based primarily on evaluations of PFDR testing of piping components, EPRI and General Electric (GE) (Tagart and Ranganath, 1992), proposed an initial set of revised criteria. These criteria were subsequently refined based on additional evaluations by the Technical Core Group (TCG) of the Advanced Reactor Corporation (ARC)[3] during the ASME Special Task Group on Integrated Piping Criteria (STGIPC)[3] deliberations.

EPRI and GE observed that, the revised seismic design criteria would provide a minimum margin of failure[4] of at least 10 on cycles and hence a margin of $\sqrt{10}$ or approximately 3 on load. Subsequently, based on analytical extrapolations to cases not tested, lower margins (approximately equal to 2) were obtained but deemed acceptable during STGIPC and ARC/TCG reviews. Independent assessments of these same PFDR component test data which utilize the STGIPC and ARC/TCG extrapolation procedure indicate margins of less than

---

1 Research sponsored by the Office of Nuclear Regulatory Research, U.S. Nuclear Regulatory Commission and performed at the Energy Technology Engineering Center which is operated by Rockwell International Corporation for the U.S. Department of Energy under Contract Number DE-AC03-76-SF-00700.

2 The views expressed in this paper are those of the authors and should not be construed to reflect the NRC position.

3 These ARC/TCG and ASME/STGIPC evaluations are documented in reports of limited distribution.

4 The margin of failure is defined to be the ratio between the seismic load at failure and the load at the ASME Code allowable.

1.0 in some cases (Chen et al, 1995). Alternately, based on the equal reliability concept and probabilities of occurrence of Code Service Level D loads, minimum margins between 1.5 and 1.0 have been proposed (Slagis, 1991).

In this paper a different approach for establishing appropriate seismic margins is presented. This approach is based on considerations of current seismic PRA evaluation methods. Specifically, margins are established such that piping will not control the plant High-Confidence-Low-Probability-of-Failure (HCLPF) seismic capacity, and consequently the current seismic PRA methodology can be maintained.

In the following, only the framework of this different approach is provided. Additionally, judgement has been exercised in a number of places in establishing the appropriate margin. A more detailed development of this approach is planned for a later date. This future development will rely less on judgement and/or include sensitivity studies. However, in view of ongoing reviews, these results are presented in the interest of timeliness.

MARGIN DEVELOPMENT

The required seismic margin for a piping system will depend upon the required seismic margin for the plant as a whole for some damage state such as seismic induced core damage. Currently, the NRC has requested that all existing plants define their plant seismic margin in terms of the High-Confidence-Low-Probability-of-Failure (HCLPF) seismic capacity. The seismic margin for the design basis Safe Shutdown Earthquake (SSE) is then defined by the ratio R_H, between the HCLPF and SSE ground motion levels, i.e.,

$$R_H = \left(\frac{HCLPF}{SSE}\right) \quad (1)$$

Initially, the HCLPF seismic capacity was defined as the level at which one had approximately 95% confidence of less than about 5% probability of failure. However, in NUREG - 1407 (US NRC, 1991a) the NRC has suggested that this HCLPF seismic capacity can be approximated as the 1% composite (mean) probability of failure. With this approximation, one does not have to separate their estimates of variability into "uncertainty," and "randomness," but can work with a single composite (mean) fragility curve which defines mean probability of failure versus ground motion level. For simplicity, this composite (mean) approximation of the HCLPF capacity will be used herein.

After anchorage, seismic-interaction, or other significant deficiencies identified as a result of a careful seismic walkdown of a plant have been corrected, seismic probabilistic risk assessments (PRA) and seismic margin studies conducted on existing nuclear power plants have reported in ERI/NRC 94-502 (Energy Research, Inc., 1994) that:

$$\text{For Existing Plants: } R_H \geq 1.25 \quad (2)$$

In addition, as a target goal for the standardized Advanced Light Water Reactor (ALWR) designs, the NRC has suggested:

$$\text{For ALWRs: } R_H > 1.67 \quad (3)$$

These levels for R_H will be used herein to establish the required seismic margin for piping systems.

In addition, based upon the pre-1994 ASME Code Addenda seismic design criteria, seismic PRA and seismic margin studies have concluded that piping is not a significant contributor to the seismic risk of a nuclear power plant and does not control the HCLPF seismic capacity of the plant. A desirable goal is to not liberalize piping seismic design criteria so much that piping begins to control the HCLPF seismic capacity of the plant.

However, because of the large number of at-least partially independent piping segments[5], piping will control the plant HCLPF capacity unless piping has substantially less than a 1% probability of failure at the plant HCLPF capacity. In order to provide reasonable assurance that piping will not control the plant HCLPF capacity, it is suggested that the probability of failure P_{P_f} of a piping segment be limited to about 0.1% at the plant HCLPF seismic capacity level (the results of a sensitivity study in which P_{P_f} has been varied from 0.02% to 0.5% are presented later in the paper), i.e.:

$$P_{P_f} < 0.1\% \quad \text{at Plant HCLPF Level} \quad (4)$$

Derivation of Required Seismic Margin for Structures and Other Components

The ratio of the median (50% probability of failure) seismic capacity of any component to the design basis SSE can be defined by a median factor of safety $F_{50\%}$ given by:

$$F_{50\%} = \left(\frac{Median}{SSE}\right) \quad (5)$$

---

[5] A piping system can be thought of, from a capacity point of view, as made up of a large number of segments containing at-least partially independent failure mode locations whose failure could lead to failure of the piping system.

This median factor of safety can be defined by the product of the median capacity factor of safety $F_{C_{50\%}}$ and the median response factor of safety, $F_{R_{50\%}}$,i.e.:

$$F_{50\%} = (F_{C_{50\%}})(F_{R_{50\%}}) \tag{6}$$

The variability [logarithmic standard deviation or approximate coefficient of variation (COV)], β, on the factor of safety can be obtained by adding capacity and response variances (β_C^2 and β_R^2, respectively) or:

$$\beta^2 = \beta_C^2 + \beta_R^2 \tag{7}$$

where β_C and β_R include both randomness and uncertainty.

Since the plant HCLPF is defined at the 1% probability of failure, then assuming a log-normal distribution:

$$R_H = F_{50\%} e^{-2.326\beta} \tag{8}$$

where (-2.326) is the standardized normal coefficient corresponding to the 1% probability of failure.

Investigations (Kennedy and Short, 1994) of Code minimum ultimate, Code limit-state, or Code Service Level D capacities indicate a capacity factor $R_{C_{1\%}}$ greater than unity on the 1% non-exceedance probability (NEP) capacity. Thus:

$$F_{C_{50\%}} \geq R_{C_{1\%}} e^{2.326\beta_C} \tag{9}$$

Furthermore, it is assumed that response analyses conducted in accordance with the Standard Review Plan (SRP, US NRC, 1991b) are aimed at defining the seismic response at the 84% NEP. However, because of either excess conservatism in some aspects of the SRP response requirements, or because of excess conservatism introduced by the analyst, most response analyses achieve a response factor $R_{R_{84\%}}$ greater than unity when compared to a probabilistic computed 84% NEP response. Thus:

$$F_{R_{50\%}} \geq R_{R_{84\%}} e^{\beta_R} \tag{10}$$

By combining Equations (6) through (10):

$$R_H = (R_{C_{1\%}})(R_{R_{84\%}})(f_\beta) \tag{11}$$

where:

$$f_\beta \geq e^{\beta_R - 2.326(\beta - \beta_c)} \tag{12}$$

However, for structures and most other components, the factor f_β will be very close to unity over the likely range of β_C from 0.2 to 0.4 and β_R from 0.2 to 0.3 as is shown in Table 1. Therefore, R_H may be closely

approximated by:

$$R_H \approx (R_{C_{1\%}})(R_{R_{84\%}}) \tag{13}$$

Thus, to estimate a seismic margin ratio R_H , defined as the ratio of the HCLPF/SSE capacity, it is sufficient to establish the capacity factor $R_{C_{1\%}}$ above the 1% NEP capacity and to establish the response factor $R_{R_{84\%}}$ on the 84% NEP response. Equation (13) can then be used to estimate R_H without having to explicitly estimate the variabilities β_C and β_R. Equation (13) defines the essence of the Conservative-Deterministic-Failure-Margin (CDFM) method for finding the HCLPF seismic capacity.

For structures, and other compact components, it is estimated that:

$$R_{C_{1\%}} = 1.25 \text{ to } 1.9 \tag{14}$$

The lower estimate of 1.25 corresponds to brittle failure modes where there is negligible inelastic energy absorption such as a welded connection, or out-of-plane shear failure of a concrete member without shear reinforcement. The high estimate is for ductile failure modes with significant inelastic energy absorption capability defined as the ratio by which linear elastically computed responses can exceed the ultimate capacity for oscillatory, limited duration dynamic events.

Based upon reviews of design basis response analyses, for structures and other non-distribution system components it is estimated that:

$$R_{R_{84\%}} = 1.0 \text{ to } 2.0 \tag{15}$$

At low elevations for structures founded on rock, $R_{R_{84\%}}$ = 1.0 for response analyses which satisfy the SRP. Higher values often occur at higher elevations and for structures embedded in soil. However, there is no consistency in these higher values so that they cannot be counted upon to reduce the required capacity factor $R_{C_{1\%}}$. It is judged that all that can be counted upon is:

$$R_{R_{84\%}} \geq \begin{cases} 1.0 \ \text{(Existing Plants)} \\ 1.25 \ \text{(ALWR)} \end{cases} \quad (16)$$

The estimate that $R_{R_{84\%}} \geq 1.25$ for standardized ALWR plants is based on the design seismic responses for these plants being an envelope of the responses computed for many sites and thus being conservative for any particular site.

Thus, combining Equations (13), (14), and (16) with the plant seismic margin R_H goals defined by Equations (2) and (3), one obtains for the minimum required $R_{C_{1\%}}$.

Plant Type	R_H	$R_{R_{84\%}}$	$R_{C_{1\%}} = (R_H / R_{R_{84\%}})$
Existing Plants	1.25	1.00	1.25
ALWR	1.67	1.25	1.33

In conclusion, for structures and compact components, it is believed that an adequate capacity factor $R_{C_{1\%}}$ on the 1% probability of failure capacity is in the range of 1.25 to 1.33. However, it is not believed that this is an adequate capacity factor for distribution systems such as piping which have a large number of segments with at-least partially independent failure locations.

Derivation of Required Seismic Margin for Piping and Other Distribution Systems With Large Numbers of Segments

As summarized in Equation (4), it is recommend that the probability of failure P_{P_f} for piping and other multiple segment distribution systems be limited to about 0.1% at the plant HCLPF level. To achieve this lesser probability of failure, Equations (11) and (12) must be modified for piping as follows:

$$R_{CP_{1\%}} = \left(R_H / R_{RP_{84\%}} \right) (f_{\beta P})^{-1} \quad (17)$$

where:

$$f_{\beta P} = e^{\beta_{RP} + X_{CP}\beta_{CP} - X_{fP}\beta_P} \quad (18)$$

$X_{CP} = 2.326$ (1% NEP)
$X_{fP} = 3.090$ (0.1% NEP)

and the subscript P refers to piping specific estimates.

For distributed, i.e., extended, piping systems, the capacity variability β_{CP} and possibly the response variability β_{RP} are typically larger than those tabulated in Table 1 for structures and other compact components. In addition, the coefficient X_{fP} of 3.090 in Equation (18) for $f_{\beta P}$ associated with 0.1% NEP is larger than the corresponding coefficient of 2.326 in Equation (12) for f_β associated with 1% NEP. Table 2 presents the resulting estimates for $(f_{\beta P})^{-1}$ for piping systems. Note that $(f_{\beta P})^{-1}$ is nearly constant over the estimated typical ranges of β_{CP} and β_{RP}. The mean value $(f_{\beta P})^{-1}$ is 1.42 with a small COV of 0.07. It is recommended that $(f_{\beta P})^{-1}$ be taken at about the 84% NEP level given by:

$$(f_{\beta P})^{-1} = \overline{(f_{\beta P})^{-1}} \ (1 + COV) = 1.50 \quad (19)$$

For piping the response margin $R_{RP_{84\%}}$ is highly variable and is often excessively conservative particularly at high elevations in structures at frequencies near resonance with the structure frequencies. It is judged that:

$$R_{RP_{84\%}} = 1.0 \text{ to } 4.0 \quad (20)$$

Here again, at low elevations for structures founded on rock, $R_{RP_{84\%}} = 1.0$ for response analyses which satisfy the SRP. Therefore, no credit for the potentially excess but highly uncertain conservatism in $R_{RP_{84\%}}$ should be taken. It is recommended that $R_{RP_{84\%}}$ be set equal to $R_{RP_{84\%}}$ from Equation (16).

Thus, from Equations (17) and (18):

Plant Type	$(R_H / R_{RP_{84\%}})$	$R_{CP_{1\%}}$
Existing Plants	1.25	1.875
ALWR	1.33	2.0

In conclusion, it is recommended that an adequate capacity margin $R_{CP_{1\%}}$ on the 1% probability of failure piping capacity is in the range of 1.875 to 2.0. This recommendation is based on the goal of limiting the probability of failure of a piping segment P_{P_f} to about 0.1% at the plant HCLPF seismic capacity level. It was judged that this goal is sufficient to provide reasonable assurance that piping will not control the plant HCLPF seismic capacity. In order to determine the sensitivity of the capacity factor $R_{CP_{1\%}}$ to the selected probability of failure P_{P_f} at the plant HCLPF level, this probability of failure was varied from a credible lower bound of 0.02% to a credible upper bound of 0.5% following the same approach as described above. The results of this sensitivity study are as follows:

P_{P_f}	$(f_{\beta P})^{-1}$	$R_{CP_{1\%}}$
0.50%	1.13	1.4 to 1.5
0.10%	1.50	1.875 to 2.0
0.02%	2.02	2.5 to 2.7

Over this range of credible P_{P_f} values, $R_{CP_{1\%}}$ should be selected to lie within the range of 1.5 to 2.5 centered on the recommended value of 2.0.

However, it may not be possible to define the 1% probability of failure piping capacity from a limited number of component seismic tests. Therefore, it may be necessary to define the required capacity margin $R_{CP_{Y\%}}$ in terms of some higher Y% probability of failure. At a Y% other than 1%, the desired stability of $(f_{\beta P})^{-1}$ shown in Table 2 for Y = 1% will not be so well achieved. For example at Y = 5% the Coefficient of Variation (COV) increases to 0. 12 versus 0.07 at Y = 1%. Even so, this COV is still reasonably small. For Y > 1% the largest $(f_{\beta P})^{-1}$ will always occur at the largest typical β_{CP} of 0.6. Therefore, at some other Y% probability of failure, the required capacity margin $R_{CP_{Y\%}}$ can be slightly conservatively estimated from:

$$R_{CP_{Y\%}} = (R_{CP_{1\%}})\, e^{\alpha \beta_{CPmax}} \tag{21}$$

where:

$\beta_{CP_{max}} = 0.6$

$\alpha = 2.326 - X_{CP_{Y\%}}$

Y	$X_{CP_{Y\%}}$	α	$(R_{CP_{Y\%}} / R_{CP_{1\%}})$
1%	2.326	0	1.00
2%	2.054	0.272	1.175
3%	1.881	0.445	1.30
5%	1.645	0.681	1.50

Thus, if $R_{CP_{1\%}}$ = 2.0 is deemed sufficient for piping as per the prior recommendation, then:

Y	$R_{CP_{Y\%}}$
1%	2.00
2%	2.35
3%	2.60
5%	3.00

The required capacity margin for piping on the Y% probability of failure seismic capacity can be specified by any of these $R_{CP_{Y\%}}$ values. Thus, for the 1% to 5% probability of failure capacities, the required seismic capacity margin should range from about 2.0 at 1% NEP capacity to 3.0 at the 5% NEP capacity.

CONCLUSIONS

The framework of an approach for determining and estimating appropriate seismic margins in nuclear plant piping was presented. This approach is based on the assumption that the probability of failure of a segment of a piping system be limited to about 0.1% of the plant HCLPF seismic capacity level. For this probability of failure, piping will not control the plant HCLPF capacity and current seismic PRA methods can be maintained.

It was estimated that for 1% to 5% probability of failure capacities, the required seismic capacity margin should range from about 2.0 at 1% NEP capacity to 3.0 at the 5% NEP capacity.

REFERENCES

Chen, W. P., Jaquay, K.R., Chokshi, N.C., and Terao, D., 1995, "An Assessment of Seismic Margins in Nuclear Plant Piping," Transactions of the 13th International Conference on Structural Mechanics in Reactor Technology J.D. Riera, General Editor, Editoria da Universidade, Universidade Federal do Rio Grande do Sul, Porto Alegre, Brasil, Vol. 3, pp. 507-512

Energy Research, Inc., 1994, "An Approach to Seismic Scope Re-Assessment for Individual Plant Examination of External Events", Technical Report ERI/NRC 94-502

EPRI 1994, "Piping and Fitting Dynamic Reliability Program", Technical Reports TR-102792-V1 through V5, Electric Power Research Institute, Palo Alto, CA

Guzy, D., 1988, "Piping Design Criteria and Research, Current NRC Activities in Dynamic Design" Nuclear Engineering and Design, Vol. 107, pp. 161-167.

Kennedy, R.C and Short, S.A., 1994, "Basis for Seismic Provisions of DOE-STD-1020", Lawrence Livermore National Laboratory and Brookhaven National laboratory Technical Report UCRL-CR-111478, S/C B235302, BNL-52418, Livermore, CA, Upton, NY

Slagis, G.C., 1991, "Basis of Current Dynamic Stress Criteria for Piping," Bulletin 367, Welding Research Council, New York, NY

Tagart, Jr., S.W. and Ranganath, S., 1992, "Proposed Dynamic Stress Criteria for Piping," ASME PVP 237-1, pp. 133-137.

US NRC, 1991a, "Procedural and Submittal Guidance for the Individual Plant Examination of External Events for Severe Accident Vulnerabilities", US Nuclear Regulatory Commission Technical Report NUREG-1407, Washington, DC

US NRC, 1991b, "Standard Review Plan for the Review of Safety Analysis Reports for Nuclear Power Plants, LWR Edition", US Nuclear Regulatory Commission Technical Report NUREG-0800, Washington, DC

Table 1

Typical Values for Factor f_β Structures and Other Non-Distribution System Components

β_C	f_β	
	β_R = 0.2	β_R = 0.3
0.2	1.01	0.93
0.3	1.06	1.01
0.4	1.09	1.07

Table 2

Typical Values for Factor $f_{\beta P}$ for Piping Systems

β_{CP}	$(f_{\beta P})^{-1}$		
	β_{RP} = 0.2	β_{RP} = 0.3	β_{RP} = 0.4
0.3	1.24	1.37	1.56
0.4	1.29	1.37	1.52
0.5	1.35	1.40	1.52
0.6	1.43	1.46	1.54

PVP-Vol. 340, Seismic Engineering
ASME 1996

On the seismic design of piping for fossil fired power stations

by L Lazzeri(Ansaldo Energia)

Via N Lorenzi 8 , 16152 Genova, Italy

ABSTRACT :The seismic design criteria are briefly reviewed : the importance of the yielding phenomena on the seismic response is presented . The decisive importance of ductility is confirmed by the field observations . The ductility causes reduction in the response with flattening of the peaks . Some analyses are performed on several piping systems in static equivalent conditions with ZPA loading . Such analyses assume some ductility in the system . Problems are found only for very flexible systems

1.Introduction

The aim of this paper is to address the problem of the seismic design of piping systems in conventional power stations. Three basic considerations should be made:

a) the design of piping in nuclear power stations has been traditionally based on the following characteristics: dynamic elastic analysis, very low damping, very conservative procedures in the evaluation of the seismic loads (UNRC) ; papers against this practice have been indeed published by Svenson et al (1968) , Lazzeri (1993)

b) the design of piping for conventional power stations is generally based (with very few exceptions) on the consideration of vertical loads only ; some lateral rigidity can be present due to the analysis of the consequences of waterhammer (the most frequently calculated dynamic load) ; very seldom some consideration is given to the lateral seismic load and some lateral bumper or some other device limiting the lateral displacement has been added ; an excerpt of a seismic specification used for the design of a coal power station in an extremely seismic area in South America is shown in appendix A.

c) the field experience shows that even very limited damage occurred to piping after even very catastrophic seismic events, even if no consideration has been given in the design to such loads , as shown by Silver et al (1985) ;possible exceptions were generally easily explained with very poor design (in general inadequate ductility) ; a very large testing program sponsored by EPRI in the eighties showed that the design according to a) is definitely overconservative , as shown by Tagart et al (1987, 1988), Onesto etal (1986) , DeVita (1988)

Hence the experimental findings both in the field and in the laboratory seem to confirm that a piping system can in reality survive very severe events and suggest that ductility is the main parameter in explaining such a behavior : if the system is adequately ductile its capacity of surviving even a catastrophic seismic event is indeed very large

A few examples of a fully plastic dynamic calculation for piping are indeed present in the open literature , however the calculations are so expensive and time consuming that a piping design is likely to provide a little extra equipment (e.g. . bumpers or guides) rather than perform such an analysis .

Lazzeri et al (1987, 1988,1989,1992) have published a series of papers dealing with this problem and providing a simplified method to deal with such phenomena . The model is revisited herein trying to describe in a semiquantitative manner the behavior of a piping system subject to a seismic excitation. The analytical model contains a few arbitrary although quite reasonable assumptions , however it shall be shown that the picture obtained by means of this method is clear enough and allows some general conclusions

2.On the nature of the seismic excitation

Seismic loads are characterized by the following:

a) the load is dynamic in nature and due to the interposing presence of the building (and soil structure interaction) is generally concentrated around a few quite distinctive frequencies (the building frequencies); some broadening of the peaks may be made in the design to account for uncertainties in the definition of the building eigenfrequencies ; such procedures lead to amplification factors sometimes even bigger than ten with respect to ZPA (Zero Period Acceleration), due to the very low damping postulated in the analysis; a typical example is shown in fig 1 . Moreover it should also be mentioned that the procedure itself completely ignores the counteraction of the mass of the piping with respect to the mass of the building ; this effect as discussed by Lazzeri et al(1979,1990) and by Sackman et al (1979) may lead to a substantial reduction of the peak amplification

b) the load possesses a large number of cycles and ratcheting may play an important role due to the superposition of the dynamic load to a constant static load (weight ,pressure etc.); nevertheless experiments show that shakedown occurs very frequently

c) even under extreme conditions the piping stays generally elastic and the non linear phenomena are generally confined to very small areas , possibly in elbows ; in any case the analyses show that the dominant fraction in the displacements is generally due to rotations concentrated in elbows or other similar items

d)strain hardening phenomena are indeed important (as suggested by the presence of shakedown phenomena)

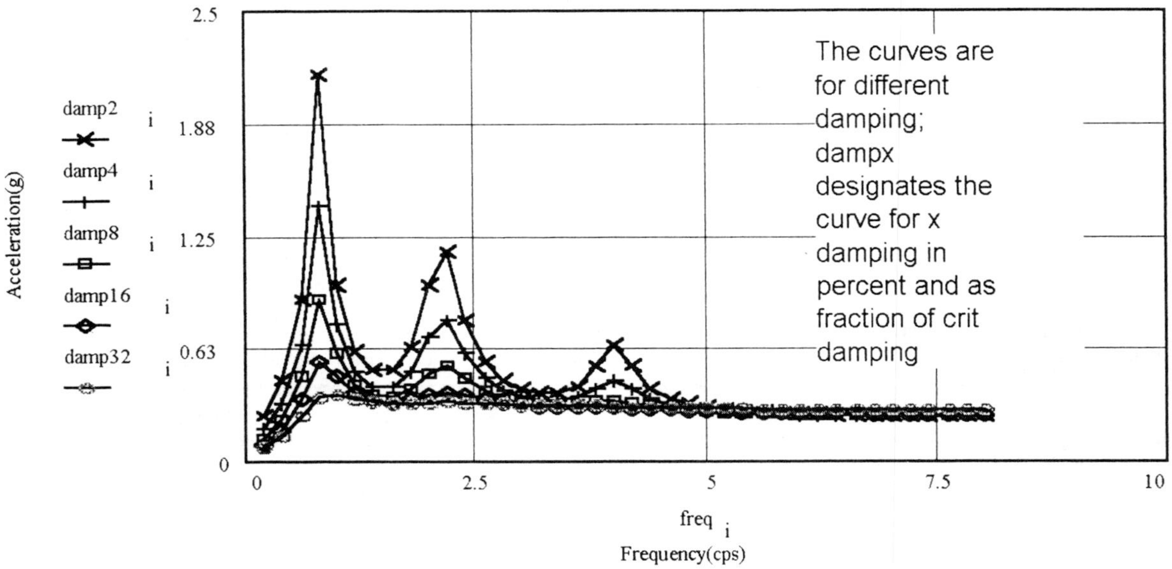

Typical FRS

fig1

A piping system is generally designed to be very flexible , in particular in areas where the design temperature can be in excess of 500 C , where the thermal stresses are indeed dominant ; in any case , if one has to account for such loads as in fig 1 , it is common practice to design the piping so that its basic frequencies are larger than the ones corresponding to the peaks , special restraints are used in order to allow the thermal 'breathing' of the piping. Cost and maintenance are affected .

However one should also try to answer the following question : what happens if the piping has a frequency in the ' peaks ' region so that the elastically calculated loads are indeed very large ? The answer using the classical procedure is obvious; there is no way the piping can sustain such loads. However experimental findings may suggest a quite different conclusion. In particular EPRI and other tests make clear that safety margins as large as twenty or even more are found. In particular the tests described by DeVita(1988) on representative piping are extremely significant , both in the identification of the safety margins (even greater than thirty) and in the identification of the area of failure (an elbow or a tee) and in the identification of the failure mode (ratcheting) ; about the same conclusions can be reached by examining the results of the tests by Tagart et al(1987,1988) and Onesto et al (1986) . The basic findings are summarized in the following table 2.1 Some slight improvement can be found if the damping value is increased as an example from the typical values around 2% to a more realistic 5% ; however even in this case the design procedure does not address the real nature of the problem :

a) the dissipation of energy is not viscous but it is due to anelastic behavior , so that the elastic method can be regarded simply as approximate

b) the frequency of the system is not a constant but changes as a consequence of plasticity and yielding , although this may lead to progressive distortion (ratcheting) which eventually leads to failure . Consequently the survival of the system is tied to its capacity to yield under loads with a progressive softening , changing its frequency and increasing the energy dissipation , until the progressive distortion leads to eventual failure

Table 2.1

Experimental findings /1 to6/

Loading :	dynamic , seismic type , peak region
Failure :	due to progressive distortion , concentrated
Safety margin with respect to ASME3 :	very large , even more than twenty

3. A one D.O.F analog

A model dealing with the phenomena described so far should be able then to predict frequency shifting, energy dissipation and progressive distortion (ratcheting) as a function of some parameter, describing the yielding of the system. Lazzeri et al (1987,1988,1989,1992) hasve presented various papers dealing with this problem and a model has been worked out. The model is briefly reviewed hereinafter. Consider a one degree of freedom non linear system, the basic equation is:

$$\varpi^2 x + x'' - \lambda x = a(t) \qquad (1)$$

ϖ = elastic eigenfrequency

' = time derivative

x = displacement

$a(t)$ = acceleration of the support, a function of time

λ = non linear term

Some methods (e.g. Krylov Bogolyuboff(1943) , see also Pipes and Harvill (1968)) try to change the equation (1) into an equivalent equation

$$(S * \varpi)^2 + x'' + 2 * \beta * x' = a(t) \qquad (2)$$

S = frequency shift

β = equivalent damping

where the two parameters S and β are a function of the level of yielding . The basic parameter describing the yielding is assumed to be

D= ductility (the ratio between the actual displacement and the displacement corresponding to yielding)

The first analog by Lazzeri (1989) used Krylov Bogolyuboff method in order to predict frequency shift and equivalent damping for an elastic perfectly plastic system ; further refinements by Lazzeri (1988) included the use of an energy method , which was able to predict quantitatively progressive distortion (ratcheting) due to the superposition of static and dynamic forces. The method makes reference to an energetic balance among kinetic energy, dissipation and external work; for a validation reference is made to Lazzeri (1992) Typical results in terms of frequency shift , additional damping and ratcheting are presented in figs 2 The frequency shift is analyzed in fig 2a for a typical value of the strain hardening coefficient defined as the ratio between plastic and elastic coefficient. The coefficient in fig 2b represents the ratcheting coefficient per cycle normalized to the yield displacement. The coefficient is also a function of the static factor Ss (ratio of the applied static force to the value producing yield) The computed damping factor in fig 2c has been obtained by means of Krylov Bogolyuboff analog, which is perfectly equivalent to the energetic computation method by Shibata et al (1987) and Masri et al (1985) ; a 23% cutoff value is also introduced

An important factor is the type of strain hardening. Lazzeri (1989) discussed the case of both the kinematic and the isotropic strain hardening model: the tests by Tagart et al (1987,1988)would indicate the presence of evident shakedown phenomena, which can be better explained if at least a partially isotropic model is used. Another important factor to be taken into consideration is the shape of the strain hardening curve; as the piping is basically elastic even under extreme loads while all the deformation is concentrated in elbows, it is quite reasonable to use the deformation curve of an elbow as representative of the behavior of the piping itself. Data about the behavior of elbows have been collected in Lazzeri (1992), further data are available in Tagart et al (1988); all the data confirm the asymmetric nature of the loading curve, when the elbow is loaded in plane, in opening and closing mode; a typical case is shown in figs 3 , where the adimensional moment is the applied moment over first yield moment. The effect is due to the change of the elbow cross section in the plastic field. On the contrary the curve is symmetric in out of plane conditions. As the in plane curve is not symmetric some ratcheting may occur under dynamic load even if no static force is present; this is the self loading effect mentioned by Lazzeri (1987) Finally the effect of pressure can be taken into consideration assuming a yield locus consistent with the findings in fig 3 , which has been built for an elbow. The method has been discussed by Lazzeri (1987,1988,1989) , taking into account the membrane theory of shells after Novozhilov(1964) In this way the progressive distortion in the hoop direction can be modeled , as discussed by Beaney (1987) and Lazzeri (1992): at each point on the yield locus curve the flow shall take place along the perpendicular to the curve itself . Then in a pressurized pipe under alternating bending load the deformation shall progress also in the hoop direction or only in the hoop direction as found by Beaney for a straight pipe As far as the significative numbers of cycles are concerned , they are assumed to be a function of the ductility . It should be reminded that smaller amplitude cycles are not significative due to shakedown , however if the load amplitude and hence the ductility increases then more and more cycles become significative . Consequently the final ductility after the seismic event is calculated as :

$$D_t = D + N * R(D, S_s) \qquad (3)$$

D_t = total ductility at the end of the seismic event

D = ductility in each cycle

N = number of significative cycles in the event

$R(D,S_s)$ = Ratcheting coefficient

S_s = Static coefficient (ratio of applied load to yielding load)

The basic characteristics of the model are then summarized in table 3.1

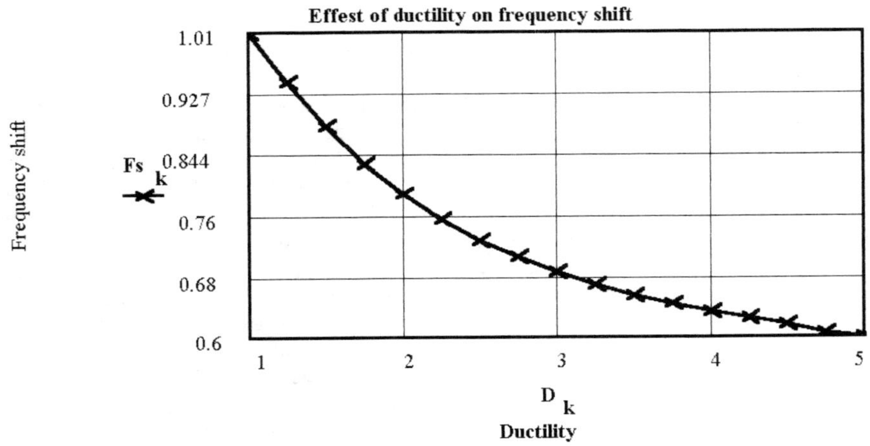

fig 2a,

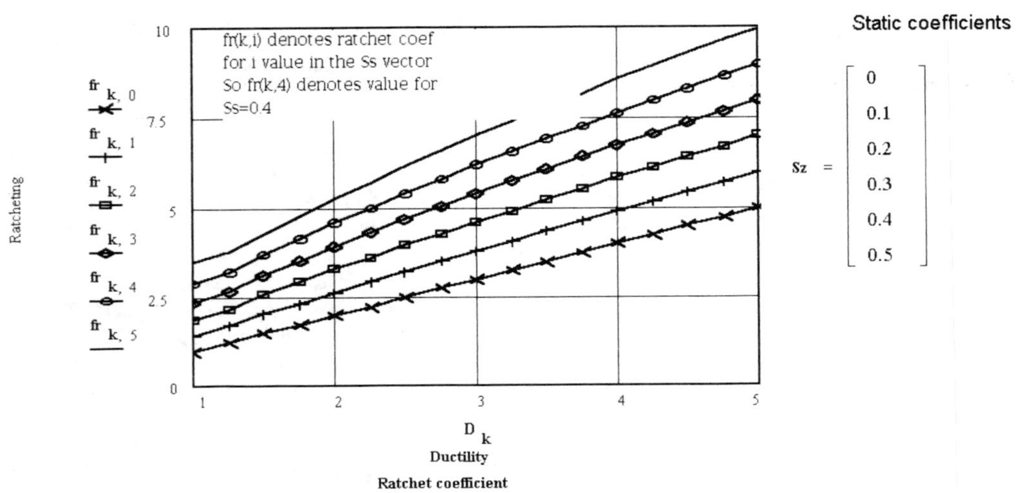

fig 2b

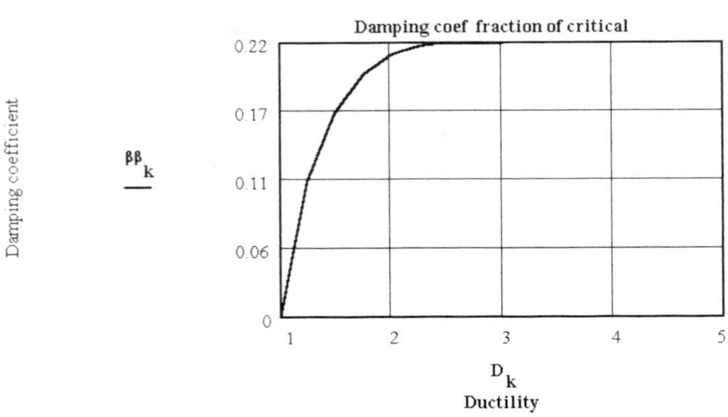

fig 2c

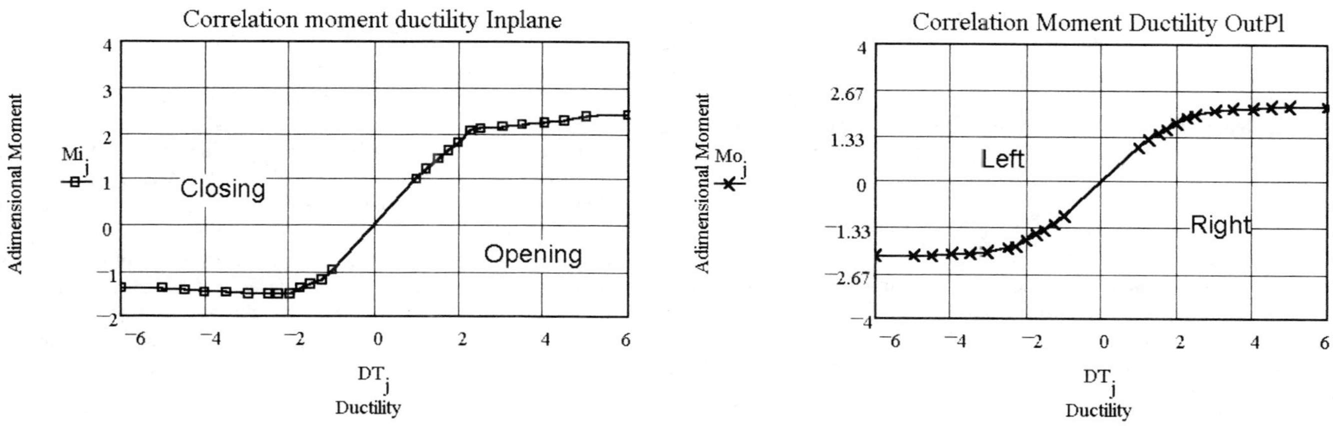

figs 3a.b

Ductility

duct = 1.2

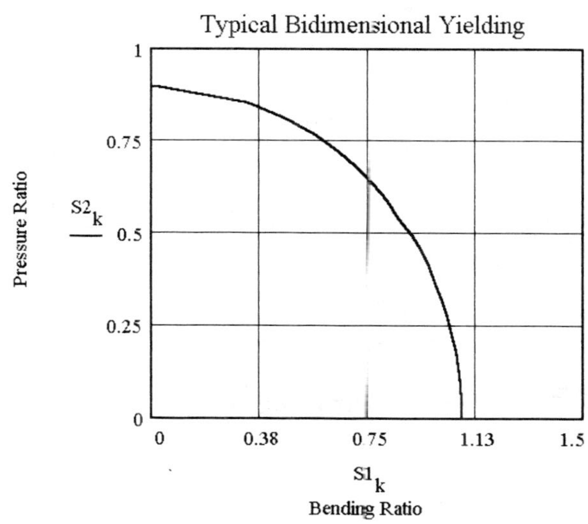

figs 3 c

Table 3.1

Basic characteristics of the analog (Lazzeri , 1987,1988,1989,1992)

frequency shift	according to energy method
equivalent damping	according to Krylov Bogolyuboff method
ratcheting	according to energy method /
final ductility	cyclic ductility plus ratcheting
strain hardening	taken into consideration (kinematic , isotropic , any combination)
Basic model	elbow like , asymmetric in loading
Other models	bilinear with-without cutoff
Pressure effect	included
Number of cycles	Six to eight according to ductility

Note : An averaging technique has been used to account for the varying effect of strain hardening on the ductility and static coefficient

However the use of the elbow like model only may be sometimes inaccurate due to the fact that once the most stressed elbow (or similar item) has yielded and its rigidity has been reduced, other components may be important . Evidently some peculiarities may be indeed present that can hardly be modeled by a one d.o.f. system. However the analysis of many cases and circumstances may be helpful.

4. Typical results

The typical flattening of an FRS as a consequence of the yielding is shown in fig 4 together with all the significant parameters. The comparison with the usual FRS is also shown. Obviously a considerable flattening of the peaks is found ; the model in figs2a predict decrease of the frequency with the increase of the ductility level ; consequently the peaks tend to be displaced towards larger frequencies with increasing ductility . This type of results, although semiquantitative, is deemed to be highly representative of the behavior of a piping under seismic loading. However even more interesting are the results presented in fig 5 , where an hypothetical piping system is supposed to have a frequency exactly in resonance with the main peak (the FRS in fig 1 is used) ; under any elastic theory its effective acceleration would be strongly amplified , the only limit to the amplification being the design damping factor . The analog in section three is used to evaluate the effective acceleration as a function of the ductility Some cases shall be examined as shown in table 4.1

Table 4.1
Calculation grid

Case	Model	Strain Hardening
A	Elbow type	Partially isotropic
B	Elbow type	Kinematic
C	Bilinear no cutoff	Partially isotropic
D	Elastic Perf Plastic	Kinematic

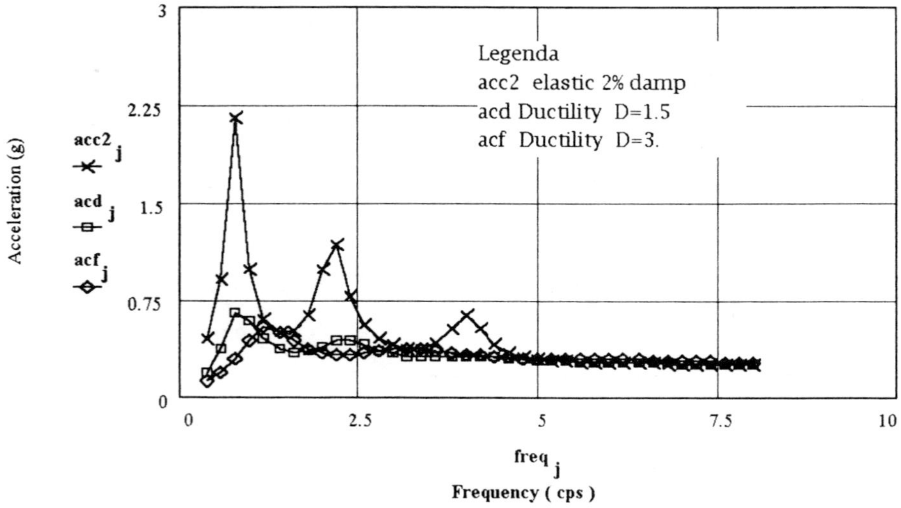

Typical FRS
fig 4

A comparison among the various results (in the case of a full resonance condition) is shown in fig 5a ; it appears that , regardless of the basic theory under examination , a substantial decrease in the acceleration is achieved with even limited values of tductility. To better understand this point it should be mentioned that , in the case of the elbow the initial elbow yielding is considered i.e. when most of the section is still elastic; under this aspect the relatively high value of the strain hardening coefficient (ratio between the plastic and the elastic coefficient) is used to model the progressive yielding of the elbow as shown in figs 3 . However even more interesting is a plot of the normalized acceleration (ratio of the actual acceleration to ZPA value) with respect to the total ductility (i.e. including the ratcheting) The data are reported in fig 5b It is clearly seen that regardless of the model used (cases A`to D) the load is totally deamplified with total values of the final ductility around 15 to 20 (with a static factor around 0.3 , which somewhat accounts for the asymmetry in the elbows loading)

A more general situation is depicted in figs6 . where a typical case has been examined for various frequencies as a function of total ductility (see formula (3)); a correlation between the adimensional load multiplier (ratio between current and reference ZPA) and ductility is presented in fig 7 . Some general conclusions can be drawn (quite similar indeed to the 'provocative' suggestion by Lazzeri (1993)

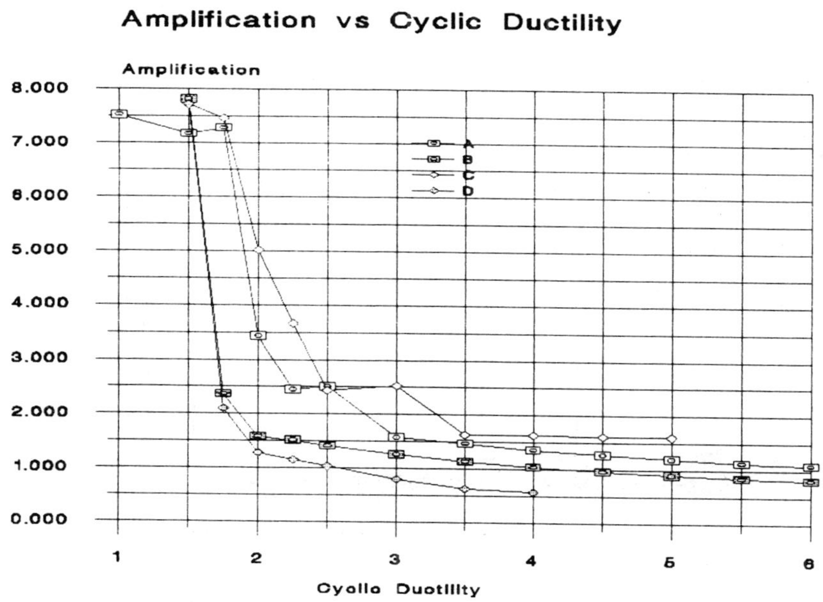

fig 5a

a) a limited amount of ductility is sufficient for a nearly complete elimination of the amplification , so that a modest increase in the effective acceleration is present ; the ductility requirements depends on the frequency , being larger in the peaks zone ; they also are decreased if the reference acceleration is increased

b) the actual shape of the acceleration vs. ductility curve is a function of the frequency; 'Duffing instability' phenomena (the increase of the response with the increase of the ductility) are present ; the most critical situation would appear to be the systems with a frequency slightly larger than the one corresponding to the main peak

In conclusion it would appear that some ductility (generally small) is necessary for a nearly complete elimination of the peaks , so that the actual design acceleration is not too different from the value corresponding to the ZPA value

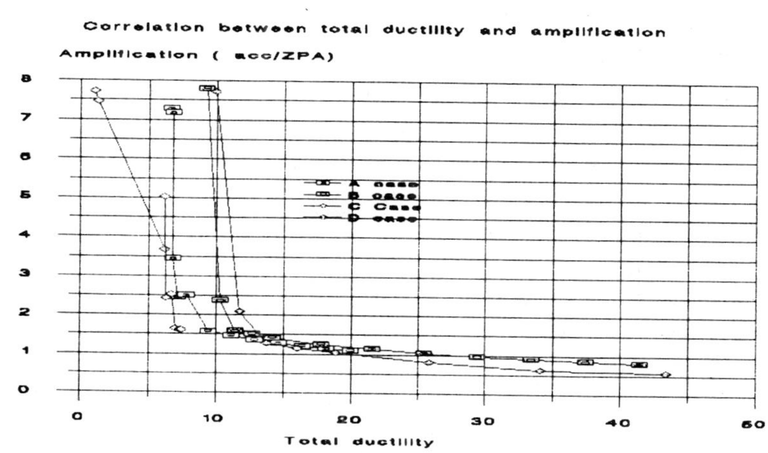

fig 5b

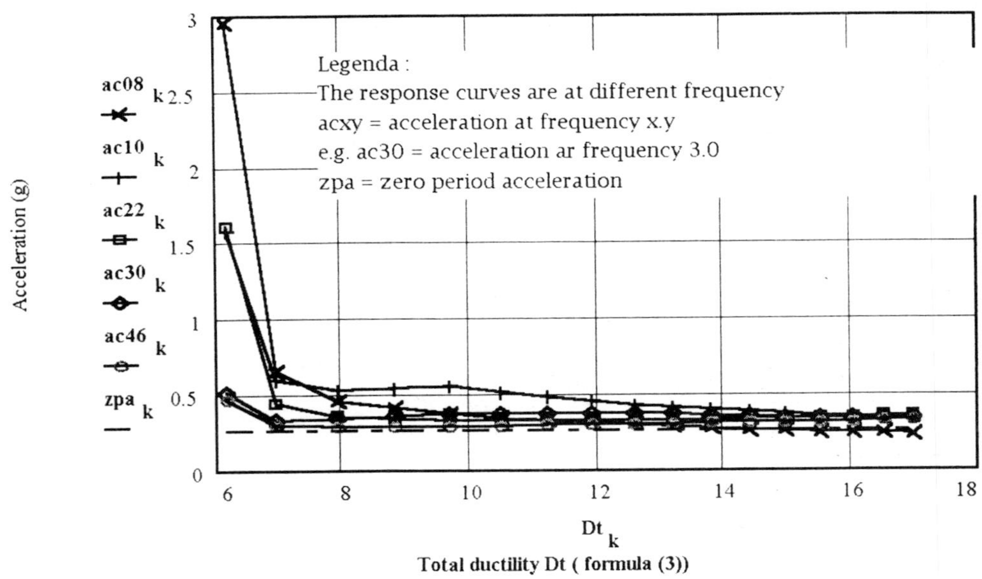

fig 6

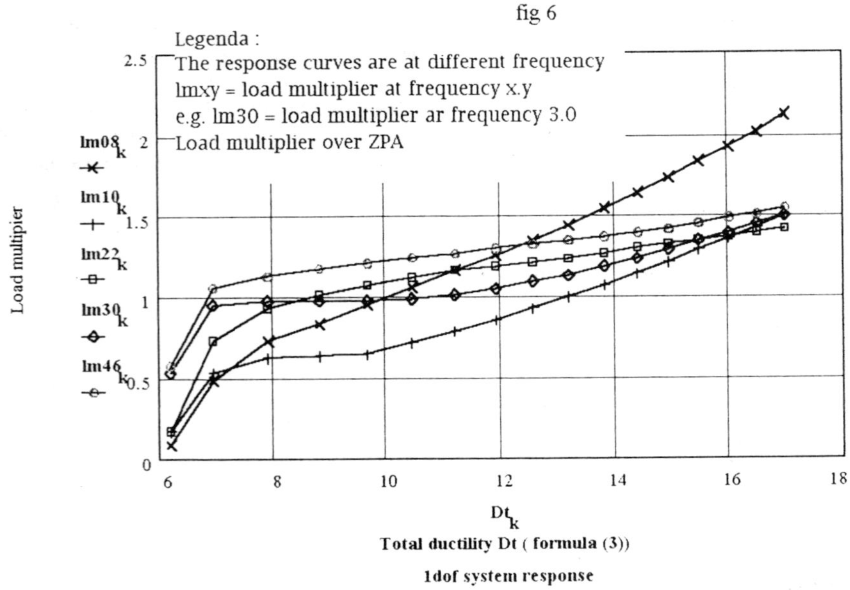

fig 7

5. Application to the design

The data presented in the preceding sections would indicate that:

a) in a fully ductile situation the design acceleration is very little amplified with respect to ZPA even at amall ductilities

b) the piping mechanical resistance should be based on the ZPA value

Then an obvious question arises: whether a piping not formally analyzed for lateral load is capable of sustaining a lateral load equal to ZPA times the weight of the piping itself. However for highly seismic zones ZPA values around 0.2g to 0.5g (including the effect of the building ductility) can be reasonably anticipated.

Several representative piping systems not specifically designed for lateral seismic loads has been examined; the basic characteristics of this analysis are reported in table 5.1. As (reference is made to figs 6,7) the ZPA level is the minimum requirement, even in the presence of a reasonable ductility level with the possible exception of very flexible systems , this level of loading was considered . In any case the analysis shall establish a base case for futher analyses

Table 5.1

	Piping representative analysis
Type	Conventional power stations : Thermal cycle
Designed for seismic	No
Level of acceleration	0.3g each horiz axis & 0.2 Vertical
Type	Static equivalent (frequency independent)
Number of cases	10
Allowables	ANSI 31.1 limits , no increase over normal service loads

The results, which are obviously to be taken with a lot of care are summarized as follows:

Table 5.2

Results of the analysis	
Within ANSI 31.1 limits	5
Within 1.5 ANSI limits	3
Within 2.0 ANSI limits	0
Beyond 2.0 ANSI limits	2

Due to the very nature of the analysis, it was somewhat arbitrarily decided that if the resulting stress is within 200% of the allowable limit of ANSI31.1 (in a manner quite consistent with the usual stress limits increase over the limits in normal conditions) , the result is acceptable and compatible with the ductility hypotheses made previously. The cases beyond such limits need additional consideration. From a detailed analysis it appears that in the cases when the stresses are well beyond the 200% mark , the part of the line where the stresses are large is generally quite limited and the line itself is very flexible (first eigenmode around 0.5 cps) Some sample problems are discussed in some detail in appendix B , however as a rule it can be reasonably stated that the problem arises near the fixed points of very long and flexible lines that are supported only vertically . A plot of seismic stress ratio as a function of the main frequency is plotted in fig 8; the seismic load is always consistent with the hypotheses in table 5 , i.e. static equivalent ; the correlation is evident. The inertia load , which is evaluated as shown in table 5.1, is supposed to be static equivalent hence no decoupling is assumed . In reality yielding shall take place and the actual frequency shall be decreased with respect to the elastic value ; in general this tends to reduce the acceleration level even with modest ductility values . It should also be noted that indeed very simple design actions might be taken in order to avoid such overloading (e.g. a snubber or a bumper to defend the overstressed areas , this procedure has been sometimes used by the author organization)

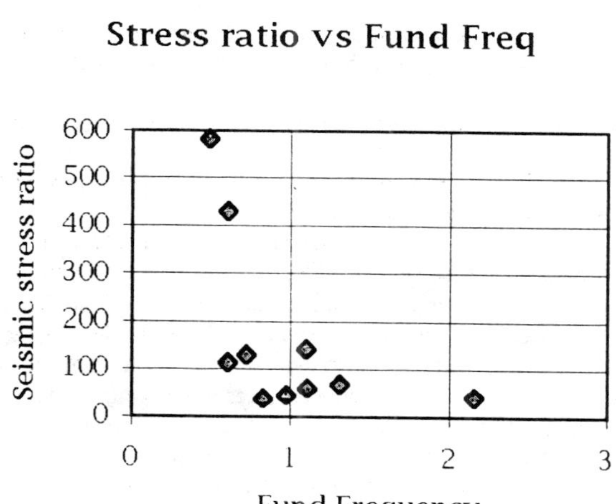

The load is constant and as shown in table 5.1 ; the FRS is neglected

fig 8

6. Conclusions

The aim of the paper has been a discussion of the seismic capability of the piping systems; an approximate but quite reasonable calculation method , based on the elastic plastic analysis of a one dof system has shown , in agreement with experimental and test data , that :

a) the dominant failure mode is ratcheting , due to the superposition of static and seismic loads

b) with even limited ductility values the resultant Floor Response spectra are practically flattened , with a small amplifications in the peaks areas and deamplification for very low frequencies
c) some typical calculations , performed on piping systems , coming from different power stations , but never specifically designed for seismic lateral loads , have shown that :
ca) the piping stay generally elastic and only limited yielding is present
cb) reasonably stiff systems are capable of sustaining loads (about 0.3 g each direction) typical of ZPA load , so that the ductility requirements are very low
cc) flexible systems are obviously much more prone to be overloaded by such loads , however the increase in the ductility causes frequency shifting and then load limitation and the actual acceleration is quite smaller so that no large overloading takes place ; however comparatively simple remedial actions can be taken

Although the examined data are too few to give quite general conclusions , nevertheless it wold appear that piping are capable to sustain seismic loads if adequate ductility reserves are present, in accordance with the field observations . This fact occurs quite often even if no special provision for seismic load is taken ; however very simple design remedial actions can be easily taken .

Further analyses are under way , in particular to confirm the validity of the mathematical approach by means of time history analyses

7. References

E M Beaney ' The response of piping to seismic loading ' 9 SMIRT Conf , K2 , Lausanne , Aug 1987

S Cecconi , V Giuliano , L Lazzeri ' On the seismic design spectra for heavy components and comparison with the usual FRS tecnique ' 5SMIRT K9/2 , Berlin 1979

V DeVita ' Seismic fragility testing of representative nuclear piping systems' Seismic engineering 1988 , PVP 144 , ASME PVP Conference , Pittsburgh 1988

N Krylov,N Bogolyuboff 'Nonlinear Oscillations ' Engl Transl, Princeton University Press , 1943

L Lazzeri ' The design of piping in plastic conditions under dynamic loads ' Int J Pres ves & Piping 30, 1987

L Lazzeri , M Scala ' Equivalent damping in piping due to local yielding ' ASME J pressure Vessel Technology , Aug 1989 , vol 111,337

L Lazzeri ' Effects of plasticity on dynamic response of piping structures ' ASME J Pressure vessel technology , vol 110 ,Aug 1988

L Lazzeri ' Simplified non linear ratchet model for dynamic analysis of pressure components ' ASME J Pressure vessel technology , vol 114 Aug 1992

L Lazzeri, M Scala , G Viti , M Agrone 'Design considerations for the analysis of piping in dynamic conditions with moderate ductiliy values ' PVP 127 , pg 207 ,SanDiego, 1987

Lazzeri ' Refined models for the analysis of the elastic plastic behaviour of pipeline components under dynamic loadings' PVP 113 , PVP meeting , Pittsburgh , 1988

L Lazzeri ' The design of piping in plastic conditions under dynamic loads ' Int J Pres Vessel & Piping , 1987

L Lazzeri ' The design of piping in plastic conditions under dynamic load ' Intern J Press Vessel and piping , vol 30 ,1987,p 351-365

L Lazzeri 'System ductility appraisal by means of a similitude rule' PVP 113 , PVP meeting , Pittsburgh , 1988

L Lazzeri ' Seismic analysis criteria for piping ' ASME PVP Meeting 1993

L Lazzeri G Viti ' Heavy interaction problems in Floor response spectra ' Res Mechanica 30 (1990)

S F Masri , A H Hadjan ' On the problem of equivalent damping estimate from experimental data' 8 Smirt Conf , K18/5 , Brussels , Aug 1985

V V Novozhilov ' The theory of thin shells ' 2nd ed ,1964

A T Onesto, W P Chen , V DeVita ' Six inches diameter pipe seismic fragility test' Proc 14th Water Reactor Safety Eng meeting VSNRC , DEC 1986

L Pipes, L Harvill ' Applied mathematics for engineers and physicists' McGraw Hill Editors ,1970

J L Sackman , J M Kelly ' Equipment response spectra for nuclear power plant system' 5SMIRT K9/1 , Berlin 1979

H Shibata , Y Shiraki , Y Hajimura ' Test and evaluations about damping characteristics of piping systems ' 8 SMIRT Conf , K18/4 , Brussels ,Aug 1985

M Silver , G Hardy , Y Tang , P Smith ' Piping performances during and after earthquakes ' ASME PVP Meeting NewOrleans 1985 PVP 98-6

P O Svensson G C Slagis F S Braun ' Seismic design of piping systems in the flexible range ' ASME paper 83-PVP-68

SW Tagart , JK Tang , HL Hwang , KL Merz , DJ Guzy ' Seismic analysis and testing of piping components ' ASME PVP meeting SanDiego ,June 1987 ,PVP 127

SW Tagart , JK Tang , HL Hwang , KL Merz , DJ Guzy, V DeVita 'Seismic anaysis and testing of piping systems and components ' Seismic engineering 1988 , PVP 144 , ASME PVP Conference , Pittsburgh 1988

US Nuclear Regulatory Commission , Regulatory Guides , 1980

Acknowledgements : the analyses described in appendix B have been performed by E Ferrarese and K.Schultz

Appendix A : Typical seismic criteria for a conventional power station

The basic criteria , typical of a plant in a very severe seismic area are summarized as follows

a) horizontal seismic coefficient : depends from the frequency and type of structure ; however an indicative value around 0.2g can be found , which incorporates a reduction factor , a function of the type of structure and in particular of its ductility reserve ; the analysis method is somewhat similar to Uniform Building Code

b) for dynamic analyses a spectrum is suggested , which is a function of the basic frequency as indicated in fig A1 ; the factor in fig A1 is applied to a basic acceleration (0.4g in this case) and then again reduced by means of a ductility factor , about 5 for ductile structures ; as buildings tend to be characterized by frequencies around a few cps, again a value around 0.2 g is typical . Obviously the rigorous application of the modal analysis and response spectrum is intended to avoid local amplifications as due to local structural deficiencies

In general the static method is deemed as sufficient (with a few exceptions as an example the boiler structure) and it is applicable to all the buildings . For heavy components the only requirement is the anchorage verification , with the load calculated assuming a design acceleration equal to the acceleration calculated in the building at the level of the anchorage (which does not include the above mentioned reduction factor , a function of ductility) and a modest amplification factor equal to 1.2 ; values of such acceleration as large as 0.6 to 1.0 g can be anticipated . Provisions for heavy equipment (decoupling) and flexible mountings are contemplated . The electrical equipment is designed according to the dynamic method or to experimental qualification procedures . No special provisions are contemplated for the piping

It is interesting to note that in analyzing the shear at the base of heavy equipment the ductility reduction factor is not applied , although the ductility obviously limits the effective acceleration ; this may be due to the fact that such reduction may be applied , but it is an obvious precaution to take care that an oversizing of the building , and so an increase in the acceleration , does not jeopardize the equipment and its anchorage

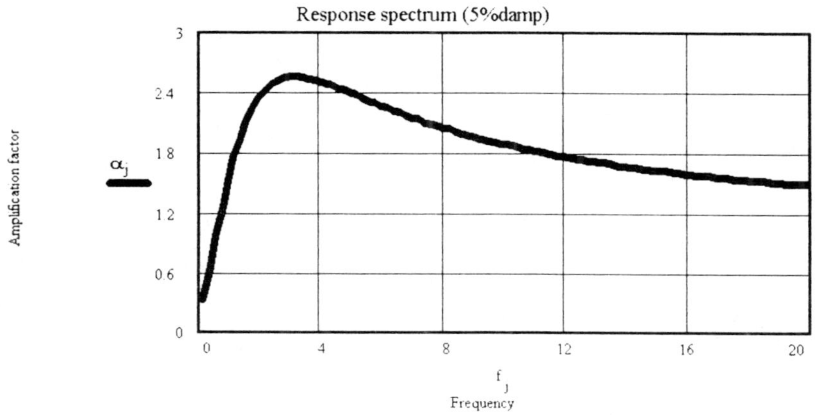

fig A1

Appendix B : Some results about seismic loads on conventional piping

The present section analyzes in some detail one of the cases where according to the criteria described in tables 5.1 and 5.2 overstressing occurs . The geometry and the displacement pattern in the first eigen mode is shown in fig B1 ; the corresponding frequency is about 0.6 cps; in the lower left corner region overstress occurs on the elbow. It is evident that the piping has no adequate lateral support in the overstress region in particular , hence the seismic loads have to be transferred to the fixed points at the end of the line

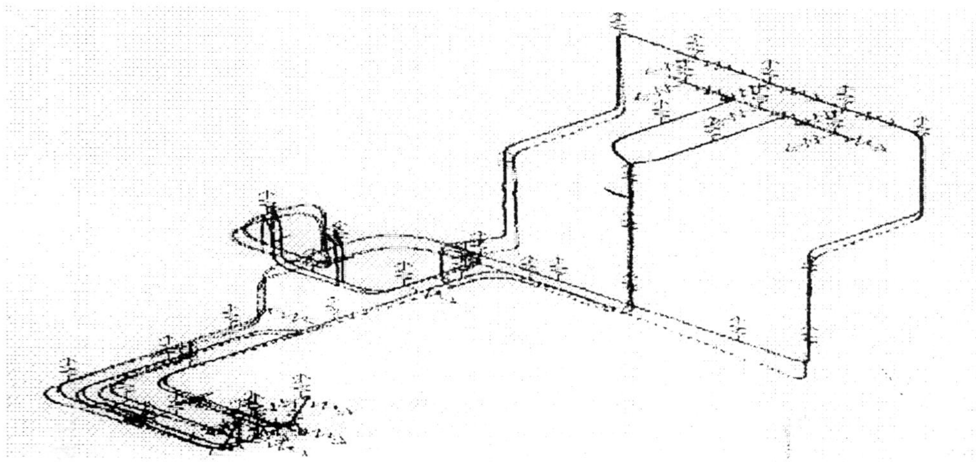

fig B1

Obviously the actual change of seismic load on the piping as a consequence of frequency shift due to softening largely depends from the actual shape of the seismic load , however assuming the seismic spectrum in fig 1 and hence assuming that the main vibration mode is at the left of the peak it is shown that substantial reduction of the acceleration can be obtained with modest ductility values as shown in fig B2 , relative to the low frequency range . It is also interesting to give an indication of the data obtained on the sample cases as shown in table B1

Table B1

Case	Main frequency	Overstress factor	Notes
1	0.487	5.85	Yielding limited
2	0.722	1.30	
3	0.602	1.14	Yielding localized
4	2.154	0.41	
5	0.973	0.45	
6	0.604	4.31	Yielding limited
7	1.101	1.42	Yielding localized
8	1.307	0.68	
9	0.826	0.38	
10	1.104	0.60	

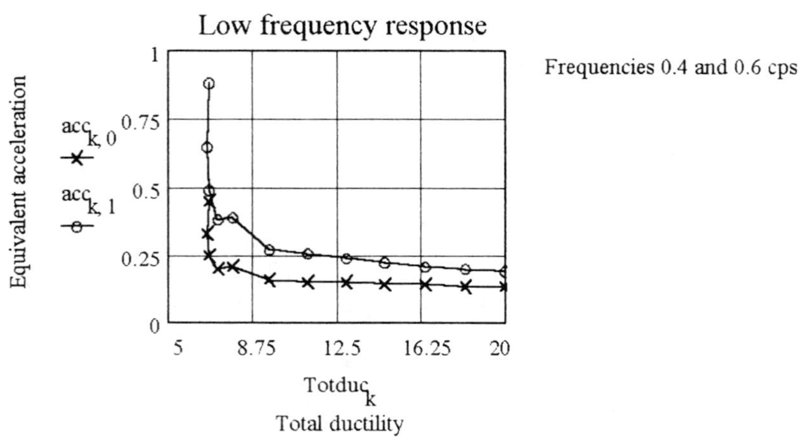

fig B2

Appendix C : ANSI B31.1 stress limits

The usual stress limits according to ANSI B31.1 concerning the primary (i.e. not including such loads as the thermal ones nor any increase for abnormal conditions) loads is :

$$0.25pd / t + 0.75iM / Z \le \sigma_{all} \quad \text{(C1)}$$

p = pressure d = diameter t = thickness

i = stress intensification factor

M = resultant moment

$Z = 0.25\pi t d^2$ section modulus

σ_{all} = allowablestress

σ_{all} = lesser of / 4 of the ultimate and 2 / 3 of the yield stress

Then basically the section is fully elastic under the normal loads , if the allowable is increased by a factor two , then the section is only at initial yielding and well within the section collapse capability . It is also evident that the computation method using elastic stress distribution is somewhat pessimistic

PVP-Vol. 340, Seismic Engineering
ASME 1996

FINITE ELEMENT SIMULATION OF PIPE DYNAMIC RESPONSE

Gerry C. Slagis
G C Slagis Associates
Walnut Creek, California

Richard W. Litton
PMB Engineering Inc.
San Francisco, California

ABSTRACT

Nonlinear finite element dynamic analyses of the response of a pipe span to controlled-displacement, sinusoidal vibration have been performed. The objective of this preliminary study is to compare strain and acceleration response data to those generated by Beaney in the Berkeley Nuclear Laboratories experiments. Results for an unpressurized, 5 Hz, carbon steel pipe are in good agreement with the experiments. Hence, it appears that analytical simulation will be useful to assess seismic margins. Recommendations for additional studies are provided.

The analyses confirm the test results -- dynamic response is greatly attenuated by material plasticity. Analytical strains and accelerations are about 30% higher than test data. There are several possible explanations for the differences. To assess the effect of frequency on response, the length of the pipe span was increased. Analysis of the longer, 2 Hz, pipe span shows significantly greater cyclic strains than the 5 Hz span at the same input excitation levels.

INTRODUCTION

There have been a number of high level dynamic tests of piping. Testing done by Beaney at the Berkeley Nuclear Laboratories was summarized and evaluated by Slagis [1]. Displacement controlled, sinusoidal vibration tests were performed on one-inch to four-inch diameter pipe segments. Straight pipe spans of 5 to 10 Hz fundamental frequency, unpressurized and pressurized, carbon steel and stainless steel, were subjected to sinusoidal excitation at resonance of as much as 5g input for hundreds of cycles without failure by collapse or fatigue.

In order to fully understand these test results and the impact on seismic margins, analytical simulations would be very beneficial. By analysis, the effects of geometry, frequency, and earthquake time histories can be assessed. As the initial step in analytical simulation, finite element studies of a single straight pipe span were performed. An unpressurized, carbon steel pipe was selected for this preliminary study. The advantages of evaluating unpressurized carbon steel is that there is no influence of hoop ratcheting on the dynamic response, and hardening effects should not be significant.

TEST RESULTS

One of the early tests performed by Beaney [2] was on carbon steel straight pipe, 25.4 mm (1.0 in) outer diameter, 2.64 mm (0.10 in) pipe wall, and 3530 mm (139 in) length. The fundamental frequency was about 5 Hz. The test fixture is shown in Figure 1. Axial strains at the center of the span are measured with a strain gage, and the acceleration at the center is measured with an accelerometer. This particular straight pipe is the one chosen for study.

The yield stress of the pipe was reported as 209 MN/m^2 (30.3 ksi) with no hardening until strain reaches 1%. However, in later experiments, Beaney found a large variation of yield strength (a factor of two) in different test specimens. Elastic damping is estimated to be 0.5%.

For this experiment, Beaney reported strain response as shown in Figure 2 and acceleration response as shown in Figure 3. These are two different tests with different test specimens. The test procedure is not clearly specified. Apparently, for the strain response as a function of frequency, the test is run continuously. The test starts at the lowest level of input excitation, and the forcing frequency is swept. Then, the input level is increased, and the frequency is swept again. There is some ramp-up of input excitation (probably 3 cycles of loading), and each segment is run for a certain time (perhaps 2.5 seconds or 12 cycles). For the acceleration response, the test is apparently run continuously with discrete increases in input level. It is assumed that there is some ramp-up of input excitation and the excitation is held for a

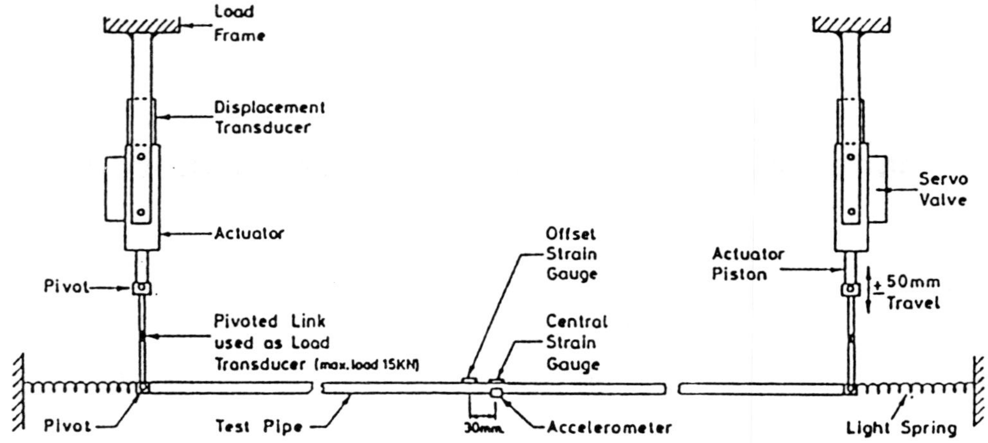

FIGURE 1. TEST SETUP [REF 2]

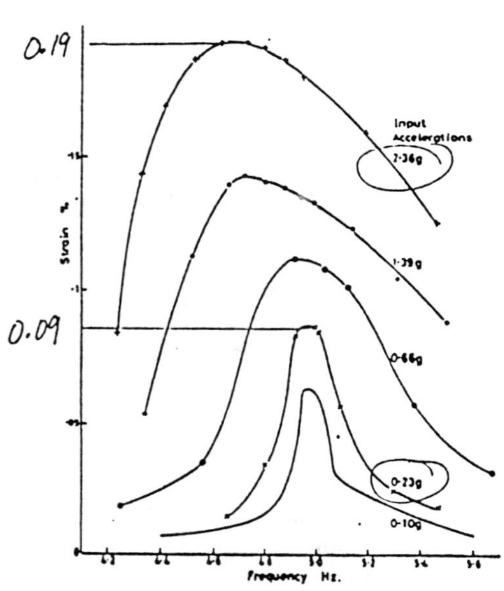

FIGURE 2. STRAIN RESPONSE [REF 2]

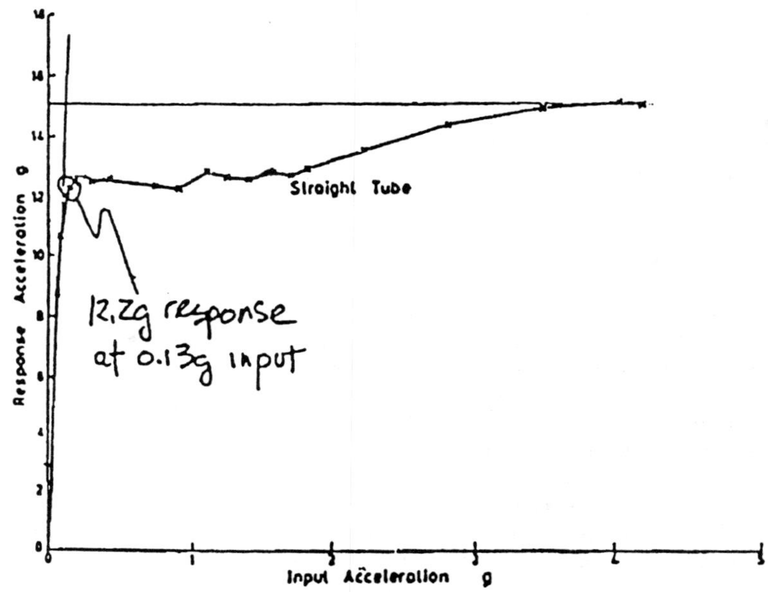

FIGURE 3. ACCELERATION RESPONSE [REF 2]

certain time. There is one major difference for the acceleration response. The forcing frequency was tuned to the inelastic response frequency. The maximum shift in frequency reported by Beaney for this particular pipe was about 8%.

Strain and acceleration data was sampled at an unspecified rate. Then, the raw data was fitted to a sinusoidal shape for the reported values. It is apparent from the Beaney reports that the strain amplitude reported does not include the inelastic strain from permanent displacement (sag) of the pipe. The sag varies with response time.

The next experiments conducted by Beaney [3] tested different size pipes. Acceleration response, as a function of frequency, was determined for a larger size pipe as shown in Figure 4. Although this data is not for the same size pipe as used in Figure 2, the data is shown to demonstrate the difference in "shape" between the strain data and the acceleration data. The acceleration data is much flatter than the strain data.

FINITE ELEMENT ANALYSES

The analytical simulation was performed using a nonlinear, finite element, structural analysis program, CAP, which stands for Capacity Analysis Program. This program was developed by PMB Engineering for the offshore industry, and the program was recently modified for analysis of piping systems. For the analyses reported here, the nonlinear beam element was used. The program generates the moment curvature relationship for the beam considering material yielding. The material behavior is elastic-perfectly plastic.

The Beaney straight pipe (25.4 mm outer diameter, 2.64 mm wall, 3530 mm long, unpressurized, carbon steel) was modeled with fifteen nonlinear beam elements. There were more elements near the center of the span then at the ends. The yield stress was set at 30 ksi. Elastic damping was specified as 0.5%. Pinned end conditions are modeled so that there is no axial restraint.

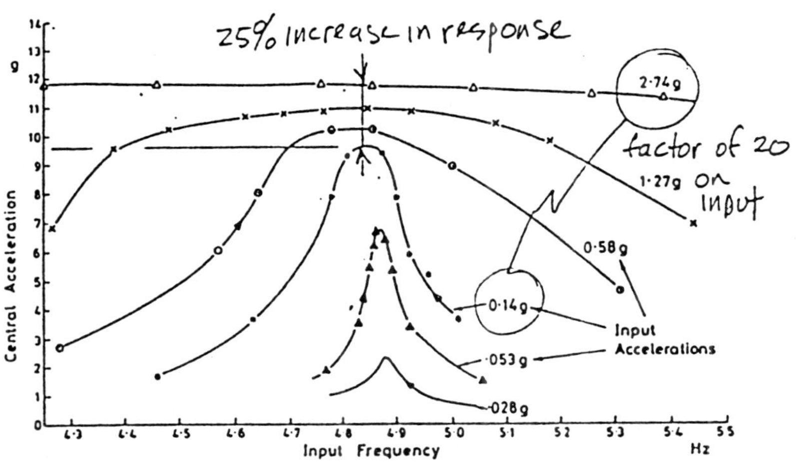

FIGURE 4. ACCELERATION RESPONSE VS FREQUENCY [REF 3]

The frequency analysis gave a slightly higher frequency than the 5 Hz test frequency. The total length of the beam was increased by 2% to get the same 5 Hz frequency.

Both the frequency sweep analyses and the response for increased input excitation analyses were performed. Results are reported and compared to the Beaney test data in the next section. There are several possible reasons for the differences. The elastic response is very sensitive to damping levels and input characteristics. Amplification factors for sinusoidal input with low damping are very high. The actual material yield strength of the test specimen is not known. The actual input history in the experiments is not known. An exact replication of a sine wave as the input to the test specimen is highly unlikely. The dwell time during each test segment is also not known.

For the inelastic analyses, ten cycles of input loading were applied with the amplitude building up to the test level in 2-1/2 cycles. The 7-1/2 full cycles were sufficient to produce the maximum response.

FINITE ELEMENT RESULTS

Strain Response

The finite element simulation was performed with each input level as a single continuous loading. Each load level was run as a new loading condition with no preexisting inelastic strain. In the Beaney test, all load levels were run on the same specimen continuously. Cyclic strain amplitudes as determined by analyses are shown in Figure 5 for comparison to the Beaney data shown in Figure 2.

The strain response curves are similar, but strains are higher in the finite element model. For the 2.36g input level, the maximum cyclic strain amplitude is 0.25% compared to the test value of 0.19%. The broadening of the curves with increased input level is indicative of the effect of plasticity on yield. The shifts in frequency with plasticity are almost the same. The peak strain response at 2.36g input is at a frequency of about 4.6 Hz. This is an 8% shift from the elastic

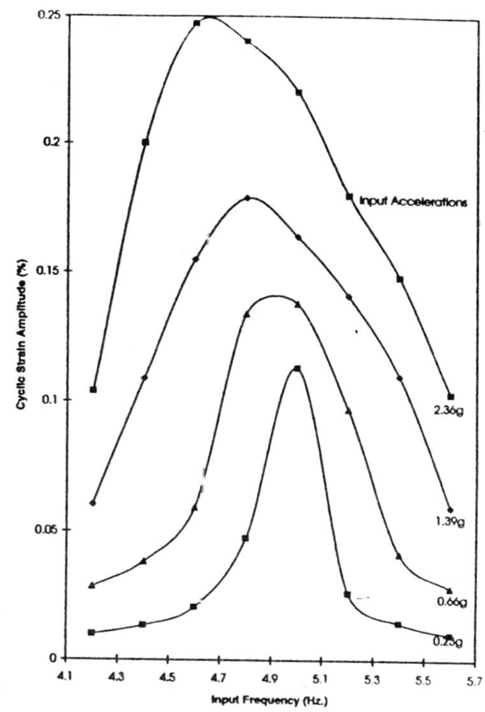

FIGURE 5. FEA STRAIN VS FREQUENCY

frequency of 5 Hz. The increase in strain with input level demonstrates the attenuation of response with plasticity. For a 10 times increase in input (2.36g vs 0.23g), the strain increases by 2.3 times (0.25% vs 0.11%). The increase in the experiment was 2.1 times (0.19% vs 0.09%).

Acceleration Response

The acceleration response for the same set of frequency analyses is shown in Figure 6. Beaney did not present a corresponding curve for this particular pipe geometry, but did report the response for a larger size pipe (Figure 4). The acceleration response shows similar frequency behavior as strain response, but the attenuation is even more drastic. For the 10 times increase in input, the acceleration increases by only 1.5 times (17.8g vs 12.2g). Note that at the 2.36g input level, the acceleration response is much flatter with frequency than the strain response.

In another test of another specimen, Beaney subjected the pipe span to an increasing level of sinusoidal input. At each input level, the frequency was tuned to the peak response frequency. Acceleration response as reported by Beaney is shown in Figure 3. In the finite element simulation, the input frequency is maintained at the elastic response frequency of 5 Hz so some difference in response is expected. The response is shown in Figure 7. The finite element results do not show the distinct flattening in the 0.2g-1.8g input range as the experiments. And the maximum response at 4g input is higher (19g vs 15g or 27%). The analysis results confirm the attenuation of response with plasticity. For a 16 times increase in input (4g vs 0.25g), the acceleration response increases by only a factor of 1.6 (19g vs 12g).

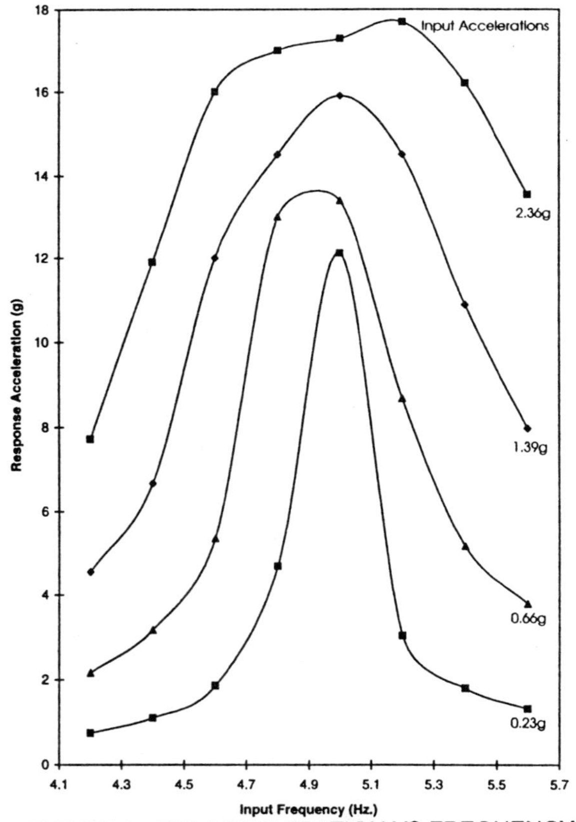

FIGURE 6. FEA ACCELERATION VS FREQUENCY

A time history plot of acceleration response is shown in Figure 8, and the displacement response, in Figure 9. The acceleration plot indicates rapid changes as a plastic hinge forms and then unloads. The displacement plot indicates the sag from deadweight increasing with each cycle of loading.

The influence of weight on deflected shape is shown in Figure 10. At the lower input level (0.66g), the deflection upward is less than the deflection downward. At the higher input level (2.36g), where sag is significant, the deflection upward never reaches the initial position.

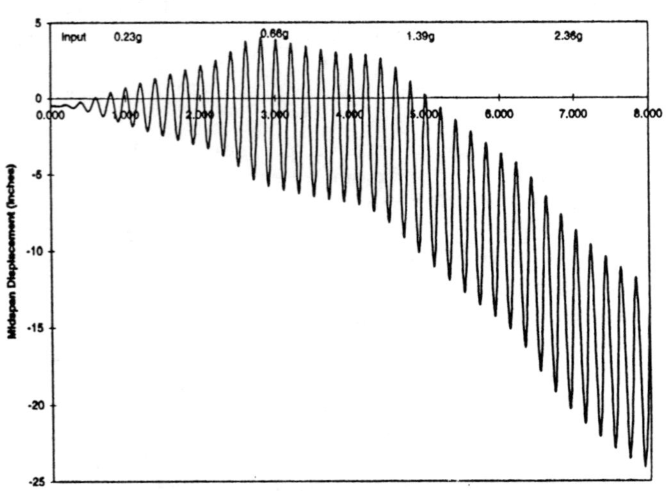

FIGURE 9. FEA DISPLACEMENT TIME HISTORY

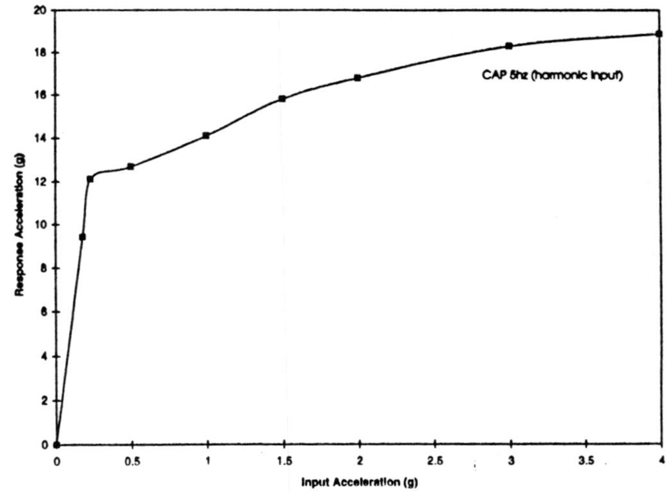

FIGURE 7. FEA ACCELERATION RESPONSE

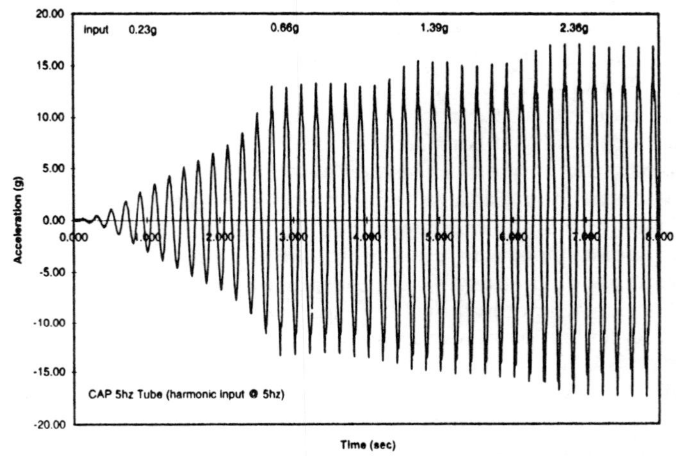

FIGURE 8. FEA ACCELERATION TIME HISTORY

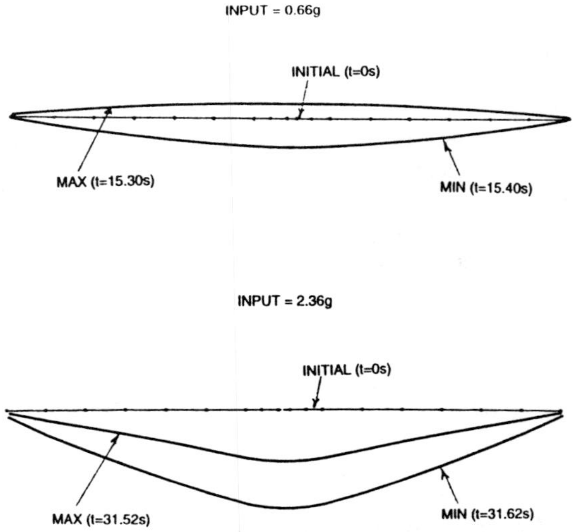

FIGURE 10. DEFLECTED SHAPES

DISCUSSION

In comparing results, it is important to keep in mind that both the experimental results and the analytical results are approximations to reality. The experimental results are approximate because of limitations in instrumentation, limitations in the control of the loading system, variability in material properties, and possible human error. The analytical simulations are approximate because of limitations in the applied mathematical algorithms, limitations in using discrete modeling for a continuous system, and possible human error.

The finite element results are comparable to the experimental results, but there are certain differences. Strain amplitudes are larger, and accelerations are larger. Acceleration response is not as flat in the 0.2-1.8g input range. There are several possible explanations -- a difference in yield strength, material cyclic strain hardening, the continuous nature of the Beaney tests, the sampling rate on the experimental data, and the input forcing function. The last two are expected to be the significant factors. The strain and acceleration data in the experiments were taken at an unknown sampling rate, and then curve fit to a sinusoidal. If the sampling rate is not high enough, the Beaney reported values will be lower than actual. This is very likely in the case of the acceleration, as indicated by the analysis time history plot (Figure 8). Also a strain gage of finite length (the length is not known) will average the strain if there is a gradient.

Although the experiment was designed to produce a sinusoidal input displacement, it is difficult to obtain a true sinusoid. Any deviation from a pure sinusoidal input will lead to a smaller response. To explore the influence of input time history, a modified input was developed. Figure 11 compares the effective dynamic loading for a pure sinusoid and a loading function that has the same maximum acceleration but is more peaked. A fourier decomposition of the "nonharmonic" loading indicates that the first component is 35% less than for the pure sine wave input.

Using the "nonharmonic" input resulted in analytical strain response in close agreement to the experimental work (for an input frequency of 5 Hz) as shown in Figure 12. For acceleration response with the nonharmonic input, accelerations flattened out up to a 1.4g input similar to the test data as shown in Figure 13.

Theoretically, a 0.23g sine input should yield the pipe, but the Beaney test results (see Figure 2) indicate a strain level well below yield. This suggests that the input was not a pure sine wave and that the test results underestimate the true sinusoidal response.

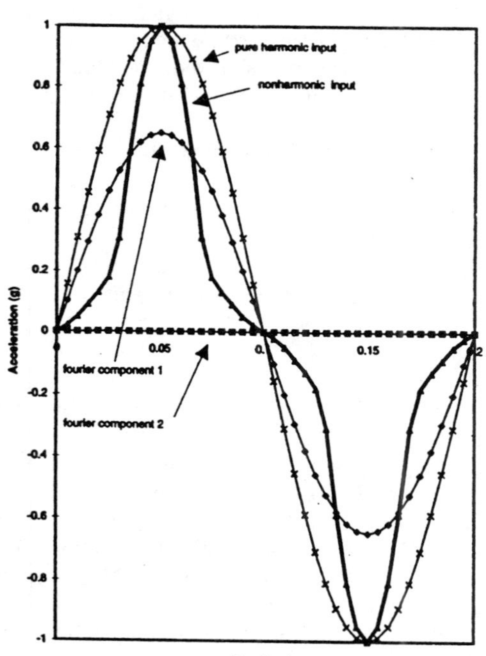

FIGURE 11. MODIFIED INPUT HISTORY

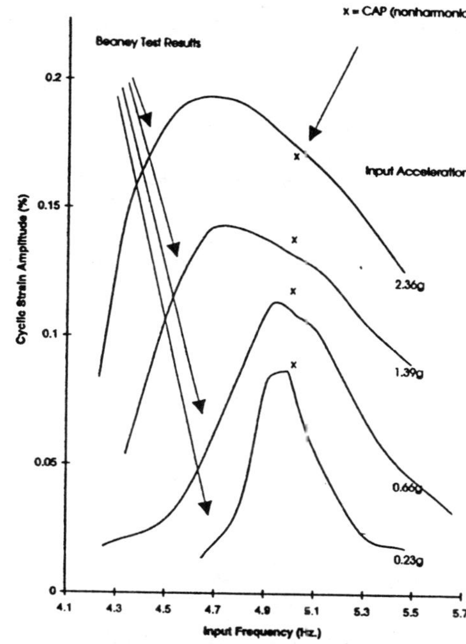

FIGURE 12. MODIFIED INPUT STRAIN RESPONSE

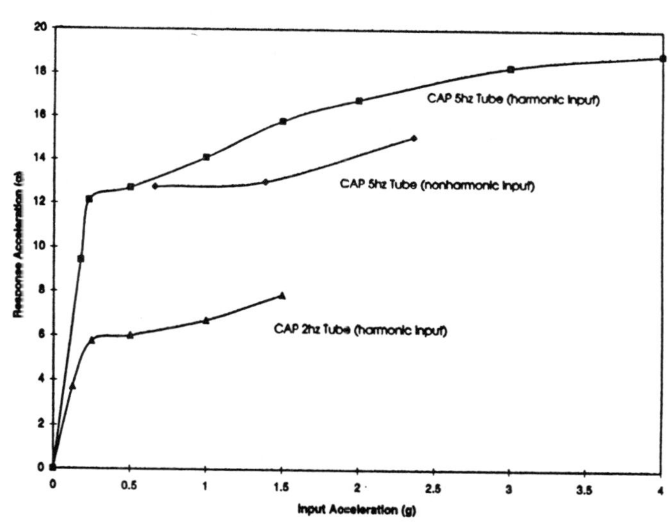

FIGURE 13. MODIFIED INPUT ACCELERATION RESPONSE

EFFECT OF FREQUENCY

Beaney [4] developed an energy balance method for predicting dynamic response and compared predictions with the experimental data. The energy response method predicts a higher response as the frequency of the pipe span is decreased. To investigate the influence of frequency on response, a finite element analysis was performed. The same beam as previously analyzed was lengthened to 5635 mm to obtain a fundamental frequency of 2 Hz.

The acceleration response of the 2 Hz span as compared to the 5 Hz span is shown in Figure 13. The behavior is similar: the acceleration magnitude is less on the 2 Hz model since yielding starts at a much lower acceleration level. But notice the strain response as shown in Figure 14. The strains in the 2 Hz span are much larger than the strains in the 5 Hz span for the same input level.

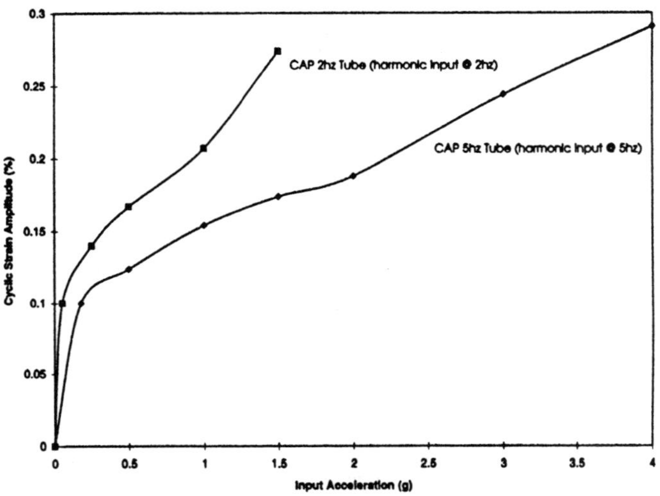

FIGURE 14. STRAIN VS SPAN FREQUENCY

CONCLUSIONS

Finite element analyses confirm the basic dynamic response characteristics as demonstrated by test. Response is greatly attenuated by material plasticity. As the input level increases, there is very little increase in acceleration response. Strains increase at a greater rate than acceleration. For a 5 Hz system, as the input increases by 16 times, the acceleration response increases by only 1.5 times. Strain response increases by 2.3 times.

A 2 Hz pipe span has a larger strain response at the same input level than a 5 Hz system. This frequency effect could be significant for seismic design considerations.

Differences between experiment and analyses are to be expected for the following reasons:

a. potential problems in determining the peak values from the instrumentation
b. difficulty in achieving a true sinusoidal input in an experiment
c. material hardening
d. the loading sequence, and
e. differences in yield strength.

Considering the above uncertainties, the analytical results are considered to be in good agreement with the experimental data.

Ideally, a useful analytical procedure will provide modestly conservative results with a well defined bias. For this preliminary study, there is good agreement in the trends between analyses and experiments, and the analytical results are more conservative.

RECOMMENDATIONS

Additional analytical studies are recommended to investigate the following characteristics of dynamic response:

1. comparison of earthquake time history response versus sinusoidal response
2. influence of different earthquake time histories
3. effect of span frequency on seismic margin
4. response of pressurized carbon steel spans
5. response of unpressurized and pressurized stainless steel spans

REFERENCES

1 G.C. Slagis, "Experimental Data on Seismic Response of Piping," PVP-Vol. 312, ASME, 1995, pp. 27-40.

2 E.M. Beaney, "Response of Tubes to Seismic Loading," CEGB Report TPRD/B/0605/N85, January 1985.

3 E.M. Beaney, "Response of Pipes to Seismic Excitation - - Effect of Pipe Diameter/Wall Thickness Ratio and Material Properties," CEGB Report TPRD/B/0637/N85, July 1985.

4 E.M. Beaney, "Response of Pipes to Seismic Excitation - - Theoretical Response Prediction," CEGB Report TPRD/B/0919/R87, May 1987.

PVP-Vol. 340, Seismic Engineering
ASME 1996

SIMULATION OF RATCHETING IN STRAIGHT PIPES USING ANSYS WITH AN IMPROVED CYCLIC PLASTICITY MODEL

Tasnim Hassan*, Yimin Zhu† and Vernon C. Matzen‡
Center for Nuclear Power Plant Structures, Equipment and Piping
North Carolina State University
Raleigh, North Carolina, 27695-7908
USA
FAX: (919) 515-5301 or (919) 515-7908
Voice: *(919) 515-8123, †(919) 515-6277, ‡(919) 515-7736
email: *thassan@eos.ncsu.edu, †ymzhu@eos.ncsu.edu, ‡matzen@eos.ncsu.edu

ABSTRACT

Ratcheting has been shown to be a contributing cause of failure in several seismic experiments on piping components and systems. Most commercial finite element codes have been unable to simulate the ratcheting in those tests accurately. The reason for this can be traced to inadequate plasticity constitutive models in the analysis codes. We have incorporated an improved cyclic plasticity model, based on an Armstrong-Frederick kinematic hardening rule in conjunction with the Drucker-Palgen plastic modulus equation, into an ANSYS user subroutine. This modified analysis code has been able to simulate quite accurately the ratcheting behavior of a tube subjected to a constant internal pressure and axially strain controlled cycling. This paper describes simulations obtained from this modified ANSYS code for two additional tests: (1) a tube subjected to constant axial stress and prescribed torsional cycling, and (2) a straight pipe subjected to constant internal pressure and quasi-static cyclic bending. The analysis results from the modified ANSYS code are compared to the experimental data, as well as results from ABAQUS and the original ANSYS code. The resulting correlation shows a significant improvement over the original ANSYS and the ABAQUS codes.

NOMENCLATURE

$\underline{a}$ = current center of the yield surface in deviatoric stress space
E = modulus of elasticity
H = plastic modulus
$J_2 = 1/2\, \underline{s}.\underline{s}$
N = number of loading cycles
n =Ramberg-Osgood hardening parameter
$\underline{s}$ = deviatoric stress tensor
$\underline{\alpha}$ = current center of yield surface in stress space
$d\underline{\varepsilon}^p$= plastic strain increment tensor
ε_x = axial strain
ε_{xm} = maximum axial strain in a cycle
ε_θ = circumferential strain
$\underline{\sigma}$ = stress tensor
σ_0 = size of yield surface
σ_x = axial stress
σ_y = Ramberg-Osgood yield parameter
σ_θ = circumferential stress

INTRODUCTION

A question that continues to generate discussion in the nuclear piping area concerns the primary mode of failure in piping components and systems subjected to seismic excitation — is it collapse as assumed in the 1992 and earlier ASME Codes (ASME, 1992) or is it ratcheting as indicated by several recent sets of experiments? The 1995 Code (ASME, 1995), which has not yet been accepted by the NRC, makes an attempt at incorporating reversed dynamic loading (implying ratcheting behavior) into the Code based on a set of experiments performed in the EPRI/GE Piping, Fitting and Dynamic Reliability (PFDR) Program (Tagart et al., 1988; Chen et al., 1995).

As part of an ongoing project at the Center for Nuclear Power Plant Structures, Equipment and Piping at North Carolina State University, we have reviewed numerous tests on piping components and systems (Zhu and Matzen, 1993; Matzen et al., 1995). We concluded that ratcheting is, in fact, a significant contributing factor in the failure of components and systems subjected to high level seismic excitation. We also concluded that existing commercial finite element codes cannot consistently simulate ratcheting behavior and, furthermore, that the reason for this can be traced to inadequacies in the plasticity constitutive models used. Another part of our Center project has been the implementation of an improved five-parameter cyclic plasticity model composed of the Armstrong-Frederick kinematic hardening rule and Drucker-Palgen plastic modulus equation into a user subroutine in ANSYS.

The modified ANSYS code was first verified using a set of data from a tube subjected to cyclic axial strain and constant internal pressure (Zhu et al., 1995). That set of six material experiments on stabilized carbon steel (CS) 1026 was generated by Hassan et al. (1992) and involved a wide range of rates of ratcheting in the circumferential direction. The simulations achieved for those tests using modified ANSYS were quite satisfactory. In those tests the principal strain directions remained fixed.

In this paper we present and discuss the correlations obtained from the modified ANSYS code for two different sets of test data in which the principal strain directions change during cyclic loading. Simulations are found to be promising for both cases.

THEORY

A brief review of the cyclic plasticity theory used and how it was incorporated into ANSYS will be presented first. The constitutive model assumes cyclically stable material behavior and J_2-type plasticity in which the yield surface is represented by

$$f(\underline{\sigma}-\underline{\alpha}) = \left[\frac{3}{2}(\underline{s}-\underline{a})\bullet(\underline{s}-\underline{a})\right]^{1/2} = \sigma_0 \quad (1)$$

where $\underline{\sigma}$ is the stress tensor, $\underline{s}$ is the deviatoric stress tensor, $\underline{\alpha}$ is the current center of the yield surface in stress space, and $\underline{a}$ is the center in the deviatoric stress space. The size (σ_0) and shape of the yield surface remain unchanged during plastic loading increments. The parameter σ_0 is determined from the linear part of stable hysteresis loops obtained from an axial test.

The plastic strain increment (flow rule) is given by:

$$d\underline{\varepsilon}^p = \frac{1}{H}\left(\frac{df}{d\underline{\sigma}}\bullet d\underline{\sigma}\right)\frac{df}{d\underline{\sigma}} \quad (2)$$

where H is the plastic modulus which is evaluated using the Drucker-Palgen Model (1981). In this model H is assumed to be a function of the second invariant of the deviatoric stress tensor J_2. An expression for H can be obtained in the following form:

$$H = \left[\frac{3n}{7E}\left(\frac{3J_2}{\sigma_y^2}\right)^{\frac{n-1}{2}}\right]^{-1} \quad (3)$$

where E, n, σ_y are material parameters which are evaluated by fitting a Ramberg-Osgood curve to the upper part of a stable hysteresis loop (Shaw and Kyriakides, 1985).

A hardening rule determines the evolution of the yield surface in stress space during plastic loading increments — it can be kinematic, isotropic or a combination of the two. We use the Armstrong-Frederick kinematic hardening rule in the following form (Hassan et al., 1992):

$$d\underline{\alpha} = d\mu\left[(1-k)(\underline{\sigma}-\underline{\alpha}) - k\underline{\alpha}\right] \quad (4)$$

where k is a model parameter determined using a biaxial ratcheting experiment and $d\mu$ is determined using the consistency condition.

This constitutive model was incorporated into a customized version of ANSYS through a user subroutine. A Newton-Raphson iteration scheme was used for nonlinear material response calculations.

In the first verification of this modified code, data were obtained from tests on cyclically stabilized specimens; hence, the assumptions in the above theory were appropriate for that data set. The results to be used in the further verification were obtained from experiments on CS 1020 and stainless steel (SS) 304, which are known to exhibit cyclic softening and cyclic hardening, respectively.

TABLE 1. Constitutive Model Parameters

Material	E, ksi (GPa)	n	σ_y, ksi (MPa)	σ_0, ksi (MPa)	k
CS 1020	25,125 (173.2)	3.9	27 (186)	35 (241)	0.091
SS 304	28,500 (196.5)	7.3	32 (221)	35 (241)	0.040

We do not currently have a cyclic plasticity model which can accurately simulate ratcheting of cyclically softening and harding materials (Hassan and Kyriakides, 1994). Before carrying out this more complex step of developing such a model, we would like to examine the performance of the current modified ANSYS code in simulating the ratcheting response for these two materials.

ADDITIONAL VERIFICATION

The two sets of data to be used are (1) symmetric shear strain cycling at constant tension (Hassan and Kyriakides, 1986) and (2) displacement controlled cyclic bending at constant internal pressure (Gau, 1990). The first of these tests is at the material level, i.e. the state of stress, except near the ends, is everywhere the same. The second, a bending test, is at the structural level and the state of stress varies from point to point.

(1) Symmetric Shear Strain Cycling at Constant Tension

This set of tests involved strain symmetric cycling in shear at a prescribed constant tensile force. A schematic drawing of the test specimen and loading is shown in Fig. 1. The reduced portion of the specimen had a 1" (25.4 mm) outer diameter with a wall thickness of 0.05" (1.27 mm). In this test ratcheting occurred in the axial direction. An axial strain-shear strain response from a test on CS 1020 is shown in Fig. 2b, where we see that the nominal axial stress due to prescribed constant tensile force was 17.5 ksi (120.66 MPa) and the amplitude of the prescribed shear strain cycle was 0.8727%. It is clearly seen that ratcheting occurs in the axial direction. The maximum value of axial strain, ε_{xm}, in a cycle (at the positive shear strain peak) is plotted against the number of applied cycles in Fig. 3. After the initial transient, the rate of ratcheting (increment in strain per cycle) is almost constant which is characteristic of a cyclically stable material. Hence, it appears that the cyclic softening of CS 1020 had little effect on the ratcheting rate in the axial direction.

The finite element model used in the calculation, shown in Fig. 1, represents the central portion of the specimen where the stress is uniform. The model has two rings, each composed of thirty-two Shell 43 (4 node flat plate) elements, around the circumference. For boundary conditions, the nodes at one end were constrained against translation in the axial and in the circumferential directions, but were allowed to translate freely in the radial direction. At the other end, rotations were specified to simulate the prescribed oscillating shearing strain of the specimen. A stable axial stress-strain hysteresis curve from Hassan and Kyriakides (1992) was used to determine the constants in Eq. 3. In Fig. 3, the ratcheting rate obtained from the constitutive model (dashed line) was matched to the stable ratcheting rate in the test (solid circles) to determine the value of k in Eq. 4. The material constants determined for CS 1020 are shown in TABLE 1.

Calculated results for this test using both the modified and original ANSYS codes are plotted in Figs. 2 and 3. We see that neither the original nor the modified ANSYS codes is able to simulate ratcheting in the first cycle accurately. This is because the monotonic loading response has not been modeled separately. The rate of ratcheting in the remaining cycles is reproduced quite accurately with the modified code. The original ANSYS code yielded results in which the rate of ratcheting is overpredicted up to the fifth cycle, and then the rate gradually decreased towards shakedown. This deficiency is mainly due to the kinematic hardening rule (Prager) used by ANSYS. In Fig. 3, by comparing the constitutive model results with the original ANSYS results, we see that the incorporation of the model into the ANSYS user subroutine introduces negligible error.

The same set of constants was then used to simulate the ratcheting rate for three other experiments in the set using the constitutive model. Each of these tests were subjected to a different constant tensile force but same shearing strain cycle of amplitude 0.8727%. The results of the rate of ratcheting from these tests are plotted in Fig. 4 where we see that, as the nominal axial stress increases, the rate of ratcheting also increases. The predictions of the model are plotted in the same figure. The simulations for σ_x = 17.5 ksi (120.66 MPa) show nearly the same rate of ratcheting as the experiment since these were the data used to obtain the material constants. The prediction for the two cases with lower axial stresses are also quite good. For the higher axial stress test, however, the analysis underpredicts the rate of ratcheting. The reason for this may be that the kinematic hardening rule we use in our plasticity model assumes that both the shape and size of the yield surface remain unchanged during plastic deformation. When shearing deformation becomes significant, this assumption is no longer valid. To further improve the ability to simulate the behavior of this type of test, we need to incorporate a material model which accommodates changes in both the shape and size of the yield surface.

(2) Cyclic Bending at Constant Internal Pressure

Two sets of test data from cyclic bending experiments conducted by Gau and Scavuzzo at the University of Akron (Gau, 1990; Scavuzzo, 1993) were used to verify the modified ANSYS code at the structural level. The two we selected were from their displacement controlled series. The specimens were 1" (25.4 mm) schedule 40, SS 304 pipes 48" (1.22 m) long. A schematic drawing of their test set-up is shown in Fig. 5. Our finite element model took advantage of the double symmetry in the test — we modeled only one quarter of the pipe. We used eight elements (Shell 43) around the half-circumference, ten elements along the length between the vertical centerline and displacement load point, and eight elements from this point to the support.

The material tests needed to evaluate the constants in Eqs. 3 and 4 were not available. We used a stable hysteresis curve for SS 304 (see Fig. 6) from Alameel (1985) to determine the constants in Eq. 3. Since a biaxial ratcheting experiment to determine the constant k in Eq. 4 was not available, we used the first set of Gau's data [from a bending test with internal pressure 2000 psi (13.8 MPa) and cyclic displacement amplitude of δ = 1.0" (25.4 mm) at the rigid spreader bar] to determine the value of k. The results obtained from both the original and modified ANSYS codes and two ABAQUS analyses (digitized from Gau's dissertation) for the first test are shown in Figs. 7 and 8. Gau used two different values for E_t, the modulus of the upper curve in his bilinear model: the higher value of 5500 ksi (37.5 GPa) representing the knee portion of the curve, and the smaller value of 500 ksi (3.44 GPa) representing the flattened portion of the upper curve. All of the analyses simulate the axial strains adequately since this is a displacement controlled experiment (correlations not shown here).

In the hoop direction, however, the original ANSYS and the ABAQUS with lower E_t greatly overpredict the rate of ratcheting even though in these two analyses the plastic modulus calculations were reasonable (see Fig. 6 for hysteresis loops with lower E_t value). The actual reason for the overprediction is the kinematic hardening rule used in the codes. The rule used in the original ANSYS is Prager's rule, and in ABAQUS it is the Prager-Ziegler rule. Both of these have been shown to be incapable of simulating biaxial ratcheting (Hassan et al., 1992.) The modified ANSYS analysis and the ABAQUS analysis which uses the larger value of E_t both match the rate of ratcheting quite well, with the modified ANSYS code giving a slightly better correlation. In the modified ANSYS code the improvement is achieved by incorporating a new kinematic hardening rule. In the ABAQUS analysis the kinematic hardening rule was not modified, but the improvement in simulation was obtained in an artificial way by making the material stiffer during plastic deformation. As shown in Fig. 6 the larger E_t value resulted in an unrealistic hysteresis loop. It is unlikely that this method would simulate results from other experiments equally well.

Using the material constants obtained from the first set of Gau's tests, we predicted the results from the second. This test had an internal pressure of 3000 psi (20.7 MPa) and cyclic displacement amplitudes prescribed in steps of 10 cycles at δ = 0.25" (6.35 mm), 50 cycles at 1.0" (25.4 mm), 10 cycles at 1.5" (38.1 mm), and then 10 cycles at 1.0" (25.4 mm). The test data and simulations from both the original and modified ANSYS are plotted in Figs. 9-11. As before, the axial strains are simulated quite accurately by both original and modified ANSYS.

For hoop strains, the rate of ratcheting is greatly overpredicted by the original ANSYS simulation in the first 25 cycles, and then it stabilizes to shakedown for the remaining cycles. In the third step of the prescribed cyclic bending, the original ANSYS simulated negative ratcheting, whereas the experiment ratchets in the positive direction. The rate of ratcheting and the amplitudes of hoop strains from the modified ANSYS code simulate the test data quite well. The differences observed in the correlations can be traced to the assumptions that the material is cyclically stable and that the yield surface shape remains unchanged. Another reason for the differences may be the fact that material parameters were not obtained from tests on the pipe specimen material, but were obtained from data in the literature.

SUMMARY, CONCLUSIONS, AND FUTURE WORK

The correlations of results from the two sets of tests described in this paper — symmetric shear strain cycles at constant tension and cyclic bending at constant internal pressure — show that the ANSYS code containing the improved cyclic plasticity model is able to simulate ratcheting more accurately than the original ANSYS or the ABAQUS codes. Further improvements are needed, however, since the constitutive model assumes a cyclically stable material, which is not the case for either stainless steel or carbon steel. Modifications to the constitutive model are underway to incorporate cyclic hardening and softening.

Changing the shape of the yield surface should also be incorporated for accurate simulations, but this is more difficult and will be the subject of future research.

One difficulty in using component test data presented in the literature is that we do not have the necessary material tests to evaluate all of the five model parameters. To address this problem, we are now performing a consistent set of tests at the material level and the structural level for both straight pipes and elbows.

Following the successful verification of our improved ANSYS code for quasi-static behavior at room temperature, we will modify the constitutive model to include both time and temperature dependencies so that it can be used to simulate the seismic response of piping components and systems under operating conditions. With a finite element code capable of simulating ratcheting accurately under realistic operating conditions, it will be possible to provide a rational basis for the incorporation of ratcheting in the ASME Code. The result would be less uncertainly in piping designs, and piping systems which were both more economical and safer.

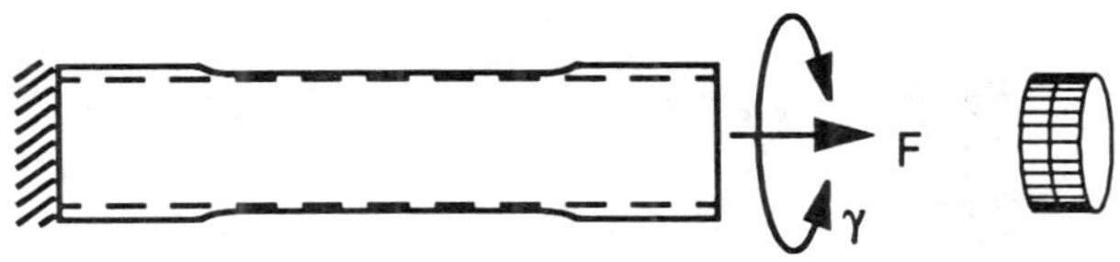

Fig. 1. Test specimen and finite element model for constant tension-cyclic shear strain tests.

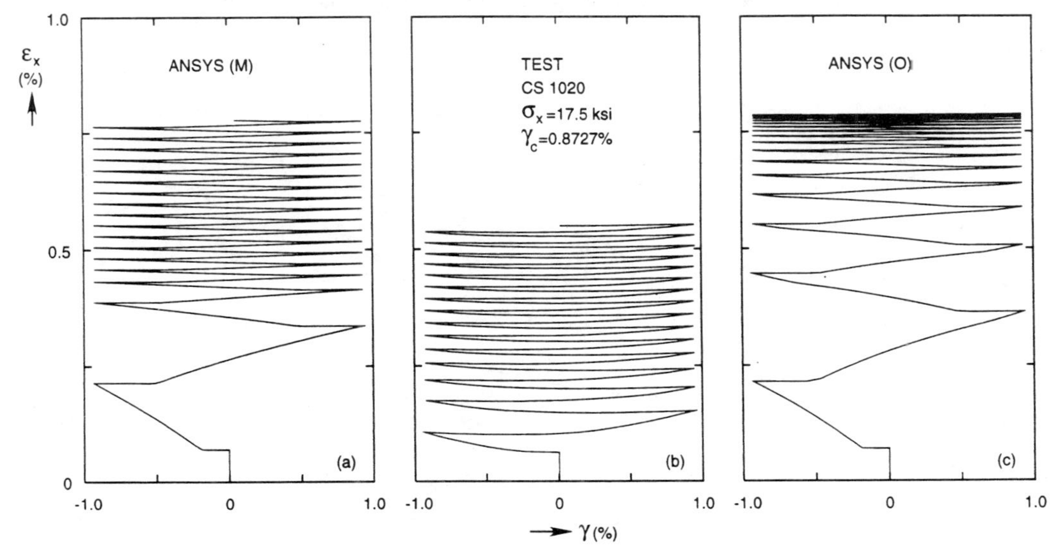

Fig. 2. Axial ratcheting in constant tension and cyclic shear strain. (a) Modified ANSYS, (b) test, and (c) original ANSYS.

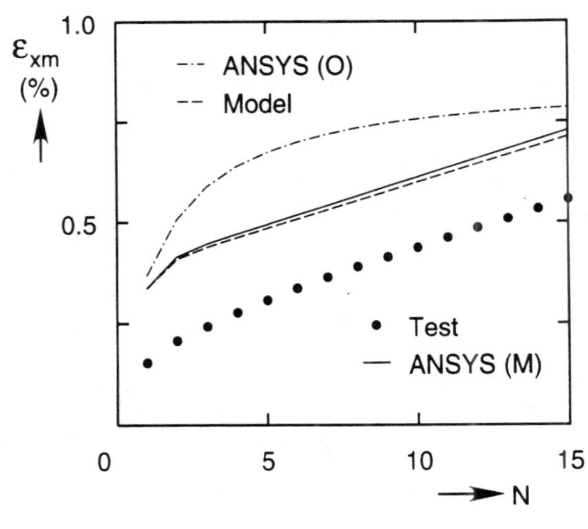

Fig. 3. Comparison of the rates of axial ratcheting from original and modified ANSYS, material model and test with constant tension σ_x = 17.5 ksi (120.66 MPa) and cyclic shear strain of amplitude γ_c = 0.8727%.

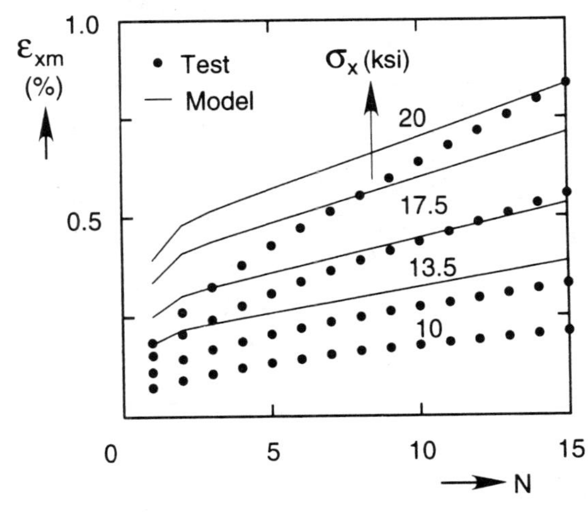

Fig. 4. Predictions of the rate of ratcheting by the material model for four tests with constant tensions of σ_x = 10 (69.0), 13.5 (93.1), 17.5 (120.7) and 20 (137.9) ksi (MPa) and cyclic shear strain of amplitude γ_c = 0.8727%.

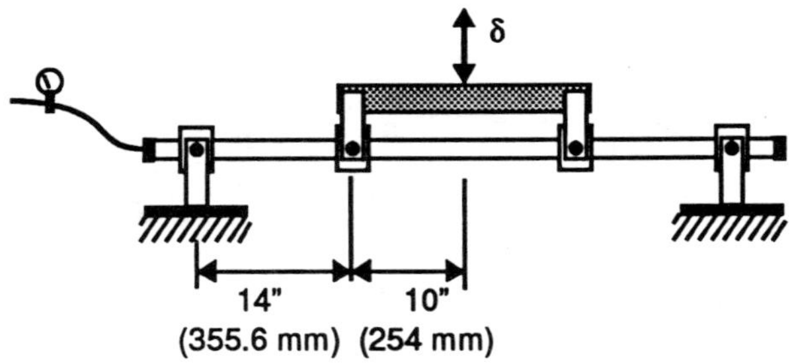

Fig. 5. Set-up used by Gau for testing straight pipes in cyclic bending and constant internal pressure.

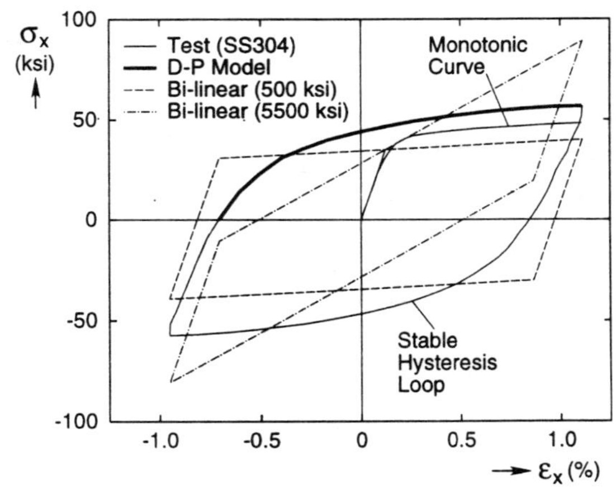

Fig. 6. Monotonic curve and stable hysteresis loop of SS 304 and models of modulus calculation in modified ANSYS (Drucker-Palgen model) and ABAQUS (Bi-linear models).

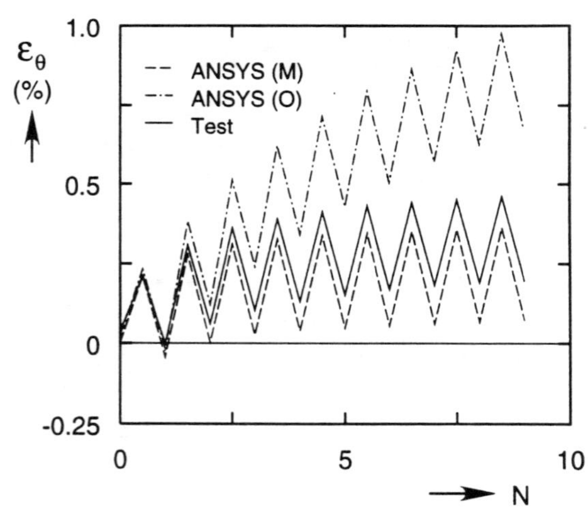

Fig. 7. Hoop strain ratcheting of a straight pipe subjected to cyclic bending [δ = ±1.0" (25.4 mm)] and constant internal pressure of 2000 psi (13.8 GPa). Comparison with analyses results from modified and original ANSYS.

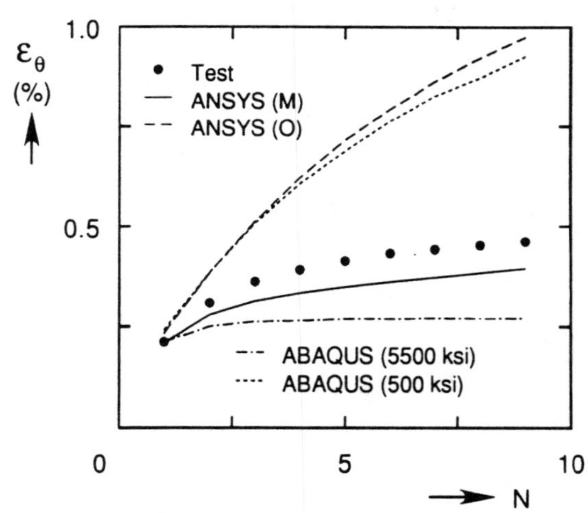

Fig. 8. Comparison of the rates of hoop ratcheting from the test in Fig. 7 and results by ANSYS and ABAQUS.

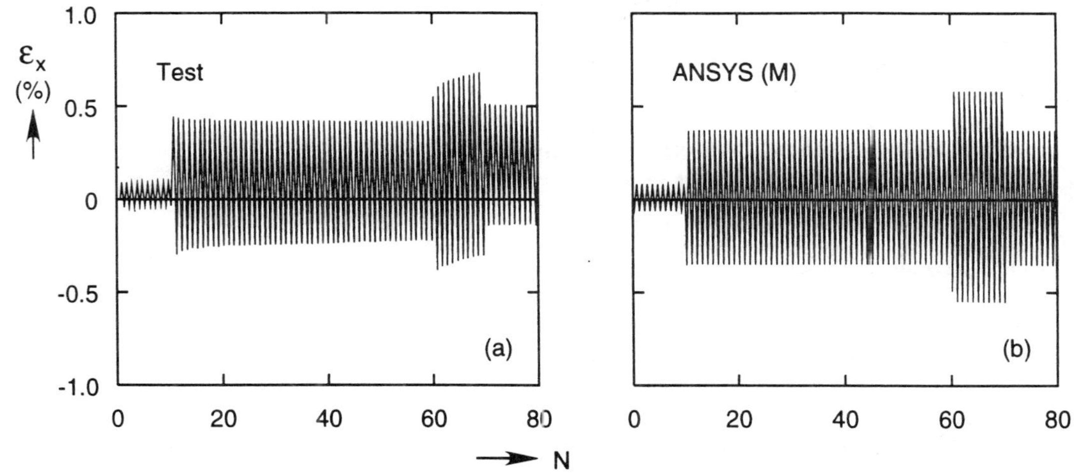

Fig. 9. Axial strain response of a straight pipe in displacement controlled cyclic bending [δ = ±0.25" (6.35 mm), ±1.0" (25.4 mm), ±1.5" (38.1 mm), ±1.0" (25.4 mm)] at internal pressure of 3000 psi (20.7 MPa). (a) Test, (b) modified ANSYS result.

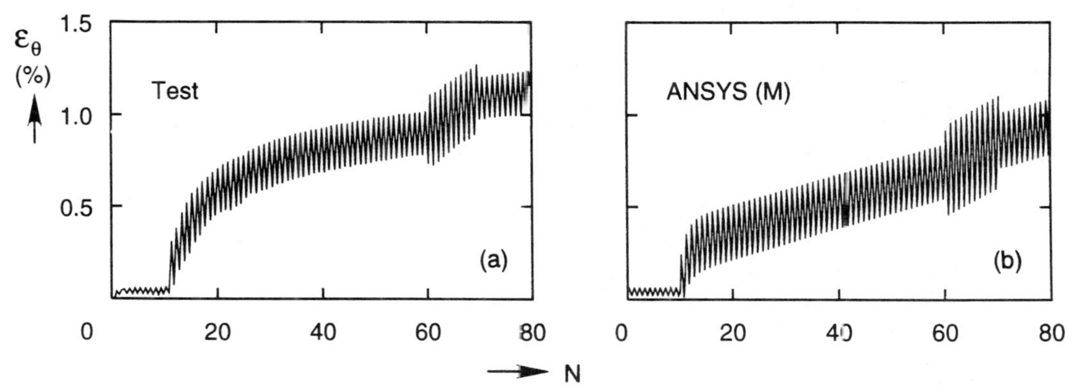

Fig. 10. Hoop strain ratcheting of a straight pipe in cyclic bending [δ = ±0.25" (6.35 mm), ±1.0" (25.4 mm), ±1.5" (38.1 mm), ±1.0" (25.4 mm)] and constant internal pressure of 3000 psi (20.7 MPa). (a) Test, (b) modified ANSYS result.

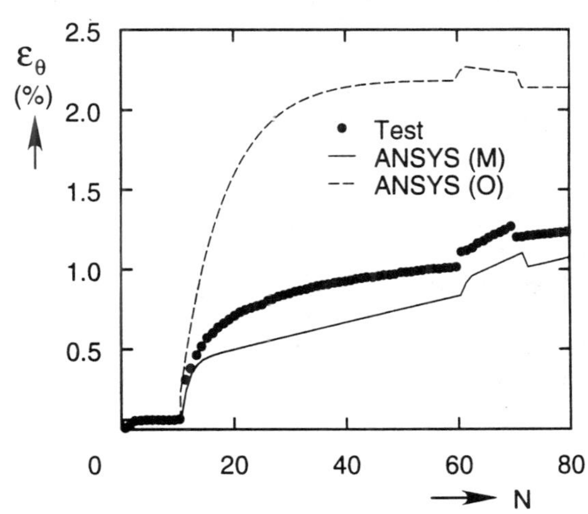

Fig. 11. Comparison of the rates of hoop ratcheting from the test in Fig. 10 and results by original and modified ANSYS.

LIST OF REFERENCES

Alameel, G. M., 1985. "Cyclic Loading of Inelastic materials: Experiments and Predictions," M.S. Thesis, University of Texas at Austin.

ASME, 1992. Boiler and Pressure Vessel Code, Section III, Division 1, Subsection NB-3600 Piping Design, The American Society of Mechanical Engineers.

ASME, 1995. Boiler and Pressure Vessel Code, Section III, Division 1, Subsection NB-3600 Piping Design, The American Society of Mechanical Engineers.

Chen, W. P., Jaquay, K. R., Chokshi, N. C. and Terao, D., 1995. "An Assessment of Seismic Margins in Nuclear Plant Piping," 13th International Conference on Structural Mechanics in Reactor Technology, Vol. III, pp. 507-512.

Drucker, D. C. and Palgen, L., 1981. "On Stress-Strain Relations Suitable for Cyclic and Other Loading," ASME Journal of Applied Mechanics, Vol 48, p. 479.

Gau, J.-S., 1990. "Elastic-Plastic Behavior of Pressurized Pipe," Ph.D. dissertation, University of Akron, Ohio.

Hassan, T., Corona, E., and Kyriakides, S., 1992. "Ratcheting in Cyclic Plasticity, Part II: Multiaxial Behavior," International Journal of Plasticity, Vol. 8, pp. 117-146.

Hassan, T. and Kyriakides, S., 1986. Data developed at the University of Texas at Austin.

Hassan, T. and Kyriakides, S., 1992. "Ratcheting of Cyclic Plasticity, Part I: Uniaxial Behavior," International Journal of Plasticity, Vol. 8, pp. 99-116.

Hassan, T. and Kyriakides, S., 1994. "Ratcheting of Cyclically Hardening and Softening Materials, Part II: Multiaxial Behavior," International Journal of Plasticity, Vol. 10, No. 2, pp. 185-212.

Matzen, V. C., Zhu, Y.-M., and Hassan, T., 1995. "Review of Experiments and Related Analyses on Piping Systems and Components," *Transactions*, 13th International Conference on Structural Mechanics in Reactor Technology, Vol. III, pp. 471-476.

Scavuzzo, R. J., 1993, "Seismic Plastic Behavior of Piping," Chapter 6 in *Technology for the '90s: a decade of progress, the ASME Pressure Vessels and Piping Division*, ed. Au-Yang, pp. 1092-1104.

Shaw, P. K. and Kyriakides, S., 1985. "Inelastic Analysis of Thin-walled Tubes Under Cyclic Bending," International Journal of Solids and Structures, Vol. 21, No. 11, pp. 1073-1100.

Tagart, S. W., Tang, Y. K., Hwang, H. L., Merz, K. L., Guzy, D. J., and DeVita, V., 1988. "Seismic Analysis and Testing of Piping Systems and Components," ASME PVP-Vol. 144, June, pp. 229-236.

Zhu, Y.-M., Hassan, T., and Matzen, V.C., 1995. 'Performance of a Commercial Finite Element Program in Simulating Ratcheting in Piping," *Transactions*, 13th International Conference on Structural Mechanics in Reactor Technology, Vol. III, pp. 489-494.

Zhu, Y.-M. and Matzen, V. C., 1993. "Reconciling Results of Seismic Tests on Piping Systems with Analysis Methods: A Summary of Existing Work," Report, Center for Nuclear Power Plant Structures, Equipment and Piping, NCSU, September.

ACKNOWLEDGMENT

The authors wish to acknowledge the Center for Nuclear Power Plant Structures, Equipment and Piping and the Department of Civil Engineering for their support of this research.

PVP-Vol. 340, Seismic Engineering
ASME 1996

THE 1995 FORUM ON APPROPRIATE CRITERIA AND METHODS FOR SEISMIC DESIGN OF NUCLEAR PIPING

Gerry C. Slagis
G C Slagis Associates
Walnut Creek, California

ABSTRACT

A record of the 1995 Forum on Appropriate Criteria and Methods for Seismic Design of Nuclear Piping is provided. The focus of the forum was the earthquake experience data base and whether the data base demonstrates that seismic inertia loads will not cause failure in ductile piping systems. This was a follow-up to the 1994 Forum when the use of earthquake experience data, including the recent Northridge earthquake, to justify a design-by-rule method was explored. Two possible topics for the next forum were identified -- inspection after an earthquake and design for SSE only.

INTRODUCTION

The objective of the forum is to have an open discussion on technical issues regarding seismic design of nuclear piping systems. Nuclear piping seismic design is a controversial issue right now because of the perceived overconservatism of the criteria. Changes to the Section III piping seismic stress limits and changes to regulatory requirements are being considered. Experimental results indicate that piping systems will not collapse in an earthquake. The focus of the session is appropriate methods of analysis, appropriate criteria, and appropriate design. Hopefully, a diverse cross-section of organizations will be represented at these discussions so that different perspectives can be debated.

These forums were started in 1992, and this session was the fourth forum. The 1995 session was attended by nineteen people. The organizations represented at the start of the session were . . . . .

A/E	2
Utilities	2
Engineering Services	3
Regulatory	2
NSSS	1
Academic	2
Vendor	2
Other	5

The forum is a discussion session. The discussion was lively and interesting. The session was focused on the earthquake experience data base. A record of the discussion follows.

SUMMARY

Less than half of the participants felt that the earthquake experience data base demonstrates that seismic inertia loads will not cause a collapse or fatigue failure in a single earthquake event in well designed ductile steel piping system without seismic restraints. The earthquake experience data base shows the capability of ductile piping to withstand large earthquakes, but there are failures reported. The root causes of the failures need to be understood. The EPRI component tests were discussed. Seismic inertia loads did cause failure in a single event in some of these tests. The stresses were much higher than allowed by the Section III code. The impact of mechanisms, such as corrosion, on earthquake capability was discussed. Enhanced inspection of susceptible areas was identified as one way to ensure earthquake capability. The need for code guidelines for inspection after an earthquake was briefly discussed. This question was then linked with the issue of design for SSE only.

FORUM DISCUSSION

The moderator, Gerry Slagis, started the session by reviewing the areas for comment, previous forums, and the topic of the day. Viewgraph #2 shows the four areas for comment -- earthquake definition, seismic input to piping, seismic response of piping, and piping stress criteria. The 1st forum generated various topics for discussion as shown on viewgraph #3. In the 2nd forum, static methods of analysis versus dynamic methods were discussed. The group conclusion (viewgraph #4) was that there is a place for both methods. Most people felt that dynamic analysis should be required for Class 1 piping. Dynamic analysis gives a better definition of the load distribution. The interesting conclusion was that the only reason to go to a static method is to use a ductility approach -- there's no benefit to the static method if you have to use elastic criteria for accelerations that are expected to occur. Another conclusion from the group was that there is no significant cost savings in using static methods for new plant design.

Design-by-rule (viewgraph #5) was the topic at the 3rd forum. Most people felt that there was no significant advantage in going to a design-by-rule approach for new plant design. Extensive training would be required. Verification is just as important as in design-by-analysis. Again, if we have to use elastic analysis with existing code stress limits to qualify the design-by-rule, there is no benefit. The only reason to go to design-by-rule is to use higher allowable stresses and justify by ductility considerations.

Possible topics for the 4th forum are shown on viewgraph #6. The topic chosen was the earthquake experience data base. Some people say that earthquake experience demonstrates that piping systems will not fail under inertia loading conditions. In the 3rd forum, David Terao took the position as shown on viewgraph #7 -- you cannot justify a design-by-rule approach from the earthquake experience data. Therefore, the focus of the 4th forum is -- What do we learn from the earthquake experience data base? To start the discussion, viewgraph #8 was shown giving the failure information from EPRI report NP-5617. 97 failures were reported in this report. Failure was a leak or break of the pipe.

Slagis -- I don't see what you can conclude from this information except that there are a large number of failures. The EPRI report concluded that seismic inertia loads do not cause failure of piping systems.

Glickstein -- How many of the failures are due to lack of piping flexibility? When the report says seismic anchor motion, is that lack of flexibility?

Geraets -- As I remember, most of the failures are due to lack of flexibility. The piping is attached to equipment, like a tank. The tank moves away and the pipe fails at the connection.

Slagis -- If you looked at the pipe you would conclude that the piping is sufficiently flexible if the equipment were anchored. [1] Many of the seismic anchor motion failures are from the equipment not being anchored and moving. Nothing would have happened if the equipment were anchored.

Glickstein -- Ductile piping or non-ductile piping?

Slagis -- Usually ductile.

Matzen -- Are there no failures in welded joints, or are welded joints included in one of the other categories.

Slagis -- I believe that failures in welded joints are included in the other categories. My reaction to the study was that every failure that was not obviously a movement problem was put into the corrosion category.

Glickstein -- Was piping designed for equipment movement?

Slagis -- No.

Thailer -- Was that because the plants were not seismically designed?

Geraets -- Yes, these were fossil, petrochem, and industrial plants.

Baliga -- Were these earthquakes equivalent to an OBE or an SSE?

Slagis -- They were high level earthquakes -- 0.2 to 0.4g zpa.

Baliga -- How many earthquakes are in the data base.

Slagis -- I'm not sure. I'd guess 10.

Slagis -- The basic question is -- Can you conclude that piping will not fail from inertia loads from this data base?

Geraets -- We have rules for pressure design of equipment. However we have failures even for equipment designed according to the rules. We still believe that the Section III rules are suitable for design of equipment and piping. Therefore, I would not conclude that because there are failures it is correct to say that inertia stresses do cause failure. Because there are failures is not a reason to disregard the experience data base as a method to qualify piping. I believe that seismic inertia stresses do not cause failure as long as the pipe is designed, fabricated and constructed to good state of the art practice. The pipe does not necessarily have to be designed for seismic.

Slagis -- Is this conclusion based on the earthquake experience data base?

Geraets -- Not only that. Also, from experience in requalifying piping that was not originally designed for seismic. We could requalify most piping systems without adding seismic supports. However, we had agreement to go above the existing code requirements using a ductility criterion. A ductility level of about 3.

[1] A different viewpoint on this statement on flexibility is provided at the end of the paper.

Slagis -- But I don't understand how you go from requalifying to say that the piping will not fail from inertia effects.

Geraets -- Based on my judgement.

Glickstein -- Is there any data to the contrary?

Slagis -- I believe the EPRI experience data indicates that piping may fail.

Glickstein -- I don't know that any of the failures are from inertia.

Slagis -- I think most of the failures under corrosion are inertia type failures. I think many of the failures under nonwelded joints are inertia type failures.

Thailer -- But we do not design piping systems for corroded piping systems, and we are not required to evaluate for corroded conditions. Until we establish code guidelines for corroded conditions, we shouldn't be concerned with those conditions.

Slagis -- If inertia effects will fail a corroded pipe, why won't inertia effects fail a noncorroded pipe? How do you make that jump?

Thailer -- Fossil plants have very little inspection programs for corrosion, and I would be concerned about corrosion failures. However, in nuclear, we have monitoring programs on the critical systems. Based on that program and reasonable detection of conditions, I believe that the systems will not fail under inertia conditions.

Geraets -- Remember the Surry accident in 1986. The pipe failed due to pressure surge. The real reason was that the pipe was completely eroded/corroded. There was no more material. Do you consider this as a pressure burst problem? Does this mean that the pressure design rules are not suitable?

Slagis -- Do we have actual data that tells us that piping systems will not fail from seismic inertia loads? Or is that a subjective judgement?

Thailer -- We have the experience in California. We had a plant, Moss Landing, within 15 miles of the Loma Prieta earthquake. The observed damage was not in mechanical areas. We saw some buckling on supports of tanks from inertia. The problems were in the electrical areas without seismic design and brittle materials. I cannot remember any breach of pressure integrity. The plant was designed for B31.1. Seismic displacements were limited by rattle type restraints. But rigorous analysis was not required. Because of this design philosophy, I believe the system survived. We permitted damage to the restraints. The main steam restraints deformed. We detuned the piping from the resonance conditions and did not allow the levels of response to build.

Dermenjian -- Since 1986, there are erosion/corrosion programs in place. We should not mix mechanisms with loadings. For seismic, we should consider inertia, seismic anchor motions, interaction, and 2 over 1 concerns -- not corrosion. Corrosion is a different mechanism. If we design for the four issues, the pipe will not fail.

Slagis -- The question of the moment is -- If you don't design for the four effects, can the piping fail in an earthquake?

Dermenjian -- I don't think anyone can accurately answer that question.

Slagis -- But that's what the EPRI report did. They said that piping will not fail under inertia loads.

Dermenjian -- We have to carefully use that report. That was a good record of the experience. Here is a report that documents the failures. As long as our design practice includes inertia, sam, interaction, and 2/1 issues, there is added assurance to demonstrate that piping will not fail.

Pecinka -- We have experience on two earthquakes (Armenia 1988 and Kozloduy 1983) with the Russian reactors. There were no failures in piping systems. The NSSS design is very different. Before seismic upgrade, the first frequency ranged from 0.36 to 0.41 Hz. With the addition of snubbers to prevent the steam generator motion, the first frequency is 2.6 to 3 Hz. The steam generator is horizontal and moved 150 mm in the earthquake [before seismic upgrade].

Glickstein -- What about all the EPRI component tests. Piping was subjected to severe input and did not fail due to inertia loads. I don't see how we can discount all the EPRI test information because someone says I don't agree with the margins calculated.

Slagis -- If you look at the EPRI component test data, there were a lot of failures in a single earthquake event. Of 30 component tests, 6 failed in a single event.

Thailer -- But, at what level?

Slagis -- At high levels.

Thailer -- Beyond the SSE requirements of any plant.

Slagis -- You can't correlate it that way. At very high stress levels. Because the EPRI component test saw a 20g zpa loading, you can't say that a piping system in a plant can take a 20g zpa load. Yes, the stress levels were very high, but there were failures in a single event. That's the problem. We were thinking that the EPRI tests were going to conclusively show that piping will not fail in a single earthquake event, but they did fail at high levels. Now we have to figure out at what level.

Thailer -- What kind of systems were these failures?

Slagis -- These were the EPRI component tests. When elbows were tested, it was correct to say that the elbow would not fail. It was not possible to fail the elbow in a single earthquake event. But when a tee was tested, it failed in a single event. It was a high loading, but when you factor in the question of actual material strength and other considerations, the loading was only about 2 times

Level D. And failure was by fatigue in one-half an earthquake event. Failure was at a weld joint subject to elastic follow-up. When you look at the EPRI component test data, you have to conclude that a failure in one earthquake is possible. And the worse case configuration was not tested.

Thailer -- Going back to the earthquake experience data base, I don't see any root cause determination for any of the failures. Before we make any general statements about what the data means, we need to better understand the root cause of failure.

Slagis -- I think that is also what David Terao's position was. We have to understand what the data means before we can make any conclusions. In the Northridge earthquake, there was a piping steam line that failed during the earthquake. It was patched, but the root cause of failure has not yet been determined. There are several other aspects of the earthquake data base that need to be considered. Earthquake effects vary greatly with location. The levels at the plant are estimated; I don't think there was seismic monitoring at the plants. Another aspect of the experience data base is that the piping is installed in ductile steel structures that are not designed for seismic events. The building is going to protect the piping system. I think that the levels of seismic load on the piping are nowhere near the levels we are required to design for in nuclear. Nuclear plants are designed so robust that the response is going to be close to elastic, and the energy will be transmitted directly to the piping rather than absorbed.

Thailer -- I want to comment on potential failure problems. My feeling is that, in an earthquake event, certain socket welds have a higher probability of failure than butt welds. Because of geometry, potential vibration, and the joint can't be inspected. There may be other similar mechanisms. We have to understand the potential mechanisms. If we have a program that assures that there are no underlying conditions prior to an earthquake, then the piping will withstand the earthquake.

Wais -- At PVRC we have been testing socket welds and have not been able to fail them in fatigue in less than 100 cycles at extremely high stress levels. We did not test the condition with cracks from vibration welds.

Thailer -- We can have a subset of socket welds with no vibration conditions. Those systems will likely not fail. We need to concentrate on those that are more susceptible, and develop NDT techniques for evaluation.

Dermenjian -- Socket weld capability depends on the arrangement, the location, and poor design. We have solved the vibration problem on socket welds by design. The small drain and vent lines are tied back to the main line all the way to the valve.

Pecinka -- The probability of failure for primary piping from seismic has been calculated by Lawrence Livermore to be around 10^{-10}.

Yukawa -- Do we need guidelines in the code for what to look for, what to study, and what to inspect after an earthquake event? Guidelines would be different for nuclear than fossil.

Slagis -- NRC has published a draft regulatory guide on inspection after an earthquake.

Yukawa -- Now you need to get into the details -- socket welds versus butt welds versus elbows, etc.

Dermenjian -- Before you decide on inspection, you need to decide if you are going to design for both OBE and SSE.

Glickstein -- I assume that the utility will do some level of inspection after an earthquake, but I don't know if it is necessary to quantify the process by rules.

Thailer -- We have an emergency plan that specifies the pipe walkdowns after an earthquake based on the level of earthquake. It's a graded approach -- the extent of the walkdown varies depending on the earthquake magnitude. At the lower levels, it's a visible exam for obvious physical damage. At OBE, we shut the plant down and look more extensively especially for the nonseismic designed equipment.

The chairman called for final comments.

Thailer -- We should leave it up to the individual organization on how or what to inspect after an earthquake. It would be better to develop enhanced pre-earthquake criteria in Section XI, not Section III. Include enhanced inspection for potentially vulnerable areas.

Glickstein -- There is information that suggests that piping will not fail from seismic inertia loads. To me, an inertia failure is total collapse or excessive deformation. Most of the failures in the EPRI component tests are fatigue or ratcheting. It may have occurred in one seismic event; but I don't classify that as inertia type load.

Dermenjian -- The preexisting condition of the piping system should be taken into account when you talk about failure due to seismic inertia loads.

A vote was taken to see if there was a consensus.

Does the earthquake experience data demonstrate that inertia loads will not cause a collapse or fatigue failure in a single earthquake event in well designed ductile steel piping system but without seismic restraints?

Agree	9
Disagree	4
Not voting	6

The chairman (Geraets) asked the moderator (Slagis) on his position. The moderator doesn't vote.

Slagis -- I don't see how you can reach that conclusion. I disagree.

COMMENTS AFTER THE MEETING

Mr. Glickstein commented on the question of whether the piping that failed in the earthquake experience data base had sufficient flexibility. If each piece of equipment is anchored to the same foundation (no relative motion), then the piping may have sufficient flexibility. However, if the stiffness of the equipment foundation varies, such that the anchored equipment could generate relative motion, then the piping may not have sufficient flexibility.

The fourth annual forum on . . . . .

Appropriate Criteria and Methods for Seismic Design of Nuclear Piping

Co-Chair	Luc Geraets
Co-Chair	T. H. Liu
Moderator	Gerry Slagis

Viewgraph #1

Four areas for comment

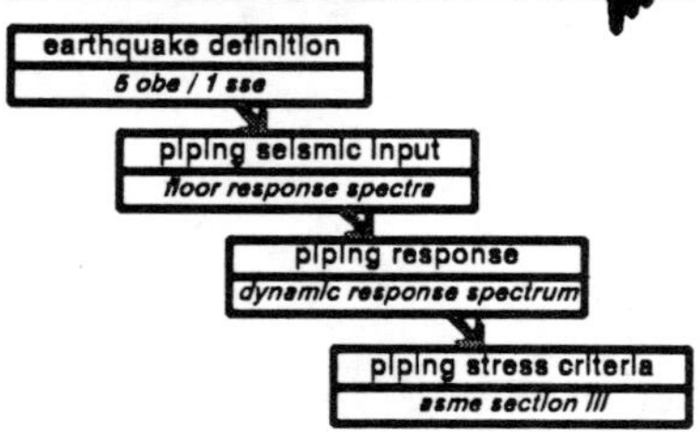

Viewgraph #2

first annual forum

- function is not a concern
- section III is too conservative
- experience data should be used
- sse only design
- floor rs are too conservative
- ism is not pratical
- reliability based design
- supports with gaps
- integrate flaw evaluation and risk assessment into design
- special criteria for low level sites
- leak-before-break for all piping
- no need to evaluate axial stesses
- ductility and stress allowables
- loca may control design
- static versus dynamic

Viewgraph #3

static versus dynamic

- there's a place for both
- dynamic -- better prediction of load distribution
- but static is also acceptable
- dynamic for Class 1
- if static, go to UBC ductility approach
- no significant cost difference for new plant design

Viewgraph #4

design-by-rule

- no significant advantage for new plant design
- extensive training will be required
- verification is as important as in design-by-analysis
- too conservative if use 1992 Edition stress limits
- need ductility considerations (UBC approach)

Viewgraph #5

topics for today

- ductility considerations
- new SIII seismic stress limits
- earthquake experience data base

Viewgraph #6

"you cannot justify design-by-rule from earthquake experience data"

Viewgraph #7

EPRI NP-5617

Recommended Piping Seismic-Adequacy Criteria Based on Performance During and After Earthquakes

	Power Plants Piping Failures
seismic anchor movement	15
corrosion	7
interaction	3
non-welded joints	36
supports	11
internal equipment	15
buried	5
miscellaneous	5
total	97

Viewgraph #8

EQUIPMENT SEISMIC QUALIFICATION

Introduction

Yong Zhao
Battelle Memorial Institute
Columbus, Ohio

David C. Ma
Argonne National Laboratory
Argonne, Illinois

The qualification of equipments for seismic service is a critical element to ensure that components can survive in a dynamic environment. This chapter contains six technical papers presented at the 1996 ASME PVP Conference that discuss issues, methods and solutions related to seismic qualification of equipments in a nuclear power plant.

The first paper written by Bhargava describes an approach for seismic qualification of valves by establishing bounding amplification factors from which amplified response spectra and peak acceleration values can be obtained. McElhaney and Staunton have examined the historic use of generic component reliability data, its sources and its limitations for check valve performances. An issue related to performance of synthetic grease as the motor operated valve stem lubricant is described by Gleeson and Cline. This subject should be of great interest to nuclear plant operators.

A PRA based seismic margin assessment of evolutionary light water reactors is presented by Ali and Bagchi. Liang Han et al have performed a seismic test on lower part model of Chasnupp steam generator and have achieved a good agreement with the analytical results. Both tests and analysis are also performed by Gu, Li and Peng in their work to establish the discharge force of pressurizer safety relief valve under three water seal volumes.

On behalf of the Seismic Engineering Committee, the contributing editors of this chapter would like to express their deep appreciation to the authors and reviewers of their contributions.

PVP-Vol. 340, Seismic Engineering
ASME 1996

SEISMIC RESPONSE FOR QUALIFICATION OF VALVES AT NUCLEAR PLANTS

Divakar Bhargava
Virginia Power
5000 Dominion Blvd.
Glen Allen, VA 23060

Abstract

Power-operated valves constitute about a quarter to one-third of all safety-related components at a typical US nuclear plant. While extensive work has been done in the industry to quantify amplification factors for components such as electrical cabinets, little guidance is available to estimate amplification factors that can be applied to floor or piping anchor seismic response spectra to obtain valve responses. This paper establishes bounding amplification factors from which amplified response spectra and peak acceleration values can be obtained for the seismic qualification of valves.

Parametric analyses of three representative nuclear piping systems consisting of different pipe sizes and support configurations are performed by the time-history method. Parameters such as the shape, frequency and energy content of the floor seismic spectra, peak shifting, size, weight and natural frequencies of the valves, support locations, and modal and spectral damping values are varied. Based on this study, amplified response spectra at the valve location for a variety of conditions are developed for each system. Amplification factors are then obtained by comparing spectra at the valve with the floor spectra. The peak seismic accelerations, which can be used in a valve's static analysis, are automatically obtained from these results. They are, by definition, the zero period accelerations of the amplified response spectra at the valve.

Current Practice

For seismic qualification of components located in distributed systems, e.g. valves, the prevailing design practice at many nuclear plants is to specify some arbitrarily high accelerations at the valve center-of-gravity (c.g.), say, 3g in each horizontal direction and 2g in the vertical direction for Safe Shutdown Earthquake, together with deadweight and other operating loads. Stress and deflection responses of weaker valve components from these seismic loads are determined and combined using the probabilistic square root of the sum of squares method or the 100-40-40 rule to account for the three spatial components of earthquake. These seismic accelerations are based on past collective industry experience. Therefore, they can be used in performing qualification when no further reduction is needed, even when the floor spectral peaks are high, i.e., greater than 3g. If the use of such high seismic accelerations is unacceptable for qualifying the valve, then a refinement may be required. A better estimate of seismic accelerations can be obtained from existing or new response spectrum analyses which are typically performed for valve-piping systems.

However, when seismic amplified response spectra (ARS) at a valve are needed to perform a dynamic analysis of the valve or for comparison with existing or new test spectra which may be the case for motor-operated, solenoid-operated, and air-operated valves, piping response spectrum analyses can not provide the solution. A simple, practical and technically acceptable method does not currently exist, and the development of amplified response spectra for the qualification of a valve generally requires a specific, case-by-case evaluation.

Parameters of Interest

The frequency contents and amplifications in the acceleration response spectra at a valve location in a piping system will depend upon both the spectra at the piping anchor locations near the vicinity of the valve and the dynamic characteristics, i.e., natural frequencies of the piping/valve system, damping, mode shapes, and participation factors of the piping-valve assembly. The dynamic characteristics of the system depend upon factors such as the piping layout, location of supports, and size, material, weight, and flexibility of piping and valve. The influence of these parameters is significant in the calculation of amplification factors which can be applied to floor or piping anchor spectra to obtain the required ARS at the valve or valve-operator locations.

Analysis

Three representative piping systems from a pressurized water reactor (PWR) nuclear plant were selected for the parametric analysis described in this paper. They are:

- Containment Steam Generator Blowdown System - 3" line with two air-operated valves
- Component Cooling Pump Suction System - 6" line with one air-operated valve
- Containment Spray System - 8" line with two motor-operated valves.

Mathematical models were developed for each system consisting of pipe, elbow, beam, and stiffness matrix elements. All computational analyses were performed using STARDYNE computer code (Reference 4). The first step was to select the in-structure or floor response spectra at the pipe support or anchor locations in the vicinity of the valves. Two sets of spectra were evaluated for the parametric study - site specific spectra based on a modified NRC Regulatory Guide 1.60 ground response spectrum shape, and median centered spectra based on the NUREG/ CR-0098 shape. The peak ground accelerations (pga) were 0.12g horizontal and .08g vertical for both sets. Floor spectra were ±15% peak-broadened and enveloped to obtain the ARS in each orthogonal direction. Therefore, further analyses with ±15% peak-shifting to account for uncertainties were not required. The remaining steps in the analysis of each of the three piping systems were as follows:

- Develop statistically independent artificial or synthetic time-histories in each direction to closely match the target floor spectra. A comparison of the floor spectrum with the spectrum obtained from the corresponding synthetic time history in one horizontal direction is shown in Fig. 1. It is noted that the zero period acceleration (ZPA) of the ARS from the synthetic time history was somewhat higher than the ZPA of the floor ARS. The power spectral density developed from the time-histories confirmed that the energy contents in the frequency range of interest were adequate.
- Perform modal extraction analysis of the pipe-valve model. Frequencies, mode shapes and participation factors obtained from this analysis were stored. Modal damping values, based on the energy principle, were also calculated and stored.
- Perform uniform support excitation modal time-history dynamic analysis with synthetic time histories representing three orthogonal components of earthquake. The damping values for the components in the piping system, and the spectral damping of the floor ARS were varied in these analyses to account for two conditions - design basis of the plant, and seismic margin evaluations which permit higher, realistic damping values. ASME code case N411 damping, although part of the design basis, was not used in these time-history analyses. Specific damping values which were selected are discussed in the detailed analyses of each of the three systems.
- Generate ARS from the acceleration time-history responses at the c.g. of the valves in the system, in each direction for the same spectral damping as the ground spectra.

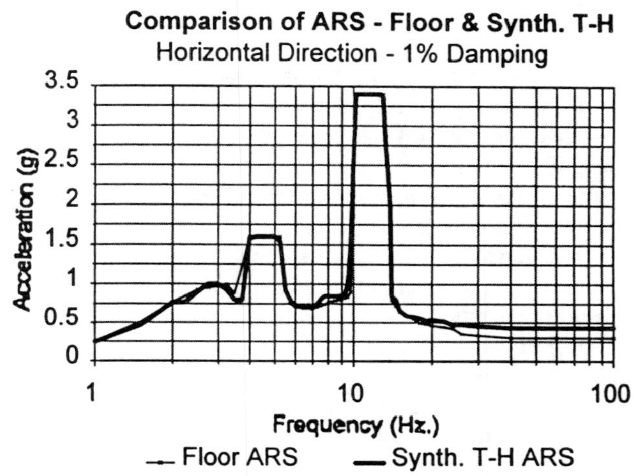

Figure 1

These steps were repeated and response spectra at the valve c.g. were calculated for several parametric cases for each system. Tables were prepared to determine amplification factors at each applicable spectral frequency as follows:

Amplification Factor = (Spectral value from the ARS at valve location)/(Spectral value of pipe anchor ARS)

Steam Generator Blowdown System

This system is located inside a PWR containment at a high elevation. Spectra at two air-operated valves in each orthogonal direction were developed for this system. The mathematical model of this system is shown in Fig. 2.

Piping Model - Steam Generator Blowdown System

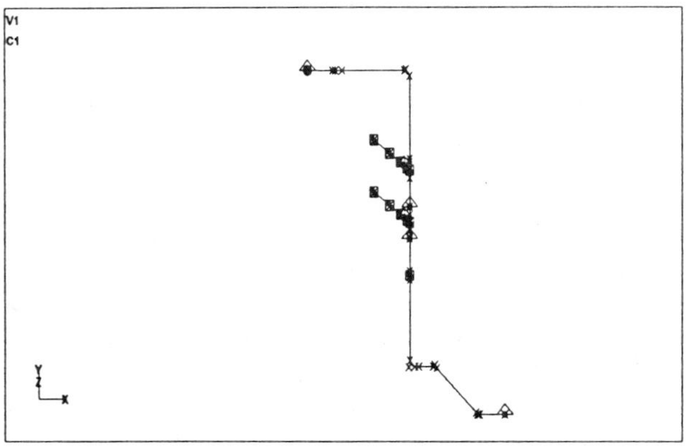

Figure 2

Eight parametric variations in valve natural frequencies, support configurations, floor response spectrum shape, and damping values were evaluated for this system. For the first five cases, synthetic time-histories were developed from 1% damped floor spectra. ARS from the response time-histories at the two valves were also developed at 1% damping. Composite modal damping values in the dynamic analysis of the first seven cases were based on 1% damping for piping, 2% for valves and 5% for supports. The dominant mode of this system varied between 1.9 hz. to 2.3 hz. for the various parametric cases. The dynamic analysis included all modes of the system up to 80 hz. since several modes exhibited significant participation factors and high mode shapes at the valve location. The spectral values from the floor spectra and the valve ARS in each direction were compared at each significant frequency and at the ZPA. Amplification factors were calculated from these comparisons.

Case no. 1

In the first model selected for this system, the valve Strong axis was oriented in the vertical direction. Valve operators for both valves were not supported independently.

Case no. 2

In this model, the stronger axis of the valve was changed to be at 45 degrees to the horizontal plane. This altered the valve natural frequency and resulted in a slight change in the frequency of the first dominant mode . Valve operators for both valves were not supported independently.

Case no. 3

This case was similar to case no. 1 except an additional support was added to the pipe near the valve. It was observed that although the natural frequency of the valve itself was about 11 Hz., the dominant modal frequency of the piping-valve system was 2 Hz., and a large peak in the response spectra at the valve c.g. occurred at this frequency. The comparison of floor spectra and the response spectra at the c.g. of one valve in one horizontal direction is shown in Fig. 3.

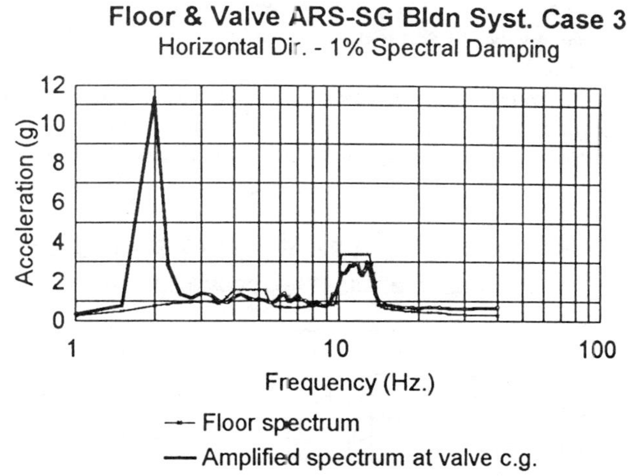

Figure 3

Case no. 4

This case was similar to case no. 3 except the valve operators for both valves were independently supported. As expected, lower amplification factors were obtained for this case.

Case no. 5

The pipe-valve model was identical to case no. 3, except that the horizontal floor spectra from which synthetic time histories were developed contained a broadened peak at the dominant mode of the system, i.e., at 2 hz. This caused a high spectral peak in the valve ARS at 2 hz. The comparison of floor and valve spectra for this case is shown in Fig. 4.

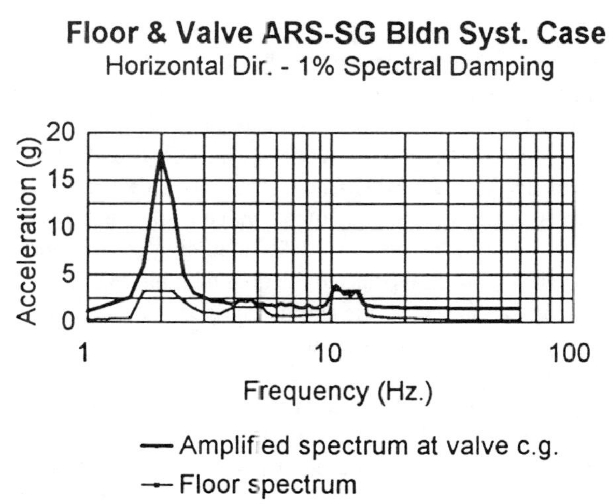

Figure 4

Case no. 6

This case was identical to case 2, except that the floor spectra from which synthetic time histories were developed, and the ARS at the two valves from the response were at 5% spectral damping.

Case no. 7

The model for this case was the same as case no. 3. The floor spectra and the response spectra at the two valves were at 5% spectral damping. The comparison of floor and valve response spectra for this case is shown in Fig. 5.

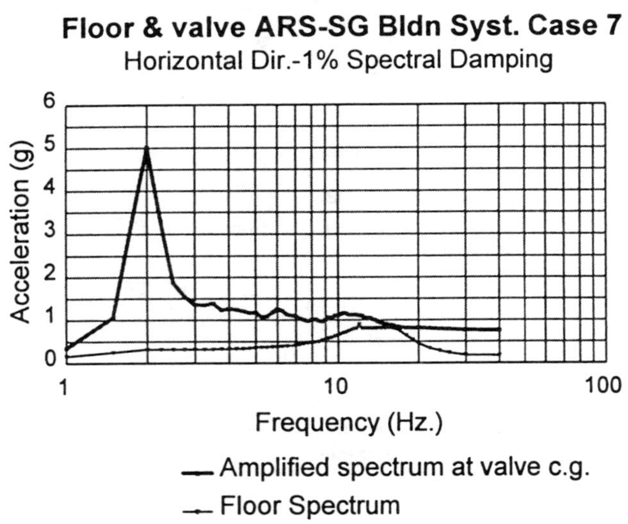

Figure 5

Case no. 8

This case was similar to case no. 7 except that 5% modal damping for the entire piping-valve system was used..

Results from the analysis of the SG Blowdown system

For each of the eight parametric cases, tables were prepared to compare the floor spectra with the response spectra at the two valves in each orthogonal direction. Amplification factors were then calculated at each frequency. The peak accelerations at both valves, i.e., the ZPA values of the valve response spectra were always found to be less than 1.5g, including case 5 where the piping anchor spectral peaks at the dominant mode of the system were 3.3g. As expected, the amplification factors were the largest when the valve operator was not independently supported and the spectral damping was 1%. The lower modal and spectral damping in the first five cases and the lower modal damping values selected for the first seven cases were representative of the plant design basis. On the other hand, the use of 5% spectral damping in the last three cases and 5% modal damping for the entire system in the last case are considered to be more realistic.

As observed from Figures 3, 4 and 5, sharp peaks were observed in the valve response spectra at frequencies at which the dominant modes of the system occurred. These frequencies were substantially lower than the valve's natural frequency. About 15% peak-clipping was performed to reduce the amplification factors in the resonant range.

In summary, the amplification factors for this system were significantly high in the resonant range of the piping system, as compared to the factors in the ZPA range. Variations in input spectra, modal damping, and spectral damping affected the amplification factors significantly. However, changes in the valve natural frequency did not have much impact on the responses at the valves. From the parametric study of the steam generator blowdown system, the following was concluded:

When lower damping values must be used, e.g., to meet the design basis of the plant, the bounding amplification factors to get the ARS at valve location in any orientation are:

Resonant range of piping anchor ARS = 12.0
Rigid range/ZPA of piping anchor ARS = 5.0

If realistic damping values for components, i.e., about 5% can be justified, then the bounding amplification factors are:

Resonant range of piping anchor ARS = 5.0
Rigid range/ZPA of piping anchor ARS = 3.0

In the intervening frequency range, i.e., between the resonant and rigid ranges, these amplification factors could be averaged to obtain a smooth ARS curve at the valve location.

Component cooling pump suction system

This system is located outside the PWR Containment wall but is anchored to the Containment external structure and also to an auxiliary structure. It contains one air-operated valve. The mathematical model of this system is shown in Fig. 6 below.

Piping Model - Component Cooling System

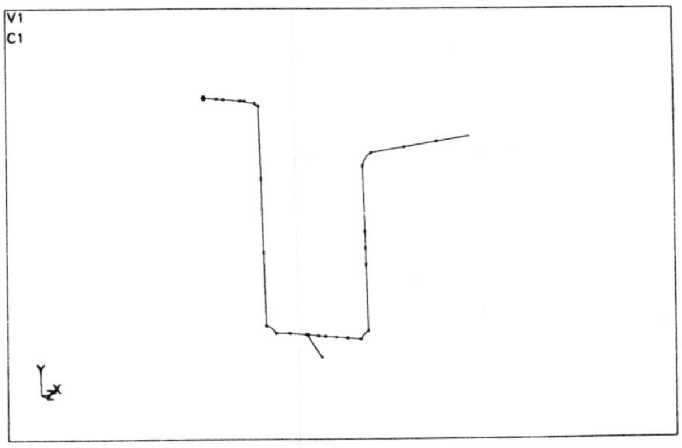

Figure 6

Five parametric variations in support configuration, ground spectral shape, and damping values were evaluated for this system. The dynamic analysis included all modes of the system up to 100 hz. since several modes had significant participation factors and high mode shapes at the valve location. The dominant mode for this system for the first case was at 10.6 Hz. The composite modal damping values in the analysis were based on 1% damping for piping, 2% for valves, and 5% for supports. Floor spectra and the ARS at the valve were at 1% spectral damping. The comparison of the floor and valve response spectra in one horizontal direction for this case is shown in Fig. 7.

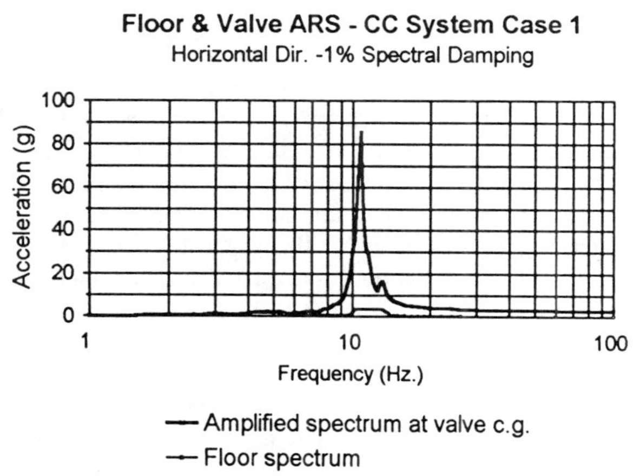

Figure 7

In the second case, the dominant mode of the system was lowered to 6.6 hz. by modifying the location, weight and stiffness parameters of the supports and the valve in the system. The comparison of the floor and the valve ARS in one horizontal direction for this case is shown in Fig. 8.

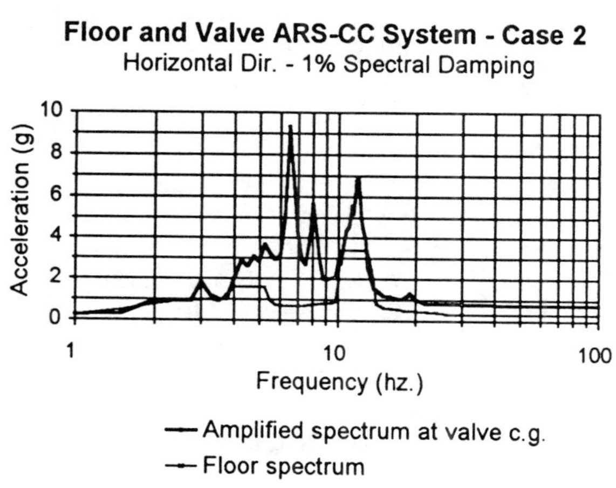

Figure 8

The models for the third and the fourth cases were the same as the first and the second cases respectively, however, the floor spectra and the ARS at the valve were at 5% spectral damping. The last case was the same as the first case, but with 5% modal and spectral damping.

Results from the analysis of the Component Cooling System

As is clear from Figure 7, the sharp peak for the first case of this system indicated the maximum amplification factor of 25 at 10.6 hz., which was the first mode. This amplification is still lower than the sinusoidal $1/2\beta$ or 50, at 1% damping. A 20% peak-clipping was performed for to reduce the amplification factors in the resonant range. The bounding amplification factors from the analyses of all the cases for this system were as follows:

Design Basis of the plant with low damping values:
Resonant range = 20
Rigid range = 9.0

Realistic 5% modal and spectral damping values:
Resonant range = 6.5
Rigid range = 4.5

Containment spray system

This system is located outside the PWR Containment wall and is anchored to the Containment external structure and to an auxiliary structure. It contains two motor-operated valves. The mathematical model of this system is shown in Figure 9.

Piping Model - Containment Spray System

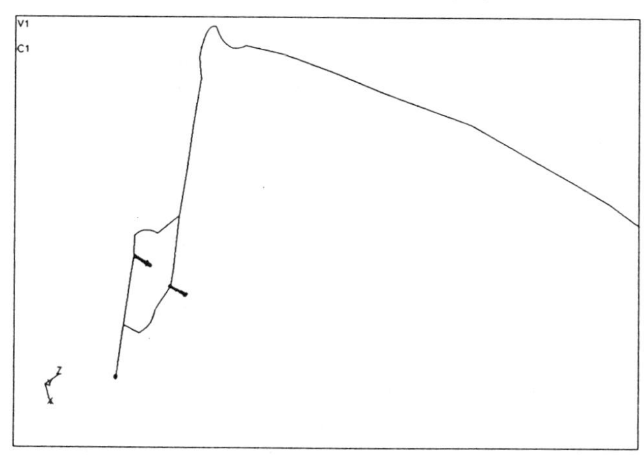

Figure 9

Five parametric variations in support configuration, ground spectral shape, and damping values were evaluated for this system. These cases were similar to the cases which were analyzed for the

Component cooling system. The dynamic analysis included all modes of the system up to 100 hz. frequency since many modes had significant participation factors and high mode shapes at the valve location in this range. Peak broadened floor spectra were used. The first mode of this system was at 7.1 hz., but the eigenvectors were small at both motor-operated valves. The comparison of the floor and valve response spectra in one horizontal direction is shown in Fig. 10 for the first case which used lower modal damping and 1% spectral damping.

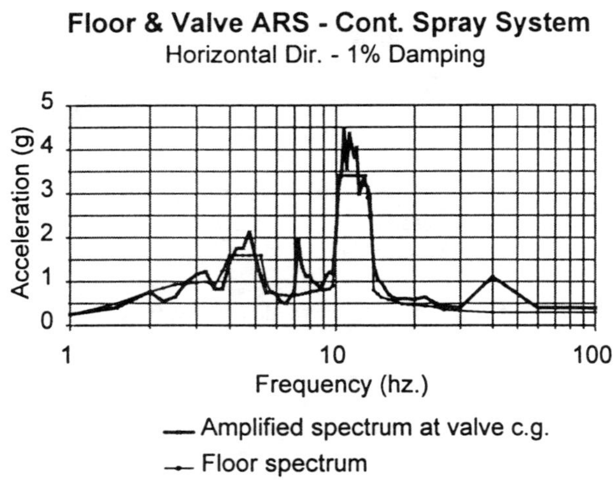

Figure 10

Results from the analysis of the Containment Spray system

From the various case evaluated for this system, it was noted that the system resonances in the vicinity of the valves were in the rigid range of the floor ARS. Thus, significantly lower amplification factors for calculating valve ARS were exhibited by this system, as compared to the Steam Generator Blowdown system or the Component Cooling system.

Conclusions

To obtain seismic amplified response spectra and peak seismic accelerations at a valve or valve-operator location, bounding amplification factors were developed based on parametric evaluations of three representative nuclear piping systems. These amplification factors depend on the modal properties of the entire system and the component and spectral damping parameters. When the dynamic characteristics of pipe/valve systems are not readily available, use of bounding amplification factors based on the analyses of these three systems is recommended. These factors are developed for two separate damping parameters. They should be applied to the envelop of piping-anchor spectra at the location of the closest pipe-supports in the vicinity of the valve.

When lower damping values, i.e., 1% to 2% of critical damping need to be used, e.g., to meet the design basis of the plant, the recommended factors are:

- Resonant range of piping anchor ARS = 15.0
- Rigid range/ZPA of piping anchor ARS = 7.0

If realistic damping values for components, i.e., about 5% can be justified, then the recommended amplification factors are:

- Resonant range of piping anchor ARS = 6.0
- Rigid range/ZPA of piping anchor ARS = 3.5

In the intervening frequency range, i.e., between the resonant and rigid ranges, these amplification factors can be averaged to obtain a smooth amplified spectrum curve at the valve location.

It is noted that for the Component Cooling system, the amplification factors were higher than those recommended above. This was because of resonance. The spectral peaks were sharp and peak-clipping was considered justified. Also, because of nonlinearities inherent in such systems, displacements at the valve would limit the amplification in spectral accelerations. Therefore, the envelop of all amplification factors obtained from the various parametric cases analyzed in this paper was not performed.

From the cases evaluated in this paper it was observed that piping systems filter the floor spectra which may even be broad band, to yield a few sharp resonances. Therefore, dynamic test procedures of Reference 7 for seismic and BWR hydrodynamic loads are justified and single frequency tests at intervals of 1/3 octave in the frequency range of interest can be performed using the peak accelerations at the valve c.g. When such testing or a static analysis of the valve is required, the bounding amplification factor for the rigid or ZPA range of the pipe anchor spectra can be used to obtain the required peak seismic acceleration at the valve-operator c.g.

A more accurate determination of seismic amplified response spectra for a valve can be obtained if the dynamic characteristics of the system are available, or by performing case-by-case, detailed dynamic analyses of the piping/valve system. Also, more insights regarding amplifications in the floor spectra to obtain required spectra at a valve can be obtained by performing and enveloping the results from additional piping analyses with more parametric variations in the future.

References:

1. IEEE Standards 344-1975 and 382-1980
2. ASME Section III
3. EPRI Report NP-6041, Revision 1, A Methodology for the Assessment of Nuclear Plant Seismic Margin
4. Stardyne Computer Code, 1995, Research Engineers Inc.
5. NRC Regulatory Guide 1.60
6. Newmark, N. M. And Hall, W. J., Development of Criteria for Seismic Review of Selected Nuclear Power Plants, NUREG/CR-0098
7. Bhargava, D, et. al, Dynamic Qualification Test Procedure for BWR Valves, ASME PVP, 1982

PVP-Vol. 340, Seismic Engineering
ASME 1996

SEISMIC MARGIN ASSESSMENT OF EVOLUTIONARY LIGHT WATER REACTORS

Syed A. Ali
Office of Nuclear Reactor Regulation
U.S. Nuclear Regulatory Commission*
Washington, D.C.

Goutam Bagchi
Office of Nuclear Reactor Regulation
U.S. Nuclear Regulatory Commission
Washington, D.C.

ABSTRACT

The objectives of the U.S. Nuclear Regulatory Commission (NRC) staff's review of the evolutionary light water reactors (ELWR) probabilistic risk assessment (PRA) are drawn from *10 CFR Part 52* (Reference 1), the Commission's Severe Reactor Accident Policy Statement (Reference 2) regarding future designs and existing plants, the Commission's Safety Goal Policy Statement (Reference 3), the Commission approved positions concerning the analyses of external events contained in SECY-93-087 (Reference 4), and NRC interest in the use of PRA to help improve future reactor designs. In general, these objectives have been achieved by the ELWR PRAs and the NRC staff's review. The staff's applicable regulation for the analysis of external events for the ELWR PRAs is as follows:

The probabilistic risk assessment required by *10 CFR 52.47(a)(1)(v)* must include an assessment of internal and external events. For external events, simplified probabilistic methods and margins methods may be used to assess the capacity of the standard design to withstand the effects of events such as fires and earthquakes. Traditional probabilistic techniques should be used to evaluate internal floods. For earthquakes, a seismic margin analysis must consider the effects of earthquakes with accelerations approximately one and two-thirds the acceleration of the safe-shutdown earthquake (SSE).

INTRODUCTION

Lessons from past risk-based studies indicate that seismic events can be important potential contributors to core damage (Reference 5). However, the estimates of core damage frequencies for seismic events continue to have considerable uncertainty. The NRC staff has concluded in Reference 5 that for ELWRs, seismic events can best be evaluated using simplified margins methods similar to those developed for existing plants.

The staff further concluded in Reference 5 that the ELWR plant designers can best determine the seismic capability of the plant through a combined approach that takes advantage of the strengths of both PRA and margin methods. The major difference between a seismic PRA and the proposed PRA-based margins approach is that the latter does not convolute fragility curves with hazard curves. The PRA-based margins approach measures the robustness of the plant to withstand earthquake ground motions as a multiple of the SSE. This method has the advantage of eliminating the need to deal with uncertainty in the hazard curves for the site and identifies potential design-specific vulnerabilities to seismic events. This assessment of seismic margin demonstrates that the plant design is robust and can resist seismically induced ground motions likely to cause severe accidents. It should be noted that the proposed siting rule and the associated regulatory guides incorporate probabilistic methods of determination of site specific ground motion for safe shutdown earthquakes not likely to be exceeded about $1x10^{-5}$ per reactor year. This frequency is based on the median reference probability for exceeding the SSE for currently licensed sites based on the Appendix A to *10 CFR Part 100*. Consequently, a margins study has an implicit assurance that the core damage frequency due to earthquakes is low enough to be inconsequential.

The NRC staff requires that each ELWR plant designer perform a PRA-based margins analysis to identify the vulnerabilities of their design to seismic events. The plant designers construct plant logic models covering the various systems that could be used to prevent core damage by modifying the plant PRA models for internal events to include logic important in considering seismic failures. The models do not include data from site-specific or generic seismic hazard curves. The designers then determine all important accident sequences using the event trees and fault trees. In addition, the designers determine the value of the minimum high confidence, low probability of failure (HCLPF) for the plant by determining the HCLPF values for the important structures, systems, and components (SSCs) for each accident sequence. The HCLPF values determined in this manner can be used to determine the robustness of the plant to resist seismic events beyond design basis without core damage.

* Views expressed are those of the authors and do not represent the position of the agency.

In general, the value of the plant HCLPF should be at least 5/3 times the design ground motion peak ground acceleration. In addition, HCLPF calculations indicate which components and systems limit the seismic capability of the plant.

PRA-BASED SEISMIC MARGINS ANALYSIS

ELWR standard plants are designed to withstand a 0.3 g peak ground acceleration (pga) SSE motion characterized by a standard response spectrum shape such as that defined in Regulatory Guide 1.60. Since the analyses used in designing the capability of SSCs to withstand the SSE have significant margin in them, it is expected that a plant built to withstand the SSE actually will be able to withstand a much larger earthquake. A PRA-based margins analysis systematically evaluates the capability of the designed plant to withstand earthquakes without suffering core damage, but does not estimate the CDF from seismic events. The margins analysis is simply a reasonably conservative method for estimating the "margin" above the SSE, that is, how much larger than the SSE an earthquake must be before it compromises the safety of the plant.

The capability of a particular SSC to withstand beyond-design-bases earthquakes is characterized by the value of the peak ground acceleration (g-level) at which there is a high confidence that the particular SSC will have a low probability of failure (HCLPF). The HCLPF capacity of a particular SSC corresponds to the earthquake level at which, with high confidence (95 percent), it is unlikely (probability less than 5 percent) that the SSC will fail. An HCLPF value for the entire plant is determined by finding the lowest sequence HCLPF that leads to core damage. It is a measure of the capability of the plant to withstand beyond-design-basis earthquakes without suffering core damage. The plant HCLPF value, which is assessed from the SSC HCLPF values, has units of acceleration. HCLPF values are computed using the conservative deterministic failure margin (CDFM) approach presented in EPRI-6041-SL (Reference 6).

HCLPF values are calculated for structures and components using the relationship

$$HCLPF = A_m * \exp[-1.65(\beta_r + \beta_u)]$$

in which:

A_m = the median peak ground acceleration corresponding to 50 % failure probability

β_r = the logarithmic standard deviation for the structure or component capacity due to randomness

β_u = the logarithmic standard deviation for the structure or component capacity due to uncertainty

To determine the mean risk, the variability due to randomness and uncertainty can be combined into composite variability $\beta_c = \sqrt{\beta_r^2 + \beta_u^2}$.

The HCLPF can then be approximated as

$$HCLPF = A_m * \exp(-2.3\beta_c)$$

Using a typical value of β_c = 0.4,

$$A_m = 2.5\ HCLPF$$

This HCLPF is usually denoted as $HCLPF_{50}$ corresponding to an earthquake response spectrum which has 50 % probability of being exceeded at any frequency. The HCLPF value computed using the EPRI CDFM approach (Reference 6) is denoted by $HCLPF_{84}$. This corresponds to an earthquake response spectrum which has only 16 % probability of being exceeded at any frequency. The relationship between these two HCLPF values is given by

$$HCLPF_{84} = HCLPF_{50} \exp(\beta_{csa})$$

β_{csa} = the composite variability associated with the spectral shape

Using a typical value of 0.18 for β_{csa} , $HCLPF_{50}$ can be estimated as

$$HCLPF_{50} = HCLPF_{84} / 1.2 = CDFM / 1.2$$

Therefore the median peak ground acceleration can be estimated as

$$A_m = 2.1\ CDFM$$

The NRC has indicated (Reference 4) that it expects that a plant truly designed to withstand a 0.3 g SSE should have a plant HCLPF at least 1.67 times the SSE (i.e., 0.5 g). The PRA-based seismic margin analyses show that the evolutionary standard plants designs meet and exceed the 0.5 g HCLPF value expectation.

DOMINANT SEISMIC ACCIDENT SEQUENCES

For ELWR designs, event and fault trees developed for the internal events PRA are modified to accommodate seismic events. In this way the random failures and human errors modeled in the internal events portion of the PRA are captured in the seismic analysis. In the current design certification applications, the underlying assumption that earthquakes exceeding the SSE will happen less than once in a thousand years is used to exclude random failures or human errors with probability less than 1×10^{-3}. This is because the combination of seismic events having acceleration higher than 0.5 g with random failures lower than 1×10^{-3} would result in estimates of CDF much less than 1×10^{-6}/year. The modified event and fault trees are merged, and cutsets are generated for all sequences that lead to core damage. These cutsets are of two kinds. One kind contains only seismic failures (i.e., without any random failures or human errors). The other kind contains random failures and/or human errors (truncated at 1×10^{-3}) in addition to seismic failures. In "quantifying" these cutsets, the HCLPF values of the seismic events (instead of mean values of failure probabilities) are used, while the probabilities of random failures and human errors are the same as for the internal events PRA. Evolutionary reactors designers have used the CDFM approach for calculating HCLPF values for components and structures and the minimum/maximum and convolution approach for the sequence and plant-level HCLPF calculations. In the minimum/maximum approach, if there is an "OR" sequence where the failure of any individual SSC would cause core damage, the lowest individual SSC HCLPF is taken as the sequence HCLPF value. If there is an "AND" sequence where the failure of all SSCs would cause core damage, the highest individual SSC HCLPF is taken as the sequence HCLPF value. The convolution method may more appropriately (in a pure mathematical sense when a margins analysis is performed) take into account the fact that when one combines HCLPFs, one is working with the tails of probability distributions.

In the light of the new probabilistic seismic margin hazard estimate methodology as discussed in draft Regulatory Guide DG-1032, the reference probability of SSE is 1×10^{-5}/year, and therefore, 1.67 times the SSE should be considered an extremely rare probability.

The CDFM method contains a set of guidelines (e.g., how to modify ground response spectra, damping, material strength, and ductility) that if followed will result in an acceptable estimate of an SSC's HCLPF. The PRA-based SMA, performed by the evolutionary reactors designers, identify the sequences that lead to core damage. In all of these sequences, offsite power is lost.

The NRC staff has identified those sequences that are the "dominant" contributors to the plant HCLPF value. The word "dominant" appears in quotes to emphasize that the use of this terminology in the context of a seismic margins study should not be taken in the same way as for a conventional PRA. Although these sequences (and associated cut sets) dominate the HCLPF values for the plant, the margins approach does not permit a determination that these are the dominant contributors to seismic risk in a probabilistic sense. However, as discussed earlier, the selection of the margins earthquake of 1.67 times the SSE ensures that seismically induced core damage is not credible.

The NRC has focused on the insights that can be drawn from the evolutionary reactors SMA. Because of the larger uncertainty in hazard curves and because seismic PRA results are dominated by the tails of the site hazard and SSC fragility curves, it is expected that if a seismic PRA had been performed, it would have been a low absolute value, but one with a CDF perhaps comparable to the CDF from other significant events. This is particularly true for the evolutionary reactors designs since the estimated CDF from internal events has been driven down so far by design changes/improvements. As it is, the evolutionary reactors designs should be better able to resist high seismic events than most, if not all, existing nuclear power plants east of the Rocky Mountains because of its built-in margins.

CONCLUSIONS

The PRA-based seismic margins approach allows one to determine which plant features are important to the plant-level HCLPF and the redundancy or diversity available in achieving that HCLPF. In order to make this determination, each sequence that can lead to core damage on the seismic event trees is examined. For ELWR designs, none of the sequences has a seismic-only HCLPF of less than 0.5*g*. The sequences were examined to determine if lowering the HCLPF value of a single SSC (to a much lower HCLPF value) or increasing the failure rate of a single system (to a much higher failure rate) would result in a plant HCLPF of less than 0.5*g*. A review of the cut sets showed that most of the important sequences require at least two failures of SSC with HCLPFs above 0.5*g*.

One of the reasons for performing an uncertainty analysis is to display the range of values within which the results of an analysis could reasonably be expected to fall. The use of a PRA-based SMA inherently makes use of the breadth of information being considered. This is because HCLPF values can be thought of as the g-level at which one has 95 percent confidence that less that 5 percent of the time the equipment will fail (i.e., we are dealing with the tails of the curves). It was not found necessary to combine (use convolution) a seismic hazards analysis with equipment fragilities, since hazard curves have a large uncertainty which reduces their value in helping to make judgments about the seismic risk. From seismic PRA analyses, it is clear that uncertainties in the hazard curves would dominate the uncertainties in equipment/structure fragilities. For the evolutionary reactors PRA-based SMA, no uncertainty analysis was performed because uncertainty is directly reflected in the choice of the ground motion level used in the margins analysis. Also, since the margins method does not result in either CDF or risk results, importance analyses were not performed. Evolutionary reactors designers did, however, perform sensitivity analyses to evaluate the effects of changes in certain assumptions used in the SMA. These sensitivity analyses covered two areas. The first addressed the sensitivity to changes in the HCLPF values of certain key SSCs. The second addressed the sensitivity to changing the site conditions from rock to various soil types.

Sensitivity analyses were also performed for the effects of siting the plant on one of a variety of soil sites (versus the base case of a rock site). For all cases, the HCLPF values for the dominant sequences and for the plant were basically unchanged. Some effects were observed for high HCLPF sequences (above 0.85 g), but the staff considers none of these effects significant.

REFERENCES

1. *U. S. Code of Federal Regulations*, Title 10, "Energy," Part 52, "Early Site Permits; Standard Design Certification; and Combined Licenses for Nuclear Power Plants."

2. U.S. Nuclear Regulatory Commission, "Policy Statement on Severe Reactor Accidents," *Federal Register*, Vol. 50, p. 32138, August 8, 1985.

3. U.S. Nuclear Regulatory Commission, "Safety Goals for the Operations of Nuclear Power Plants," *Federal Register*, Vol. 51, p. 28044, August 4, 1986.

4. U.S. Nuclear Regulatory Commission Staff Requirements Memorandum on SECY-93-087," Policy, Technical, and Licensing Issues Pertaining to Evolutionary and Advanced Light-Water Reactor Designs," July 21, 1993.

5. U.S. Nuclear Regulatory Commission," Policy, Technical, and Licensing Issues Pertaining to Evolutionary and Advanced Light-Water Reactor Designs," SECY-93-087, April 2, 1993.

6. Electric Power Research Institute, "A Methodology for Assessment of Nuclear Power Plant Seismic Design," NP-6041-SL, August 1991.

PVP-Vol. 340, Seismic Engineering
ASME 1996

RELIABILITY ESTIMATION FOR CHECK VALVES AND OTHER COMPONENTS

K. L. McElhaney
R. H. Staunton
Oak Ridge National Laboratory
Oak Ridge, Tennessee

ABSTRACT

For years the nuclear industry has depended upon component operational reliability information compiled from reliability handbooks and other generic sources as well as private databases generated by recognized experts both within and outside the nuclear industry. Regrettably, these technical bases lacked the benefit of large-scale operational data and comprehensive data verification, and did not take into account the parameters and combinations of parameters that affect the determination of failure rates. This paper briefly examines the historic use of generic component reliability data, its sources, and its limitations. The concept of using a single failure rate for a particular component type is also examined. Particular emphasis is placed on check valves due to the information available on those components. The Appendix presents some of the results of the extensive analyses done by Oak Ridge National Laboratory (ORNL) on check valve performance.

BACKGROUND

Historically, an analyst attempting to quantify component reliability for the purpose of assessing the impact of component failure has been limited to a number of sources of published "classic" generic data, and the analyst's own collection of reference material/data (NUREG/CR-6186, 1994). These "classic" sources, including two NUREGs and an Institute of Electrical and Electronics Engineers (IEEE) Standard, have historically been used (combined with the analyst's own references) as input into those assessments.

NUREG-75/014, "The Reactor Safety Study," had as its goal a risk assessment as opposed to a reliability analysis (NUREG-75/014, 1980). Hence, order of magnitude accuracy was tolerated in its quantified results. The study used a range approach (a standard statistical and probabilistic modeling technique where a range of values is determined such that the "true" value has a high probability of lying within it) because of the inadequacy of the data available at the time. Failure rates were derived from reference handbooks, reports, operating experience, and nuclear power plant experience. Primary shortcomings of the estimates are the poor quality of data, which required broad ranges, and the modest number of total failures contained in the data (only 102 valve failures).

The Licensee Event Report (LER) data summary, NUREG/CR-1363, was based on five years of LERs pertaining to valves (NUREG/CR-1363, 1982). As with the Reactor Safety Study, the report assumed a constant failure rate for the valves over the total number of demand hours. The primary shortcomings of using LER data and associated operating history are that not all valve faults and maintenance/part replacement actions are recorded and available. In general, the effects of maintenance actions, programs, and retrofits are unknown.

The LER data summary report was limited to valves in engineered safety feature systems and primary relief and safety valves, which was the only data found of sufficient quality. Failures and command faults were assumed to be random events, and no common-cause events were considered separately. The operational demands for the valves included those related to testing. The test demands were assumed to be as specified in Section XI of the ASME Boiler and Pressure Vessel Code. Operational demands were added to test demands for safety and relief valves since only one test demand was assumed for them. The number of operational demands used was estimated.

IEEE Standard 500-1984 is another generic source of valve reliability estimates which was prepared to be used by either nuclear systems reliability analysts or design engineers (IEEE, 1983). In addition to design and test planning purposes, the standard was also intended to be used for Probabilistic Risk Assessment (PRA) studies.

Drawing primarily from nuclear and applicable industrial data obtained from NUREGs, EPRI reports, and its own surveys, the IEEE Standard 500 represented the most rigorous attempt at the time (the standard was reaffirmed in 1991 with minor changes) to create a comprehensive and defendable source of reliability estimates for a wide variety of nuclear components.

The estimates provided in IEEE Standard 500 were developed using a feedback system for the communication of data and information among a group of experts. This process is an iterative one that uses expert opinion to consider both data and independent expert opinion to produce a product. Valve data was obtained through questionnaires and visits to nine nuclear units. Experts recorded maintenance records pertaining to a five year period. As in the earlier studies, the analysts probably did not have access to accurate operating time data, complete valve cycling data, and information pertaining to various part replacements and other preventive maintenance actions since complete records are generally not kept on these types of data. Assumptions were almost certainly made in these areas, although this detail is not provided in the standard itself.

SINGLE FAILURE RATE VS. A FAILURE RATE RANGE

Reliability analysts have been accustomed to using generic failure rates available from various sources. It has been generally recognized that failure rates vary according to the type of component. Failure rates have been made available for motor operated valves (MOVs), pumps, check valves, and other components. Occasionally, failure rates have even been broken down by failure mode (e.g., failure to open) and sometimes even by specific valve type, such as for a swing check valve. In general, however, no effort has been made to estimate failure rates for any component type considering such important parameters as service environment/application, failure mode, size, age, specific design features, manufacturer, maintenance practices, inservice testing practices, etc. Recent studies by ORNL (McElhaney, 1995a) have shown that some significant differences exist in relative failure rates for check valves according to, among other factors, their size, system of service, analysis period, and design features (e.g., *specific valve type*). For example, in 1991 check valves > 10 in. in size were 1.2 times more likely to fail than valves in any other size group. Check valves installed in HPCI, suppression pool support, RCIC, and diesel starting air systems were three to four times more likely to fail *significantly*[1] than were valves installed in most other systems. This result was also found to be heavily influenced by other factors, such as valve design, operating conditions, and testing practices. For example, most of the valves in the diesel starting air system were small lift check valves that failed due to the presence of foreign material, and most of the failures occurring in the HPCI and suppression pool support systems involved stuck closed vacuum breaker valves. Three of the four systems were exclusive to BWRs, a result that could lead to false conclusions regarding BWR performance until it is recognized during further analysis that BWRs may simply have superior testing practices and are therefore detecting more failures.

The point is, there is no "correct" answer to a question such as "what is the failure rate for check valves?" The answer depends upon the scope of the question: the more specific the question (i.e., the greater the number of parameters specified), the more accurate the answer. A much more accurate estimation can be established, for instance, for the failure rate of "***stuck open 10 in. Anchor Darling** model 1234 swing check valves in the feedwater systems in Westinghouse pressurized water reactor (PWR) plants during 1990-1995*," than for "*check valves in feedwater systems,*" assuming that the population of valves in service during the period of interest (corresponding to the given set of parameters) and the number of failures is known. It should be clear, then, that perhaps the best that can be done is to establish a range of failure rate values based upon the nature of the data itself and the set of analysis parameters considered. The analyst may be able to determine a worst case and a best case scenario by reviewing the failure data available, calculating appropriate cross-tabulations with the known parameters, and using the valve population data to calculate failure rates. In some cases this range may be relatively small, while in others it may vary by orders of magnitude. Again, the width of this range would be dependent upon both the data and the set of constraints given. A sensitivity analysis might be performed for the end application to determine just what effect a given failure rate range might have on the analysis.

ORNL DATABASE AND CALCULATION OF ABSOLUTE FAILURE RATES

ORNL has established performance databases for check valves, MOVs, air-operated valves (AOVs), main steam isolation valves (MSIVs), and pumps. These databases were established by downloading, reviewing, screening, and characterizing thousands of nuclear component failure records from the Institute of Nuclear Power Operations' (INPO's) Nuclear Plant Reliability Data System (NPRDS) database. For check valves, the review and screening processes eliminated incipient failures, external leakage failures, and failures not involving internal parts degradation. (A "failure" for purposes of this study was defined as a degradation of one or more valve functions.) Failure records remaining for analysis were then characterized according to specific parameters inherent in the NPRDS database, such as system, nuclear steam supply system (NSSS) vendor, valve size, valve age, manufacturer, and plant age. Other parameters not included in NPRDS were also characterized by reviewing the failure narratives. These included failure mode, failure area, extent of degradation, general and specific detection methods, and normal system operating status. Many of the ORNL failure characterizations were reviewed and verified by cognizant plant personnel, resulting in an increased level of confidence in the data. In some cases, input from the plants actually resulted in changes to the failure narratives/characterizations. Databases for MOVs, AOVs, MSIVs, and pumps were developed in much the same way as for check valves. These component performance databases also included information on the installed component populations so that relative failure rates could be calculated. The Appendix provides a more detailed review of some of the results of the ORNL check valve studies.

[1] I.e., significant in terms of degradation of one or more component functions. Examples of *significant* failures would include those involving valves that were stuck open, stuck closed, or had loose or broken parts. Other failures which would not have resulted in loss of functionality would include, for example, leak test failures.

The primary benefits of the ORNL-developed component performance databases include their incorporation of many years of actual operational data, consistency of data characterization and development methodology, and reproducibility. Their intended use has been to enable the identification and trending of any significant correlations of failure rates with plant age, component age, or other parameters while considering the effects of the number of components in service during the analysis period (i.e., population effects). NUREGs developed using these databases have contained hundreds of charts and tables showing relationships among the studied parameters while also providing calculated *relative failure rates*.[2]

It must be understood, however, that although NPRDS provides reasonably accurate information on, as well as a count of, "failure" incidents, it does not provide data on, for example, valve operating time or number of valve actuations (demands), which are essential for the derivation of failure rates. The concept of component age is also uncertain in light of the type of maintenance performed - often amounting to comprehensive replacement of internal parts. These factors, along with differences in reporting practices among plants/utilities, make it difficult to utilize even the best available operational data to establish absolute failure rates. The concept of extent of degradation to the component must be addressed also; a failure of a valve to pass a leak rate test should not necessarily be given the same weight as a failure involving a detached disc or broken hinge pin in calculating the valve's "failure rate."

Again, the analyst is left with no choice except to establish a failure rate range given the best data and largest number of parameters available. Even with its limitations, the data derived from the NPRDS database through review, screening, and characterization is the best representation of years of nuclear operating experience available. A methodology for studying and assessing the effect that such factors as design, size, age, maintenance, and service conditions have on component performance must ideally be based on data such as that characterized from NPRDS by ORNL from nuclear power plant operations. Such a methodology might involve the following steps:

1. Devise a means of sorting component operational data to contrast the factors of interest. This may be a simple sort, or it may involve considerable research, depending on whether the essential data field(s) can be sorted and are complete, reliable, and/or even present in the database.

2. Perform a careful review of all possible factors that may affect performance to determine how many other variables can be eliminated in the data sort being performed. One approach may be to perform additional data sorts of "suspect" factors to see if they are indeed biasing the results. For instance, if one utility is known to maintain the pertinent valves in a significantly different manner, these valves might be excluded from consideration. [It must be kept in mind, however, that the larger the number of parameters reviewed (both for inclusion and elimination), the more complicated and time-consuming the analysis. The analyst must weigh the time and costs involved vs. the confidence level required for the results.]

3. Evaluate the results of the data showing the effects of the factors of interest and establish a failure rate range by identifying best and worst case performance scenarios.

4. Consider the cases of markedly contrasting data carefully. For example, are the differences in performance due to some obvious factor such as a particular design feature or abrasive system fluid? Assess the strength of the factor causing the contrasting data to gain insight into how effective the factor is in causing the observed data.

5. If possible, review the reliability estimates for the component based on historical data and generic sources utilized in past studies.

6. If necessary, modify the historic failure rate estimates based on results of 2 through 4. Make generous use of expert opinion in performing the modification and subsequent peer review.

An example of the calculation of absolute failure rates using NPRDS data and a relatively specific set of analysis parameters can be found in the analysis of accumulator discharge check valve performance (McElhaney, 1995b). This study reviewed the operating history of 231 PWR accumulator discharge check valves during the period 1984-1992. During this period, only 18 failures were recorded in the ORNL check valve performance database. Only 7 of these failures were characterized as *significant* in terms of degradation to the valve itself. Although NPRDS does not contain specific information on valve operating time, the number of valve-years of service of the valves included in the study could be estimated using the available component inservice date. The number of accumulator discharge check valve failures was then divided by the number of valve-years of service for this component during the study period to arrive at an overall failure rate in terms of valve failures per valve-years.

For example, for *accumulator discharge check valves* from 1984 through 1992:

[2] Calculation of a *relative failure rate* involves the employment of a normalization process to account for both component population and service life. This process involves dividing the number of failures for a given category within a field by the number of component-years of service for the analysis period for that category. The first step in the normalization process is to determine the overall failure rate for all components (e.g., check valves) in a given analysis period. This is determined by dividing the total number of failures characterized by the total number of component-years of service. The result is the "normalizing value" which is applied to individual category failure rates to determine the relative failure rate, which is dimensionless. Normalization provides a good indication of how a particular category (e.g., > 10 in. check valves) within a field compares with other categories in the field. A relative failure rate of unity indicates that the particular category's failure rate is equivalent to the failure rate of the population as a whole. A value of less than unity indicates that a particular category fails at a rate less than the population as a whole, while a value of greater than unity indicates a failure rate greater than the general population. The normalization process accounts for variables such as population size or service life, which have been shown to influence the number of component failures, and allows comparison across a field with values that are less likely to be misinterpreted or misapplied.

Overall failure rate for all ORNL-characterized failures
= 18 failures / 1894 valve-years = 0.0095/year

Overall failure rate for all ORNL-characterized *significant* failures
= 7 failures / 1894 valve-years = 0.0037/year

Overall failure rate for valves to accomplish their accident mitigation function (failure to open)
= 0 failures / 1894 valve-years = 0 / year.

It could then be concluded that even this relatively specific set of parameters for valves with an extremely good performance record represents a range of failure rate values. (The Appendix contains a more detailed discussion of the analysis of this particular check valve application.)

CONCLUSIONS

The generic failure rate estimates given for nuclear systems components available from certain classic generic sources should be reviewed carefully in consideration of additional, more specific data since they may not necessarily represent realistic industry operating experience. Most estimates do not consider such important parameters as component design features or operating environment. The resulting single generic values given, for example for "check valve failure rates" can therefore be inaccurate and/or misleading.

A more realistic approach to component failure rate estimation should involve the utilization of as much actual operating data as possible. More accurate values can be obtained by utilizing industry performance databases such as those developed at ORNL and by providing as specific an analysis scope as possible. In other words, the greater the number of component analysis parameters specified, the more accurate the failure rate estimation. Even then, the accuracy of the estimation depends upon the accuracy and completeness of the performance data available and the methodology used to estimate the component reliability. In most cases, the best estimate of component performance can probably only be established by a bounding range of values for the best and worst case performance scenarios. In the case of utilization of these estimates in current estimates of failure impacts, a sensitivity analysis might be in order to determine the relative contribution of component failure rate vs. risk importance as factors in determining the impacts. This tends to take on more meaning today as the nuclear industry heads toward "risk-based" inservice testing intervals. If the failure rates used to develop risk change as the testing intervals increase, then more data will be needed so that new risk values can be calculated. A feedback loop will become necessary to ensure overall confidence in the analyses.

REFERENCES

Casada, D.A. and Todd, M.D., "A Characterization of Check Valve Degradation and Failure Experience in the Nuclear Power Industry - 1984-1990 Failures," NUREG/CR-5944, Vol. 1, September 1993. Prepared by Oak Ridge National Laboratory for the U.S. Nuclear Regulatory Commission.

IEEE Standard 500-1984, The IEEE Guide to the Collection and Presentation of Electrical, Electronic, Sensing Component, and Mechanical Equipment Reliability Data for Nuclear Power Generating Stations, December 1983.

McElhaney, K.L, "A Characterization of Check Valve Degradation and Failure Experience in the Nuclear Power Industry - 1991 Failures," NUREG/CR-5944, Vol. 2, July 1995. Prepared by Oak Ridge National Laboratory for the U.S. Nuclear Regulatory Commission.

McElhaney, K.L., "An Evaluation of PWR Safety Injection Accumulator Tank Discharge Check Valve Performance," ORNL/NRC/LTR-95/22, August 1995. Letter report prepared by Oak Ridge National Laboratory for the U.S. Nuclear Regulatory Commission.

NUREG-75/014, "Reactor Safety Study - An Assessment of Accident Risks in U.S. Commercial Nuclear Power Plants, USNRC report WASH-1400, Appendix III, June 1980.

NUREG/CR-1363, "Data Summaries of Licensee Event Reports of Valves at U.S. Commercial Nuclear Power Plants, USNRC report, October 1982.

Staunton, R.H., et al., "Valve Reliability for Passive Advanced Light Water Reactors," NUREG/CR-6186, manuscript completed December 16, 1994. Prepared by Oak Ridge National Laboratory for the U.S. Nuclear Regulatory Commission.

Appendix

ORNL Check Valve Studies

FAILURE ANALYSIS AND RESULTS

Original ORNL Analysis - 1984–1990 Data

The original analysis performed by ORNL consisted of a review of nearly 5000 check valve failure records from the Institute of Nuclear Power Operations' (INPO's) Nuclear Plant Reliability Data System (NPRDS) database entered through May 1990 (Casada and Todd, 1993). Preliminary manual review of the failure narratives resulted in elimination of those failures occurring prior to 1984 (when significant improvements in NPRDS reporting requirements were implemented), as well as the exclusion of those events not involving valve internal parts damage (e.g., external leakage), valves incorrectly coded as check valves, nonfailures, and minor seat leakage. The resulting analysis database consisted of 1227 failure records occurring during the time period 1984 through 1990.

Failure records remaining for analysis were then characterized according to specific parameters inherent in the NPRDS database, such as system, nuclear steam supply system (NSSS) vendor, valve size, valve age, manufacturer, and plant age. Other parameters not included in the database were also characterized by reviewing the failure narratives. These included failure mode, failure area, extent of degradation, general and specific detection methods, and normal system operating status. The ORNL analysis focused on failures involving degradation of valve internal parts. A "failure" for purposes of this study was defined as a degradation of one or more valve functions (e.g., failure to open, failure to close, loose/broken parts).

After the failure records were manually reviewed, characterized, and recorded in database format, exhaustive data manipulations and cross-correlations were performed. Calculations were performed using data normalized according to the population of check valves existing during the analysis period to enable the establishment of relative failure rates. The Nuclear Industry Check Valve group (NIC) also provided a forum for utility review of much of the characterized data through actual site visits and interviews, further substantiating the validity of the analysis.

Analysis Updates

Desiring to gather additional information and to identify any performance trends, the NRC requested ORNL to update the original failure analysis for two additional study periods, 1991 and 1992. Analysis for 1991 has been completed, while that for 1992 is currently in progress. A somewhat augmented approach was undertaken for the 1991 study, building from experience gained from the 1984–1990 analysis. An important difference in the two studies was the incorporation of specific valve type data for 1991 (e.g., swing, lift/piston, tilting-disc check valves), since this parameter was recognized as being perhaps the most significant in terms of failure correlation (McElhaney, 1995a).

Another notable difference in the analyses for 1984–1990 and for 1991 was the filtering process used to exclude valves from further analysis. While both studies excluded failure reports involving no internals degradation, nonfailures, and noncheck valves, the 1991 (and subsequent) analysis included most failures involving seat leakage. As a result, more "moderate" type failures were included in the 1991 study. Although this approach somewhat invalidates direct comparison between the two periods, it was felt that a better representation was gained of actual check valve performance. Comparison of data based on "significant" failures for the two periods is considered to be valid.

The first data available relative to specific valve type for 1991 resulted in a failure distribution of 35% lift/piston check, 33% swing check, 12% unknown valve type, and the remainder composed of tilting-disc, duo/double disc, stop, in-line, and other. (It was not possible, however, to establish relative failure rates, since information does not yet exist regarding the population distribution by specific valve type.) Insight gained from this data, however, includes correlations such as valve type vs. failure mode and failure cause.

ACCUMULATOR DISCHARGE CHECK VALVES

A detailed analysis was also undertaken by ORNL to evaluate check valves in an application that had been reported to have experienced good performance and reliability (McElhaney, 1995b). The purpose of the analysis was to examine the potential for relaxation of current requirements for inservice testing for check valves in this particular application based on industry performance history. Accordingly, 231 PWR safety injection accumulator tank discharge check valves were identified and their performance history reviewed from 1984 through 1992. During this time period, only 18 failures involving these valves were recorded in the ORNL-characterized database derived from NPRDS. Of these, only seven failures were characterized as being significant in terms of extent of degradation (to the valve), and no failures were reported that would have affected any valve's ability to perform its forward flow function in the event of its actuation during an accident.

Methodology

In an effort to more accurately assess the performance of check valves in this specific application, a characterization of information available from NPRDS was performed. Since the engineering record associated with each component included in the NPRDS database does not list the component's specific application, it was first necessary to review plant Final Safety Analysis Reports (FSARs) to determine which valves (according to location and component identifier) were actually accumulator discharge check valves. FSARs from 39 randomly chosen PWR units were reviewed, resulting in the identification of 231 check valves serving an accumulator discharge function. Of the 231 valves identified, 142 were at Westinghouse plants, 61 were at CE plants, and 28 were at B&W plants.

After identification of the 231 accumulator discharge check valves, a review of their failure history was initiated. Using the ORNL check valve database, 18 failures of accumulator discharge valves were identified. The criteria for failure in this evaluation was the same as that employed in the previous studies; i.e., a "failure" was defined as a degradation of one or more valve functions. Failures were again evaluated in terms of their effects on the valves themselves, not in terms of any resulting plant or system effects.

Overall failure rates

The overall failure rates for accumulator discharge check valves are presented in Fig. A1. Overall failure rates are presented for all failures and for significant failures only (i.e., failures characterized as being significant in terms of extent of degradation of the valve). Failure rates are presented for all check valve failures and for accumulator discharge check valves only. These rates were calculated as follows, using failure data from NPRDS as filtered and characterized by ORNL (*Note that these failure rates were derived for comparison purposes only. Their magnitude is affected by many factors, including NPRDS reporting practices, ORNL filtering methodology, and sample size for the valves included in the accumulator discharge study. They should not be considered for use in probabilistic or risk-based applications without further evaluation.*):

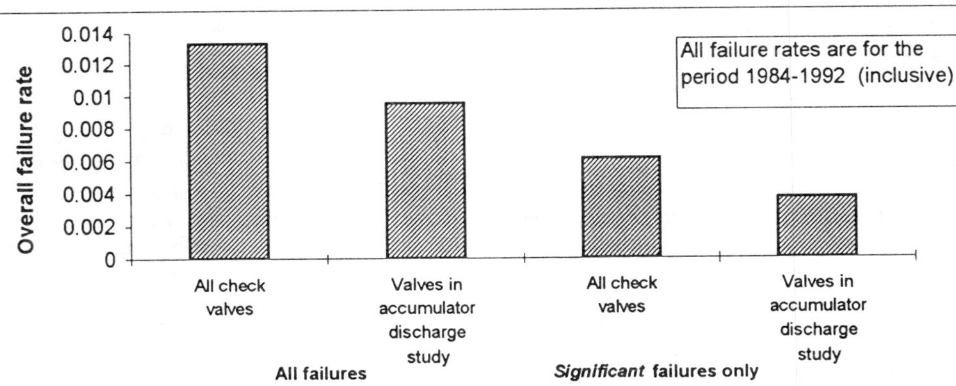

Figure A1 Overall Failure Rates For Check Valves In Accumulator Discharge Study vs. All ORNL-Characterized Check Valve Failures

Overall failure rate for all ORNL-characterized check valve failures (1984-1992)
= No. failures / No. valve-years
= 2065 failures / 155,498 valve-years = 0.0133/year

Overall failure rate for all ORNL-characterized significant check valve failures (1984-1992)
= No. significant failures / No. valve-years
= 949 failures / 155,498 valve-years = 0.0061/year

Similarly, for accumulator discharge valves in the study,

Overall failure rate for all ORNL-characterized failures
= 18 failures / 1894 valve-years = 0.0095/year

Overall failure rate for all ORNL-characterized significant failures
= 7 failures / 1894 valve-years = 0.0037/year

Overall failure rate to accomplish the accident mitigation function
= 0 failures / 1894 valve-years = 0 /year

Figure A1 indicates that the valves in the accumulator discharge check valve study exhibit failure rates lower than both the ORNL-characterized check valve failure population for all failures (both moderate and significant in terms of extent of degradation) and for significant failures only.

Conclusions

Of the 231 accumulator discharge check valves identified, only 18 failures were recorded in the ORNL database from 1984-1992, inclusive. Of these, only seven failures were characterized as being significant in terms of their extent of degradation (to the component itself). Also, of the seven significant failures, three were attributable at least in part to human error. The other four significant failures involved either broken or missing parts. None of the reported failures should have had any effect on the forward flow functionality of the valves, had they been called upon in the event of an accident. This result, in addition to the absence of any stuck closed failures, indicates that in terms of ability to perform their required accident functions, accumulator discharge check valves have been extremely reliable.

PVP-Vol. 340, Seismic Engineering
ASME 1996

SEISMIC RESPONSE ANALYSIS AND TEST OF CHASNUPP STEAM GENERATOR LOWER PART MODEL

Liang,H., Jinkang,X., JingXian, Z., Yinbiao,H., Peizhu, W.
Shanghai Nuclear Engineering Research and Design Institute
Shanghai, People's Republic of China

ABSTRACT

The seismic response analysis and test of CHASNUPP steam generator lower part model has been performed. The lower part model is consisted of a tube sheet, 441 U-shaped rods (modeled on U-shaped tubes), 9 tie rods, a tube bundle wrapper, a lower shell and some position bolts between the lower shell and the wrapper. The analytical and experimental data show that the steam generator lower part model is more stiffer than the previous pure tube bundle model (Han, 1994).

The movement of the lower part model appears in a systematic mode in earthquake condition. The fundamental natural frequency of this model is higher than that of the tube bundle model and lower than that of the lower shell. The global frequency components are absolutely dominant. The local frequency components are insignificant. The experimental data are in good agreement with FEM results. The effectiveness of mathematical model is verified. The effects of the contained water in the lower shell and the added water of the rod are both considered. The research work is more practical in reactor engineering.

1. INTRODUCTION

The steam generator for CHASNUPP is an improved steam generator for Qinshan NPP unit 1. In research work of seismic response analysis and test of the steam generator lower part model for CHASNUPP, its modifications of structure are as follows:

(1) installation of three groups of anti-vibration bars in stead of two groups,
(2) decreasing of the support plate span as scale ratio,
(3) consideration of the stiffness effect of the lower shell,
(4) increasing of the zero period accelerations of applied response spectra.

For modeling the main characters of seismic response and manufacturing model conveniently, adequate simplification of the lower shell model structure is made. Firstly, the thicknesses of the lower shell and tube-sheet are thinner than the scaled thicknesses. Secondly, a flat top head is used as the closure head of the model. Thirdly, modeling of the steam generator lower head, separators, the conical and cylindrical shells are not considered. The stainless steel rods are instead of Incaloy-800 heat transfer tubes. The U-shaped rod pitch is enlarged and the number of rods is decreased. The diameters and thicknesses of tube-sheet and support plate , the span of the support plate, the diameter of rod and dimensions of anti-vibration bars are modeled as scale ratio of 1:4.4. Two photographs of the steam generator lower part model are shown in Figure 1-a and 1-b.

2. SEISMIC RESPONSE ANALYSIS

2.1 Simplified Mathematical Model

The mathematical model of the lower part model is simplified as a row of six rods with different U-bend radii and constrained by three anti-vibration bars. Firstly, the perforated support plate and the tube-sheet are considered as equivalent solid plate (Slot, 1971). Then they are treated as equivalent beams (Timoshenko, 1940). At the two ends of equivalent beam for support plate, except the rotation about horizontal axis z, the displacements in three directions of x, y and z and rotations about horizontal axis x and vertical axis Y are constrained. The equivalent tube-sheet beam is clamped at both ends, its six degrees of freedom are constrained. At the contact point between rod and tube support plate, the horizontal displacements of the rod and the plate are equal, but the vertical displacements and rotations of the rod and the support plate may be inconsistent. The constraint conditions are represented as equation (1).

$$U_{ix} = U_{jx}$$
$$U_{iz} = U_{jz} \qquad (1)$$

Where U denotes displacement, subscripts i and j denote node numbers of rod and tube support plate at contact point, x and z denote the x and z direction of reference coordinates.

At the contact point, the out-off plane displacements of rod and anti-vibration bar are equal, the other displacements and rotations may be inconsistent, the constraint condition is represented as equation (2).

$$U_{mz} = U_{nz} \qquad (2)$$

Where U denotes displacement, subscripts m and n denote node numbers of rod and anti-vibration bar at contact point, z denotes z direction of reference coordinates.

The longitudinal section plane of the lower part model is considered as x-y plane of global coordinates. The added mass of rod is considered for the lower part model only. The added mass ratio are calculated as per (Chen, 1975). Six springs in series simulate a tie rod. The anti-vibration bars are hinged in in-plane and free in out-off plane.

Three tension-compression springs (No. 7, 8 and 9 in Table 3) at the bottom points of anti-vibration bars simulate the constrain conditions.

2.2 Seismic Loadings

The safe shutdown earthquake (SSE) acceleration spectra in two horizontal directions and one vertical direction are used for both analytical analysis and experimental test. The corresponding damping value of 3% is selected according to (R. G. 1.61).

The frequency intervals for input data are selected according to (NUREG-0800, 1981). Two-fold amplified accelerations and two-fold broadened frequency bands corresponding to SSE response spectra at 15m elevation of internal structure for CHASNUPP containment are used both for the seismic analysis and experimental test.

3. ORIGINAL DATA AND ANALYTICAL RESULTS

3.1 Original Data

The main structural dimensions and material properties for the lower part model are shown in Tables 1 to 3.

Table 1 Original data of rod

Outer diameter(cm)	0.5
Inner diameter(cm)	0
Young's modules (MPa)	1.96×10^5
Poisson ratio	0.3
Mass per unit length (kg/cm)	1.53×10^{-3}

Table 2 Original data for tube-sheet, support plate, anti-vibration bar & lower shell

	Tube-sheet	Support plate	Anti-vibration bar	Lower shell
Cross section area (cm^2)	9.00	0.54	0.26	204.2
Shear area (cm^2)	9.00	0.54	0.26	204.2
Torsional moment of inertia (cm^4)	21.18	0.1499	0.0375	215741
Moment of inertia (cm^4)	2.43	0.1458	0.0367	107870.5
	18.75	0.00405	0.000867	107870.5
Young's modulus $\times 10^5$ (Mpa)	591.6	520.2	55.95	2.667
Poisson ratio	0.2903	0.2975	0.1411	0.30
Density (kg/cm^3)	7.95×10^{-3}	7.384×10^{-3}	7.867×10^{-3}	0.320×10^{-3}

Table 3 Spring constants ($N/cm \times 10^6$)

Spring No.	1	2	3	4	5
Spring constant	1.18	0.59	0.394	0.296	0.236
Spring No.	6	7	8	9	
Spring constant	0.197	10^{-3}	10^{-3}	10^{-3}	

3.2 Computer Code and Analytical Results

The tube, beam, spring elements and constraint equations of computer code ANSYS are used. Where 3-D straight and curved tube elements are used for rod, Spring elements are used for the tie rods. The straight tube elements are used for the lower shell. Considering the accuracy requirements for frequencies and responses (NUREG-0800, 1981), 355 3-D straight tube elements, 120 3-D curved tube elements, 126 3-D beam elements and 21 tension-compression spring elements are used. The total number of nodes is 608. Both cases of the pure tube bundle model and the lower part model with water are considered in calculations. In the second case, not only the mass and the stiffness of the lower shell and 441 rods are considered, but also the added mass of the rod and the contained water in the lower shell are considered. Six representative rods simulate the mass and stiffness of 441 rods. The first normal mode of the pure tube bundle model is shown in Figure 4. The first normal mode of the lower part model is shown in Figure 5 . The calculated natural frequencies for the pure tube bundle model are 15.62, 33.68, 34.28, 42.78, 62.96 and 81.30 Hz. And the calculated natural frequencies for the lower part model with water are 28.26, 32.29, 40.29, 41.85, 59.30 and 76.6 Hz. For identifying the dynamic behavior of the single rod, the lower shell and the pure tube bundle model, the calculation results of their natural frequencies are shown in Table 4.

Table 4 Calculated natural frequencies for rod and lower shell (Hz)

	Lower shell	Rod(R303)	Rod(R159)	Rod(R15)
1	168.16	146.47	153.70	153.19
2	1052.76	153.69	153.73	153.21
3	2943.96	154.14	153.93	153.52
4	5770.16	154.78	154.23	153.94

The calculated displacements and stresses under SSE loadings for the pure tube model and the lower part model with water are shown in Table 5 and 6.

Table 5 Calculated rod displacements of the pure tube model and the lower part model water (mm)

Node	Ux	Uy	Uz	U
38	0.214	0.084	0.079	0.243
	(0.241)	(0.094)	(0.031)	(0.261)
125	0.112	0.043	0.312	0.334
	(0.139)	(0.053)	(0.124)	(0.194)
127	0.110	0.038	0.319	0.340
	(0.136)	(0.046)	(0.117)	(0.185)
443	0	0	1.180	1.180
	(0)	(0)	(0.642)	(0.642)
587	0	0	1.089	1.089
	(0)	(0)	(0.373)	(0.373)
443	0	0	1.453	1.453

Note: data in brackets denote displacement of the lower part model

Table 6 Calculated stresses in rod of the pure tube bundle model and the lower part model (MPa)

Node	29	116	203	290	369	440
Effective stress	4.54	5.77	13.51	24.94	39.89	81.10
σ_e (MPa)	(4.88)	(4.70)	(9.00)	(16.21)	(29.34)	(43.72)

Note: data in brackets denote displacement of the lower part model

4. SEISMIC RESPONSE TESTS

Both seismic response tests of the pure tube bundle model and the lower part model with water are performed. During tests the accelerometers and strain gauges are used . The first group of nine accelerometers and eleven strain gauges on rod and tube support plate and the second group of six accelerometers and ten strain gauges on rod and lower shell are used in the seismic response tests and shown in Figures 6 and 7.
The water proof measure is considered for accelerometers and strain gauges to insure the accuracy of determined data.

4.1 Exploratory Tests

Before seismic response tests , the exploratory tests are performed. The selected swept frequency range is 0 to 100 Hz. Accelerations of 0.1g and 0.2g for white noises are excited to identify the non-linear effects of the models under earthquake loadings. The natural frequencies and corresponding damping ratios at measured points are shown in Tables 7 and 8 respectively.
The natural frequencies of the first mode and the third mode of the tube bundle model are determined with the data of 16.992, 79.492 Hz in X direction , but the natural frequency of the second mode is not appeared in the test . And 15.625, 43.164, 85.330 Hz in Y direction are determined as shown in Table 7. The determined natural frequencies of the lower part model are 27.865, 37.630, 49.659 Hz in X direction and 29.102, 38.412, 50.992 Hz in Y direction respectively, as shown in Table 8.

Table 7 Natural frequencies and corresponding damping ratio of tube bundle model under 0.2g white noises

Direction	Location	T1	T2	T3	T4	T5	T6
X	f1(Hz)	16.992	16.992	16.992	16.992	16.992	16.992
	ξ	0.0313	0.0312	0.0306	0.0296	0.0282	0.0278
	f2(Hz)	___	___	___	___	___	___
	ξ	___	___	___	___	___	___
	f3(Hz)	79.492	79.492	79.492	79.492	___	79.297
	ξ	0.0285	0.0281	0.0281	0.0283	___	0.0211
Y	f1(Hz)	16.221	15.820	15.625	15.625	15.625	15.625
	ξ	0.0257	0.0336	0.0336	0.0349	0.0336	0.0297
	f2(Hz)	43.459	43.164	43.164	43.359	43.164	42.578
	ξ	0.0189	0.0245	0.0248	0.0238	0.0234	0.0257
	f3(Hz)	86.328	82.227	84.180	86.328	84.180	83.789
	ξ	0.0061	0.0095	0.0126	0.0160	0.0242	0.0212

Table 8 Natural frequencies and corresponding damping ratio of the lower part model with water under 0.2g white noise

Direction	Location	T1	T2	T3	T4	T5	T6
X	f1(Hz)	29.297	28.711	29.297	28.711	25.781	25.391
	ξ	0.1074	0.1118	0.0947	0.0983	0.1046	0.1554
	f2(Hz)	37.695	37.695	38.086	38.086	38.086	36.133
	ξ	0.0363	0.0362	0.0431	0.0631	0.0558	0.0482
	f3(Hz)	48.438	48.438	50.195	51.563	___	___
	ξ	0.1327	0.1327	0.1099	0.1213	___	___
Y	f1(Hz)	___	___	___	___	___	___
	ξ	___	___	___	___	___	___
	f2(Hz)	38.477	38.477	38.281	___	___	___
	ξ	0.0282	0.0282	0.0288	___	___	___
	f3(Hz)	___	___	___	51.593	50.391	___
	ξ	___	___	___	0.0529	0.0114	___

4.2 Determined Stresses

Determined stresses of the pure tube bundle model and the lower part model at the measured points are shown in Figure 6 and 7 and listed in Table 9

Table 9 Determined stresses of tube bundle model and lower part model with water

Location	T10	T11	T12	T13	T14	T15	T16	T18	T19
Determined strain ε(μ)	100.62	93.10	99.85	78.75	43.72	55.59	29.80	52.41	49.30
Determined stress σ_e (MPa)	19.72	18.25	19.57	15.44	9.57	10.90	5.84	10.27	9.66
Calculated stress σ_e (MPa) (tube bundle model)	21.23	___	___	21.23	3.89	___	___	___	3.89
Determined strain ε(μ)	128.69	93.43	152.67	130.44	82.82	99.94	32.81	88.97	125.95
Determined stress σ_e (MPa)	25.22	18.31	29.91	25.57	16.23	19.59	6.43	17.44	24.69
Calculated stress σ_e (MPa) (lower part model)	12.93	___	___	12.93	4.17	___	___	___	4.17
Determined strain ε(μ)	225.2	248.5	342.1	346.1	192.9	245.7	110.6	224.3	285.6
Determined stress σ_e (MPa)	44.15	48.71	67.04	67.84	37.80	48.16	21.68	43.95	55.98
Calculated stress σ_e (MPa) (Fragility test of lower part mode)	25.86	___	___	25.86	8.34	___	___	___	8.34

5. DISCUSSION AND CONCLUSION

(1) The input response spectra simulate well the earthquake loadings for both models (see Figures 3).

(2) The coherence factors are very small (see Figure 8), it is indicated that the combined effects of single direction earthquake loadings are consistent with three direction input earthquake loadings.

(3) The fundamental frequency components are dominant, the second and the third frequency components have different contributions at different measue points for both models under earthquake loadings (see Tables 7& 9 and Figure 9). The determind frequencies of the tube bundle model and the lower part model with water are in good agreement with the corresponding calculated results. The determined stresses are with the same order of magnitude as the corresponding calculated results. The maximum stresses appear at the tangency of bend and straight part of rod.

(4) The fundamental frequencies of both models exited by white noises with different accelerations have no significant effects. The corresponding global non-linear effect of both models is insignificant . The global dynamic behavior can be basically obtained by linear seismic response analysis.

(5) In comparison between the lower part model with water and the pure tube bundle model, the fundamental natural frequency of the former is increased by 64 %. It is indicated that the structural modifications can effectively strengthen the anti-seismic ability of tube bundle.

(6) Because the rod pitch is enlarged, added mass of the rods for both analytical and experimental results are insignificant.

(7) The effect of impact between rod and the tube support plate and real damping value of the rod will be considered in the future.

Reference

1. Liangbi, H., Fugao,Z., and Jinkang, X., 1994, " Seismic Analysis and Test of Model Steam Generator Tube Bundle" CNIC-00828, SNERDI-0024, China Atomic Energy Press (in Chinese).
2. Slot, T. and O'Donnel , W.J., 1971, " Effective Elastic Constants for Thick Perforated Plates with Square and Triangular Penetration Pattern ", JEI, Trans. ASME, Vol. 93, No. 4.
3. Timoshenko,S., 1940 , " Theory of Plates and Shells ", Mcgraw- Hill Book Company, Inc..
4. Chen, S. S., 1975, " Vibration of Nuclear Fuel Bundles", Nuclear Engineering and Design, Vol. 35, pp399-422.
5. " Damping Values for Seismic Design of Nuclear Power Plants", Regulatory Guide 1.61, U.S. Atomic Energy Commission, October 1973, Rev. 1, July , 1981.
6. NUREG-0800 (formerly NUREG-75/087), July, 1981, " Standard Review Plan" , 3.7.1, U. S. Nuclear Regulatory Commission, Office of Nuclear Reactor Regulation.
7. "Experimental Report of Seismic Response for Model Steam Generator Tube Bundle", 1995, Tong Ji University and Shanghai Nuclear Engineering Research and Design Institute.
8. ANSYS, Swanson Analysis Systems, Inc. Revision 4.4.

Figure 1-a photograph of tube bundle model

Figure 1-b photograph of lower part model

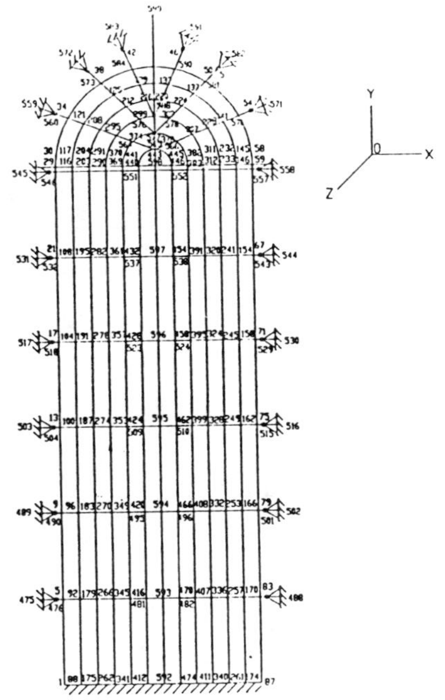

Figure 2. mathematical model

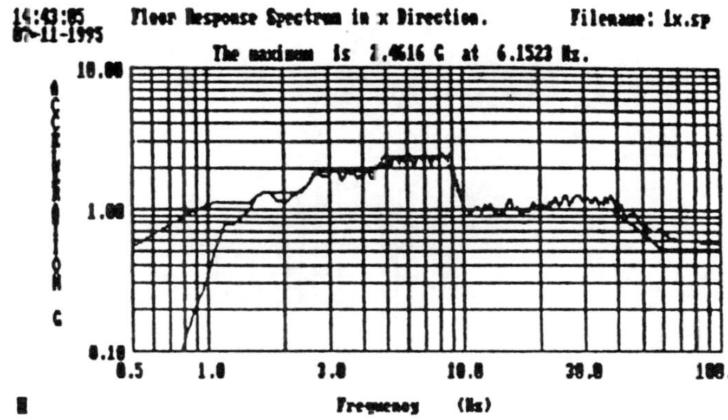

a) x-direction

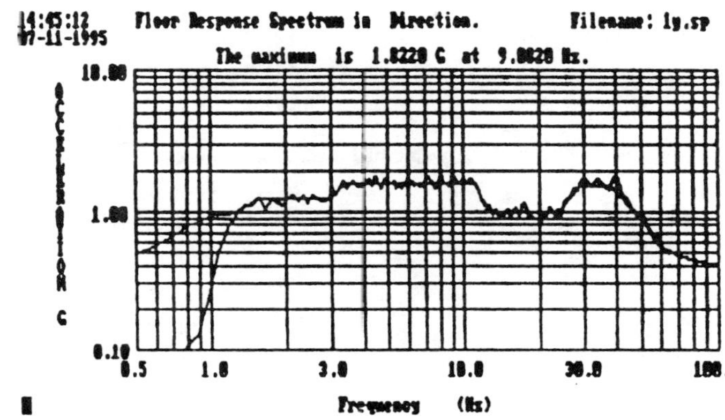

c) z-direction

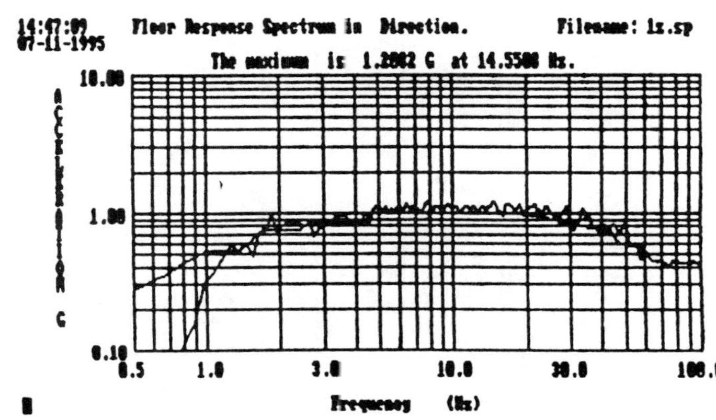

b) y-direction

Figure 3. acceleration response spectra

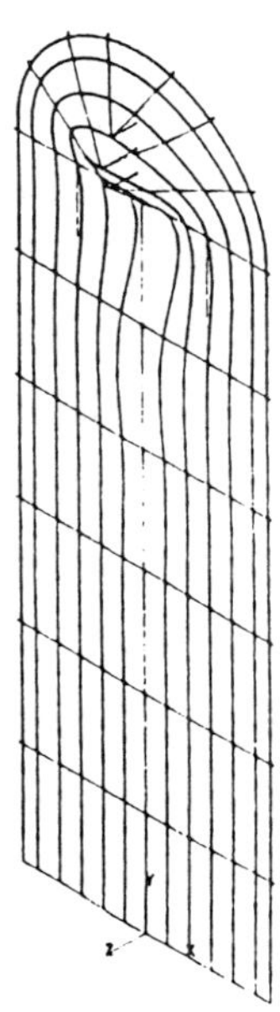

Figure 4. first normal mode of pure tube bundle model

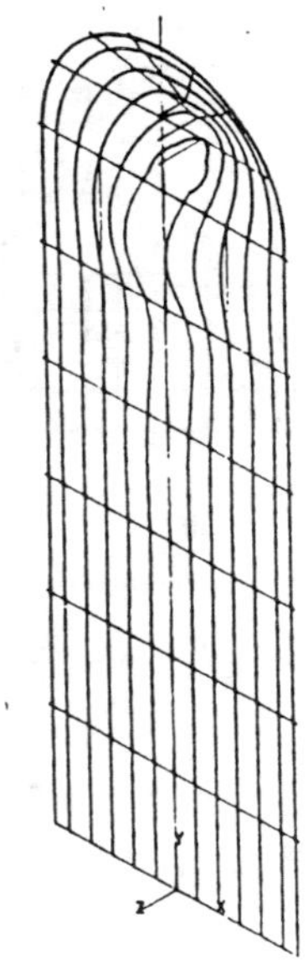

Figure 5. first normal mode of lower part model

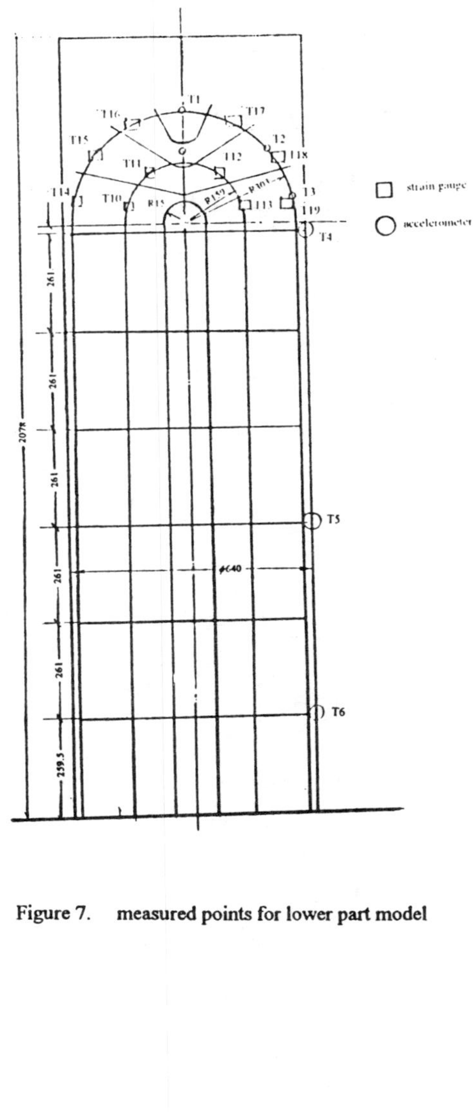

Figure 7. measured points for lower part model

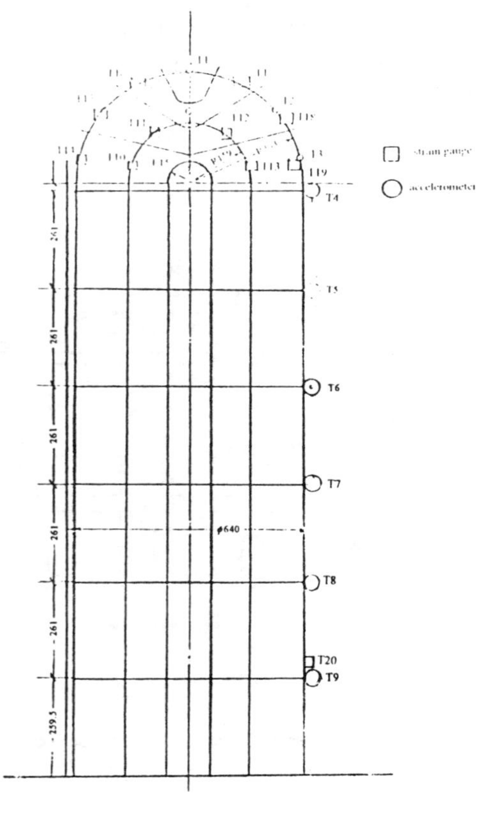

Figure 6. measured points for pure tube bundle model

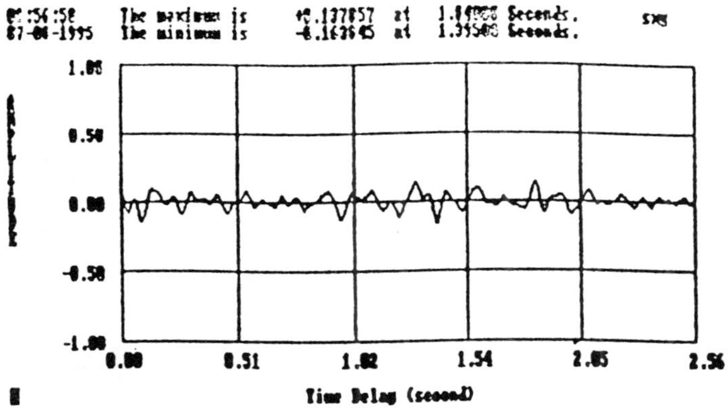

a） X与Y

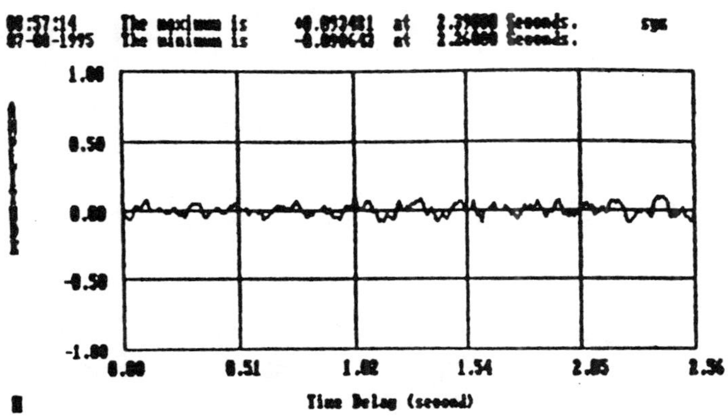

b） Y与Z

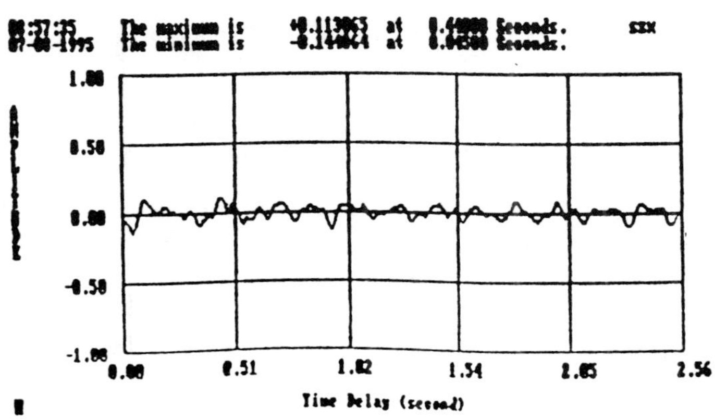

c） Z与X

Figure 8. correlation functions of artificial earthquake waves

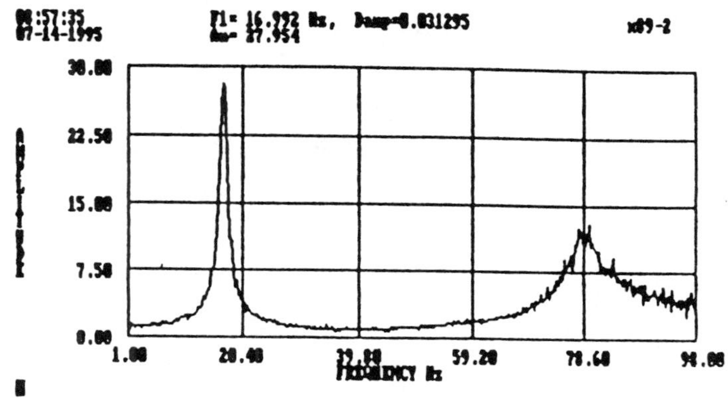

a). x direction for pure tube bundle model

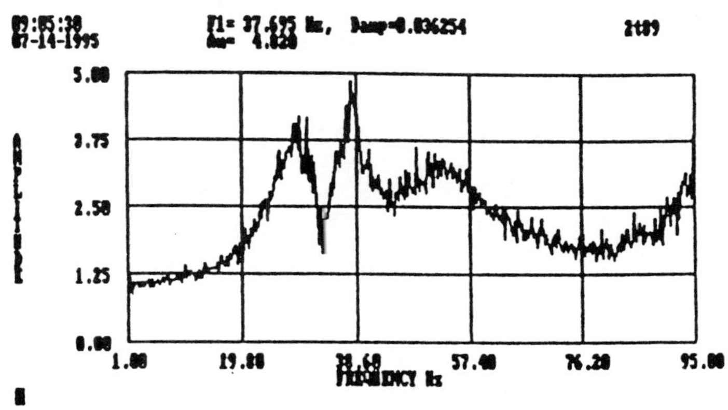

b). x direction for lower part model with water

Figure 9. transfer functions excited by white noise for measured point T1

PVP-Vol. 340, Seismic Engineering
ASME 1996

Performance of Two MOV Lubricants

James B. Gleeson
Battelle Memorial Institute
Columbus, Ohio 43201, USA

Timothy Cline
Duke Power Company
Charlotte, NC 28021, USA

ABSTRACT

Recent testing at Battelle has shown that a synthetic grease (Mobil 28) provides both improved friction coefficient and lower rate-of-loading (ROL) characteristics. Duke Power nuclear plants currently use an antiseize (N5000) as a motor operated valve (MOV) stem lubricant but they are considering changing to Mobil 28. However, it is not practical to completely remove all the residual antiseize from the stem threads when changing to a new lubricant. The testing performed during this study was to determine the friction coefficient and ROL effects for Mobil 28 when applied to a valve stem with residual N5000. This testing demonstrated how mixtures of Mobil 28 and N5000 perform under varying MOV conditions.

Grease performance was determined from standardized ASTM tests and from simulated MOV tests. Standardized grease tests were performed to determine if an incompatibility exists between the two greases and to determine the sensitivity of particular ratios of these greases mixed together on relative performance. The simulated MOV tests determined the friction coefficient and ROL effects for each grease and various mixtures of the two greases. Battelle's MOV test stand provides a means to test valve actuators and stems with representative valve load profiles and to accurately measure the actuator performance. These tests were conducted at a high torque switch setting on a Limitorque SMB-0. Stem-stemnut pairs with known rate-of-loading (ROL) effects ranging from approximately zero to 30% were tested. Stem temperature, number of strokes, and stemnut material were varied to establish their influence on grease performance.

Data from these experiments were compared with results of previous tests which used different stem lubricants. The results with Mobil 28 yielded consistent, reduced ROL effects. These results agreed with the trend predicted by a squeeze film model of the interaction at the stem-stemnut interface. The model shows that the basic rheological properties of the lubricant, the thread composite surface roughness, and the thread geometry all have a significant influence on the magnitude of ROL effects. However, under certain conditions, N5000 produced unexpected results that could not be modeled.

BACKGROUND

The actuator thrust output during motor-operated gate and globe valve closures is typically limited by use of a torque switch. In a typical actuator, torque is applied by an electric motor through a gear train to rotate a stemnut. The stemnut, in turn, drives the valve stem in translation, thus opening or closing the valve. The magnitude of sliding friction at the stem-stemnut interface depends on a number of parameters relative to the proportion of shear of boundary films and hydrodynamic films. Such factors include: lubricant reactivity with the stem and stemnut materials (including EP or other additives), surface roughness, aging of the lubricant, load, lubricant viscosity, operating temperature, and lubricant creep or bleeding, which influence the availability of lubricant in the critical contact region. Using experimental data on stemnut friction measurements for different load, speed, and viscosity conditions, Ham and Ryan (Ham and Ryan, 1932) demonstrated that very little hydrodynamic effects exist at the stem-stemnut interface. Nakashima (Nakashima et al., 1991) reached a similar conclusion by demonstrating that stemnut wear coefficients fall in line with those expected from boundary lubricated components.

However, results of previous MOV testing at nuclear plants (Black, 1990; Dogan and Hosler, 1992) and in laboratories (Steele et al., 1992) have shown that the stem load-time history can affect the thrust output of torque switch controlled MOVs. The effect has been called "rate-of-loading" or "load-sensitive behavior" and has generally been attributed to changes in the stem-stemnut friction coefficient. These changes in the friction coefficient affect the efficiency with which operator output torque is converted to stem thrust.

The term "rate-of-loading" (ROL) is used to describe the variation in MOV output thrust between a static and a valve closure at differential pressure (DP) and is defined as the percent change in thrust from static to DP conditions. For this report, ROL is always evaluated at torque switch trip (TST) as:

$$\text{ROL \%} = \frac{\text{static thrust - DP thrust}}{\text{static thrust}} * 100 \qquad (1)$$

The ROL phenomenon is a concern because in-plant testing of MOVs to verify that sufficient thrust is developed at TST is often conducted under conditions which are different from those under which the MOV is required to function. For example, valves are often tested under "static" conditions of zero flow and zero differential pressure (DP). As the valve is closed under static conditions, the gate or globe seats and the stem load increases rapidly (> 100 kip/s). However, when the valve is required to close against design basis conditions (high flow and high DP), the stem load will develop more slowly. Under DP conditions, the torque switch may trip at a value of stem thrust which is significantly lower than in the static test. Therefore, evaluation of data from *in situ* tests needs to consider the potential thrust differences which may occur between static tests and valve closures under design basis conditions.

A temporary reduction in stem-stemnut friction coefficient during valve seating was documented in experiments studying the ROL effect (EPRI, 1995). The hypothesis presented was that rapid loading of the thread contact area can result in a thin pressurized lubricant film that temporarily supports the stem load. Depending on the properties of the stem-stemnut surfaces, the lubricant rheology, and the loading rate during seating, this friction coefficient transient may last up to the full seating duration (several hundred milliseconds). During this time, the effective friction coefficient is significantly lower than for pure boundary lubrication. This squeeze-film phenomenon was proposed as the principal factor related to the ROL effect.

Recent testing at Battelle (Gleeson, 1995) has shown that Mobil 28 provides both improved friction coefficient and lower ROL characteristics. Duke Power nuclear sites currently use Felpro N5000 as a MOV stem lubricant but they are considering changing to Mobil 28. However, it is not practical to completely remove all the N5000 from the stem threads when changing to a new lubricant. Therefore, Duke Power carried out a study to establish the appropriateness of a new stem lubricant with residual N5000 on the stem.

TEST DESCRIPTION

MOV Test Facility. The testing was conducted at Battelle on specially constructed MOV test stands. The facility was designed to provide a means to test valve actuators and stems with simulated load profiles and to accurately measure the actuator performance under these conditions. Figure 1 shows a exposed view of a MOV test stand. The principal elements of the facility are the support frame, the simulated valve yoke, the hydraulic loading system, and the actuator and stem. The simulated valve yoke and top and bottom plates form the structural frame used to react the loads developed by the actuator. It was designed to be representative of a rigid valve/yoke assembly. Load profiles simulating valve operation are applied by a hydraulic cylinder attached to the lower part of the stem. The load is controlled by a closed loop servo-hydraulic system, so that pre-determined load profiles can be applied. The support frame also includes a simulated valve seat against which the stem can be loaded. The stiffness of the seat can be changed by adjusting the position of the stops and by substituting different stop beams.

The stems that were tested were typical of hardware used in nuclear service. The stems were made of 17-4PH stainless alloy and had standard ACME threads. These stems had known rate-of-loading effects ranging from approximately zero to 30% (Gleeson, 1995). Stemnuts were fabricated using standard industry methods from Limitorque manganese bronze stemnut "blanks" or Rotork-specified aluminum bronze material. All the tests were performed using a Limitorque SMB-0 motor operator. The custom aluminum bronze stemnuts were used in order to reproduce the stem-stemnut interface conditions of Rotork operators.

Instrumentation. The principal measurements for these tests were actuator output thrust, actuator output torque, and detection of torque switch trip

(TST). Actuator output thrust and torque were measured using a torque/thrust cell (TTC) mounted between the actuator and the simulated yoke. The combined nonlinearity, hysteresis, and crosstalk errors for the TTC was less that 0.4%. Torque switch trip was detected with a custom electronic circuit which sensed frequency changes in the motor contactor holding coil current. The contactor holding coil current was also recorded.

It is important to measure torque switch trip precisely because TST is used as a standard reference point for comparing MOV performance. The limiting factor in resolving data at TST was the sampling frequency of the data acquisition system. The sampling frequency used for the tests was 500 Hz which allowed a minimum resolution of 200 lbs for the thrust at TST (assuming a 100 kip/sec loading-rate during seating). This relates to an uncertainty of ±0.5% on ROL measurements. In any case, data at TST are always plotted as two points that define the uncertainty range due to the sampling rate.

Typical Operation and Types of Tests. Operator performance tests were run in the direction of closing a valve. Tests were typically started by energizing the actuator with the stem positioned two to four inches from the seat to achieve about 15 seconds of running at load prior to seating. The operator torque switch was set for each stem to achieve a final thrust output of approximately 22 kip from a 0.5 kip running load static test. The stem was then driven in the open direction by the actuator with a slightly lower compressive load on the stem. Data were recorded in the closing direction from 1 sec. before motor actuation until after the motor stopped. Stems were cleaned and lubricated prior to each new grease mixture tested but they were not lubricated between individual tests.

Instead of trying to simulate actual valve stem load histories in detail, a few basic load profiles were used. The two profiles which provided the most information were a constant running load (static test) and a linearly increasing load (ramp test). In the static test profile, the stem load remains constant with time until the stem contacts the seat, whereupon stem load increases rapidly. The ramp load profile produces an increasing stem thrust over the stroke. The ramp rate was constant at approximately 2 kip/sec for these tests. The stroke length and the initial ramp value were set to cause the torque switch to trip before contacting the seat. This load profile simulates a DP test and the loads are sufficiently high to generate a stem-stemnut friction coefficient representative of design basis conditions. Figure 2 shows two representative ROL tests; one for a stem-stemnut that displays no ROL effects (Stem a with Mobil 28) and one with considerable ROL effects (Stem h with N5000).

The behavior of a MOV at torque switch trip is affected by the instantaneous load and also by the load earlier in the stroke. To document these effects, each test series was conducted by increasing the running load in succeeding static tests, e.g., 0.5, 1, 3, 5, 7, 10, and 13 kip (up to the maximum load allowed by the torque switch setting), For stems that display ROL effects, each increase in running load results in a decrease in the output thrust and this trend continues until the output thrust approaches the thrust level developed from a ramp test. These static tests were followed by a ramp test to establish the DP performance, and to calculate the ROL effect.

Another phenomenon related to valve stem lubrication is called the memory effect (Steele et al., 1992). This is manifested when the stem friction coefficient is determined by the load history from a previous valve stroke. MOV studies at INEL have shown that two low-load static tests are sufficient to erase memory effects (if they are exhibited). In order to account for possible memory effects, two low-load static tests (0.5 kip) were added to each test series followed by another ramp test. If memory effects are involved they would cause a significant difference in the friction coefficient between the two ramp tests.

A typical test series consisted of static tests with increasing running load in succeeding tests, a ramp test, two low-load static tests, and another ramp test. A test series, then, consisted of up to 11 individual tests which were repeated for varying parameters of stem geometry, stemnut material, stem temperature, lubricant mixture, and number of MOV cycles between each test series.

DISCUSSION OF RESULTS

Standardized grease tests. Standardized grease tests were performed to determine if a chemical incompatibility exists between Felpro N5000 and Mobil 28. These tests should also indicate if particular ratios of these greases mixed together might result in a reduction of performance. Five combinations of N5000 and Mobil 28 were tested in the following ratios : 100:0, 90:10, 50:50, 10:90, and 0:100 (by weight). Five standardized tests were performed on each proportioned grease mixture. The oil separation or bleed test (Federal Test Method Std. No. 791C) measures the percent weight loss of oil drained through a wire screen cone at

100°C for 30 hours. The dropping point test (ASTM D566) determines the temperature at which a grease melts or passes from a semisolid to a liquid state. The four ball wear test (ASTM D2266) determines the relative wear preventing properties of lubricants by rotating a steel ball under load against three stationary balls and measuring the diameter of the wear scars on the stationary balls. The worked penetration test (ASTM D217) determines the consistency of a grease after 60 strokes by penetration of a cone, measured in tenths of a millimeter. This test is also used to determine the grease grade classification, where 310-340 (tenths of a millimeter) is defined as grade 1. The roll stability test (ASTM D1831) determines the change in consistency of a grease after 2 hours of milling at 165 rpm and 50g load followed by a cone penetration test.

The standardized grease test results are graphically summarized in Figure 3. The general appearance of the grease mixtures indicated that N5000 and Mobil 28 are miscible. Also, wear (steel-on-steel), roll stability, and worked penetration are not influenced by mixing these greases in any proportion. Furthermore, when Mobil 28 is added to N5000, particularly for mixtures of 50 percent or greater ($\geq$ 50% Mobil 28), grease bleed and dropping point performance improve significantly. These results suggest that MOV performance should only improve with the addition of Mobil 28.

Friction Coefficient and ROL Effects. There is a fundamental relationship between the valve stem friction coefficient (μ) at TST and the resulting thrust at TST. ROL occurs when the friction coefficient drops significantly during seating in static tests. For example, Stem h displays ROL effects and its μ transient when lubricated with N5000 is shown in Figure 4. This figure shows friction coefficient as a function of time for a static and a ramp test. Figure 4a clearly shows the trend for low load static tests to undergo a (temporary) reduction in friction coefficient. This trend has been similarly demonstrated for all stems tested that demonstrate an ROL effect. A friction coefficient of 0.03 (at TST) is significantly below a value reflecting boundary lubrication. In contrast, the ramp test demonstrates a consistent friction coefficient of approximately 0.14 (a value expected for this boundary lubrication condition).

Load prior to seating is a factor in determining the magnitude of ROL for stem-stemnut pairs that demonstrate this effect. This is why a typical test series was conducted by increasing the running load in succeeding static tests, e.g., 0.5, 1, 3, 5, 7, 10, and 13 kip (some series were truncated because high ROL necessitated a low torque switch setting). For each increase in running load, the output thrust decreased for stem-stemnut combinations that demonstrated ROL. This trend continued until the output thrust approached the level obtained from a ramp (DP) test. The reduced thrust and its respective ROL effect correlate to a change in the stem-stemnut friction coefficient during seating. Stem h displays high ROL and Figure 4a clearly showed a transition to a low friction coefficient at torque switch trip for low static running loads conditions. Figure 5 is a plot of friction coefficient at TST and the respective ROL as a function of running load for Stem h lubricated with N5000 and with Mobil 28. As the running load is increased, the μ at TST approaches a peak that is equivalent to the ramp (DP) friction coefficient. Figure 5a shows that seating of the valve from a low load (static) condition causes departure from an expected (boundary lubrication) friction coefficient. This figure also shows that the lubricant significantly affects the magnitude of friction coefficient transitions and, hence, the magnitude of the ROL effect. In Figure 5b, low ROL is identified with low load static tests because the highest thrust output is used as the reference for calculating ROL.

Typically, the friction coefficient, both running and during seating (i.e., μ at TST), is constant for stems that do not demonstrate ROL effects. As discussed, μ at TST varies considerably as a function of the load prior to seating for stems that demonstrate ROL effects. However, even stems that display ROL effects usually have repeatable friction coefficients, both running and at TST, for tests run at the same condition. Figure 6 summarizes the friction coefficient results from three successive test series. These tests used Stem h lubricated with N5000 or Mobil 28 and each succeeding test series was preceded by 50 additional valve strokes. As shown in Figure 6a, the maximum running μ was repeatable and constant (independent of running load) for Stem h lubricated with Mobil 28. The maximum running μ also agrees with the friction coefficient at TST from a ramp (DP) test. However, the maximum running friction coefficient for Stem h lubricated with N5000, shown in Figure 6b, had a significantly different trend. With N5000, only the low load running μ agreed with the friction coefficient from a ramp (DP) test (but at a slightly higher level than Mobil 28). It can also be seen from Figure 6b that the maximum running μ was increasingly higher for static tests at a load of 3, 5, and 6 kip.

The variability in running friction coefficient (comparing similar static tests from different test series) is related to a valve stem lubrication phenomenon called the memory effect. This is manifested when the stem friction coefficient is determined by the load history from a previous valve stroke. Two ramp tests were run in order to account for possible memory effects. The first ramp followed the highest static load test (high load history) and the second ramp test was preceded by two low load static tests (low load history). If memory effects are involved, they would cause a significant difference in the friction coefficient between two, otherwise equivalent, ramp tests. Figure 7 shows the ramp test friction coefficients from three successive test series using Stem h lubricated with N5000, where the second and third test series were preceded by 50 additional valve strokes (at 4 kip). Obviously, there is a significant variation in the running μ and the μ at TST in these ramp tests. This effect was not observed with Mobil 28 nor with mixtures of Mobil 28 and N5000 with > 50% Mobil 28. Nevertheless, the occurrence of this effect may be random and does not become more likely with the number of strokes (which could be interpreted from Figure 7). Another example of the memory effect is shown in Figure 8, which shows the ramp test friction coefficients from three test series using Stem hR (Stem h with a different stemnut) lubricated with N5000. It can be seen from Figure 8 that the memory effect appears to be a minimum after 50 strokes and hence, not clearly increasing with increasing strokes. In any case, memory effects were only observed with N5000.

Mixtures of N5000 and Mobil 28 on Stem h. One of the principal objectives of this study was to determine the friction coefficient and ROL effects for Mobil 28 when applied to a valve stem with residual N5000. The initial MOV stem-stemnut tests used the same ratios of grease mixture defined for the standardized grease tests. Friction coefficient and ROL were calculated to document changes attributed to each particular grease mixture.

Stem h was tested first because it displays the largest ROL effect measured in a laboratory environment. Figure 9a summarizes the relative running friction coefficients measured as a function of grease mixture; Figure 9b summarizes the relative μ at TST. These plots display ramp data at loads $\geq$ 2 kip in order to represent DP performance. Each lubricant condition (percent N5000 in grease mixture) consists of data from three separate test series (again, where succeeding test series are preceded by 50 additional valve strokes at 4 kip). Certainly, N5000 lubrication displays significantly higher friction coefficients (both running and at TST) compared to Mobil 28. The curve fits plotted in Figure 9 represent a linear fit of the maximum and minimum friction coefficients to describe their variation as a function of the percent grease mixture. Most important, though, is the reduction in frictional variability associated with Mobil 28. The frictional variability associated with N5000 is problematic because it increases the uncertainty in predicting MOV performance. The variability in friction coefficient (comparing similar ramp tests from different test series) is primarily due to the memory effects displayed by N5000. This effect is reflected in the calculated ROL, as represented in Figure 9c. The general trend for ROL appears to roughly follow a linear relationship between the baseline performance of N5000 and Mobil 28. The reduction in variation in ROL can be seen from the relative magnitude of the error bars in Figure 9c (highest static thrust compared to lowest ramp thrust and lowest static thrust compared to highest ramp thrust).

High Temperature Performance of N5000 and Mobil 28 on Stem h. Stem h was tested at 225°F to simulate the performance of MOVs that operate at elevated temperatures. This stem was gun-bored (from the top of the stem) so that a 6 inch long cartridge heater could be centered axially inside the stemnut when the stem was positioned 4 inches off the seat. Two thermocouples were mounted on the stem surface just below the stemnut and were used to control and to record the stem temperature. This arrangement simulated MOV heating assuming that the stem was the primary conduction path. This modification allowed the stem temperature to be continuously controlled throughout the high temperature testing.

Figure 10a summarizes the high temperature, relative running friction coefficients measured as a function of grease mixture on Stem h; Figure 10b summarizes the relative μ at TST. Each lubricant condition consists of ramp data (at loads $\geq$ 2 kip) from three separate test series (where succeeding test series are preceded by 50 additional valve strokes at 4 kip). Again, N5000 lubrication displays significantly higher friction coefficients compared to Mobil 28. For N5000, the differences in maximum and minimum friction coefficients (both running and at TST) are much greater at 225°F compared to those measured at ambient conditions. Figure 10c shows the calculated ROL for Stem h, at 225°F. Again, the general ROL trend for mixtures of these two lubricants at 225°F follows a proportional relationship between the baseline performance of N5000 and Mobil 28. For both lubricants, the average ROL is reduced compared to ambient conditions. However, the variation in ROL with N5000 is significantly greater at 225°F compared with the ambient ROL, as can be seen from the magnitude of the error bars in Figures 9c and 10c.

Overall Performance of N5000 and Mobil 28. MOV testing established the baseline performance of N5000 and Mobil 28 valve stem lubrication under various load conditions. These tests also determined the influence on MOV performance for particular mixtures of these two greases. A summary of the ambient friction coefficients and respective ROL effects for Stem h was given in Figure 9. Subsequent testing on Stems a, i, and hR measured the MOV lubricant performance for N5000, for Mobil 28, and for a 50:50 mixture of the two greases. For each stem, the trend for friction coefficient for 50:50 mixtures generally followed a linear relationship between the baseline performance of N5000 and Mobil 28. Figure 11 summarizes the range of running and TST friction coefficients for these stems (again μ data is taken from ramp tests). Consistently, N5000 lubrication displays much higher friction coefficients and greater variability (for both running and at TST) compared to Mobil 28. The frictional variability with N5000 is problematic because it increases the uncertainty in predicting MOV performance. In any case, the frictional variability associated with Mobil 28 was negligible.

The influence of load and loading-rate on ROL is hardware-specific and varies with different stem and stemnut combinations. Figure 12a shows the measured ROL for Stems a, i, hR, and h using Mobil 28 and N5000. For each stem the general ROL trend for mixtures of the greases appeared to roughly follow a proportional relationship between the baseline performance of N5000 and Mobil 28. In all cases, Mobil 28 lubrication displayed little variation in ROL effects. The range of average ROL for the 4 stems tested was -1% to 26% for Mobil 28 compared to 16% to 38% for N5000. Stem h had the maximum variation in ROL and was shown by the error bars in Figures 9c and 10c. Figure 12b shows the measured ROL for Stems a, b, d, and h from previous tests (Gleeson, 1995) using Mobil 28 and Swepco Moly101. Although there were hundreds of tests on Stems a and h between the tests shown in Figures 12a and 12b, the ROL effects have not changed.

These results agree with the trend predicted by a squeeze film model (Gleeson, 1995) of the interaction at the stem-stemnut interface. The model shows that the basic rheological properties of the lubricant, the thread composite surface roughness, and the thread geometry all have a significant influence on the magnitude of ROL effects.

In reviewing data of over 60 stems, Stem h displayed the greatest ROL effect. It was expected, then, that large ROL effects (46%) would be measured for Stem h with N5000 lubrication. In a previous study, Stem h also showed the least improvement in ROL effects from changing the stem lubricant to Mobil 28. Similarly, Stem hR, which uses Stem h and a different stemnut (made from Rotork stemnut material) had a surprisingly small reduction in ROL from Mobil 28 lubrication. However, it is not likely that the stemnut material influenced the ROL.

CONCLUSIONS

Testing was performed to determine the friction coefficient and ROL effects for Mobil 28 when applied to a valve stem with residual N5000. Simulated MOV tests established baseline performance for each grease and for various mixtures of the two greases. This testing demonstrated how mixtures of Mobil 28 and N5000 perform with various stem temperatures, with increasing number of strokes, and with stemnut material.

- A significant reduction in ROL was observed using Mobil 28 as the stem lubricant. For the four stem-stemnut pairs tested, rate-of-loading effects ranged from approximately -1% to 26% for Mobil 28 compared to 16% to 38% for N5000. The mean reduction in ROL effects with Mobil 28 was approximately 50%.

- A significant reduction in friction coefficient was seen using Mobil 28 as the stem lubricant. Most important, though, is that Mobil 28 displayed little frictional variability.

- N5000 demonstrated frictional "memory" effects. This was manifest when the stem friction coefficient was determined by the load history from a previous valve stroke. This effect was only manifested for N5000 lubrication, where large variations in μ and ROL were measured under the same operating conditions. This variability is problematic because it increases the uncertainty in predicting MOV performance. On the other hand, Mobil 28 had no obvious memory effects. Also, grease mixtures of > 50% Mobil 28 had minimal memory effects.

REFERENCES

Black, B.R., "MOV Testing and the Rate of Loading Phenomenon", Power Engineering, Volume 94, Number 9 September 1990, pp 36-38.

Black, B.R., "Motor Operated Valve Testing and the Rate of Loading Phenomenon", Electric Power Research Institute Third Valve Symposium and Equipment Exhibition, August 20-23, 1990, Charlotte, North Carolina, Symposium Notebook, pp 271-310.

Dogan, T. and Hosler, J.F., "Thrust Capability and Rate-of-Loading Effect in Motor Operated Valves", Electric Power Research Institute Fourth Valve Technology Symposium, August 11-13, 1992, Dallas, Texas.

Gleeson, J., "EPRI MOV Performance Prediction Program; Motor Operator Separate Effects Testing", EPRI Document # TR-103256, February 1995.

Gleeson, J., "Lubricant Effect on Rate-of-Loading", Proceedings of the 1995 ASME/JSME Pressure Vessel and Piping Conference, July 1995.

Ham, C.W., and Ryan, D.G., "An Experimental Investigation of the Friction of Screw Threads", University of Illinois Engineering Experiment Station, Bulletin No. 247, June, 1932.

Nakashima, K., and Takafuji, K., "Wear of Lead Screw Nuts", Wear of Materials, ASME 1991, pp 589-592.

Steele, R., Russell, M.J., DeWall, K.G., and Watkins, J.C., "Load Sensitive Behavior in Motor-Operated Valves", NUREG/CP-0123.

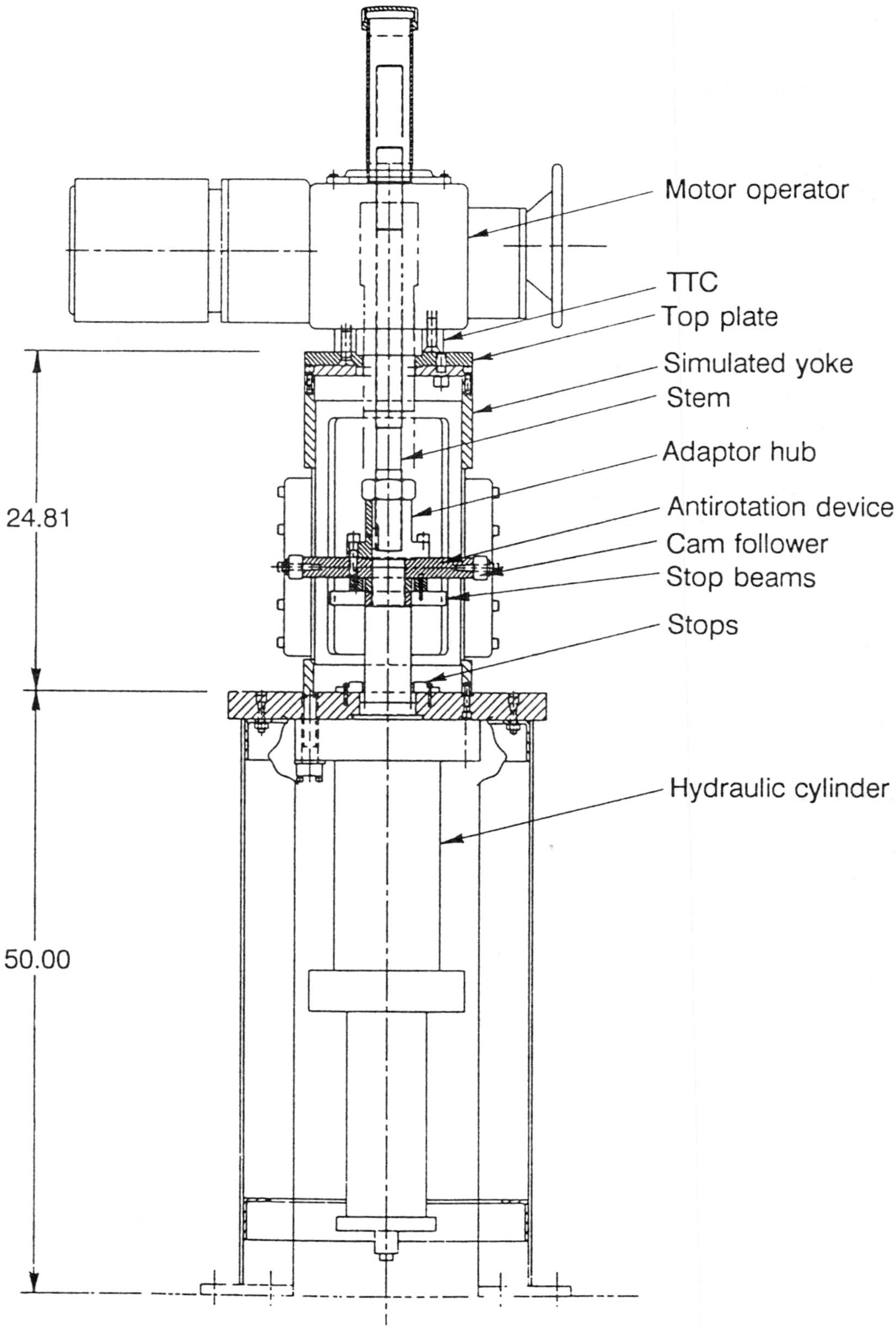

Fig. 1. Exposed view of an MOV test stand

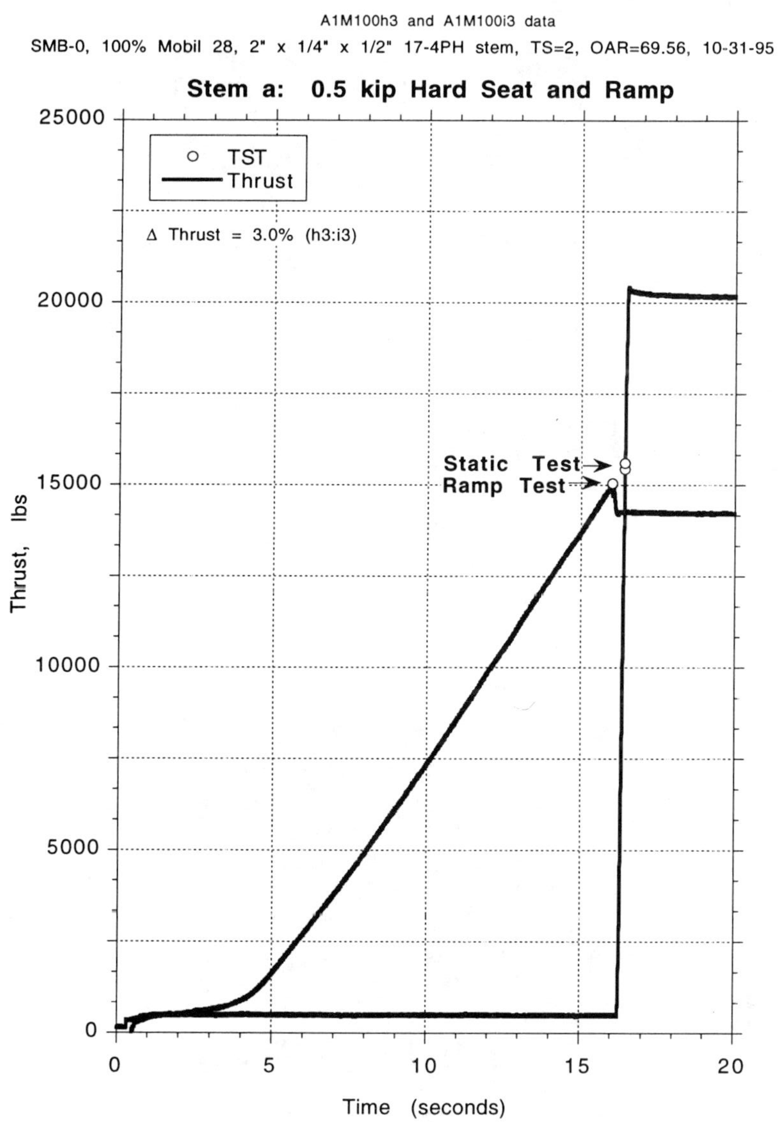

Fig. 2a. Stem h with N5000, thrust vs time for static and ramp tests

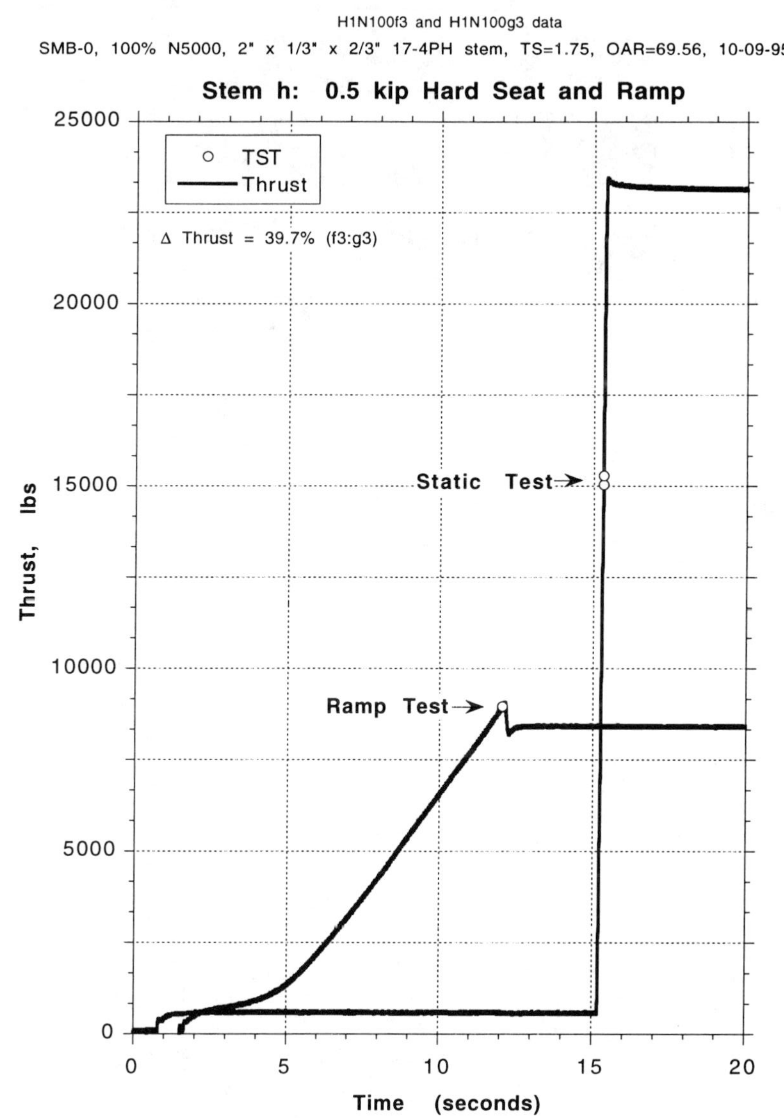

Fig. 2b. Stem a with Mobil 28, thrust vs time for static and ramp tests

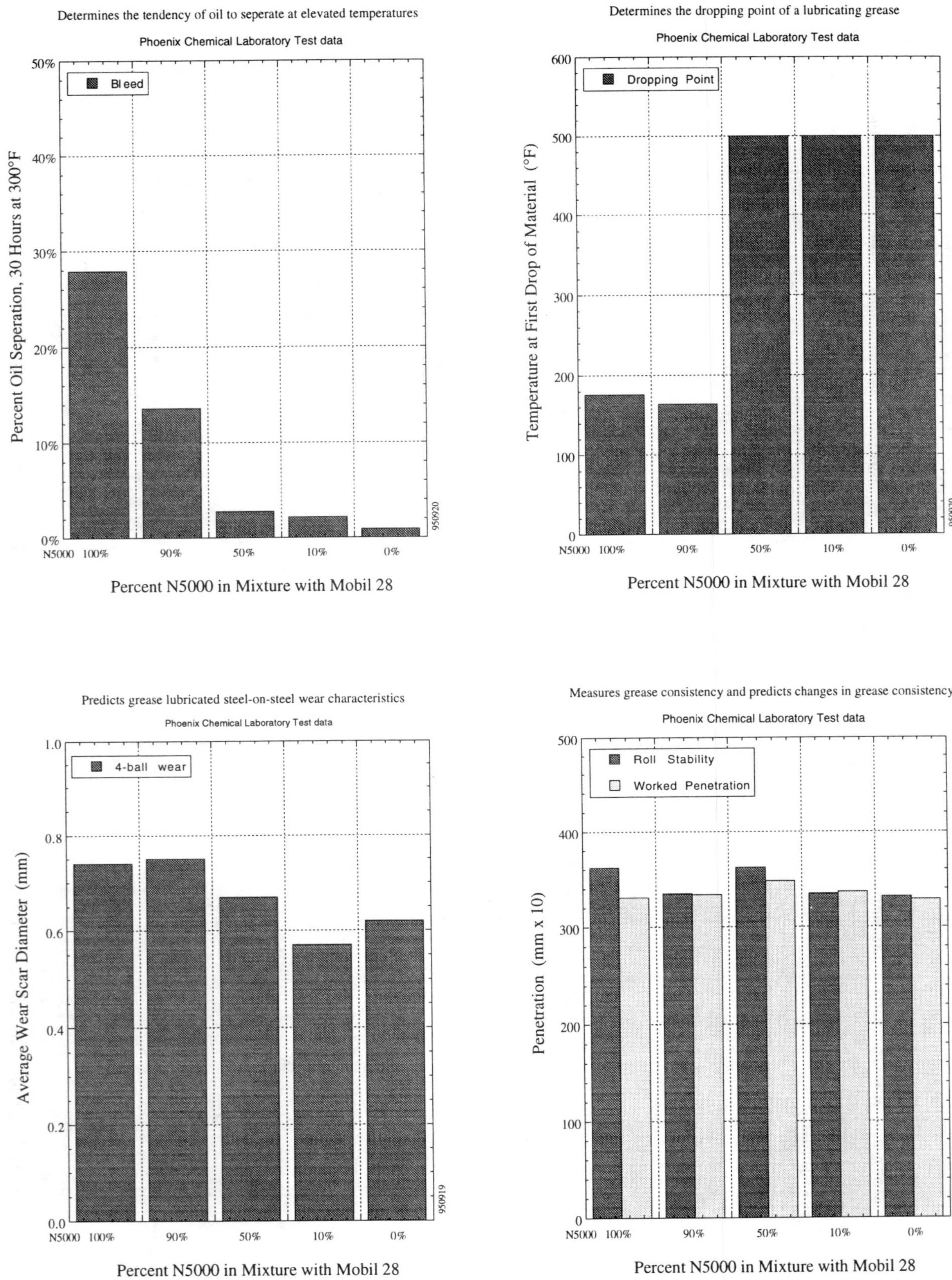

Fig. 3. Summary of standardized grease test results

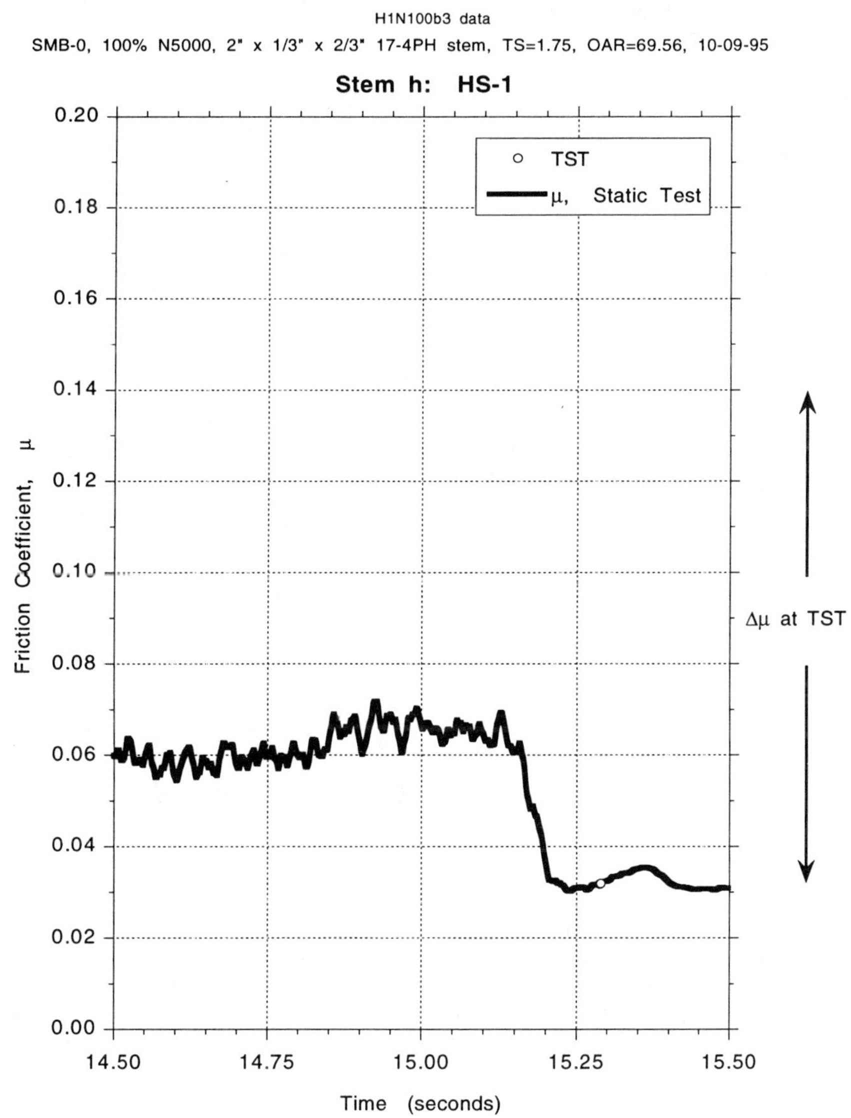

Fig. 4a. Stem h with N5000, friction coefficient vs time for a static test

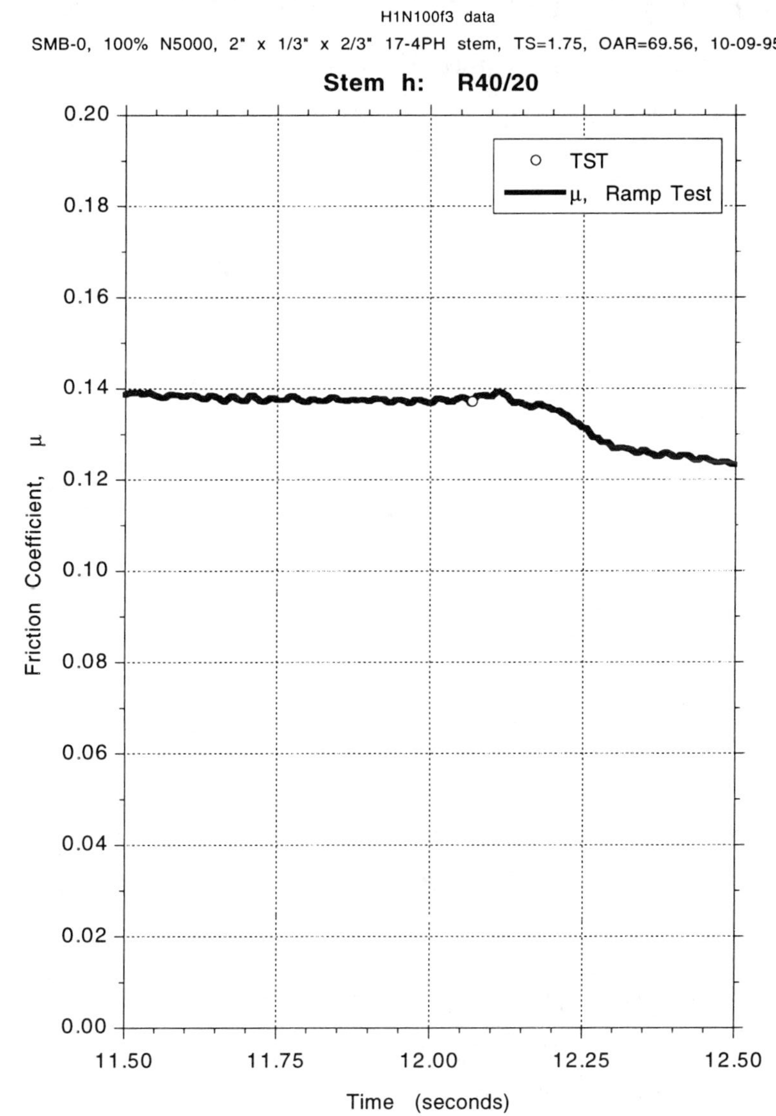

Fig. 4b. Stem h with N5000, friction coefficient vs time for a ramp test

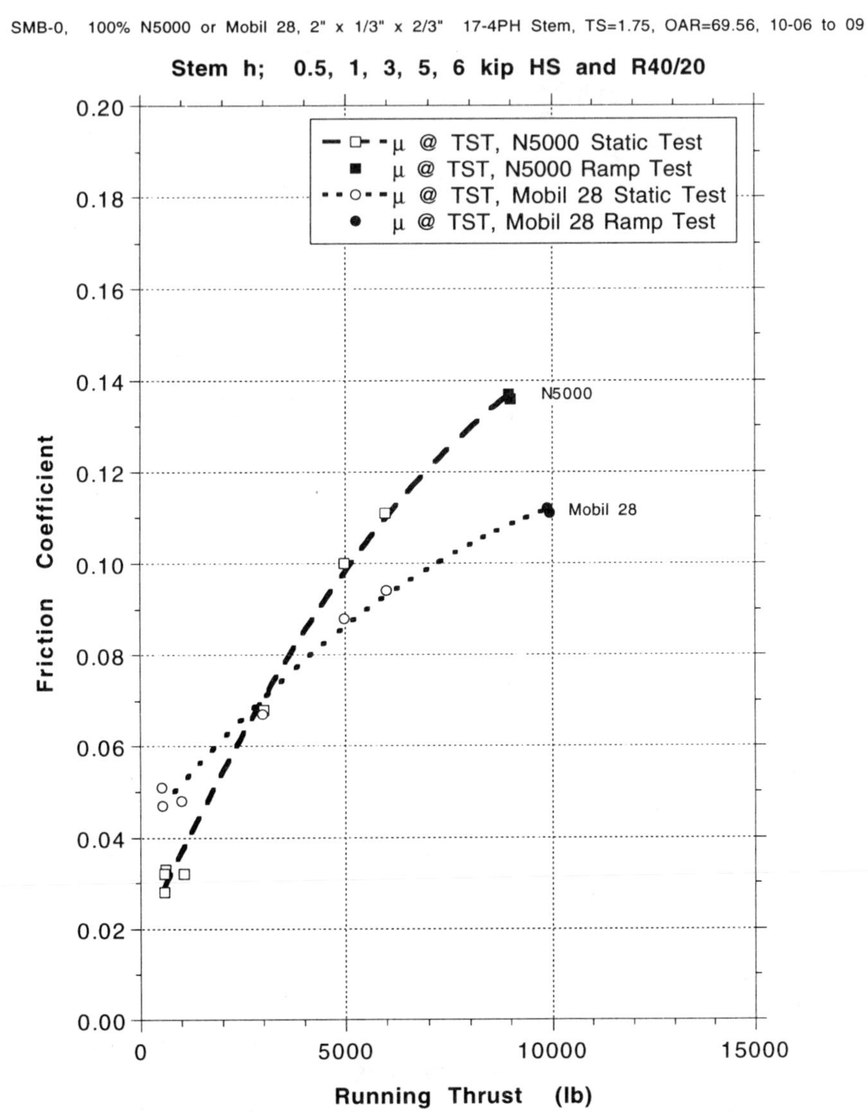

Fig. 5a. Friction coefficient vs running load for N5000 and Mobil 28 on Stem h

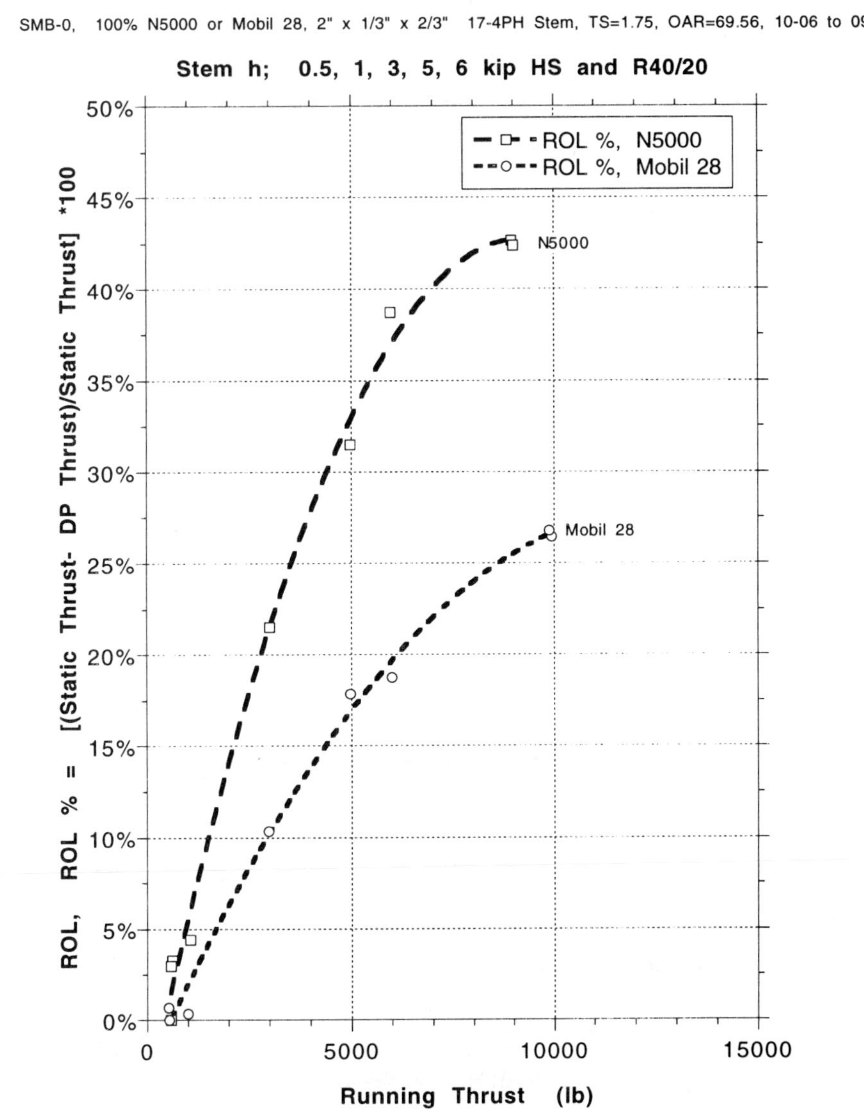

Fig. 5b. ROL vs running load for N5000 and Mobil 28 on Stem h

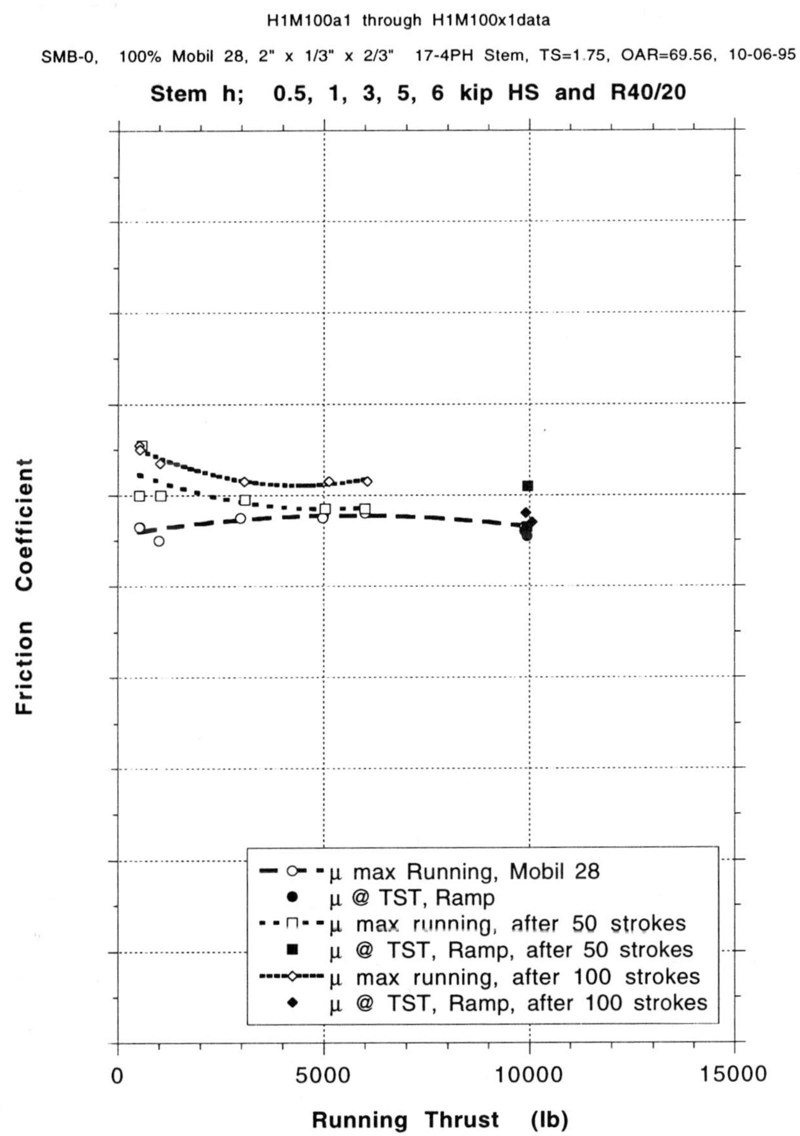

Fig. 6a. Stem h with Mobil 28, friction coefficient vs running load for three test series

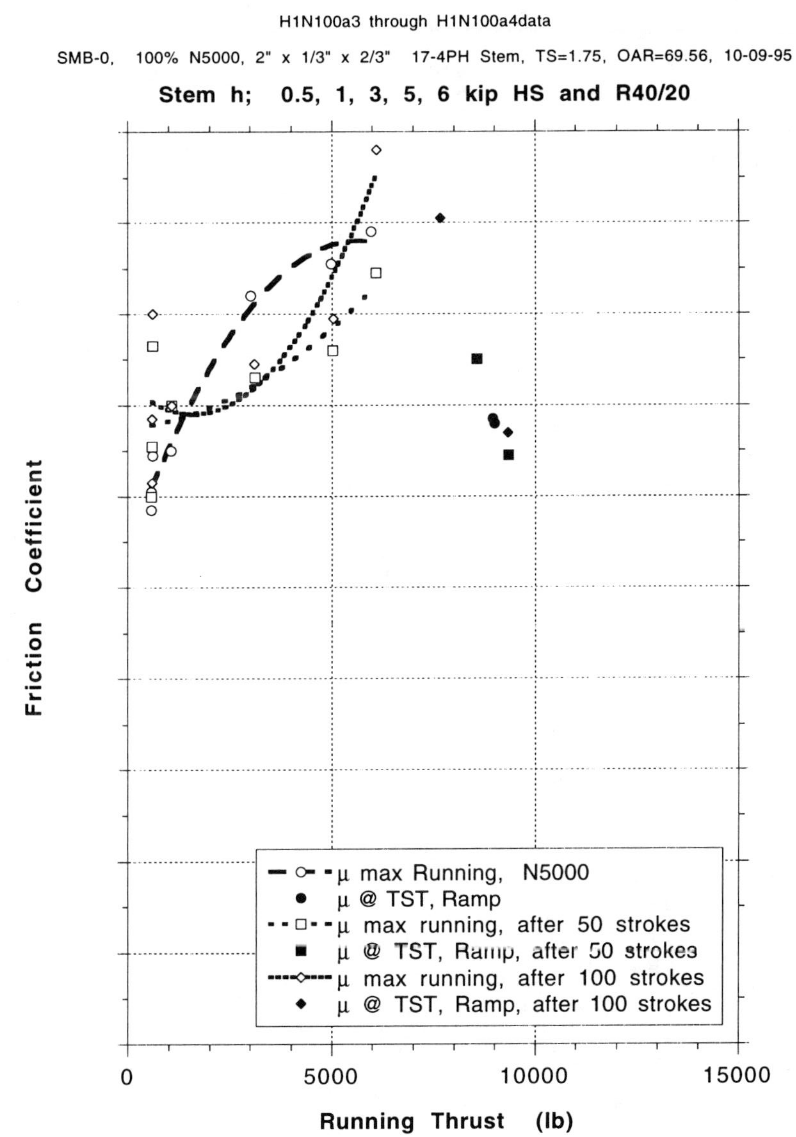

Fig. 6b. Stem h with N5000, friction coefficient vs running load for three test series

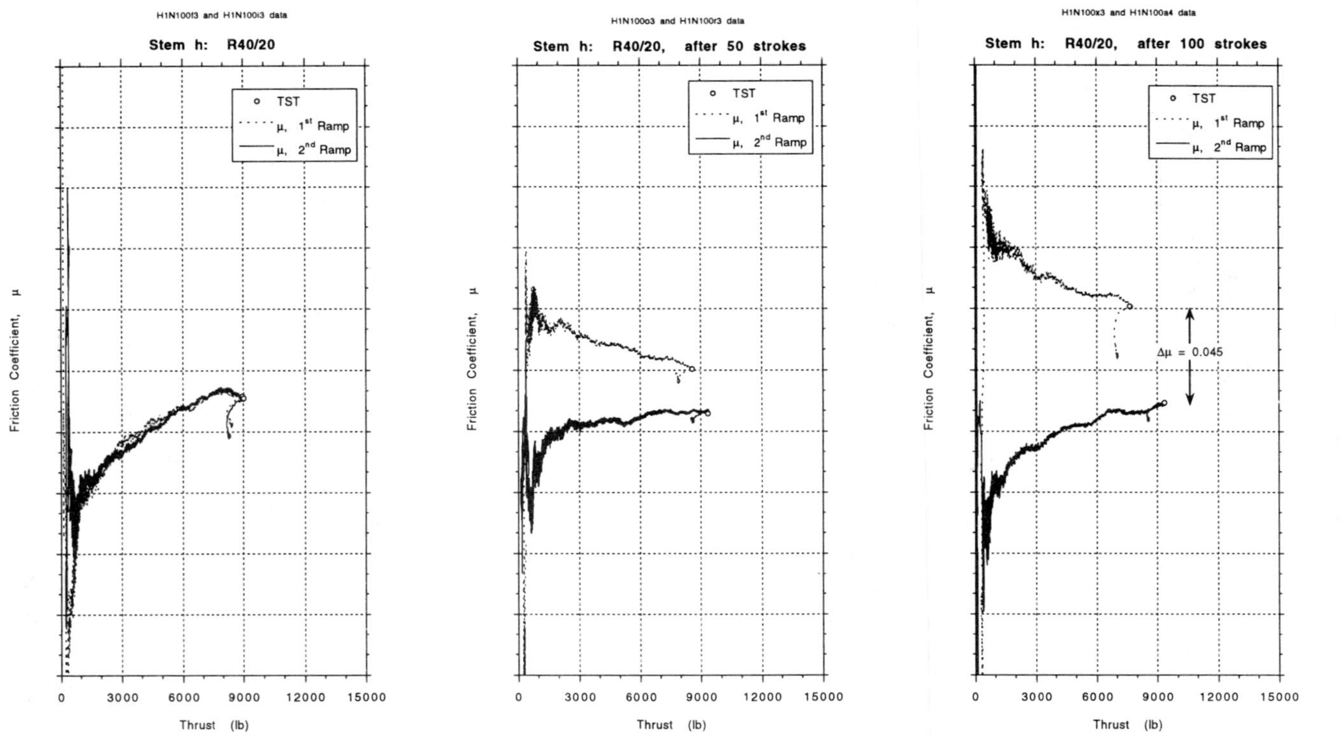

Fig. 7. Stem h with N5000, friction coefficient vs load for ramp tests from three successive test series

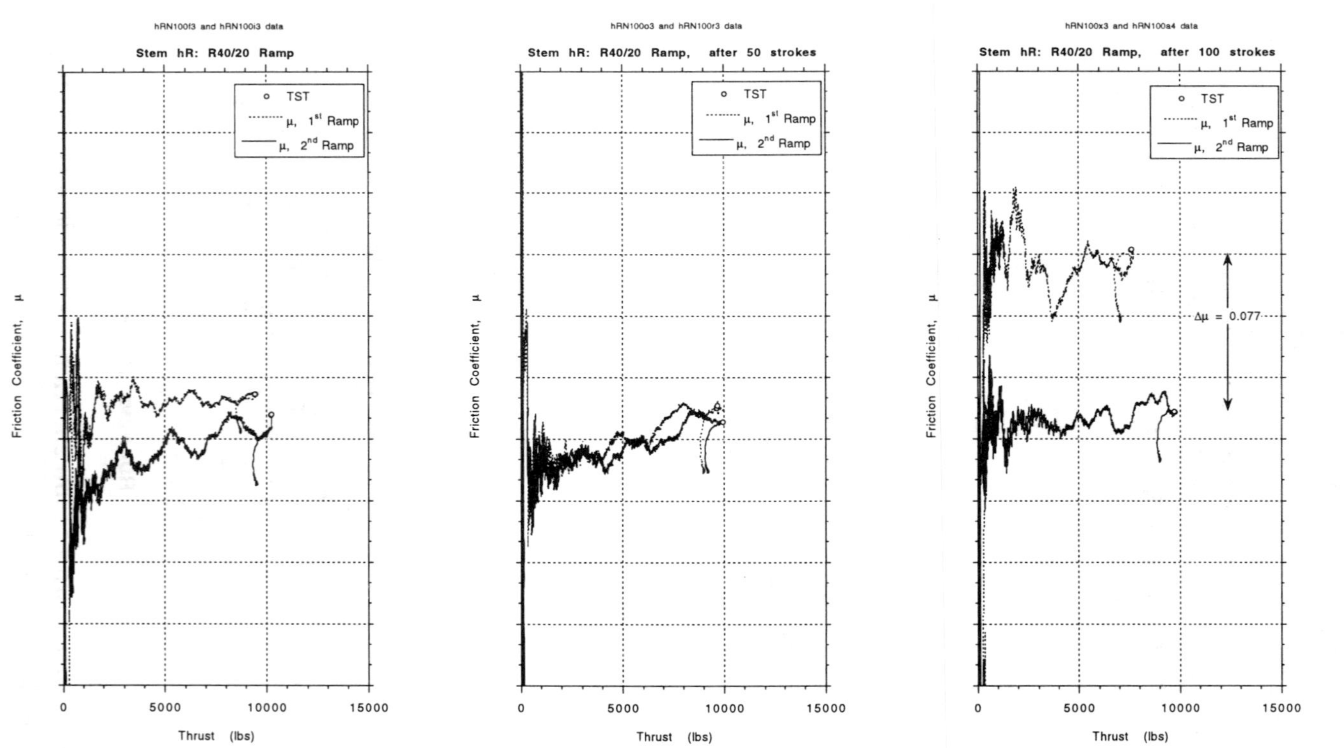

Fig. 8. Stem hR with N5000, friction coefficient vs load for ramp tests from three successive test series

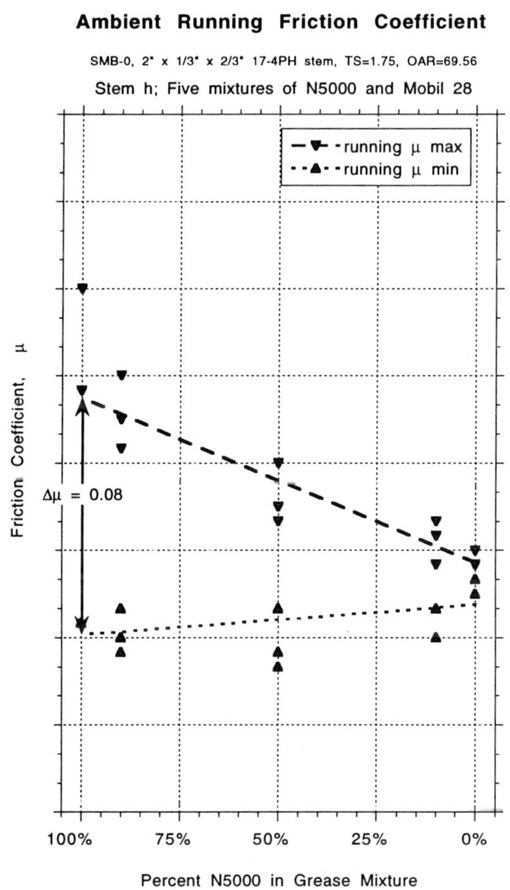

Fig. 9a. Stem h, running friction coefficient as a function of grease mixture

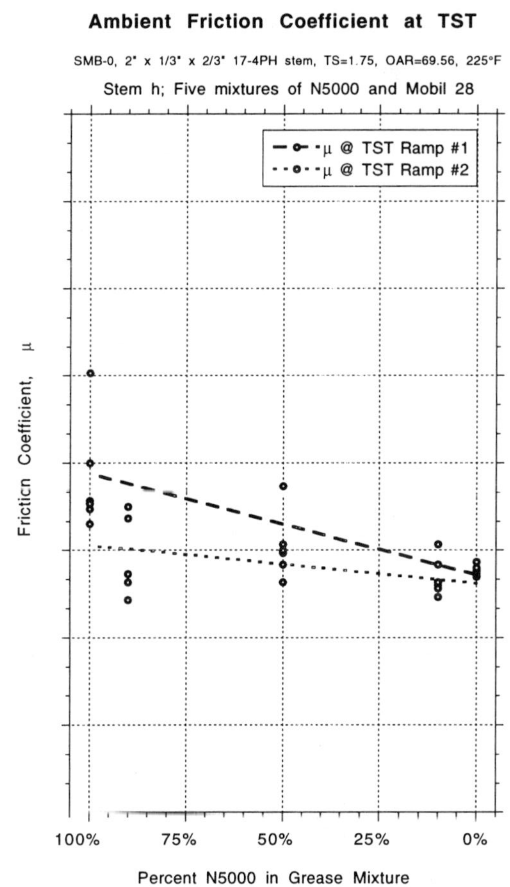

Fig. 9b. Stem h, friction coefficient at TST as a function of grease mixture

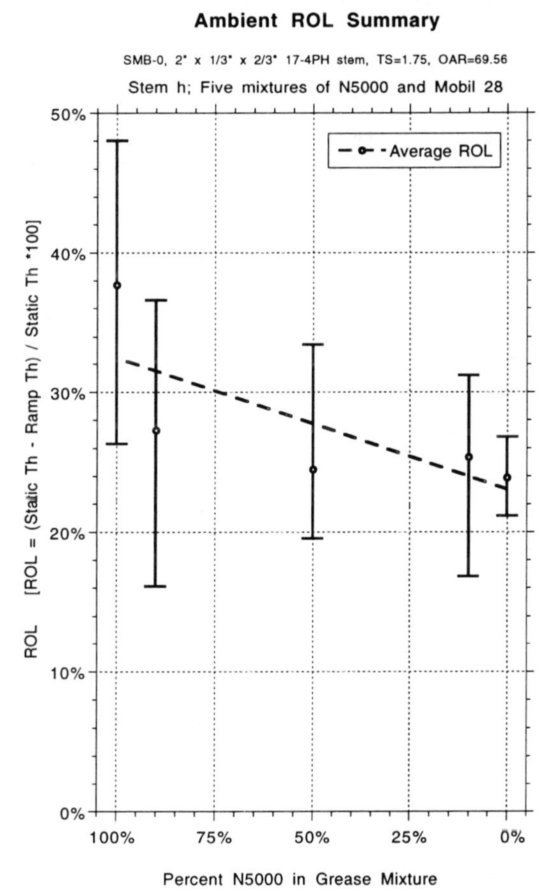

Fig. 9c. Stem h, ROL as a function of grease mixture

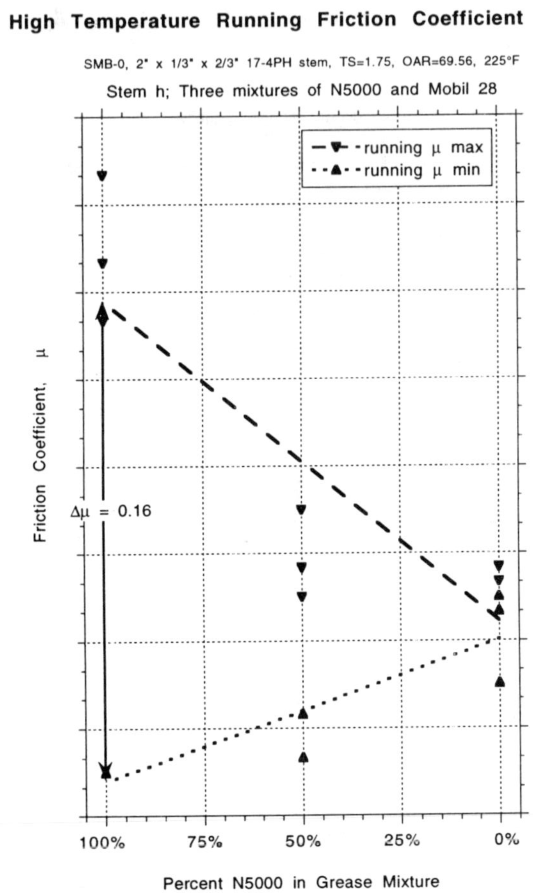

Fig. 10a. Stem h, running friction coefficient as a function of grease mixture at 225°F

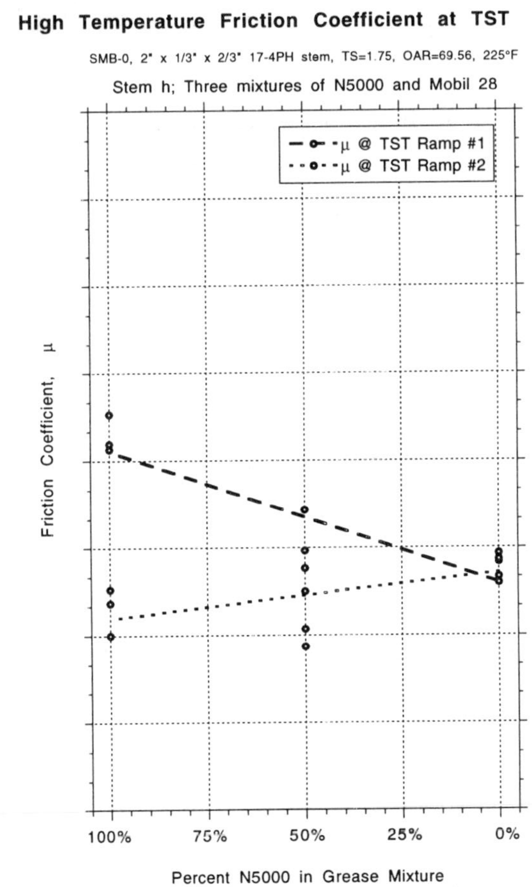

Fig. 10b. Stem h, friction coefficient at TST as a function of grease mixture at 225°F

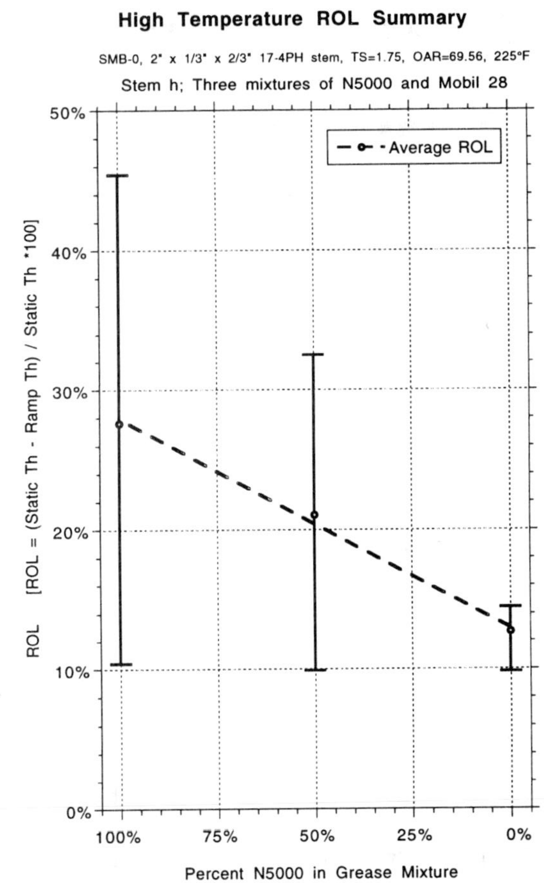

Fig. 10c. Stem h, ROL as a function of grease mixture at 225°F

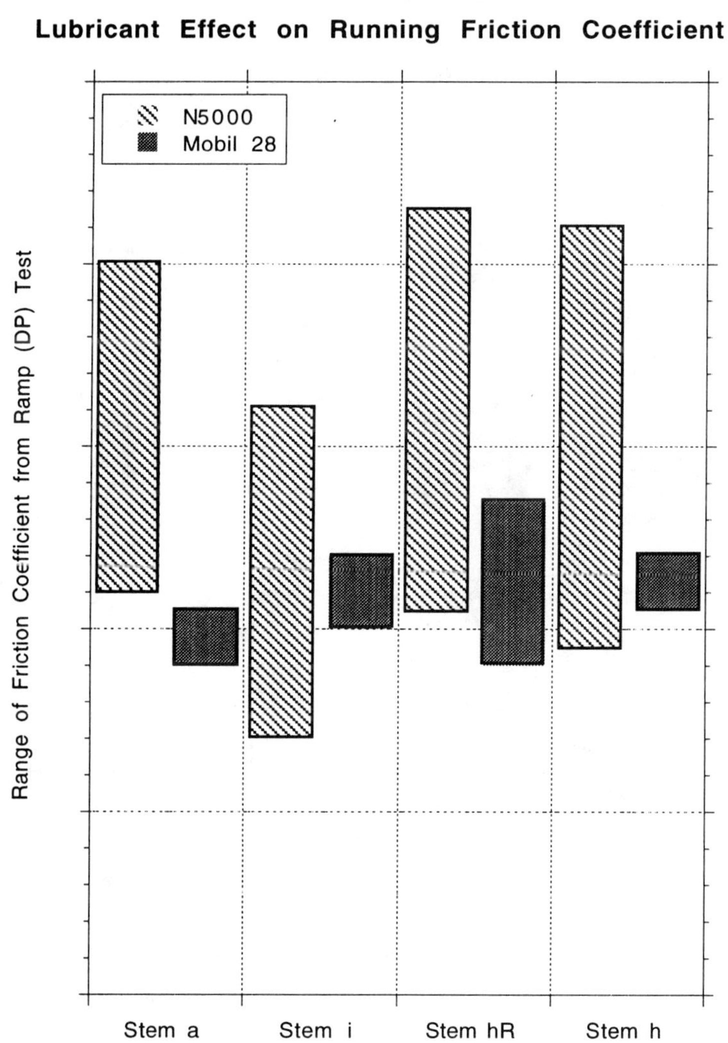

Fig. 11a. Lubricant effect on running friction coefficient

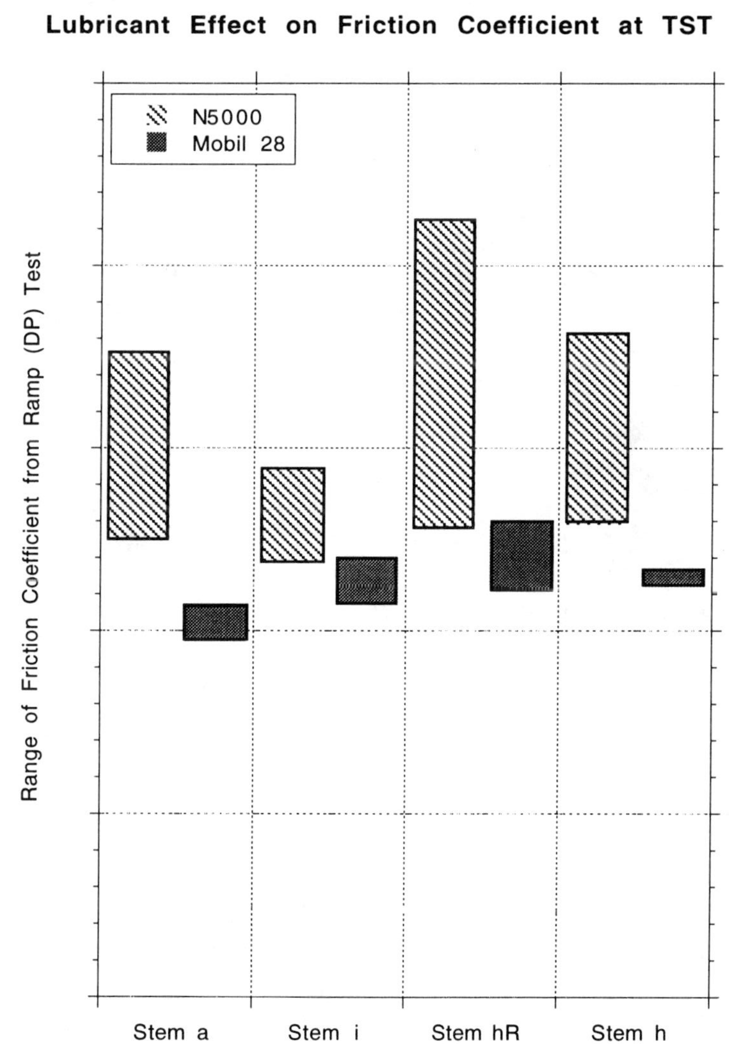

Fig. 11b. Lubricant effect on friction coefficient at TST

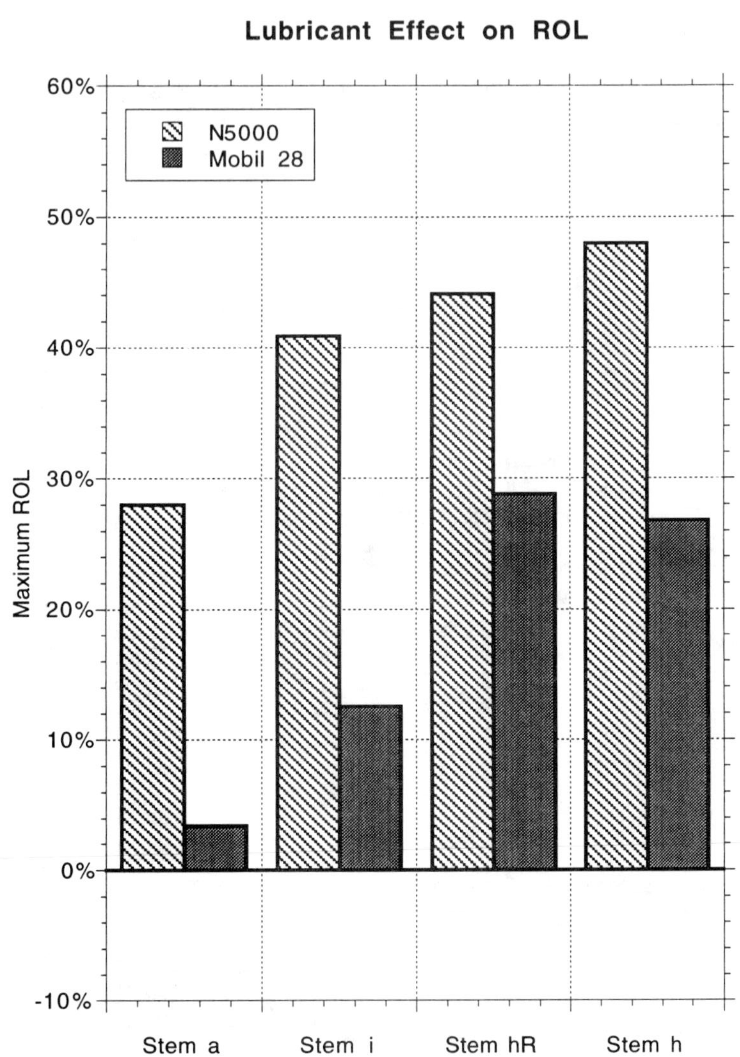

Fig. 12a. Lubricant effect on ROL

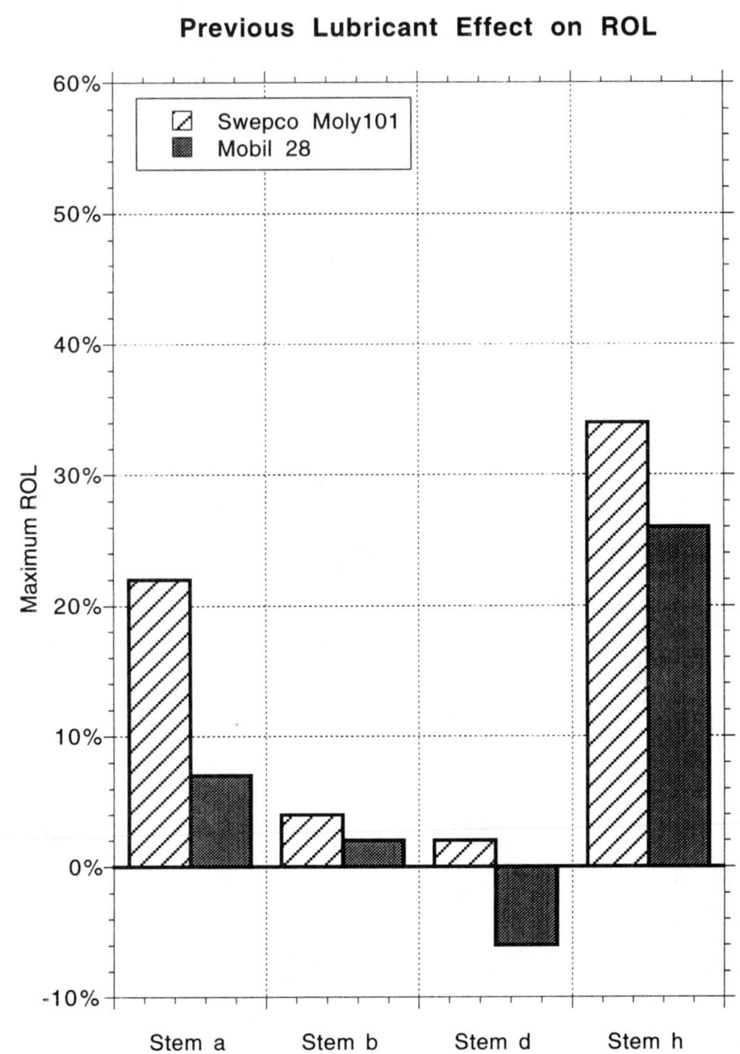

Fig. 12b. Lubricant effect on ROL from previous studies

PVP-Vol. 340, Seismic Engineering
ASME 1996

TESTING AND ANALYSIS OF PSV DISCHARGE FORCE WITH WATER SEAL

Gu Fangyu Li Pengzhou Peng Yunkang
Reactor Engineering Research Subinstitute
Nuclear Power Institute of China (NPIC)
Chengdu, Sichuan
P. R. China

ABSTRACT

Pilot—operated safety valve of pressurizer is one important equipement in NPP. It is an advanced type of safety valves and its discharge force is key parameter to the design of overpressure protection system. The discharge process can be affected by many complicated factors like volume and temperature distribution of water seal, discharge flow rate and so on.

One set of tests on discharge force have been completed under three water seal volumes and different temperature distributions by using a particular target to measure the transient dynamic response and modal analytical method to obtain the force function on the target. Meanwhile, the code Relap5/MOD2C is applied to calculate the force function. The results of tests and code analysis show that the application of Relap5/MOD2C in the calculation of PSV discharge force is successful and it has been used in Qinshan NPP.

Key words : Pilot — operated safety valve(PSV), Water seal, Discharge force

INTRODUCTION

PSV is an advanced type of safety valves. There is some condensed water, as water seal, in PSV inlet of pressurizer in NPP. Its discharge force is key parameter to design the overpressure protection system. There are several loadings which include thermal shock, internal pressure and reaction force during PSV discharge.

The discharge process is a very complex thermohydraulic phenomenon(Liu,1995) and can be affected by water seal volume, temperature distribution of condensed water, discharge flow rate, etc. Testing and analysis of PSV discharge force with water seal is a new subject and further researches are necessary to determine the discharge force correctly.

In this paper, the testing on the PSV test facility developed by NPIC and the analysis of thermohydraulic behaviors of the facility with open discharge have been finished. A particular target device was designed and installed closely in front of the discharge outlet to measure the transient dynamic response on the target disc. Model analysis method was used to obtain the discharge force function on the disc. Meanwhile, a thermohydraulic computer code Relap5/MOD2C was applied to calculate the

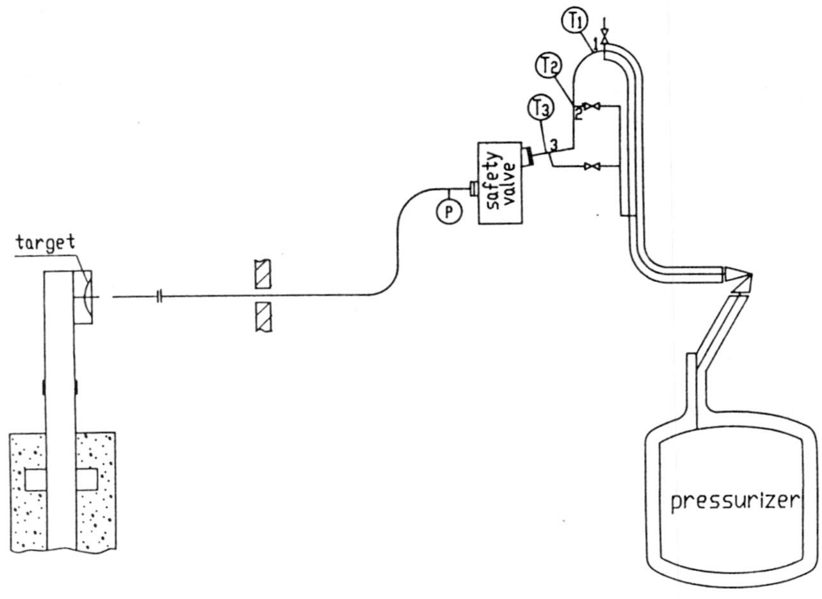

Fig. 1 PSV DISCHARGE TEST FACILITY

function.

One set of tests have been performed under three water seal volumes and different temperature distributions.

Both of the experimental and analytical results indicate that the discharge force is a function of water seal volume, condensed water temperature distribution and discharge flow rate, etc. Reasonable Selection of water seal volume and condensed water temperature in PSV design is extremely significant to reduce the influence of discharge force; In the calculation of PSV discharge process, the application of Relap5/MOD2C is successful.

TEST FACILITY AND EXPERIMENT DATA

There are two kinds of safety valve installation: open or closed discharge. For our test, the PSV discharge test facility is shown in figure 1, it is a open discharge installation and consists of a pressurizer, a PSV, a target device and connecting piping system. One pressure measuring point lies at the outlet of valve to measure the peak pressure in the discharge pipe. Besides, three temperature measuring points are located in the water seal section of the valve to measure the water temperature distribution.

The set pressure of PSV is 9. 8MPa.

The steam temperature in the pressurizer is 583K.

The minimum flow area of PSV is 1. 06 $\times 10^{-4} m^2$.

The geometry dimension of PSV inlet pipe is $\Phi 63 \times 7$mm.

The geometry dimension of PSV outlet pipe is $\Phi 76 \times 4$mm.

There is a invert U—bent in the PSV inlet where lots of steam from the pressurizer was condensed and forms the water seal avoiding leakage from PSV. However, the water seal volume and temperature distribution in the U—bent maybe affect the thermohydraulic behavior of PSV test facility seriously. The water seal effect on the thermohydraulic and mechanical performance has been studied in NPIC experimentally and analytically. Three water seal volumes are $1.847\times10^{-3}m^3$, $1.131\times10^{-3}m^3$ and $0.433\times10^{-3}m^3$ respectively in the experiment.

The target device consists of target disc, support beams and concrete base (see fig. 2a). A couple of strain measuring points are arranged at 400mm height from the base upper surface to measure dynamic response of the target during PSV discharge.

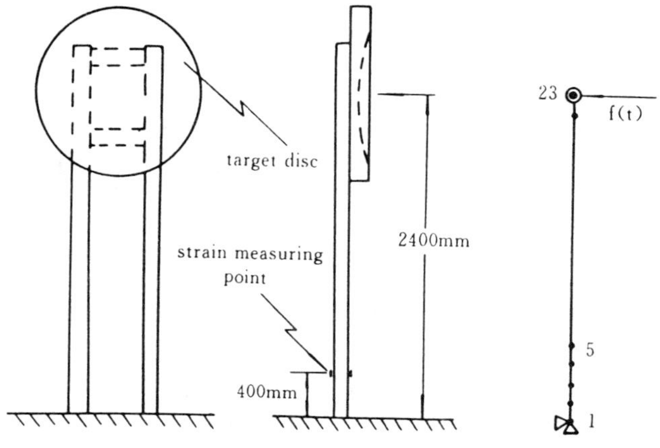

Fig. 2a, b TARGET DEVICE AND ITS FINITE ELEMENT MODEL

11 tests have been performed. The dynamic strain response at the strain measuring point, the transient pressure at PSV outlet are shown in fig. 3a, b, fig. 4a, b, and fig. 5a, b respectively for test No. 1, No. 8 and No. 11. All water temperature distribution values are listed in table 2, 3, 4.

DETERMINATION OF FORCE FUNCTION ON THE TARGET DISC

As stated above, the final experimental purpose is obtaining the time history fuction of the equivalent force applied at the target center rather than only measuring the strain response near the root of the target support beam and the transient pressure on the disc surface (Li, 1995).

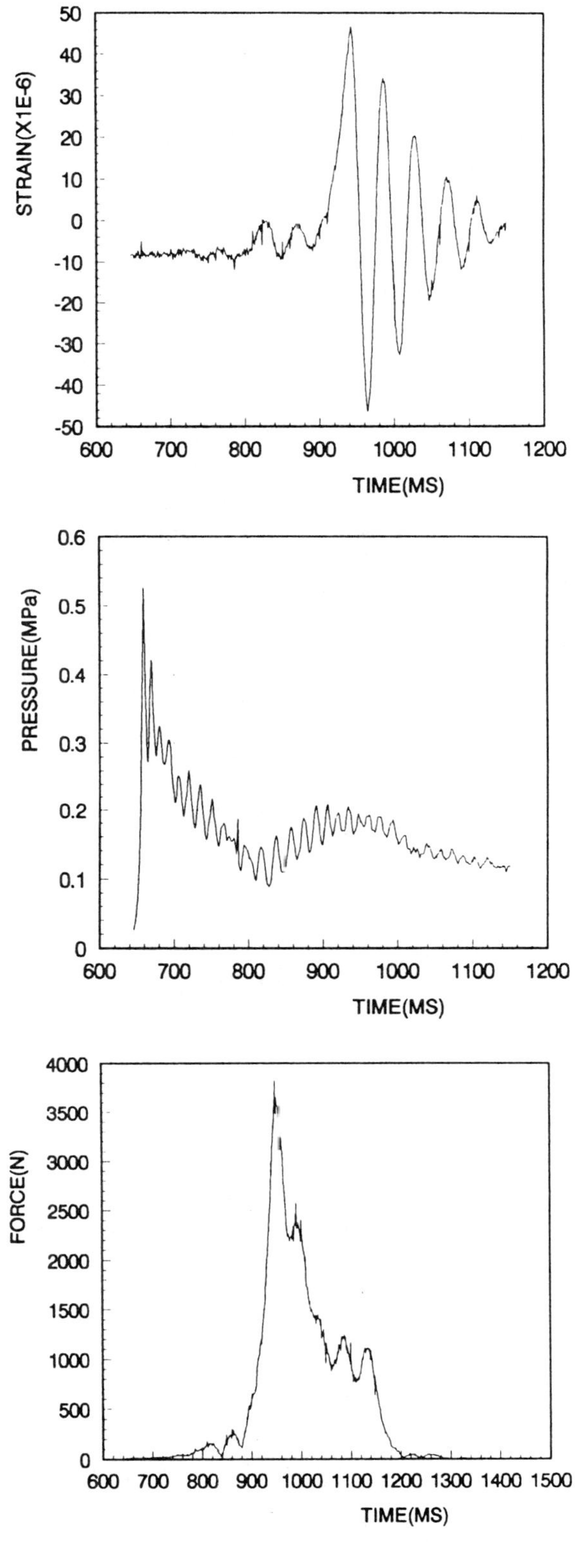

Fig. 3a, b, c PSV DISCHARGE TEST NO. 1

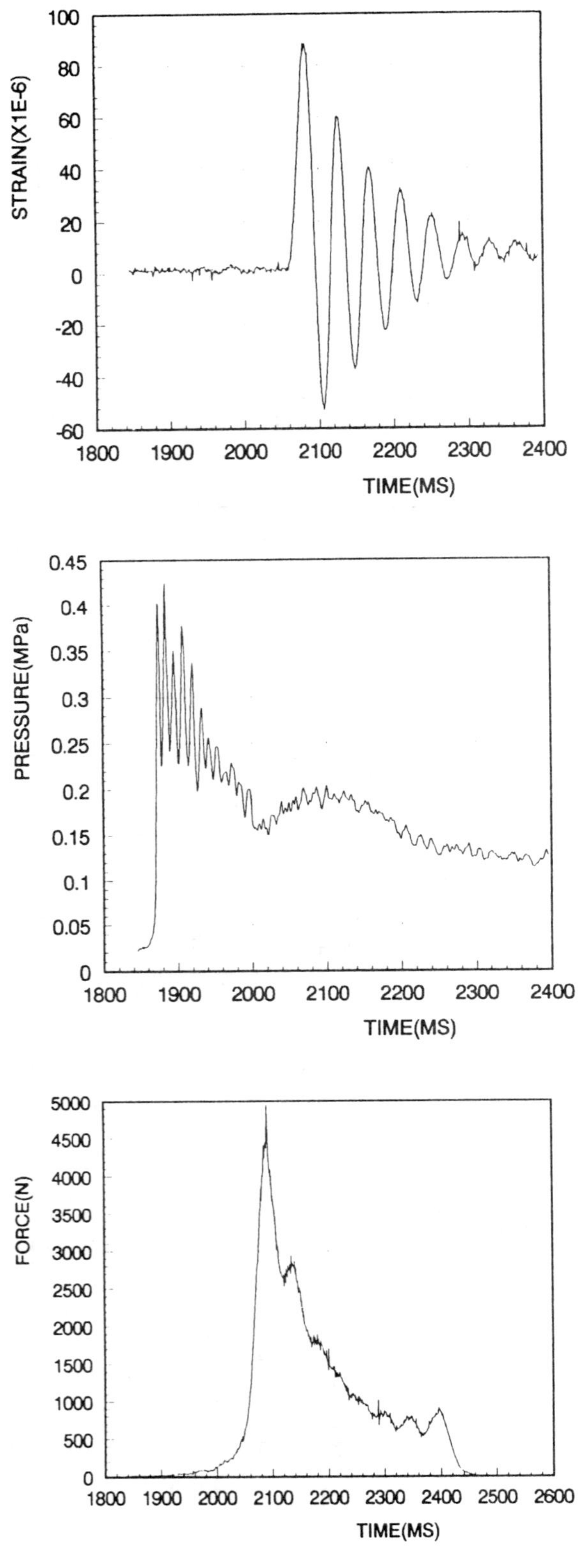

Fig. 4a, b, c DISCHARGE TEST NO. 8

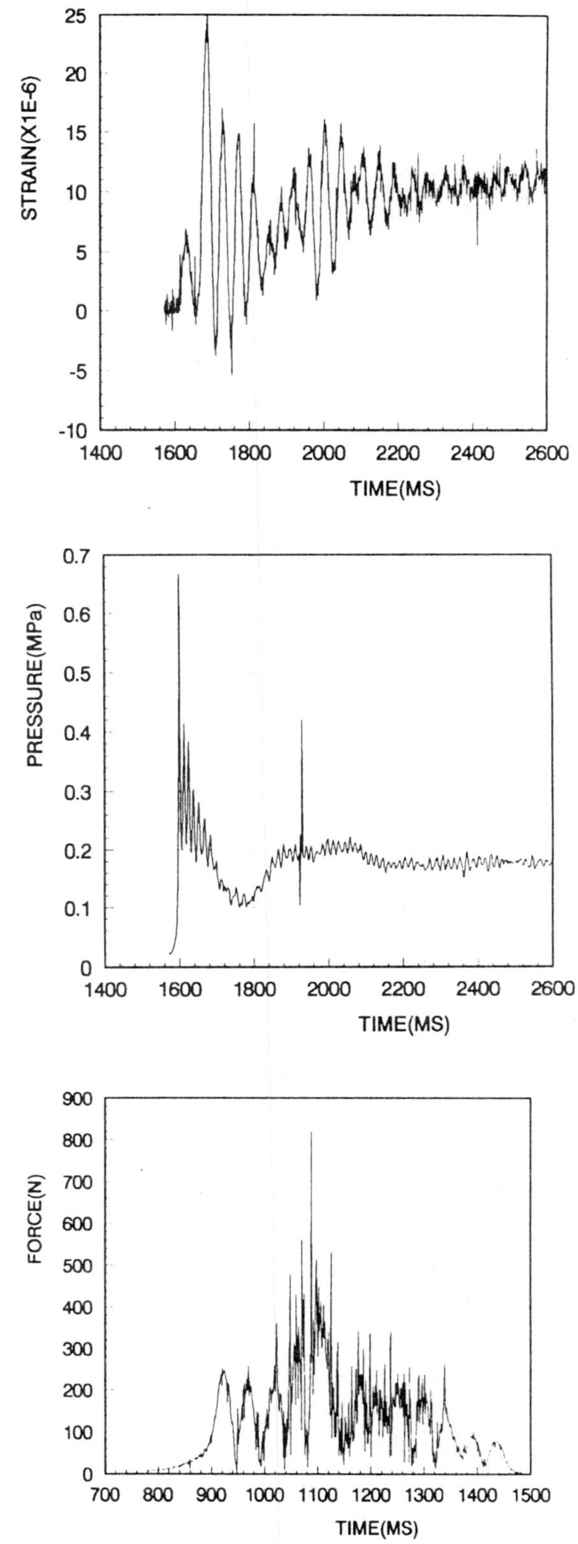

Fig. 5a, b, c DISCHARGE TEST NO. 11

Table 1 MAXIMUM VALUE OF EJECTION FORCE

NO.	TIME(MS)	MAX. VALUE(N)	STATIC VALUE(N)	STRAIN($\mu\varepsilon$)
1	946.696	4198.57	2227.68	46.8
2	1358.159	1949.62	2165.8	45.5
3	1101.991	2909.52	2299.08	48.3
4	1293.336	1623.38	1537.48	32.3
5	1122.297	1049.57	1151.92	24.2
6	1320.671	1493.85	1751.68	36.8
7	1038.73	653.82	985.32	20.7
8	2092.299	4933.59	4217.36	88.6
9	1126.983	4451.78	3384.36	71.1
10	1089.495	1394.18	1456.56	30.6
11	1990.769	1260.38	1185.24	24.9

So linear system theory of automatic control technique is used for the target structure. The ejection force on the target disc is regarded as input of the structure system and the strain reponse is considered as output of the system. When the frequency response function of the system is known, the equivalent force can be calculated surely. In this paper, modal analytical method is applied to compute the function of target structure by using the computer code ANSYS 4 • 3. The finite element model of the target structure is shown in fig2b.

The system damping ratio ξ which is needed in the calculation of the frequency response function is 3.1% by striking the disc center using hammer experimentally. The first three modal frequencies are respectively 28.67Hz, 249.26Hz and 746.00Hz without considering the torsion of the system. The frequency response function given by FEM makes no difference from the results deduced theoretically. Therefor, the design of the target device is successful and the analysis is also correct.

Some typical results of equivalent force history are respectively shown in fig. 3c, fig. 4c and fig. 5c for test No. 1, No. 8 and No. 11.

The maximum values of ejction force and its occuring time related to each test are listed in table 1 and last two columns stand for maximum values of the force transformed through static marking and strain measured at the same time. It can be found that when taking effect of inertial mass into account, the equivalent force is totally different from that given by static transform, and can only be solved as an inverse problem. The results by test and modal analysis will be compared with ones obtained by thermohydraulic analysis of the same system.

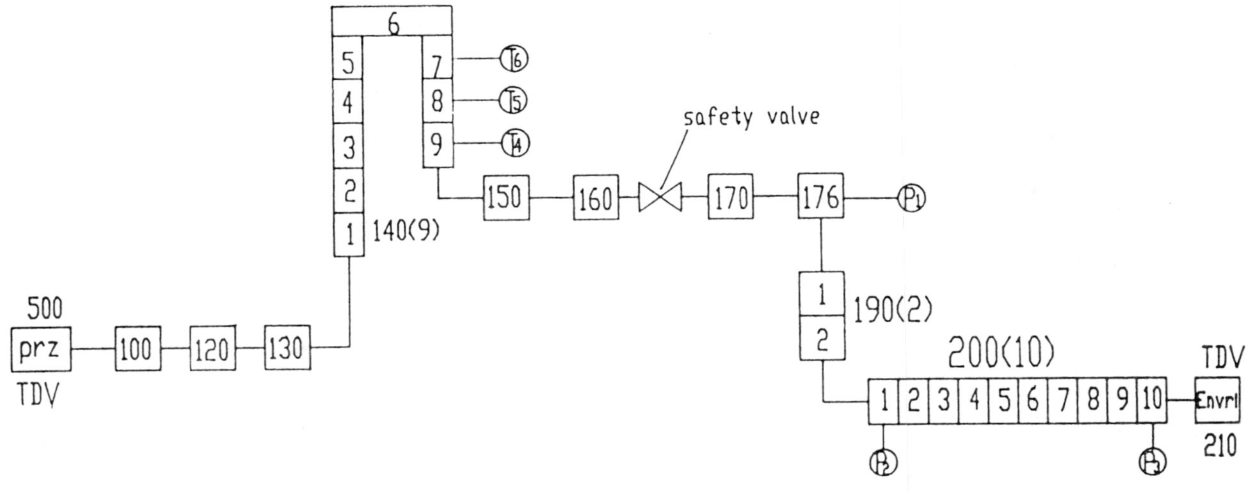

Fig. 6 PSV COMPUTATIONAL MODEL

THERMOHYDRAULIC ANALYSIS OF PSV SYSTEM

There are two kinds of PSV system, one is the installation with open discharge, and another with closed discharge. Only the open discharge system is involved in this paper, in which the condensed water in the invert U—bent is discharged directly to the atmosphere.

The Relap5/MOD2C code is used to analyze the thermohydraulic behavior of the open discharge system (Peng, 1995). The computational model is shown in figure 6. The pressurizer and the atmosphere are simulated by time — dependent volumes. the PSV is simulated by a trip valve.

Corresponding to the experimental condition, the computation by Relap5/MOD2C has been completed under three water seal volumes and three sets of temperature distribution values for each volume which are listed in table 2,3 and 4 respectively. The calculational results and related testing results are list in table 2 to 4.

A group of results are given in figure 7a to figure 7h for water seal volume $1.847\times 10^{-3}m^3$ and condensed water temperature distribution 293k, 298k, 503k.

The Comparison of results obtained by test and computation indicates that the steady results are in good agreement with each other. The peak value for target reaction force are not in so good agreement, the experimental results are within the limits specified by $\pm 50\%$ of computation results. The force decreases with the water seal temperature increase under same volume, and also decreases with the water seal volume reduction under almost same temperature distribution.

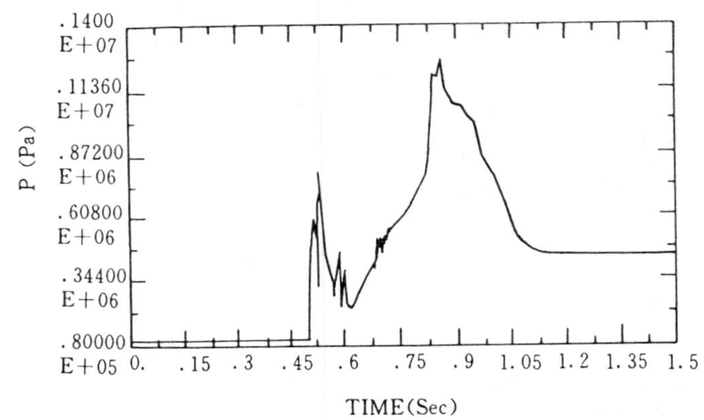

Fig. 7a THE OUTLET PRESSURE OF PSV VS. TIME

Table 2 THE RESULTS UNDER WATER SEAL VOLUME $1.847 \times 10^{-3} m^3$

Iterm \ Results		Exp.	Cal.	Exp.	Cal.	Exp.	Cal.
Water seal temperature (K)		293 298 503		313 403 558		338 453 573	
Valve outlet pressure(MPa)	peak	0.95	1.25	0.9	1.2	0.87	1.19
	steady—state	0.41	0.45	0.42	0.45	0.42	0.44
Discharge flowrate(kg/s)		1.49	1.61	1.49	1.61	1.49	1.61
Target reaction force(kg)	peak	500	980	454	440	199	380
	steady—state	95	98	95	98	95	98

Table 3 THE RESULTS UNDER WATER SEAL VOLUME $1.131 \times 10^{-3} m^3$

Iterm \ Results		Exp.	Cal.	Exp.	Cal.	Exp.	Cal.
Water seal temperature(K)		313 583 583		323 548 583		335 533 583	
Valve outlet pressure(MPa)	peak	0.8	1.15	0.79	1.16	0.85	1.12
	steady—state	0.45	0.46	0.45	0.43	0.45	0.46
Discharge flowrate(kg/s)		1.49	1.61	1.49	1.54	1.49	1.61
Target reaction force(kg)	peak	428	320	298	310	166	300
	steady—state	95	98	95	98	95	98

Table 4 THE RESULTS UNDER WATER SEAL VOLUME $0.433 \times 10^{-3} m^3$

Iterm \ Results		Exp.	Cal.	Exp.	Cal.	Exp.	Cal.
Water seal temperature(K)		533 583 583		558 583 583		563 583 583	
Valve outlet pressure(MPa)	peak	0.83	1.04	0.92	1.04	0.84	1.03
	steady—state	0.42	0.46	0.44	0.45	0.43	0.46
Discharge flowrate(kg/s)		1.49	1.61	1.49	1.61	1.49	1.61
Target reaction force(kg)	peak	297	230	152	220	129	195
	steady—state	95	98	95	98	95	98

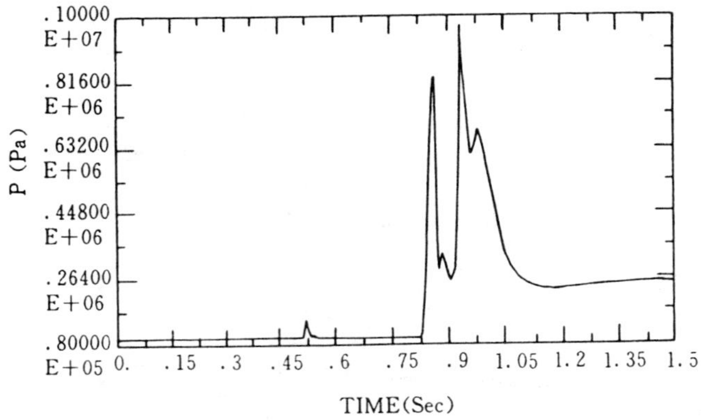

Fig. 7b THE OUTLET PRESSURE OF DISCHARGE PIPE

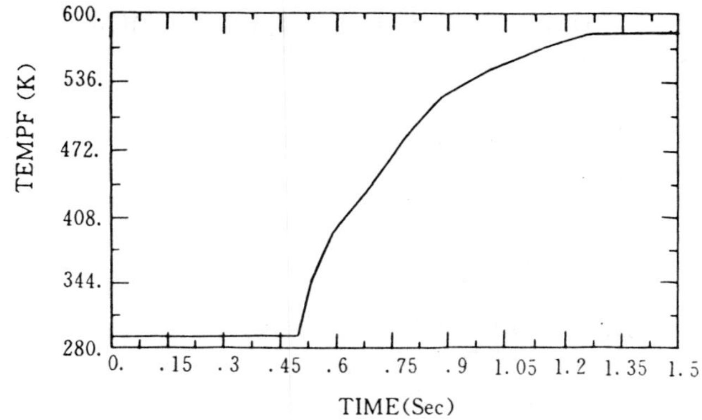

Fig. 7e THE TEMPERATURE AT MEASURING POINT 6

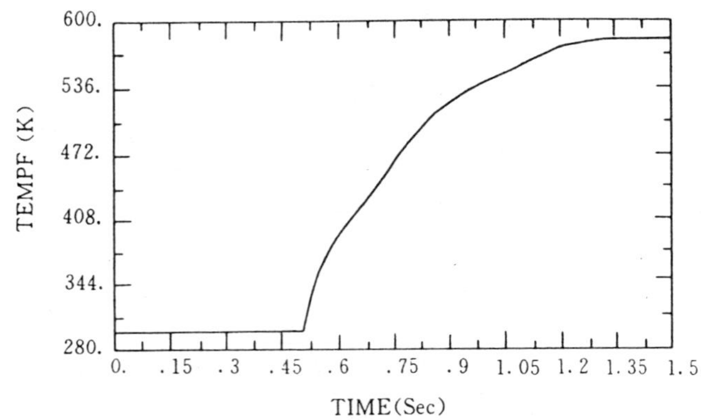

Fig. 7c THE TEMPERATURE AT MEASURING POINT 4

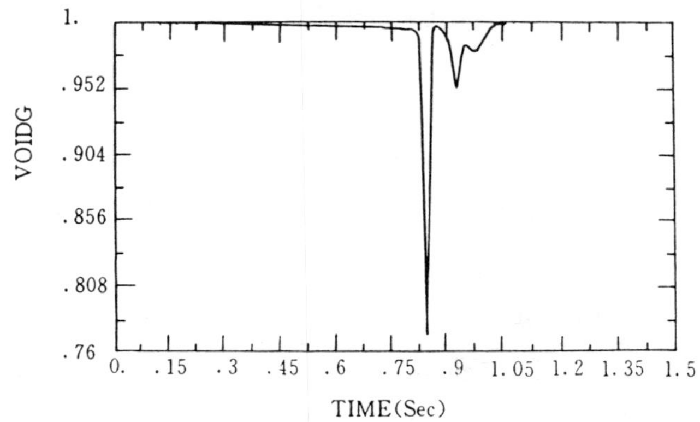

Fig. 7f OUTLET VOID FRACTION OF DISCHARGE PIPE

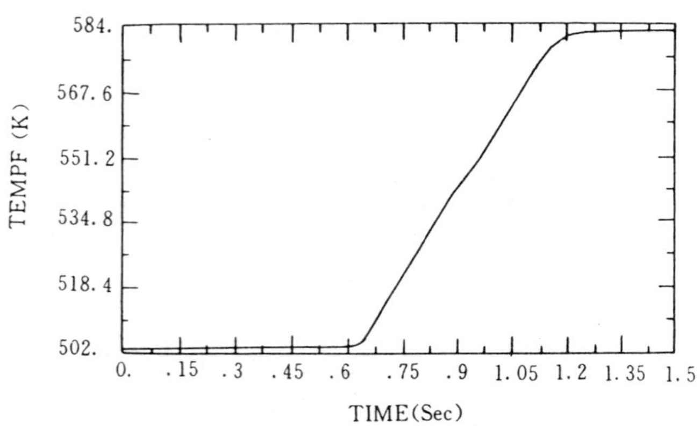

Fig. 7d THE TEMPERATURE AT MEASURING POINT 5

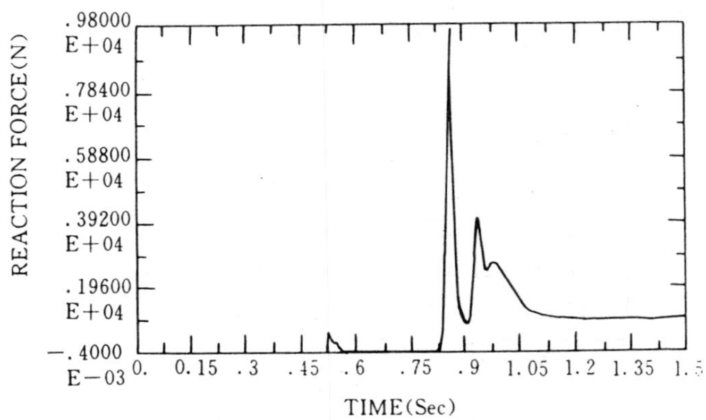

Fig. 7g THE PSV DISCHARGE FLOWRATE

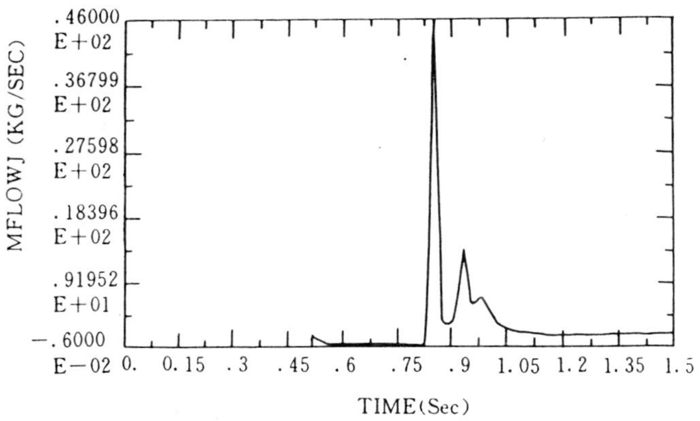

Fig. 7h THE OUTLET REACTION FORCE OF DISCHARGE PIPE VS. TIME

CONCLUSION

The conclusion from above comparison can be drawn as follows:

The technical approach in this paper is correct. Both discharge tests (including the modal analysis of the target device) and analysis of thermohydraulic behaviors by Relap5/MOD2C are successful and the results given by them make almost no difference.

The code Relap5/MOD2C can be used to predict the thermohydraulic parameters such as PSV outlet pressure, discharge pipe outlet pressure, temperatures of measuring points, gas void fraction at discharge pipe, PSV discharge flow rate, and even the reaction force at discharge pipe outlet and at the PSV itself. The code has been used in Qinshan I in China.

The water seal volume affects the thermohydraulic behavior seriously. The discharge reaction force increases with the increase of water seal volume; The water seal temperature in the U—bent of PSV is another important factor to influence the thermohydraulic behaviors of PSV discharge system. The discharge reaction force increases with the reduction of condensed water temperature.

REFERENCE

Li Pengzhou, 1995, " Calculation of Force Acting On Target During Safety Valve Discharge," Technical Report, Nuclear Power Institute of China.

Liu Shuqian, 1995, " Calculation of Leakage Rate of Pipe Crack in LBB Analysis Technology," *Nuclear Power Engineering* (Chinese), Vol. 16, No. 4, pp. 301—305

Peng Yunkang, 1995, "Study On Safety Valve System Thermohydraulic Behavior," *Computer Application in Nuclear Field* (Chinese), Vol. 27, No. 4, pp. 125—129.

SOIL STRUCTURE INTERACTION

Introduction

M. A. Saleem and **R. A. Ricciuti**
Atomic Energy of Canada Limited

This chapter contains papers that deal with soil structure interaction analysis methods, an interesting approach to establish a design soil profile for the AP600 nuclear plant standard design and paper that presents a newly developed probabilistic methodology which rigorously accounts for geotechnical and seismologic uncertainties in the quantification of seismic liquefaction risk.

The effect of soil-structure interaction is recognized to be an important consideration for the seismic design of critical structures such as nuclear power plants. SSI refers to the effects of the supporting soil medium on the motion of a structure or component during a earthquake. The SSI phenomena is a complicated process, with the response of a structure during an earthquake depending on the characteristics of the ground motion, the properties of the surrounding soil, and the structure itself.

Ricciuti, Elgohary and Usmani present the results of a study to assess the potential benefits from utilization of a state-of-the-art SSI computer program (SASSI) on the seismic response of the CANDU 6 NPP. The results predicted with SASSI are compared with the Soil-Spring method used in standard CANDU 6 plant design. The impact of soil layering, embeddment and structural damping on the seismic response have been assessed.

Ostadan, Liu, Gross and Orr present a systematic and effective approach to establish a limited set of design soil profiles from the unlimited number of site conditions that can be postulated for use in the SSI analyses of a NPP. The results were then used in the final three dimensional SSI analysis and design of the AP600 nuclear plant standard design.

Arango, Ostandan, Lewis and Gutierrez describes the application of a newly developed probabilistic methodology which rigorously accounts for geotechnical and seismologic uncertainties to predict the risk of seismically-induced liquefaction. The results of the analysis are compared with the acceptable levels of risk that have been established by the Department of Energy (DOE).

On behalf of the Seismic Engineering Technical Subcommittee, the editors of this chapter would like to thank the authors for the efforts put into the preparation of these papers as well as for the efforts of reviewers whose useful comments contributed to the quality of these papers.

PVP-Vol. 340, Seismic Engineering
ASME 1996

QUANTIFICATION OF SEISMIC LIQUEFACTION RISK (U)

Ignacio Arango

Geotechnical Services
Bechtel Corporation
San Francisco, California

Farhang Ostadan

Geotechnical Services
Bechtel National, Inc.
San Francisco, California

Michael R. Lewis

Geotechnical Services
Bechtel Savannah River, Inc.
Savannah River Site
Aiken, South Carolina

Brent J. Gutierrez

Department of Energy
Savannah River Site
Aiken, South Carolina

ABSTRACT

Explicit goals of acceptable risk for natural phenomena hazards (earthquake, extreme wind, and flood) have been established by the Department of Energy (DOE, 1994). Closely associated to the earthquake risk is the issue of seismically-induced liquefaction. Because deterministic methods currently available to answer the question to whether a site is liquefiable or not are incapable of providing a clue as to the likelihood or risk of liquefaction, the application of the criteria to a given facility requires that alternative evaluation techniques be formulated.

This paper describes the application to a nuclear facility of a newly developed probabilistic methodology which rigorously accounts for geotechnical and seismologic uncertainties. The results of the analyses are compared with the acceptable levels of risk presented by DOE. This comparison is used to emphasize the power of the methodology as a tool in the decision-making processes.

NOMENCLATURE

CSR	Cyclic stress ratio
CSRN	Normalized cyclic stress ratio
C_n	Factor to normalize standard penetration test N-Value for overburden pressure
C_q	Factor to normalize cone tip resistance
D_{bar}	Deaggregated earthquake distance
E	Earthquake
EPRI	Electric Power Research Institute
FS	Factor of safety against seismic liquefaction
g	Gravity
Hz	Hertz
LLNL	Lawrence Livermore National Laboratory
MSF	Magnitude scaling factor
M_{bar}	Deaggregated earthquake magnitude
N	SPT Blowcount
N_1	SPT Blowcount Corrected for "Overburden Pressure"
N_{60}	SPT Blowcount Corrected to 60% Energy
P	Compression Wave
PC	Performance Category
P_F	Permissible failure frequency limit
$P_E(L)$	Overall probability of liquefaction during earthquake E
P[L|E]	Conditional probability of liquefaction given E
P[E]	Probability that earthquake E occurs
P[L]	Overall Probability of Liquefaction
S	Shear Wave
SCPT	Seismic cone penetration test
SPT	Standard Penetration Test
$(N_1)_{60}$	Standard penetration test blow count
TR	Tobacco Road Formation
q_c	Cone tip resistance
$(q_c)_1$	Cone tip resistance normalized for overburden pressure
σ'_v	Vertical effective overburden pressure
τ_{MAX}	Maximum cyclic shear stress in soil
Σ	Sum over all earthquakes

INTRODUCTION

In a recent paper, Murray, et al, (1994) discuss DOE's explicit goals of acceptable risk for natural phenomena hazards, namely earthquakes, extreme winds, and floods. The acceptable risks were developed based on target probabilistic performance goals. These performance goals follow a graded approach beginning with normal-use facilities and ending with facilities involving hazardous or critical operations. Performance goals are defined by both qualitative expressions of acceptable behavior and target quantitative probabilities of exceedence/non-exceedence comparable to commercial power reactors.

The DOE requirements for earthquake mitigation were established with the intent to provide an appropriate level of seismic protection for, a) occupant and public health and safety; b) the environment; c) production and research objectives; and d) potential property losses. A graded approach was implemented by defining five performance categories (PC-0 through PC-4), each with a performance goal. Performance category 0 covers items which require no seismic criteria. Performance category 4 (the most severe) is applicable to nuclear facilities.

The quantitative performance goals are defined in terms of a permissible mean annual probability of unacceptable performance P_F (i.e., a permissible failure frequency limit). Seismically induced unacceptable performance should have an annual probability less than or approximately equal to these goals. Values of P_F for Performance Categories 1, 2, 3, and 4 are about 10^{-3}, about $5x10^{-4}$, about 10^{-4}, and about 10^{-5}, respectively. The qualitative descriptions of expected performance following design/evaluation levels of earthquake ground motions are shown in Table 1 for PC-1 through PC-4.

Earthquake ground shaking may cause the loss of strength of loose granular soils which results in settlement of buildings, landslides, failure of dams and embankments, and disruption of lifelines. This process of strength loss is called soil liquefaction. Numerous empirical relationships have been developed relating some soil characteristics which are indicative of liquefaction susceptibility and ground motions of varying intensities. Today, the relationship developed by Seed et al. in 1984 is perhaps the most widely used and accepted by the profession. This relationship, typical of the so-called deterministic models, provides a yes or no answer to the question whether liquefaction will or will not occur, and this answer is expressed numerically in terms of a factor of safety (FS).

An evaluation of the liquefaction potential for a critical facility located in Eastern U. S. was performed using deterministic qualitative and quantitative procedures. A generalized subsurface soil profile at this site is presented in Figure 1. In that evaluation, best estimate soil properties, along with the evaluation basis earthquake, were used in the analyses. Since the evaluation was based on deterministic methods, the effect of variability in soil properties and of earthquake motions on the liquefaction potential could not be integrated into the analysis. The purpose of the probabilistic evaluation presented here is to incorporate the effects of variation of the most significant soil properties and earthquake parameters in the evaluation of the annual liquefaction hazard at the site, and to compare this with the acceptable performance goals established by DOE, as referenced above. The evaluation is based on site-specific subsurface data, the Electric Power Research Institute (EPRI) seismic hazard study (EPRI, 1986 and 1993) and a recently developed methodology for liquefaction hazard evaluation.

METHODOLOGY

The recently developed Advanced Seismic Hazard/Liquefaction Evaluation (ASHLE) method (Bechtel, 1991, Ostadan et al., 1991; Arango, I. and Ostadan, F., 1995) was used for the evaluation. In this method, the probability of liquefaction at a given site is obtained from

$$P_E(L) = P[L|E]\,P[E] \qquad (1)$$

where P_E (L) is the overall probability of liquefaction during the earthquake E; P[L|E] is the conditional probability of liquefaction given that earthquake E occurs; and P[E] is the probability that earthquake E occurs. The total overall probability of liquefaction is obtained by summing over all possible earthquakes:

$$P[L] = \sum_E P[L|E]\,P[E] \qquad (2)$$

The ASHLE methodology used models for conditional probability of liquefaction developed by Liao, et al. (1988). The models were developed based on statistical analyses (logit analysis followed by likelihood maximization) of a data catalog consisting of 278 observed cases of liquefaction/no liquefaction in Holocene deposits during forty earthquakes.

The Liao model, known as the local model for silty sands, was used as the basis for this evaluation. From this model, the probability of liquefaction can be obtained once the normalized blowcount data SPT $(N_1)_{60}$, and the normalized cyclic stress ratio (CSRN) are known. The normalized cyclic stress ratio is obtained from

$$\mathrm{CSRN} = 0.65\,\frac{\tau_{MAX}}{\sigma'_v} \qquad (3)$$

where τ_{MAX} is the maximum seismic shear stress in the soil layer under consideration and σ'_v is the corresponding effective overburden pressure. The normalization of CSRN is performed so as to correct for the magnitude (duration) of the earthquake motion consistent with the approach used in the deterministic method (Seed et al., 1984).

In addition to the seismic models, the SPT blowcount corrected for overburden pressure and normalized to 60% energy, $(N_1)_{60}$, and the normalized cyclic stress ratio (CSRN) for each level of the earthquake motion (all events E in Equation 2) are required as input to the ASHLE analysis.

Seismic Models

Current baselines for seismic hazards applicable to the site are described in investigations conducted by the EPRI (EPRI, 1986) and by the Lawrence Livermore National Laboratory (LLNL, 1989) and the revisions to the LLNL study.

Both the EPRI and the LLNL mean rock spectral hazard curves at 1.0 Hz and 2.5 Hz are shown in Figure 2. As shown, while the differences between the 1.0 Hz and the 2.5 Hz hazard curves are insignificant in each study, the LLNL results predict 3 to 4 times higher hazard compared to the EPRI results. The differences in these two results have been the subject of additional ongoing studies. In this paper, only the liquefaction hazard evaluation using the EPRI data for 2.5 Hz hazard curve are presented.

Use of the 2.5 Hz spectral hazard curve for liquefaction hazard evaluation was justified based on the results of several studies, such as Costantino (1994). In that study, a series of seismic free-field ground response analyses were performed using a total of 32 time histories (real and artificial). The effect of frequency content of the input motion on the cyclic stress ratios induced in the TR soils (i.e., the soils subject to the liquefaction evaluation) was thoroughly examined using both high pass and low pass band filters on the input motion. The study concluded that "the important characteristic of the input rock outcrop motion which controls the development of the induced shear stresses in the TR soils is the low frequency content of this motion as measured by the average spectral acceleration from 1 to 3 Hz."

A point on the spectral hazard curve, shown in Figure 2, defines the annual probability of the exceedence at a given ground motion level. This probability is calculated as the summation of the occurrences of ground motion exceedences over a range of source magnitudes (M_{bar}) and distances D_{bar}. To cover the wide range of ground motion intensities (from 1 cm/sec to 90 cm/sec in Figure 2) and the associated probabilities of exceedences, a total of ten points on the EPRI 2.5 Hz hazard curve was selected. For each point, through a hazard deaggregation process, the explicit contributions to the total hazard were obtained and displayed in space as earthquake magnitude and distance interval bins. For the above M_{bar} and D_{bar} pairs, rock acceleration response spectra were developed using region-appropriate seismic data. The resulting spectra were scaled so that the 2.5 Hz spectral amplitude matched that of the corresponding hazard curve point. For this, the computer program RASCAL, Silva et al. (1987) was used. The program uses random vibration theory to calculate peak accelerations, velocity, and response spectra for specified earthquake source and propagation path parameters. Figure 3 presents the 10 spectra derived for the study. The spectra have been individually scaled so that the spectral values at 2.5 Hz maintain the appropriate spectral velocity, as defined by the EPRI mean 2.5 Hz spectral hazard curve. These spectra constitute the rock motion target spectra for which spectra-compatible time histories are generated for ground response analyses.

It may be noted that some of the target spectra in this figure are high. For example, the spectrum with the 2.5 Hz spectral velocity of 90 cm/s requires a spectral acceleration greater than 4.3g. However, the hazard level corresponding to these ground motions is about 3.5×10^{-7}, an extremely rare event.

Soil Characterization

For ground response analyses and liquefaction hazard evaluation, the soil properties consisting of the low-strain shear wave velocity, strain-dependent shear modulus and damping, and the insitu SPT blow-count data are required.

Low-strain shear wave velocity measurement from ten seismic piezo cone penetration probes (SCPTs) down to a depth of 170 feet below site ground surface, and two deep velocity (P and S) logged boreholes to a depth of about 1,000 feet below site ground surface were utilized.

A composite shear wave velocity for the ten SCPTs is shown in Figure 4. Table 2 presents the shear wave velocity, by layer, for these SCPTs, while Table 3 gives their respective layer thicknesses. Ultimately, the shear wave velocities presented in Table 2 and the layer thicknesses reported in Table 3 were used in a series of ground response analyses to represent the variability of shear wave velocity in the upper 170 feet of the subsurface profile.

To construct a complete shear wave velocity profile for ground response analyses, the deep shear wave velocity profiles were coupled with the ten shallow SCPTs. The deep and shallow profiles were randomly selected, paired and used in a Monte Carlo-type analysis.

For response analyses, site-specific modulus reduction and damping ratio vs. strain amplitude relationships were utilized.

SPT $(N_1)_{60}$ Values

The conditional probability of the liquefaction model requires the normalized SPT blow count, $(N_1)_{60}$, as a measure of the soil's resistance to liquefaction. Deterministic analyses show that the Formation in question (i.e., TR3/TR4 layer) has the lowest FS against liquefaction. Hence, the probabilistic liquefaction hazard analysis was restricted to SPT, N-values from this layer.

Available penetration resistance data for the site come from numerous boreholes with SPT, N results and SCPT probes. The borehole information consisted of two sets of data. The first set was from the original investigation performed for the facility (four underground storage tanks), while the second set was from the current study. The SCPT probes were all done as part of the current study. A rigorous quality assurance program was implemented for the current study. This included prior site-specific measurement of SPT hammer energy (measured as 60%). For the original investigations, however, documentation was not as complete. For example, the type of SPT hammer used was not recorded nor were SPT hammer energies measured. Thus, utilizing the previous SPT data required an evaluation.

The evaluation of the previous SPT data was accomplished by performing a statistical comparison of the previous uncorrected SPT data with the current uncorrected SPT data. The results showed that the previous uncorrected SPT N value data were, on the average, 1.6 times higher than the N value data of the current study. The difference suggests that the previous hammer energy was about 40%. Thus, all of the previous SPT N value data was divided by 1.6 to convert to an equivalent N_{60} value. Subsequently, all N_{60} data was converted to $(N_1)_{60}$ using the relation:

$$(N_1)_{60} = N_{60} \times C_n \tag{4}$$

where C_n equals $\left(1/\sigma'_v\right)^{0.5}$

The cone results SCPT data were also used to determine $(N_1)_{60}$. First, SCPT tip resistances (q_c) were corrected for overburden pressure using:

$$\left(q_{c1}\right) = C_q \times q_c \tag{5}$$

where, $C_q \cong 1.63 \left(\sigma'_v\right)^{-0.7}$, Seed, Idriss, and Arango (1983). Next, q_{c1} was converted to $(N_1)_{60}$ using the following relationship:

$$(N_1)_{60} = 0.285 q_{c1} - 2.1399 \tag{6}$$

This was accomplished using site-specific statistical regression analysis developed for each subsurface layer using paired data from SCPT probes and SPT boreholes located no greater than 10 feet apart.

In addition to the site-specific SPT and SCPT data, geostatistical simulations were performed to develop additional representative data for the soils. Figure 5 shows the simulated $(N_1)_{60}$ data superimposed on the measured $(N_1)_{60}$. In general, the simulated values appear to agree with the measured data both in trend with depth and in the amount of scatter, suggesting that a reasonable simulation has been performed. The measured, simulated, and combined (measured plus simulated) $(N_1)_{60}$ values for the potentially liquefiable soil layer (i.e., TR3/TR4 layer) are presented in Table 4. All three sets of $(N_1)_{60}$ values (measured, simulated, and combined) were used in the ASHLE analysis to investigate the sensitivity of the final results to the simulated values.

Cyclic Shear Strength

The conditional probability of liquefaction models developed by Liao et al. (1988) are based on the statistical analysis of field data corresponding to Holocene deposits. For application to the study site, the model was modified to reflect the increased cyclic shear strength of the soils due to age, which also includes adjustments for overconsolidation and sample disturbance. Depending upon the degree of conservatism to be adopted for the analyses, the combined effect of the above adjustment factors had a numerical value of either 1.35 or 1.9. Both of these values were used in the analyses.

Cyclic Shear Strength Magnitude Scaling Factors

The magnitude scaling factors (MSF) developed by Seed and Idriss (1982) have been widely used. They are based on laboratory test data and statistical studies relating the number of significant strong motion cycles to earthquake magnitude. However, reliability of these factors has never been confirmed. Since publication of the MSF by Seed and Idriss in 1982, several other studies have been performed. These include energy concepts (Arango, 1994), statistical evaluation (Loertscher, 1994), and seismologic evaluation (Ambraseys, 1988). They all suggest that the Seed and Idriss MSF are very conservative (overestimating the damaging effect) for small to moderate earthquakes (magnitudes less than 7) and underestimating the damaging effects for large earthquakes (magnitudes greater than 8). A comparison of the various relationships is shown in Figure 6. As shown in this figure, the curve recommended by Arango is in close agreement with that recommended by Ambraseys, both of which are more conservative than Loertscher. For the probabilistic liquefaction hazard evaluation, the magnitude scaling factors derived by Arango were adopted.

Free-Field Seismic Ground Response Analysis

Free-field ground response analyses were performed using the computer program SHAKE, (Idriss et al., 1992). In all SHAKE runs, the input motion was prescribed as outcrop at a depth of 960 feet below ground surface, corresponding to the top of rock.

Response analyses would theoretically include ten soil columns (ten SCPTs velocity profiles for the upper layers), two shear wave velocity profiles (for the deep soil layers), and ten acceleration time histories, for a total of 200 separate computer runs. To reduce this number for each input motion a total of eight soil columns were randomly selected from the set of ten soil columns: half of these were combined with one of the deep shear wave velocity profiles and the other half were combined with the other deep shear wave velocity profile. With this random selection, the variation of soil properties were considered and the number of SHAKE runs was reduced to eighty. The results of the SHAKE analyses in terms of effective CSR (defined in Equation 2) at two depths within the TR3 and TR4 layers (the critical layer for liquefaction) were obtained.

ASHLE ANALYSES AND RESULTS

Using the EPRI deaggregated hazard data corresponding to 2.5 Hz hazard curves, the normalized SPT, $(N_1)_{60}$ data (histogram, normal, and lognormal distributions), the CSRs (histograms, normal, and lognormal distributions), a series of ASHLE analyses was performed. The results are shown in Figure 7. As shown in this figure, the results are not sensitive to the use of either measured, simulated, or combined $(N_1)_{60}$ data. The results change, however, from $5x10^{-6}$ to approximately $2.5x10^{-6}$ once the shear strength factor changes from 1.35 to 1.9. The cluster of nine data points shown in each group of results in Figure 7 is due to the combination of various cases from three distributions of CSR and $(N_1)_{60}$ data (histogram, normal, and lognormal distributions) indicating that the liquefaction hazard is insensitive to statistical characterization of CSR and $(N_1)_{60}$ values.

It can also be seen with reference to the DOE criteria referred to in the Introduction section of this document that the calculated liquefaction hazard levels meet the target seismic performance goals recommended for Performance Category 4 (i.e., for nuclear facilities). Thus, the analyses show that for this site the ultimate acceptable degree of safety under natural hazards is controlled by factors other than exposure to seismic liquefaction.

CONCLUSION

The methodology developed demonstrates one of the benefits to using probabilistic analyses; that is the ability to compare the relative risk of various hazards or accident scenarios for critical facilities or components. The results can also be combined with other probabilistic studies to determine the overall risk to a facility from a seismic event, or any other postulated accident. Thus, more informed decisions regarding where evershrinking funds should be focused can be made.

REFERENCES

Ambraseys, N. N., 1988, "Engineering Seismology," Earthquake Engineering and Structural Dynamics, Vol. 17.

Arango, I., 1994, "Methodology for Liquefaction Potential Evaluation of Sites East of the Rockies," Bechtel Corporation, Technical Grant, June.

Arango, I., and Ostadan, F., 1995, "Quantification of Liquefaction Hazard and Its Application to Risk Assessment and Urban Zoning," 5th International Conference on Seismic Zonation, Nice, France, October.

Bechtel Corporation, 1991, Advanced Seismic Hazard/Liquefaction Evaluation (ASHLE), San Francisco, California.

Costantino, C. J., 1994, "Results of Recent Sensitivity Studies for the Savannah River RTF Site Subjected to the Loma Prieta and Artificially Generated Rock Outcrop Time Histories," Letter to J. Kimball from C. Costantino, January 27, 1994.

Department of Energy, 1994, Natural Phenomena Hazards Assessment Criteria, Draft DOE Standard 1023, Washington, D. C., November 1994.

Electric Power Research Institute (EPRI), 1986, "Seismic Hazard and Methodology for the Central and Eastern United States," EPRI Technical Report NP-4726-A, 11 Volumes.

Electric Power Research Institute (EPRI), 1993, "Guidelines for Determining Design Basis Ground Motions," EPRI Technical Report TR-102293, November 1993.

Idriss, I. M., and Sun, J. I., 1992, "User's Manual for SHAKE 91," University of California, Davis, November 1992.

Lawrence Livermore National Laboratory (LLNL), 1989, "Seismic Hazard Characterization of 69 Plant Sites East of the Rocky Mountains," UCID-21517, U. S. Nuclear Regulatory Commission Report, NUREG/CR5250, by D. L. Bernreuter, J. B. Savy, R. W. Mensing, J. C. Chen, B. C. Davis.

Liao, S. S. C., Veneziano, D., and Whitman, R. V., 1988, "Regression Models for Evaluating Liquefaction Probability," *Journal of Geotechnical Engineering*, ASCE, Vol. 114, No. 4, pp. 389-411.

Loertscher, T., 1994, "Magnitude Scaling Factors for Analysis of Liquefaction Hazard," Doctoral Dissertation, Department of Civil Engineering, Brigham University, Provo, Utah.

Murray, R. C., Nelson, T. A., Short, S. A., Kennedy, R. P., Chander, H., Hill, J. R., and Kimball J. F., 1994, "DOE Natural Phenomenal Hazards Design and Evaluation Criteria." In Current Issues Related to Nuclear Power Plant Structures, Equipment and Piping, Proceedings of the Fifth Symposium, Orlando, Florida, December 1994.

Ostadan, F., Arango, I., Litehiser, J., and Marrone, J., 1991, "Liquefaction Hazard Evaluation," 11th SMIRT Conference, Tokyo, Japan.

Seed, H. B. and Idriss, I. M., 1982, "Ground Motions and Soil Liquefaction During Earthquakes," Earthquake Engineering Research Institute, Monograph Series, p. 134.

Seed, H. B., Tokimatsu, K., Harder, L. F., and Chung, R. M., 1984, "The Influence of SPT Procedures on Soil Liquefaction Resistance Evaluations," Report No. UCB/EERC-84/15, Earthquake Engineering Research Center, University of California, Berkeley, California, October 1984.

Seed, H. B., Idriss, M., and Arango, I., "Evaluation of Liquefaction Potential Using Field Performance Data," Journal of Geotechnical Engineering, Vol. 109, No. 3, March, 1983.

Silva, W. J., and Lee, K., 1987, WES RASCAL Code for Synthesizing Earthquake Ground Motions, Report 24, State-of-the-Art for Assessing Earthquake Hazards in the United States, Misc. Paper S-73-1, U. S. Army Engineer Waterways Experiment Station, p. 120.

TABLE 1 QUALITATIVE SEISMIC PERFORMANCE GOALS

Performance Category	Occupancy Safety	Concrete Barrier	Metal Liner	Component Functionality	Visible Damage
1	No structural collapse, failure of contents not serious enough to cause severe injury or death, or prevent evacuation.	Confinement not required.	Confinement not required.	Component will remain anchored, but no assurance it will remain functional or easily repairable.	Building distortion will be limited, but visible to the naked eye.
2	No structural collapse, failure of contents not serious enough to cause severe injury or death, or prevent evacuation.	Concrete walls will remain standing, but may be extensively cracked; they may not maintain pressure differential with normal HVAC. Cracks will still provide a tortuous path for material release. Don't expect largest cracks greater than 1/2 inch.	May not remain leak tight because of excessive distortion of structure.	Component will remain anchored and majority will remain functional after the earthquake. Any damaged equipment will be easily repaired.	Building distortion will be limited, but visible to the naked eye.
3	No structural collapse, failure of contents not serious enough to cause severe injury or death, or prevent evacuation.	Concrete walls cracked; but small enough to maintain pressure differential with normal HVAC. Don't expect largest cracks greater than 1/8 inch.	Metal liner will remain leak tight.	Component anchored and functional.	Possibly visible local damage, but permanent distortion will not be immediately apparent to the naked eye.
4	No structural collapse, failure of contents not serious enough to cause severe injury or death, or prevent evacuation.	Concrete walls cracked; but small enough to maintain pressure differential with normal HVAC. Don't expect largest cracks greater than 1/8 inch.	Metal liner will remain leak tight.	Component anchored and functional.	Possibly visible local damage, but permanent distortion will not be immediately apparent to the naked eye.

TABLE 2
SUMMARY OF SHALLOW DEPTH SHEAR WAVE VELOCITIES (fps)

SCPT	FILL	TR1	TR2	TR3/4	DB1/3	DB4	DB5	ST	GC
Depth Range (ft)	--	0-27	27-38.6	38.6-61.5	61.5-89.2	89.2-100.7	100.7-111.1	111.1-165.3	165.3-170
SCPT1	1129	1176	1499	976	795	757	867	824	1325
SCPT5	975	1461	1302	1304	1183	977	1000	1134	1325
SCPT7	1086	1196	1110	1268	1155	1055	--	1041	1325
SCPT11	--	1085	1119	1222	1206	1092	1457	1780	1326
SCPT14	--	1086	982	1363	1076	682	--	1409	1325
SCPT18	1175	1379	1122	1308	1361	1236	1152	1178	1325
SCPT25	--	1269	1132	1132	831	747	--	1223	1325
SCPT26	1024	1331	1126	1167	951	909	800	2278	1325
SCPT29	976	1129	1112	1042	1365	1166	--	1469	1325
SCPT30	1034	1167	922	1080	1137	1087	888	953	1325

Note: SCPT - Seismic Piezocone Penetration Probe

TABLE 3
SUMMARY OF LAYER THICKNESS (ft)

SCPT	FILL	TR1	TR2	TR3/4	DB1/3	DB4	DB5
SCPT1	41	15	27	24	32	9	8
SCPT5	41	25	15	22	36	7	10
SCPT7	41	30	20	24	31	24	--
SCPT11	42	30	18	22	32	8	15
SCPT14	6	28	26	24	35	17	--
SCPT18	10	23	31	27	28	5	15
SCPT25	--	33	25	21	25	9	--
SCPT26	36	11	21	23	33	15	16
SCPT29	44	35	16	24	35	4	--
SCPT30	41	34	12	22	31	6	21

Note: SCPT - Seismic Piezocone Penetration Probe

TABLE 4
SUMMARY OF SPT N-VALUE DATA

Data Points	Location	$(N_1)_{60}$ Values
Measured (Total = 50)	Boring SPT-14	9.8, 10.4, 6.7, 7.8, 7.7, 7.7, 7.0,7.0, 8.1, 14.3, 13.6, 12.4, 11.2, 10.0
	Boring SPT-18	9.5, 17.1, 10.5, 9.3, 8.1, 8.6, 9.1
	Boring SPT-20	15.5, 8.4, 9.2, 9.8, 8.9, 8.7, 6.3, 6.9, 6.1, 8.3, 7.5, 9.5, 9.5, 13.6
	Boring SPT-23.3	9.0, 5.6, 4.8, 3.1, 4.6, 3.8, 4.5, 2.2, 2.9, 3.6, 3.6, 3.5, 4.9, 3.5, 4.8
Simulated (Total = 21)	Tanks 1 - 4	5.4, 6.6, 5.2, 8.8, 11.4, 11.2
		12.0, 15.1, 8.3, 2.9, 5.3, 7.6, 10.0
		16.9, 3.4, 1.5, 4.2, 4.1, 6.9
		5.5, 6.9
Combined (Total = 71)	Borings SPT - 14, 18, 20 & 23.3, and Tanks 1 - 4	9.8, 10.4, 6.7, 7.8, 7.7, 7.7, 7.0, 7.0, 8.1, 14.3, 13.6, 12.4, 11.2, 10.0
		9.5, 17.1, 10.5, 9.3, 8.1, 8.6, 9.1
		15.5, 8.4, 9.2, 9.8, 8.9, 8.7, 6.3, 6.9, 6.1, 8.3, 7.5, 9.5, 9.5, 13.6
		9.0, 5.6, 4.8, 3.1, 4.6, 3.8, 4.5, 2.2, 2.9, 3.6, 3.6, 3.5, 4.9, 3.5, 4.8
		5.4, 6.6, 5.2, 8.8, 11.4, 11.2
		12.0, 15.1, 8.3, 2.9, 5.3, 7.6, 10.0
		16.9, 3.4, 1.5, 4.2, 4.1, 6.9
		5.5, 6.9

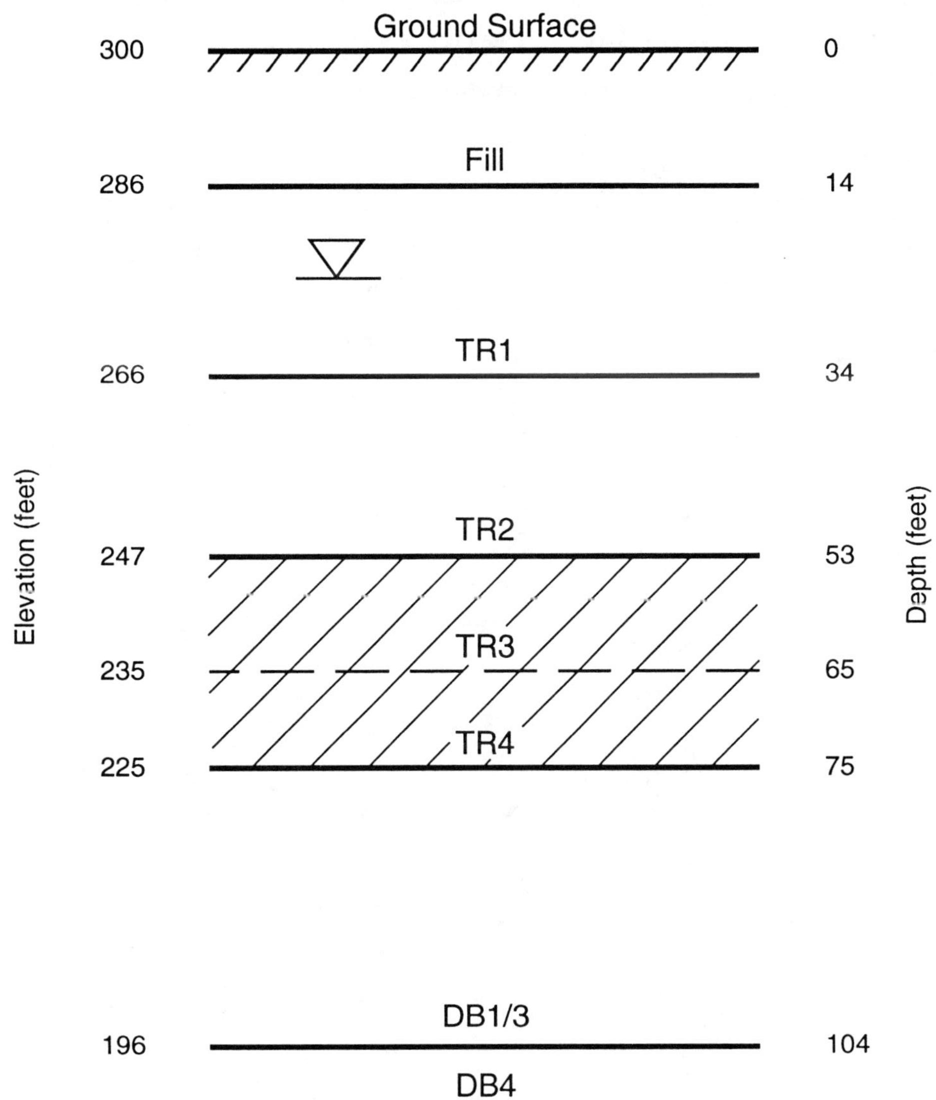

FIGURE 1 GENERALIZED SUBSURFACE SOIL PROFILE

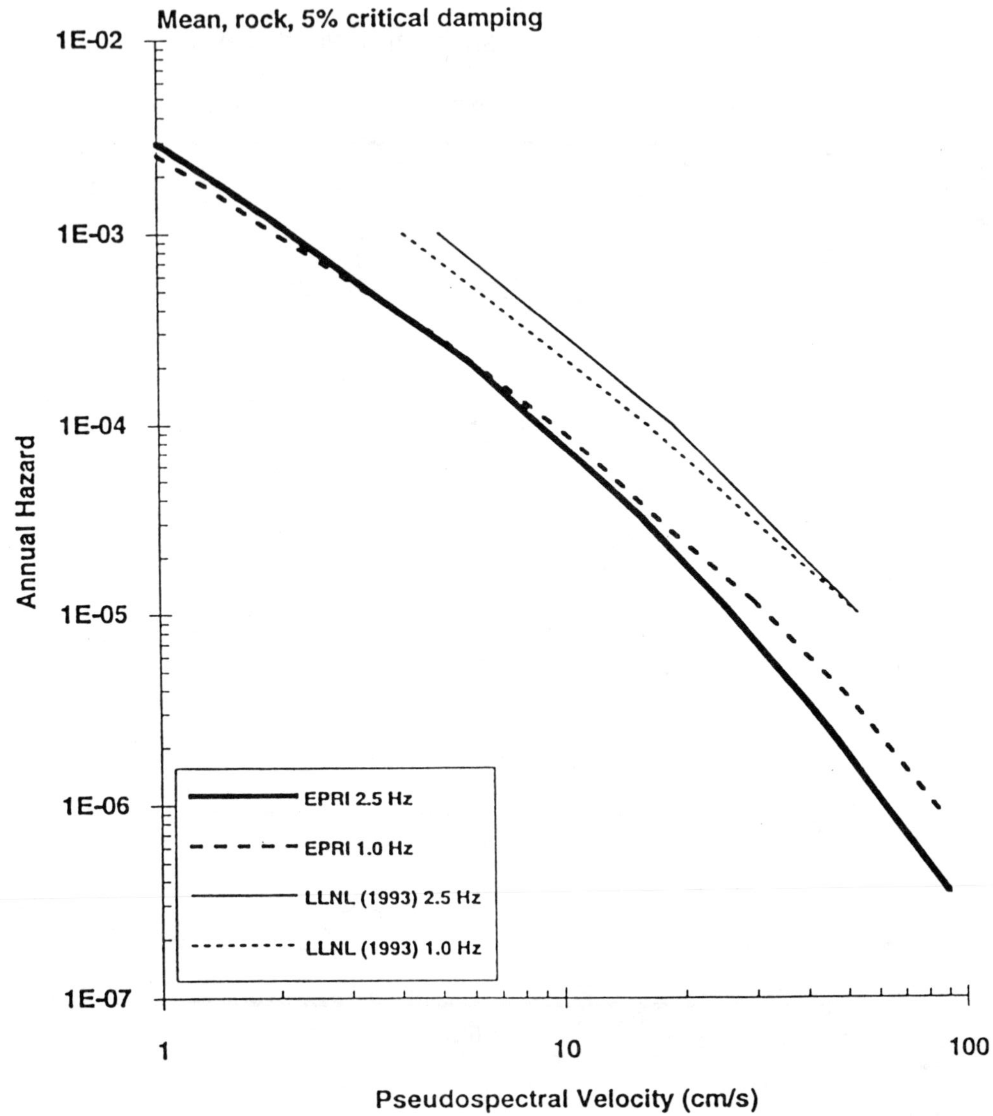

FIGURE 2 EPRI AND LLNL SPECTRAL HAZARD CURVES

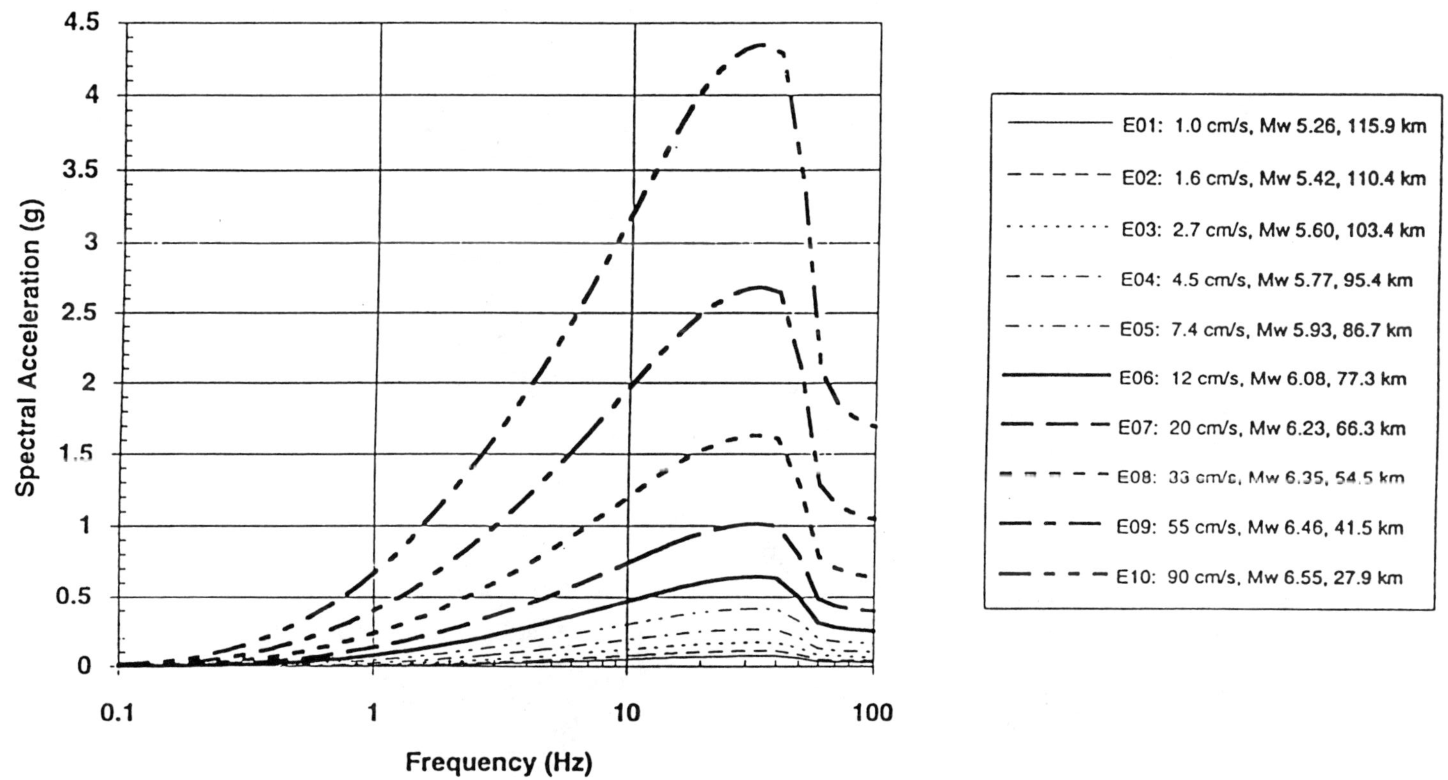

FIGURE 3 ACCELERATION RESPONSE SPECTRA (5% DAMPING) DEVELOPED FOR THE ROCK MOTION

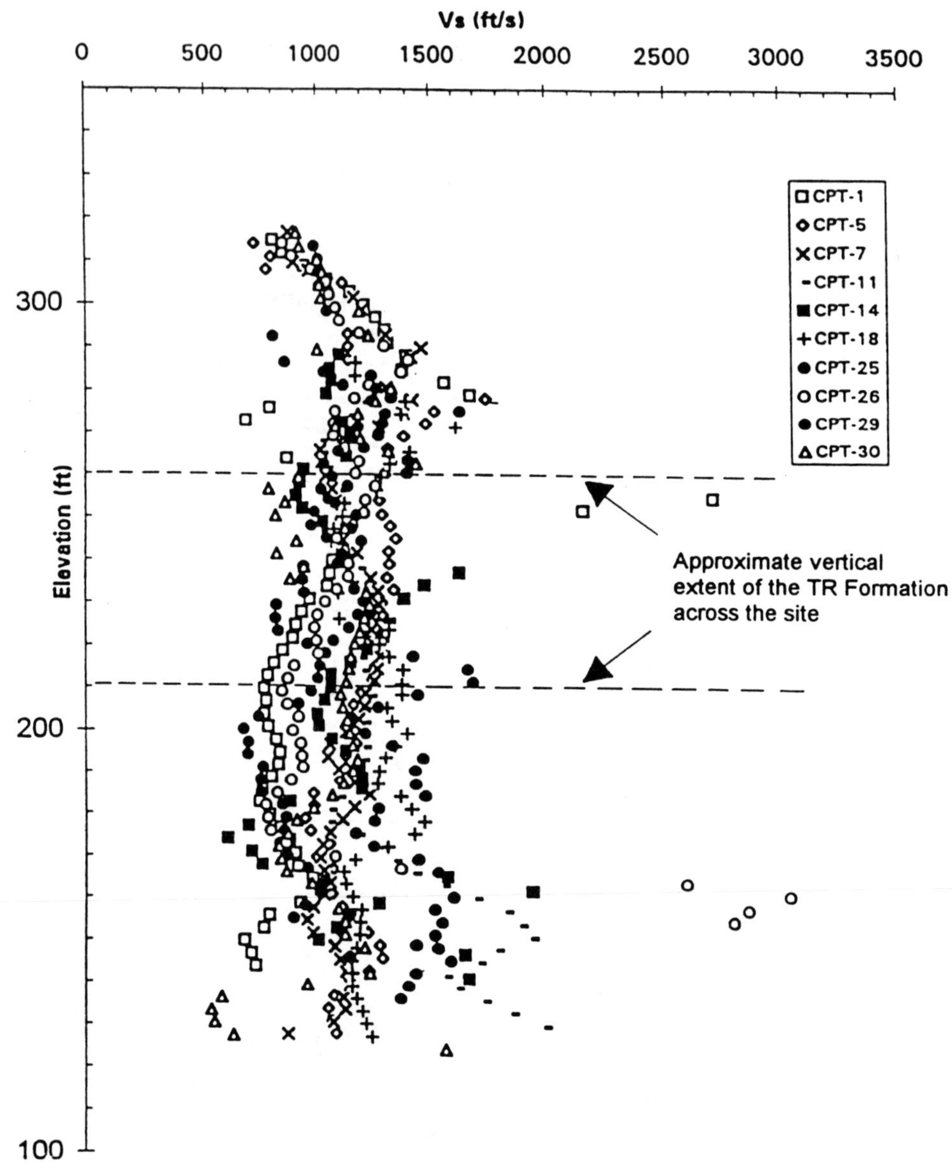

FIGURE 4 SHEAR WAVE VELOCITY PROFILES FROM SEISMIC CONE PROBES

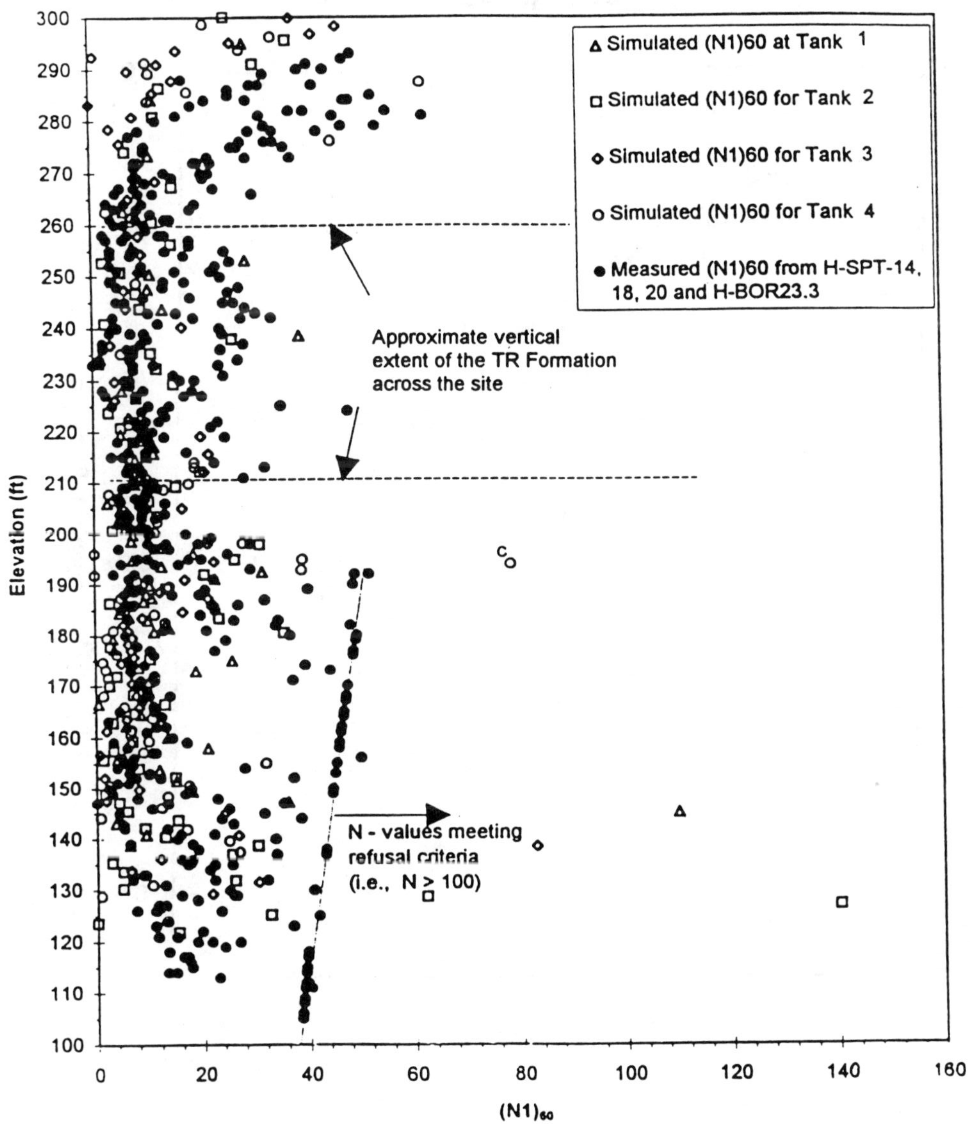

FIGURE 5 MEASURED AND SIMULATED $(N_1)_{60}$ VALUES

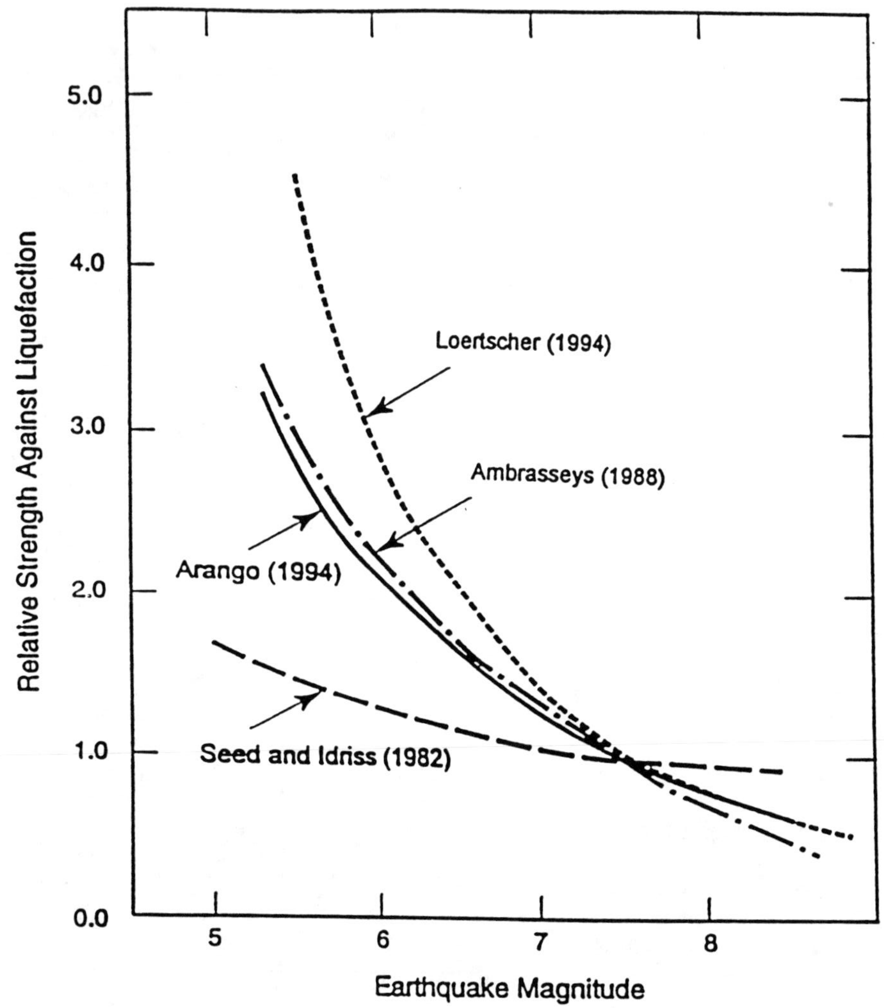

FIGURE 6 COMPARISON OF EARTHQUAKE MAGNITUDE SCALING FACTORS FROM VARIOUS SOURCES

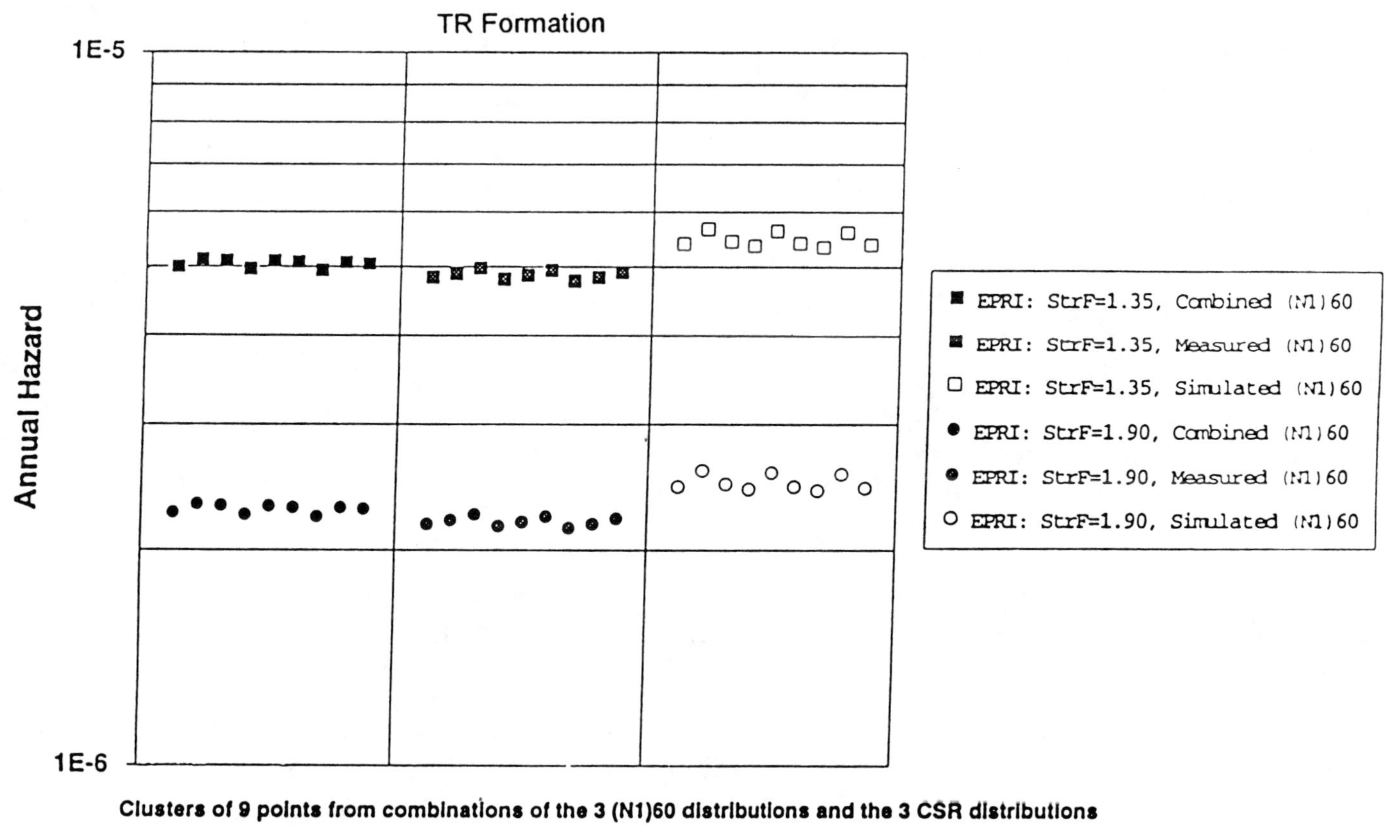

FIGURE 7 ANNUAL LIQUEFACTION HAZARD

PVP-Vol. 340, Seismic Engineering
ASME 1996

APPLICATION OF THE SASSI SOIL STRUCTURE INTERACTION METHOD TO CANDU 6 NPP SEISMIC ANALYSIS

Rick A. Ricciuti, Medhat Elgohary & S. Aman Usmani
Atomic Energy of Canada Limited
Sheridan Park Research Community
Mississauga, Ontario, Canada

ABSTRACT

The standard CANDU 6 NPP has been conservatively qualified for a Design Basis Earthquake (DBE) peak horizontal ground acceleration of 0.2g. Currently there are potential opportunities for siting the CANDU 6 at higher seismicity sites. In order to be able to extend the use of a standardized design for sites with higher seismicity than the standard plant, various design options, including the use of the SASSI [1] Soil Structure Interaction (SSI) analysis method, are being evaluated. This paper presents the results of a study to assess the potential benefits from utilization of the SASSI computer program and the use of more realistic damping ratios for the structures.

1.0 INTRODUCTION

Through their operating record in Canada and abroad, CANDU 6 Nuclear Power Plants (NPPs) have proven to be safe, reliable and competitive. Currently, there are 12 CANDU 6 units in operation and under construction world–wide. The CANDU 6 structures, systems and components are designed to sustain the effects of the seismicity requirements at a selected site. The standard CANDU 6 Nuclear Power Plant (NPP) has been conservatively qualified for a Design Basis Earthquake (DBE) peak horizontal ground acceleration of 0.2g. This is the highest DBE level for a CANDU 6 site to date. The improvement of our seismic analysis methods is an ongoing process. Concurrently new sites with different seismic conditions are being considered for CANDU 6 siting. Qualifying the CANDU 6 standard plant to a higher seismic level using the same methodology and input parameters that are currently being used may lead to unnecessary design modifications.

A rational first step for qualification of the standard plant to a higher seismic level would be the use of more realistic determination of the input design parameters (FRS and time history) for the qualification of the structures, systems and components utilizing state-of-the-art Soil-Structure Interaction (SSI) computer programs and methodology. For this purpose, the SASSI (System for Analysis of Soil Structure Interaction [1]) computer program has been investigated for its potential benefits. Secondly, in the generation of the input design parameters for the structures, a more realistic damping and treatment of uncertainties [11] that are consistent with the current state of knowledge can be used.

The purpose of this study is to assess the potential benefits from utilization of the SASSI computer program in reducing the input seismic design parameters and the potential benefits of using higher damping ratios for the structures.

2.0 SOIL STRUCTURE INTERACTION ANALYSIS METHODS

The SSI effect is an important consideration in the seismic design of a critical structure such as the CANDU Reactor Building (R/B). Several well-known methods are currently being employed in the nuclear industry for the analysis of SSI effects. These methods include the soil-spring method, the two-dimensional (2-D) finite element method (FLUSH [2]), the more sophisticated half-space continuum impedance method (CLASSI [3]). Each of these commonly-applied methods has limitations resulting from the intrinsic assumptions used in the formulation of the SSI analysis problem.

The computer program SASSI was chosen for this study because the program is best capable of rigorously analyzing 3-D SSI response for surface-supported or embedded structures with rigid or flexible foundations.

The major steps in a typical SSI analysis are schematically shown in Figure 1 and include: (1) Site response analysis; (2) Foundation scattering analysis; (3) Foundation impedance analysis; (4) Dynamic modeling of structural system and solving the SSI problem. A good description of each of these steps, the key modeling information required and the functional relationships among the SSI response parameters is given in Reference [4].

2.1 <u>Soil-Spring Method</u>

The Soil-Spring Method has been employed on the standard CANDU 6 NPP. The method is based on representation of the foundation impedances by frequency-independent soil springs and dampers which have constant coefficients. The method uses a two step solution procedure, consisting of the foundation impedance analysis and the interaction response solution. The structure is represented by either a lumped-mass stick model or a finite element model.

The method employs published formulae [5] for calculation of the soil spring and damper coefficients and are based on the assumption of a rigid massless base supported on the surface of a uniform half space. The effects of soil layering, non-vertically propagating waves and foundation embeddment are not accounted for. Seismic input motions are assumed to be prescribed on the ground surface. The method implicitly assumes that the input motions are vertically propagating seismic waves. Local site response may be performed separately. The computer program SHAKE [6] can be used in the local site response analysis to determine strain compatible soil properties to be used in the calculation of the spring and damper coefficients.

2.2 SASSI Method

The SASSI method of SSI analysis is a general 3-D finite element substructuring method using a discretized halfspace foundation model. The basic analytical method used in SASSI is the complex response method, which works in the frequency domain, and the flexible volume method of substructuring. The flexible volume method of formulation employs a finite element substructuring technique in which the structure and foundation models are partitioned as shown in Figure 2. The structure model in Figure 2c consists of finite element models for the superstructure plus the basement minus the excavated soil. The foundation model, Figure 2b, consists of the finite element model of the original site, i.e., the excavated soil for the basement is retained with the foundation soil model. With such a partitioning, the foundation medium retains the halfspace (without excavation pit) configuration which greatly simplifies the calculations of the free-field motion and the foundation impedances. The SSI interaction occurs at every interacting node within the excavated soil volume, called the flexible volume, as shown in Figure 2. The foundation impedances are calculated for each interacting node within the flexible volume, and then coupled with the structural model to form the complete SSI system. The SASSI method does not explicitly determine the scattered foundation input motions. It combines the determination of scattered foundation input motions with the solution for structural response in the single-step interaction response solution.

The flexible-volume substructuring technique used in SASSI requires several steps of analysis : (1) site response analysis which determines the free-field motions within the flexible volume resulting from the prescribed seismic wave fields; (2) foundation impedance analysis which calculates the impedance matrix associated with the interacting nodes in the flexible volume; and (3) SSI response analysis which computes the final SSI response of the assembled SSI system. A complete description of the program, its theory and capabilities is given in References [1,4].

3.0 ANALYTICAL MODELS

The reactor building consists of an internal concrete structure enclosed by a cylindrical containment wall, both supported on a common base slab. The base slab is supported on a concrete subbase (Figure 3). The internal structure houses the concrete vault, reactor and other components and systems. The containment wall supports the dome, dousing tank and other equipment.

For the seismic analysis of the reactor building a lumped mass stick model is utilized (Figure 4). The containment structure, internal structure and the reactor vault are modeled by separate sticks representing the stiffness of these structures in all three directions. The internal structure masses are lumped at major floor levels and include both structural, live load and equipment loads. The containment structure masses are lumped at several points along the height of the stick. The dousing water is considered as rigidly attached mass at the dome and the ring beam area of the containment structure. All lumped masses are considered for all three directions. In the internal structure, the beams of the model at various elevations are located at the center of stiffness. Eccentricity of floor masses with respect to the center of stiffness is also taken into consideration.

3.1 Application of Soil Spring Method

In the soil spring method, the reactor building base slab is supported on six degree of freedom springs representing the foundation media which is idealized as semi-infinite half space. The spring stiffnesses are calculated using the expressions widely published (e.g. Ref. [5]). Figure 5a shows the strain compatible dynamic rock properties for the potential site, which is layered, used in the calculation of the soil-spring stiffnesses. The effect of layering at the site was accounted for in the soil-spring determination using the methodology given in Reference [9].

Damping values for the structure are per CAN3-N289.3-M81 [7]. The damping value used for reinforced concrete internal structure and concrete vault is 5% and for the prestressed containment structure is 3%. The total soil damping is limited to 10% for rocking and 30% for translation (including radiation damping).

The Ground Response Spectra (GRS) for the standard plant site is shown in Figure 6. The peak horizontal acceleration for the Design Basis Earthquake (DBE) is 0.2g. The DBE ground motion parameters and the shape of the GRS follow the requirements of CSA CAN3-N289.3-M81. The peak horizontal DBE acceleration for the proposed site is 0.25g.

For the standard plant, the response spectrum method is used for generation of the FRS. An AECL in-house developed computer code is used for generation of the FRS. The program is based on the methodology described in Reference [8]. The program uses the results of the modal analysis of the stick model (using the computer program STARDYNE) and the GRS to generate the FRS. The FRS are generated for the three rock conditions and the results are enveloped and broadened by a minimum of +/-15%.

For the analysis of some equipment, time-history direct integration analysis was used. For this purpose, a spectrum compatible artificial time-history was generated following the matching acceptance criteria of CSA N289.3-M81.

3.2 Application of SASSI Method

The dynamic lumped mass stick models used to represent the behavior of the containment structure, internal structure and reactor vault are equivalent to the standard plant soil-spring analysis method discussed above.

In SASSI, a 3-D finite element model has been used to model the CANDU 6 base slab foundation. The brick element model of the base slab for the surface supported model is shown in Figure 7a. A total of 160 elements are used to model the foundation base slab. Stiff properties are used for the brick elements to ensure the rigidness of the foundation due to the added stiffness of the containment and internal structure walls.

To examine the effects of subbase embeddment, a second model was developed that considered the subbase to be fully embedded within the top rock layer. The foundation model for the embedded model is shown in Figure 7b. An additional 160 brick elements have been used to model the subbase.

The SASSI site model for the potential site, best-estimate rock properties, is shown in Figure 5a. The model consists of 5 layers that model the top soft rock layer, and an additional 10 layers with varying thickness, with a viscous boundary at the base to simulate the stiffer rock halfspace. To assess the uncertainty in the rock properties the best estimate soil properties are varied by +/- 50%.

It has been reported in several references [4,10] that the effect of soil layering can be substantial. The frequency dependent impedances which characterize and govern the seismic response of a given foundation can often be highly sensitive to the layered soil properties. Therefore in order to investigate the effect of soil layering on both the foundation impedances (stiffness and damping) and the seismic response of the reactor building, an equivalent uniform halfspace SASSI model has been developed and used for comparative purposes. The model is also shown is Figure 5b. This model is consistent with the soil spring model developed for a shear modulus G of 1500 MPa, which is the equivalent G for the two layer Potential site.

The seismic input used in the SASSI analysis is identical to that applied to the soil-spring CANDU 6 model. The control point for the time history input was applied at the ground surface. The input motion is assumed to consist of a vertically propagating SV-wave field. The control motion is scaled to either a maximum of .25g, the higher seismic condition site, or .20g, the standard plant DBE level.

The SASSI surface supported model of the CANDU 6 reactor building is derived from the models shown in Figures 4, 5 and 7a. The reactor building concrete base slab, which is 1.68m (5.51 ft) thick, is supported directly upon the surface of the soil profile. Embeddment effects are not considered in this model.

For the standard plant the reactor building concrete base slab is supported by a reinforced concrete subbase, which is 3.37 m thick (11.06 ft). It is assumed to be fully embedded into the top rock layer for the second SASSI CANDU 6 model studied.

4.0 ANALYSIS RESULTS

4.1 Impedance and Damping Comparison: Soil Spring vs. SASSI

The horizontal and rocking foundation impedances (stiffnesses) are shown in Figures 8a and 8b respectively for the CANDU 6 reactor building. Each figure shows the impedance as a function of frequency for three cases: 1) Soil-spring method applied to an equivalent half space site; 2) SASSI method applied to an equivalent half space site; 3) SASSI method applied to the site with explicit consideration for soil layering. In the equivalent half space method, the actual layering of the rock properties is modeled by an equivalent uniform half space using the method prescribed in Reference [9].

In the soil-spring method, the impedances are taken to be frequency independent. From the Figure of horizontal impedances, the equivalent half space impedances computed using SASSI is in good agreement with the soil spring method in the low frequency range, with the difference increasing with higher frequencies. SASSI precisely accounts for the effects of soil layering by explicit modeling of the soil layers and their respective properties. It is seen that layering results in lower impedance values and an increase in the frequency dependency (oscillation) of the horizontal impedances.

In Figure 8b, the rocking impedances are shown for the three cases. The trends are similar to the horizontal impedance results, however the rocking impedances exhibit more variability with frequency than the horizontal impedances.

Figures 9a and 9b show the horizontal and rocking damping coefficients vs. frequency plots for the three cases. Again the damping coefficients are strongly frequency dependent. The 30% translation damping ratio used in the standard plant conservatively covers the SASSI damping coefficients. However for rocking, the damping ratio of 10% is over-estimated in the frequency range of 2.5 - 13.5 Hz when compared to the layered site. The rocking mode of the CANDU 6 reactor building generally falls within the 3Hz to 5Hz range. In this range the use of a 10% rocking damping ratio is over estimated by approximately 50 %.

For the case of a single layer deep deposit site that can be represented by a uniform equivalent half-space, SASSI predicts higher damping across the entire frequency spectrum in comparison to the soil spring method and the SASSI layered site values.

These results confirm that the horizontal and rocking impedances and damping coefficients for the layered site are strongly frequency dependent. The influence of soil layering on foundation impedances and damping has been reported in References [4,10]. The effect of layering on the impedance and damping function becomes significantly important for the case of a shallow layer and for the case of a high contrast between the elastic properties of the layer and those of the underlying half space. It is reported [10] that for a soft layer resting on a stiff half-space, radiation damping is minimal for frequencies below the frequency of the top soil layer, which is approximately 9 Hz for the considered site.

4.2 Layering Effects

To illustrate the effect of soil layering and its corresponding reduction in the level of foundation radiation damping, the CANDU 6 R/B FRS at key points have been computed using the two SASSI soil models (Figure 5), a SASSI model with layered soil properties and a SASSI model using an equivalent uniform half-space soil model. The FRS at key points under a .20g peak ground acceleration DBE are shown in Figures 10a and 10b. Figure 10a illustrates the FRS for the top of the CANDU 6 R/B containment structure using best estimate soil properties. The FRS has a predominant single peak at approximately 3.8 Hz, which is the rocking mode of the reactor building. The SASSI model that includes layering has the effect of reducing the foundation radiation damping below 9 Hz, the frequency of the top soil layer. This leads to an approximate 40% increase in the peak spectral acceleration value over the SASSI value determined for an equivalent uniform half-space site for frequencies below 9 Hz.

Figure 10b shows the FRS for the 100.0 m elevation of the internal structure, where the calandria vault is supported. The two dominant peaks correspond to the reactor building rocking mode and the first mode of the internal structure at approximately 7 Hz. The FRS peak corresponding to the R/B rocking mode increases in the range of 30-40 % over the uniform half space results due to reduced radiation damping associated with soil layering.

4.3 Subbase Embeddment Effects

Accounting for foundation embeddment in the SSI analysis is generally beneficial in terms of seismic response reduction. This is because the seismic ground motion accelerations are reduced with depth below the ground surface and the increases that occur in the foundation impedance and damping coefficients. Therefore the seismic input into the structure is reduced in comparison to that of a surface supported structure. For the CANDU 6 reactor building, the subbase may be considered as fully embedded. To study the benefit of subbase embeddment, the reactor building seismic response for a SASSI model with an embedded subbase was compared with the SASSI surface supported structure response. Reactor Building FRS at key points for a .20g peak ground acceleration DBE are shown in Figure 11. Figure 11a shows the FRS for the top of the CANDU 6 R/B containment structure. The effect of subbase embeddment is to reduce the FRS peak, corresponding to the rocking mode, by approximately 15 %. The 100 m elevation of the internal structure (Figure 11b) experiences an approximate 5-10 % reduction of the spectral acceleration values beyond the rocking mode frequency. Also an approximate 10% reduction in the zero period acceleration (ZPA) is noted due to the embeddment effects.

4.4 Concrete Damping Assessment

Current state of knowledge relating to damping factors [11] suggests that the structural concrete damping factors used in the seismic response analysis can be increased to more realistic levels. The standard plant used damping factors of 5 percent and 3 percent for the reinforced concrete internal structure and the prestressed concrete containment wall respectively. The SASSI SSI analysis results above are based on the standard plant damping factors. A study has been performed to assess the benefit of using higher structural damping factors based on Reference [11] recommendations. Damping factors of 7 percent for the reinforced concrete internal structure and 5 percent for the prestressed concrete containment wall have been examined.

SASSI computed FRS for the CANDU 6 R/B at key points have been computed using the best-estimate soil properties. The unbroadened 2% damped FRS at key points for a .25g peak ground acceleration DBE are shown in Figures 12a and 12b. For comparison purposes these Figures also show the broadened standard plant FRS, determined for a 0.20g DBE level, and the potential site FRS, determined using the soil-spring method.

Figure 12a shows the FRS comparison for the top of the containment structure 136.0 m elevation in the B/D direction. It is noted the use of a 5 percent damping ratio for prestressed concrete reduces both the predominant FRS peak and ZPA levels by approximately 9% for the containment structure at this elevation. Figure 12b shows the FRS comparison for the internal structure 100.0 m elevation in the B/D direction. At lower elevations of the internal structure the benefits of higher damping are reduced. The use of higher damping factors results in a reduction of both the FRS peaks and ZPA range by approximately 4%.

4.5 Fuelling Machine Seismic Input

The CANDU 6 Fuelling Machine (F/M) is used to perform on-line refuelling of the reactor, which is a significant operational advantage of the CANDU design. Canadian safety philosophy requires that the F/M survive the Design Basis Earthquake [12]. The standard plant F/M seismic qualification methodology uses the time-history direct integration analysis technique. In this method, the reactor building time-history seismic analysis is used to generate input motion at the F/M support points, which are then used in the F/M analysis. Therefore to assess the relative benefits of using SASSI, FRS plots have been generated that compare the seismic input from the soil-spring model using time-history analysis versus that generated using the SASSI method.

Shown in Figure 13 are the FRS for the concrete internal structure at elevations 125.3 m and 100.0 m respectively. Two FRS curves are shown on each plot, corresponding to the soil-spring method and SASSI method (layered site) generated for the best estimate proposed site soil properties.

The following observations can be made:

I. As a result of the layering characteristics, and its effect to reduce the foundation radiation damping, the F/M seismic input will increase by approximately 20-30 % in the rocking mode frequency range (3 to 5 Hz).

II. SASSI will provide reductions ranging from 10-30 % in the in the F/M seismic input for frequencies in the 6 to 15 Hz range in comparison to the soil-spring method. This is a result of the frequency dependency of the foundation impedances and the fact that the foundation impedances predicted by SASSI are lower than the soil-spring method values. The internal structure response reduces for lower foundation impedances.

III. Above approximately 20 Hz there is no significant difference between the soil-spring and SASSI results.

5.0 CONCLUSIONS

This study has focused on assessing the potential benefits from utilization of the SASSI soil structure interaction analysis method in reducing the input seismic design parameters compared with that obtained from the soil-spring analysis method. In addition, an assessment of the potential benefits of using higher damping ratios for the structure has been investigated. From this study, the following conclusions are made:

a) A layered site will have strong impedance and damping frequency dependency characteristics that will play an important role in the overall seismic response of the CANDU 6 reactor building. The SASSI soil structure interaction analysis capabilities accurately addresses and captures the influences of the soil layering and therefore is recommended for application at any layered site over the soil-spring method.

b) This study has shown that under certain conditions the SASSI SSI method can reduce the F/M seismic loads significantly for higher seismicity sites.

c) Since foundation embeddment is beneficial in terms of reducing the F/M seismic input levels, a cost/benefit analysis to assess the merits of increasing the subbase depth should be considered.

d) The use of increased structural damping factors results in a reduction of the reactor building FRS in the range of 4-9%.

6.0 Acknowledgment

The support of the CANDU generic development program and especially Basma Shalaby, the program manager throughout this study is duly acknowledged. The review and endorsement of the study report, on which this paper is based, by EQE International is also greatly appreciated. Finally the review comments of J.R. Hopkins, V.K. Bajaj, J.K. Biswas, D. See Hoye, M.A. Saleem and A. Saudy is gratefully acknowledged.

7.0 REFERENCES

1. SASSI, A System For Analysis of Soil-Structure Interaction Theoretical Manual, Bechtel Power Corporation, San Francisco, California, CE994, July 30, 1988.

2. Lysmer, J., Udaka, T., Tsai, C. F. and Seed, H.B., "FLUSH-A Computer Program for Approximate 3-D Analysis of Soil-Structure Interaction Problems, " Report No., EERC-75-30, University of California, Berkeley, 1975.

3. Wong, H.L. and Luco, J.E., "Dynamic Response of Rigid Foundation of Arbitrary Shape", J. Earthquake Engineering and Structural Dynamics, Vol. 4, pp. 576-587, 1976.

4. EPRI NP-7395, "Guidelines for Soil-Structure Interaction Analysis", October 1991.

5. Richart, Jr., F.E., Hall, Jr., J.R., and Woods, R.D., "Vibration of Soils and Foundations", Prentice Hall, N.J., 1970.

6. Schanbel, P.B., Lysmer, J., and Seed, H. B., "SHAKE - A Computer Program for Earthquake Response Analysis of Horizontally Layered Sites", Report No. EERC-72-12, University of California, Berkeley, California, December 1972.

7. CAN3–N289.3–M81 "Design Procedures for Seismic Qualification of CANDU Nuclear Power Plants".

8. Duff, C.G., "Earthquake Response Spectra for Nuclear Power Plants Using Graphical Methods", CSNI Specialist Meeting on Anti-Seismic Design of Nuclear Power Plant Installation, Paris, France, December 1975.

9. Johnson, G.R., Christiano, P., and Epstein, H.I., "Stiffness Coefficients For Embedded Footings", Journal of the Geotechnical Engineering Division, ASCE Vol. 101, No. GT8, Proc. Paper 11522, pp. 789-800, August 1975.

10. American Society of Civil Engineers, Seismic Analysis of Safety Related Nuclear Structures and Commentary, ASCE Standard 4-86, September 1986.

11. Usmani, S.A., Baughman, P.D., "Recommendations For Damping and Treatment of Modeling Uncertainty in Seismic Analysis of CANDU Nuclear Power Plant", PVP Conference 1996, Montreal (to be published).

12. Biswas, J.K., "Seismic Analysis of Fuelling Machine Support Structure for CANDU6 Reactor", PVP-Vol. 256-1, pp. 215-220, Denver, Colorado, 1993.

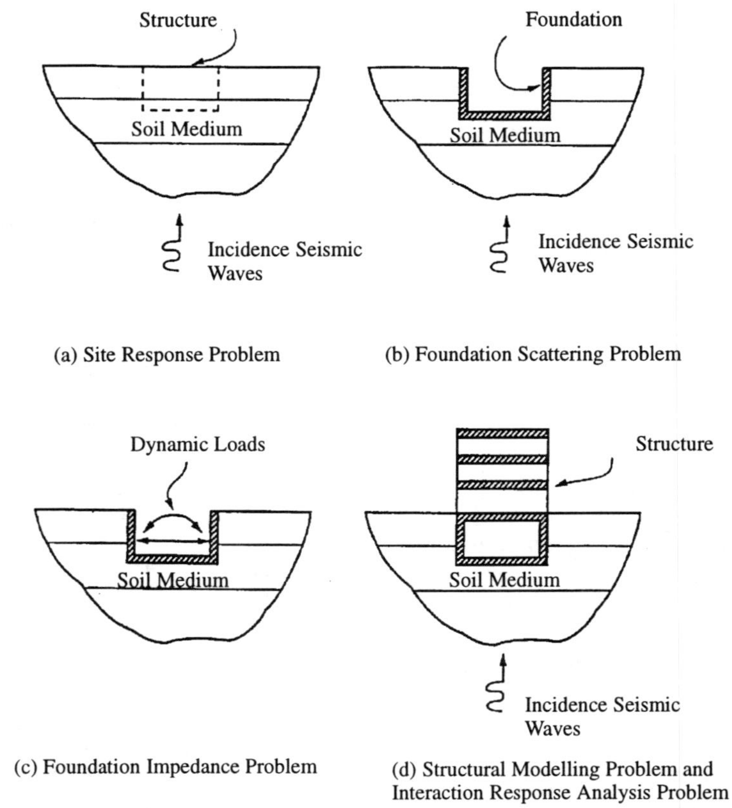

Figure 1 Major SSI Analysis Steps

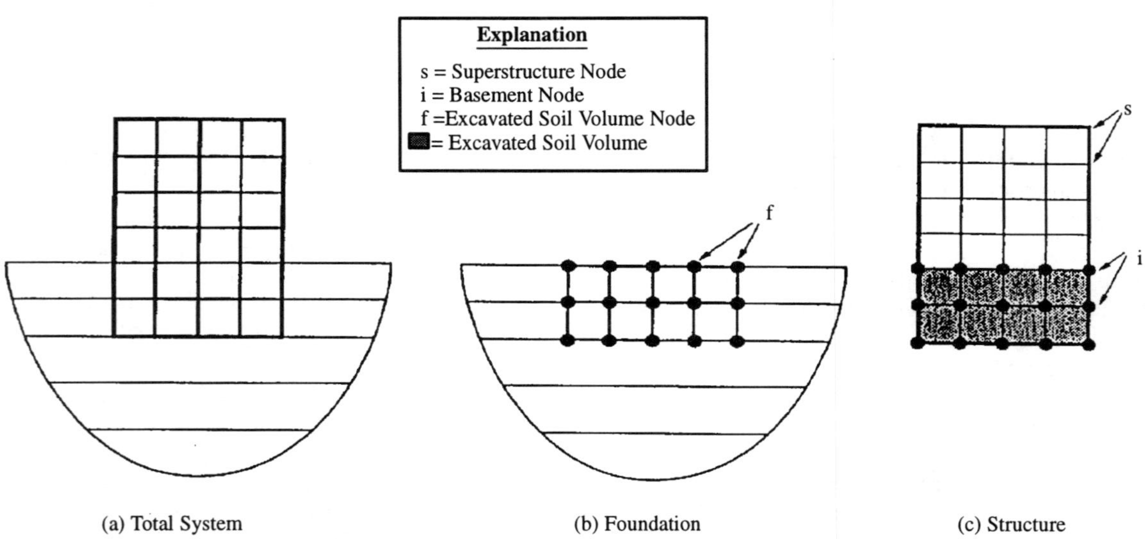

Figure 2 SASSI Substructuring Using Flexible Volume Method

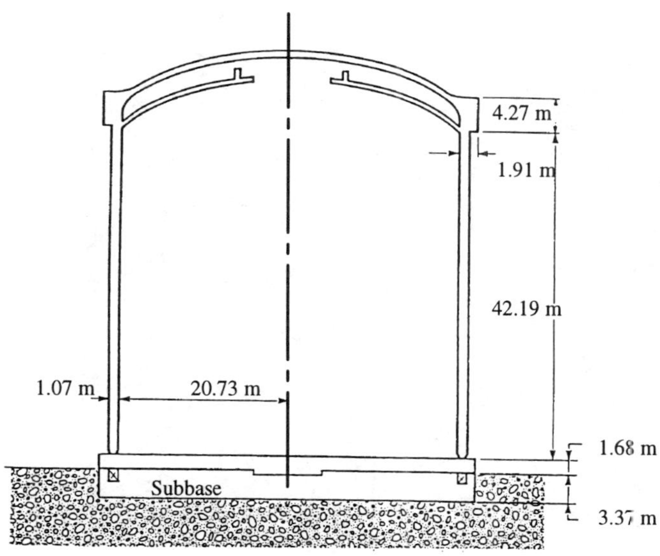

Figure 3 CANDU 6 Containment Structure Sectional View

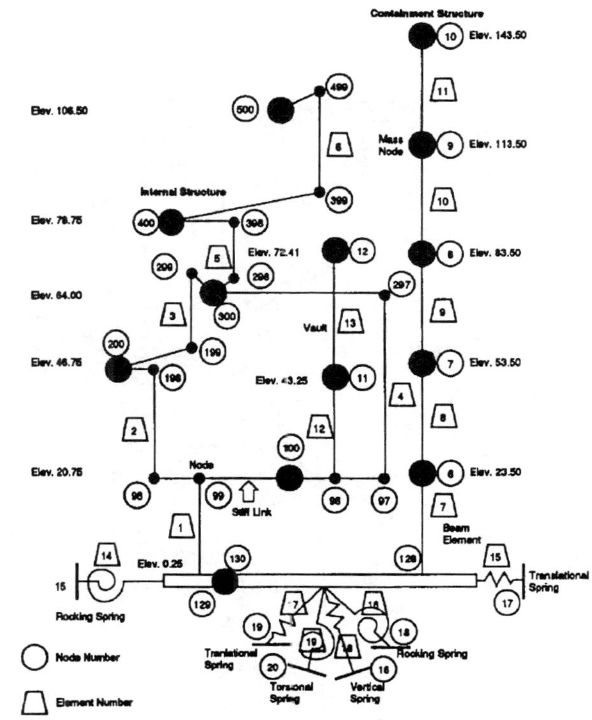

Figure 4 Schematics of the R/B 3–D Stick Model

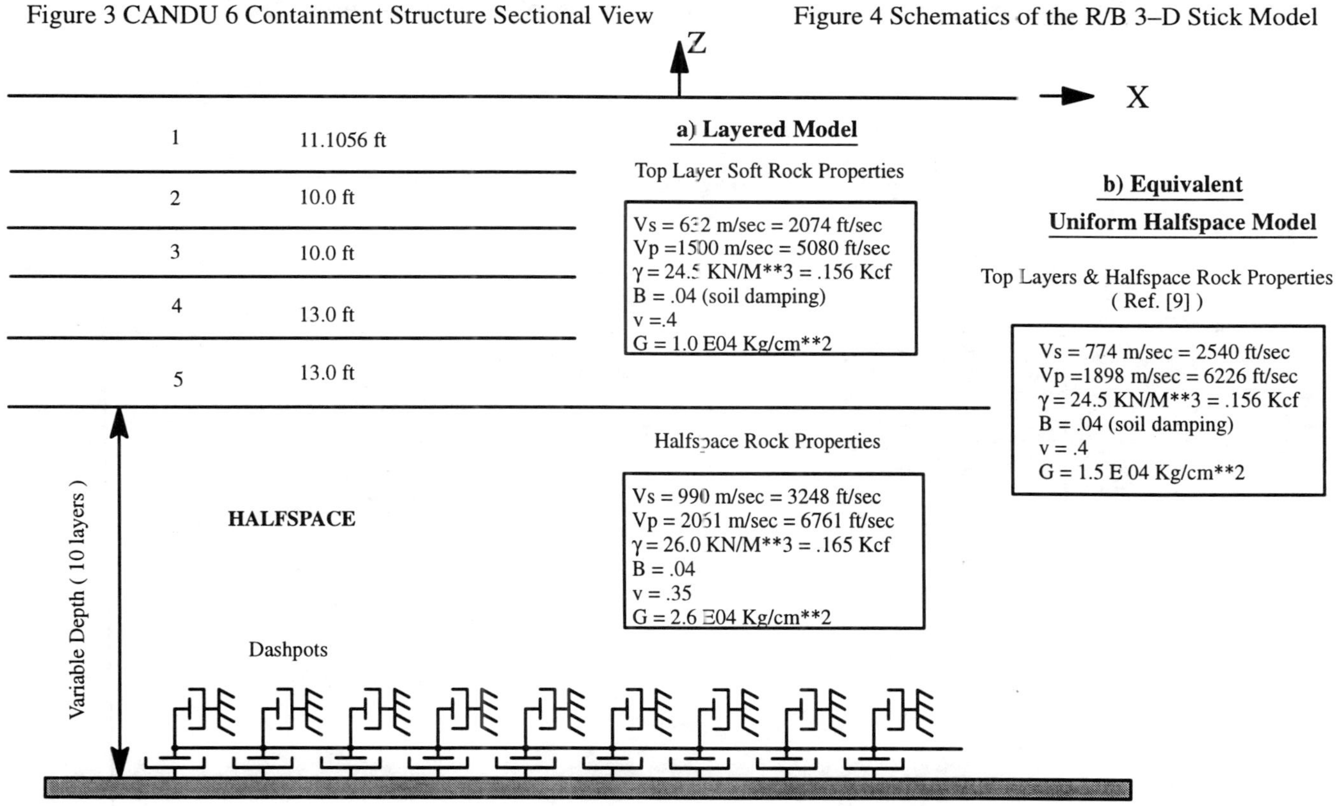

Figure 5 Potential Site – SASSI Soil Profile Model – Best Estimate: a) Layered ; b) Uniform Halfspace Models

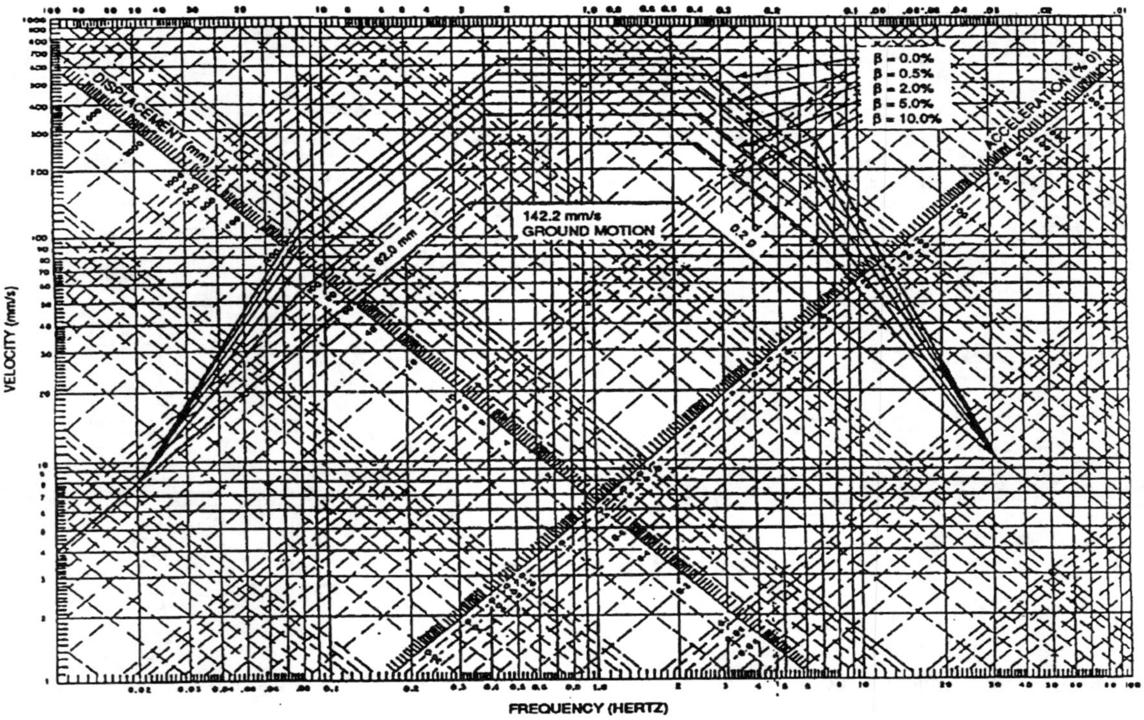

Figure 6 DBE Ground Response Spectra for CANDU 6 NPP

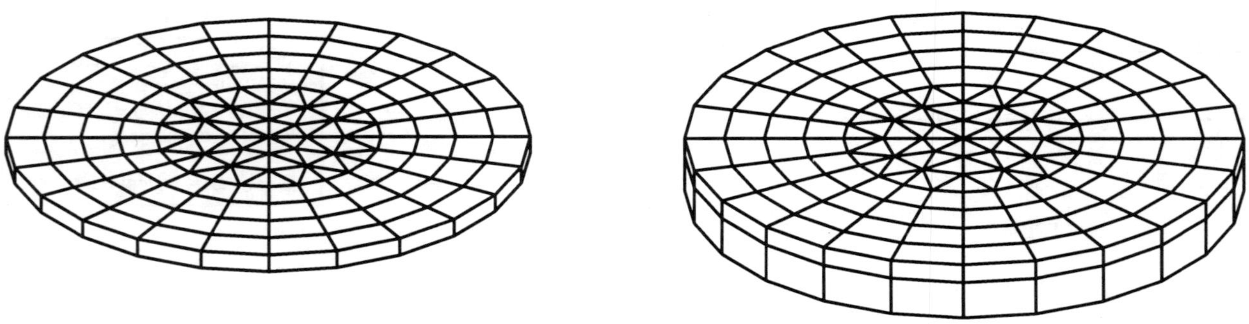

(a) Surface Supported Structure

(b) Embedded Subbase

Figure 7 CANDU 6 SASSI Base Slab Models

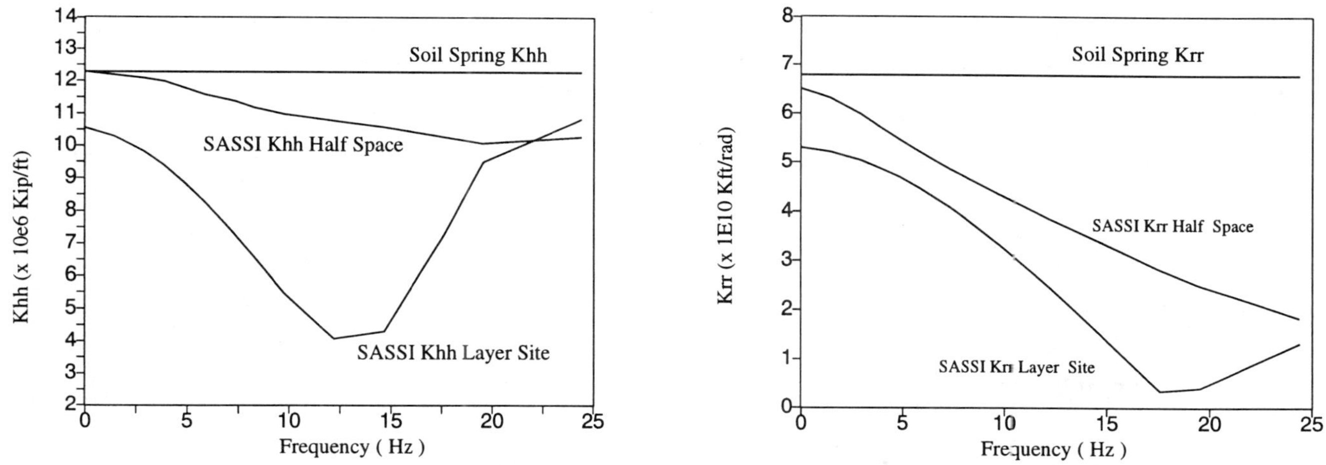

(a) Horizontal Impedance Coefficients

(b) Rocking Impedance Coefficients

Figure 8 CANDU 6 Surface Foundation Impedances

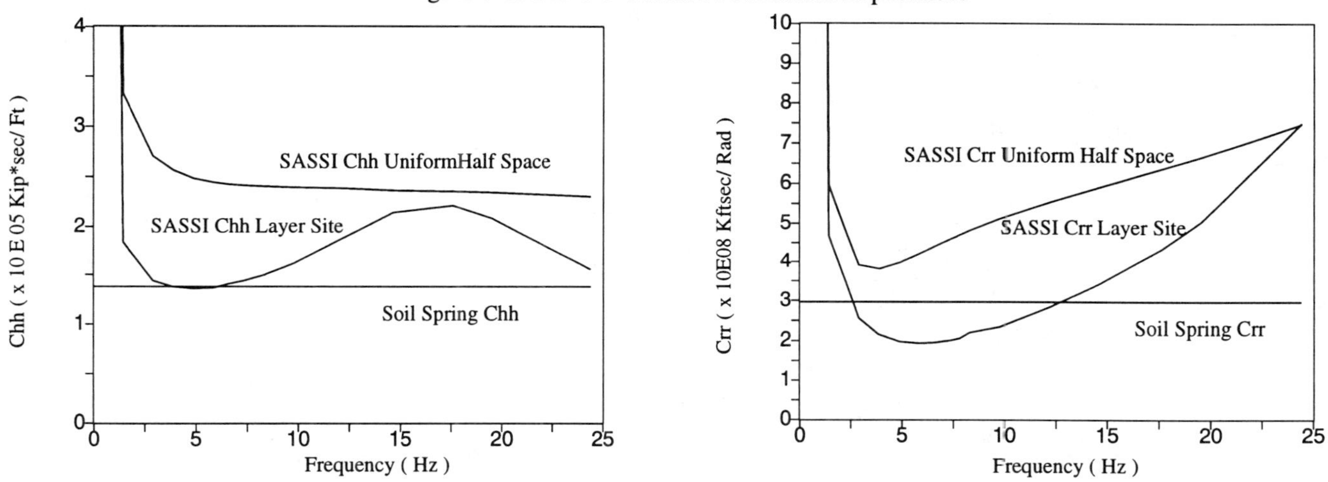

(a) Horizontal Damping Coefficients

(b) Rocking Damping Coefficients

Figure 9 CANDU 6 Surface Foundation Damping Coefficients

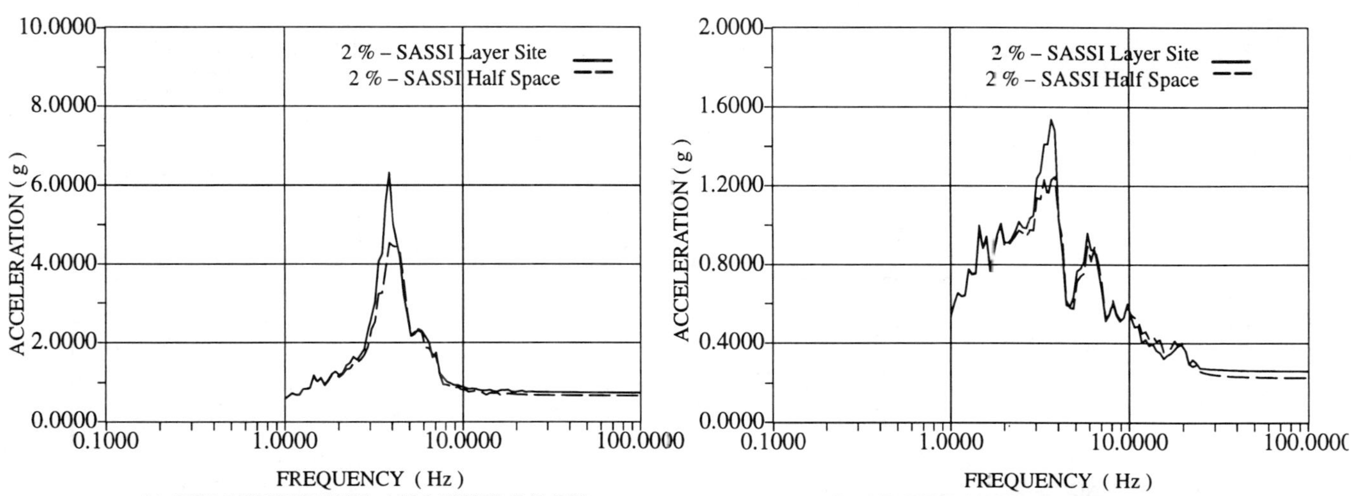

(a) TOP C/S NODE 10 EL. 136.0 M HOR B/D DIR

(b) I/S NODE 100 EL. 100 M HOR B/D DIR

Figure 10 R/B FRS: Elastic Half Space (G=1.5E04) Vs. Layered Site

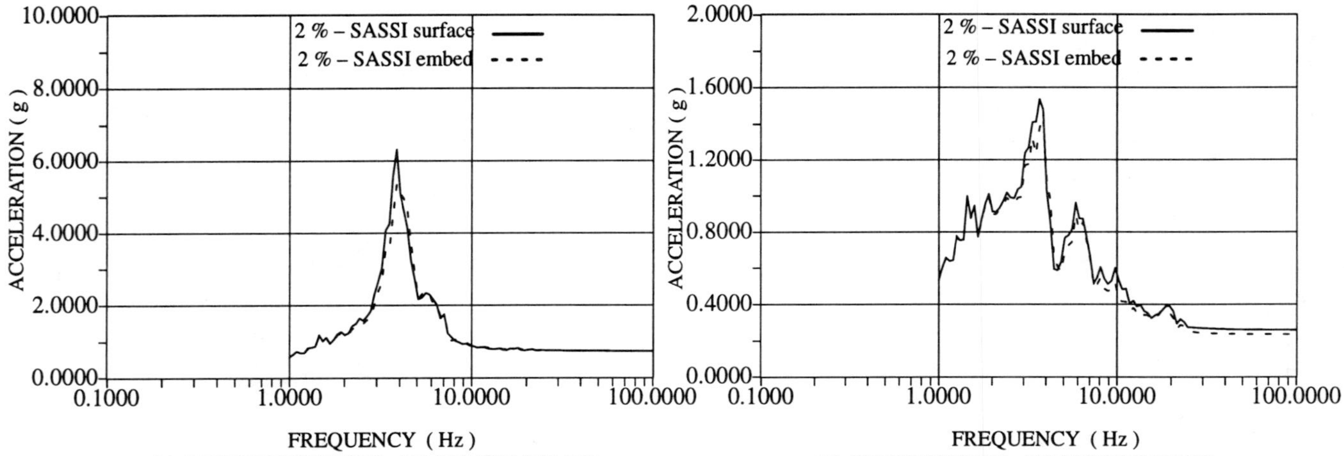

(a) TOP C/S NODE 10 EL. 136.0 M HOR B/D DIR

(b) I/S NODE 100 EL. 100 M HOR B/D DIR

Figure 11 R/B FRS Layered Site : Surface Vs. Embedded Comparison

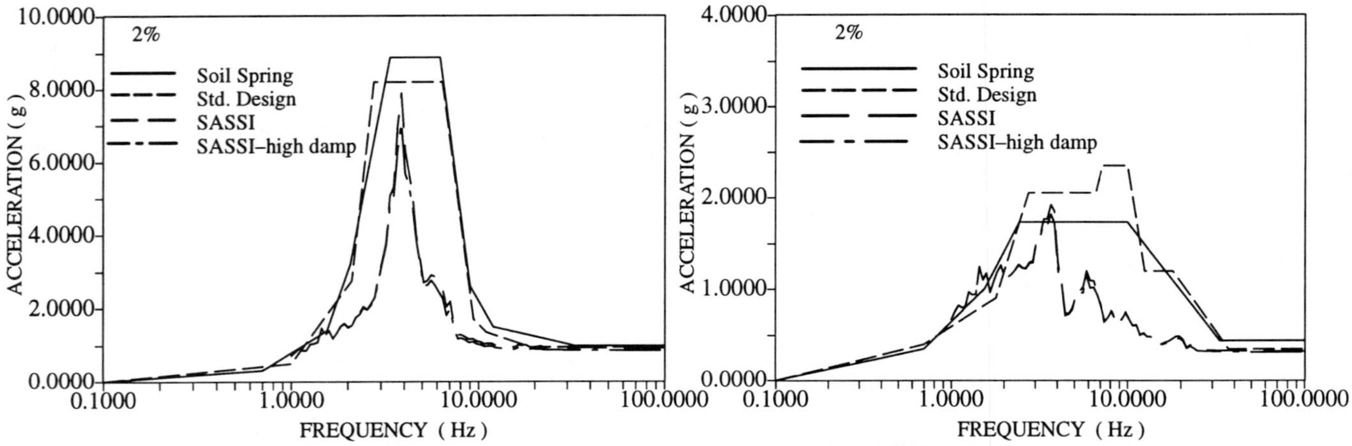

(a) TOP C/S NODE 10 EL. 136.0 M HOR B/D DIR

(b) I/S NODE 100 EL. 100 M HOR B/D DIR

Figure 12 R/B FRS Potential Site: Soil–Spring / Standard Design / SASSI / SASSI High Concrete Damping Comparison

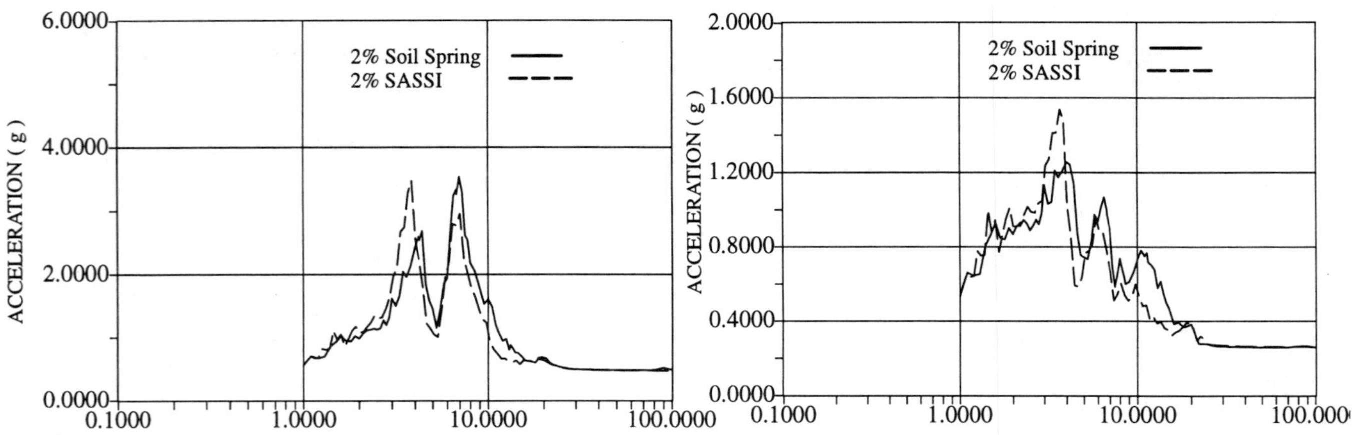

(a) TOP I/S NODE 500 EL. 125 M HOR B/D DIR – .25g

(b) I/S NODE 100 EL. 100 M HOR B/D DIR

Figure 13 R/B F/M FRS Potential Site : Soil–Spring vs. SASSI Comparison

PVP-Vol. 340, Seismic Engineering
ASME 1996

DESIGN SOIL PROFILES FOR SEISMIC ANALYSES OF AP600 PLANT STANDARD DESIGN

Farhang Ostadan

Bechtel National, Inc.
San Francisco, California

Chwen-I T. Liu

Bechtel Corporation
San Francisco, California

Karl K. Gross

Bechtel National, Inc.
San Francisco, California

Richard S. Orr

Westinghouse Electric Corporation
Pittsburgh, Pennsylvania

ABSTRACT

Future nuclear power plants are based on standard designs. For seismic qualification, a variety of subsurface conditions are considered in the design. While the soil properties play a significant role in the seismic responses, an unlimited number of site conditions can be postulated for analyses. In this paper, a systematic and effective approach is described to arrive at the design soil profiles for the AP600 nuclear plant.

INTRODUCTION

For a safer and a more economic design, future nuclear power plants are based on the Standard Design concept. One of these designs currently near completion is the Westinghouse AP600 nuclear plant. This design is suitable for a wide variety of site conditions and seismic environments.

For the AP600 design, like the design of all other nuclear structures, seismic responses are required to be developed using the soil-structure interaction (SSI) technique. This requires a priori selection of the soil parameters and site conditions representing the potential sites. Performing SSI analyses for all combinations of site conditions and parameters is a formidable task. For this reason, selection of the design soil profiles is an important step in the design which needs to be undertaken early on in the process. The design soil profiles consist of such critical soil properties and site conditions which in combination with the dynamic structural properties in the SSI analysis would lead to a conservative and yet realistic design. The choice of the design soil profiles is not only made by the complex inter-play of dynamic soil and structural properties but also by the desire for the broad applicability of the design to as many potential sites as possible.

To facilitate the selection of the design soil profiles, a total of 22 nuclear power plant sites in the Unites States were surveyed. This survey was used as a guide to establish the range of soil parameters and site conditions for AP600 SSI analyses. Observing the results of the survey, a series of seismic SSI analyses were performed using the two dimensional models of the nuclear island and the results in terms of the seismic loads and the acceleration response spectra at the key locations in the plant were evaluated and compared. These results were used to select the design soil profiles which were subsequently used in the final three dimensional SSI analyses and design.

AP600 NUCLEAR ISLAND

The AP600 nuclear island consists of four structures. The structures are the shield building, the steel containment vessel (SCV), the containment internal structures (CIS), and the auxiliary building all on the common basemat. A plan view and a cross section of the nuclear island are shown in Figures 1A and 1B, respectively. The nuclear island is embedded approximately 40 ft (from El. 100 ft to El. 60 ft).

SURVEY OF A SELECTION OF EXISTING NUCLEAR POWER PLANT SITES IN THE UNITED STATES

A survey of 22 commercial nuclear power plant sites in the United States was conducted to identify the subsurface site conditions. Figure 2 shows the geographical distribution of the plant sites surveyed. Information regarding the depth to base rock, embedment depth of the key plant structures, water table elevation, soil profile and subsurface conditions, the design motion (both SSE and OBE levels) and also the dynamic properties of the supporting soil foundation were collected and examined. Table 1 shows the plant names and the summary information. Figures 3 through 5 depict the histograms for depth to base rock, depth to water table and embedment depths of the key structures, respectively. Because of the wide range of variation of depth to base rock and depth to the water table, these two parameters were considered as variables in the selection of the design soil profiles. The histogram of the embedment depth in Figure 5 shows that the embedment depth of 40 ft is common among the nuclear plants.

Examining the subsurface conditions of the plants surveyed reveals that the soil profiles consist of layers of clay to silty and sandy clay to sand and gravel extending to shale and soft rock to competent rock at greater depth. The shear wave velocity of the soil profiles are shown in Figures 6A and 6B for profiles with velocities less than 2000 ft/sec and larger than 2000 ft/sec, respectively. The designated ID numbers (1 through 22) are identified in Table 1. The dashed lines in these two figures represent the velocities before the excavation or the properties of the backfill materials As shown in these figures, the velocities range from 1000 ft/sec to 10,000 ft/sec with the majority of the profiles in the range of 1,500 ft/sec to 3,000 ft/sec. In all cases the shear wave velocity of the in-situ materials at the basemat level exceeds 1000 ft/sec. The normalized velocity profiles normalized to the shear wave velocity at the respective base mat level in the free-field is shown in Figure 7. As shown in this figure, except for the sites where base rock is present at shallow depths, the normalized velocity profiles increase gradually with depth with the maximum ratio of approximately 2 at the depth of 240 ft. This ratio was used to select one of the velocity profiles for subsequent analysis. It should also be noted that there is no inverted velocity profile in the sites surveyed. The results of this survey was used as a guide to select the soil profiles as discussed below.

DESIGN SOIL PROFILES

The most critical soil property affecting the SSI responses is the shear wave velocity of the material. The velocity should be adjusted to account for the highly nonlinear response of soil deposits to strong ground motion.

Based on the results of this survey, the velocity profiles in Figure 8 were selected for further evaluation. The soft soil velocity profile is 1000 ft/sec at the ground surface gradually increasing to 1200 ft/sec at 240 ft depth. This velocity profile is also consistent with the lower bound of the profiles in the sites surveyed. Sites with shear wave velocities less than 1000 ft/sec at the foundation level are not considered potential sites since it is unlikely that such sites would meet the site interface requirements for the minimum bearing capacity. A lower bound to the soft soil profile was also chosen with a shear modulus of 50% of the soft soil site to account for any variation and scatter in the velocity data for such sites. The other profiles represent the variation of the soil profiles for clayey sites, deep cohesionless sites (parabolic distribution of the soil shear modulus) as well soft to firm to hard rock sites. Although the 8000 ft/sec shear wave velocity represents a competent hard rock site, a large upper bound velocity was also included in order to simulate the "fixed base" condition and capture the high frequency responses in the structures. As to the depth to base rock, depths of 0 ft (rock site), 40 ft, 50 ft, 60 ft, 80 ft, and 120 ft were selected. Water table was also varied from ground surface to the depth of 120 ft.

In order to identify the design soil profiles, a series of two-dimensional SSI analyses were performed using the computer program SASSI (Ostadan, 1988). A typical SASSI model of the nuclear island is shown in Figure 9. The structures were modeled using multiple stick beam elements. The basement, consisting of the embedded exterior walls and the basemat, is modeled using two-dimensional plane strain elements and beam elements. The design motion is based on the U. S. Regulatory Guide 1.60 spectrum scaled to 0.30g with modified amplitude at high frequencies to represent the characteristics of central and eastern Unites States earthquakes. The acceleration response spectra of the horizontal component are shown in Figure 10. The input motion was specified at the ground surface and the SSI analyses were performed for each site condition discussed above. The strain-compatible soil properties were obtained from free-field de-convolution analyses. As expected, degradation of the properties for the rock sites was found to be insignificant. The strain-compatible properties were subsequently used in the SSI analyses. The SSI results in terms of seismic loads and acceleration response spectra at key locations were obtained and compared. Typical acceleration response spectra at El. 180 ft of the nuclear island and at the crane elevation in the steel containment vessel for five soil profiles are compared in Figures 11 and 12, respectively. As shown in these figures, the high frequency responses are governed by the rock site condition. In the intermediate frequency range, responses for the medium to medium stiff soil sites are slightly larger than those for the rock site. The soft sites have no contribution to the acceleration responses in the frequency range of interest. The seismic member forces are compared in Table 2. These member forces are generally governed by the rock site condition. Such trend in the seismic responses has been observed in the study of other nuclear structures (Ostadan, 1989, 1993). With the design motion defined at the ground surface, the attenuation of motion with depth is larger for the soil sites as compared with rock sites. The softer the soil, the larger is attenuation of motion with depth resulting in a reduced effective foundation motion. This effect combined with the radiation capacity of the supporting soil medium tends to reduce the responses significantly. On the other hand the rocking vibration of the structure will be larger in the soil sites compared to the rock sites. The combined effect is that the responses are mostly governed by the rock sites except at few locations where soil sites dominate the response in a limited frequency band. The soft soil sites are expected to govern the displacement responses. However, such responses are less important in the design of the AP600 plant since all safety related systems are located on the common basemat of the nuclear island.

The 2D SSI analyses were performed systematically and the sensitivity of the results to other soil properties such as the Poisson's ratio, density, variation in nonlinear strain-dependent soil properties (sand and clay curves) were evaluated. In addition the effects of depth to base rock and depth to water table on the SSI responses were studied. Based on this study, the following design soil profiles were recommended for final 3D SSI analyses of the plant structures.

- The shear wave velocity profiles selected as design profiles are shown in Figure 13. The profiles represent the soft-to-medium soil with parabolic distribution of the shear modulus to represent deep cohesionless soil profiles, an upper bound to the soft to medium profile, the soft rock, and the hard rock profile. These profiles envelope the responses of the nuclear island throughout the frequency range of interest.

- Seismic responses increase significantly as the depth to base rock is reduced from deep to shallow depth within each soil profile. However, on the enveloping basis, the responses from the shallow rock cases are governed by the responses obtained from the rock site. For the soil sites, a critical depth to base rock was determined to be 120 ft. At this depth, the soil column frequency and the SSI frequency of the nuclear island are close enough to increase the response.

- For the soil sites, the effect of water table elevation on horizontal responses was found to be insignificant. The vertical responses, were, however, controlled by the shallow water table. This is mainly due to large compressional velocity of the saturated soil profile resulting in a reduced attenuation of P-wave motion with depth.

- The effects of other soil properties such as the Poisson's ratio, density and variation in strain-dependent soil properties (sand and clay curves) were found to be secondary effects and were bounded within the envelope of the results obtained from use of the design soil profiles.

Based on this evaluation, the design soil profiles shown in Figure 13 were subsequently used in the 3D SSI of the nuclear island. The responses were enveloped and used for the design.

SUMMARY AND CONCLUSION

Future nuclear power plants are based on the standard design concept. These standard designs are evaluated for a variety of subsurface conditions and licensed in advance in order to reduce cost and construction time and to increase safety. The AP600 design is near completion at this time. The broad applicability of this design is supported by selection of limited site conditions from virtually unlimited number of possibilities for seismic soil-structure interaction analyses. The selection process is based on studies of the effects of complex interaction between the dynamic soil and structural properties. A careful examination of the dynamic structural properties and the SSI effects as observed from the simplified SSI analyses has led to only a handful of site conditions that need to be considered for the complete SSI analyses. The steps outlined above describe a systematic and an effective approach for selection of the design profiles.

REFERENCES

Ostadan, F., 1988, "Theoretical, Validation and User's Manuals for the Computer Program SASSI," Bechtel Corporation, San Francisco, California

Ostadan, F., et al., 1989, "Effect of Site Soil Properties on Seismic SSI Response of Deeply Embedded Structures," ASCE Foundation Engineering Congress, Evanston, Illinois, June.

Ostadan, F., et al., 1993, "Seismic Design of ABWR and SBWR Standard Plants," ICONE2, San Francisco, California.

TABLE 1

SUMMARY OF COMMERCIAL NUCLEAR POWER STATIONS SURVEYED

Plant ID Number	Nuclear Plant and Unit(s)	Location (State)	Type	CB or TB Embedment Depth (Ft)	Depth to Rock (Ft)	Backfill Under CB or TB Foundation (Ft)	Depth to Water (Ft)	ZPA (g)	
								SSE	OBE
1	Hope Creek, U-1	New Jersey	BWR	39-CB	1500-2000	10	0	0.2	0.1
2	Callaway, U-1	Missouri	PWR	30-CB	50	28	15	0.2	0.12
3	Palo Verde, U-1,2,3	Arizona	PWR	50-CB	334	None	44	0.2	0.1
4	Arkansas Nuclear One, U-2	Arkansas	PWR	30R-CB	30	None	11	0.2	0.1
5	Calvert Cliffs, U-1,2	Maryland	PWR	46-CB	2500	None	35	0.15	0.08
6	Wolf Creek, U-1	Kansas	PWR	40R-CB	20	None	3	0.12	0.06
7	Grand Gulf, U-1,2	Mississippi	BWR	47½-TB	>185	None	47½	0.15	0.075
8	Millstone, U-2	Connecticut	PWR	53R-CB	53	None	19	0.17	0.09
9	Farley, U-1,2	Alabama	PWR	60R-CB	59	None	5	0.1	0.05
10	Vogtle, U-1,2	Georgia	PWR	66-CB	950	12	55	0.2	0.12
11	Hatch, U-2	Georgia	BWR	55-TB	4000	None	55	0.15	0.08
12	Susquehanna, U-1,2	Pennsylvania	BWR	31R-CB	30	None	0	0.10	0.05
13	Monticello	Minnesota	BWR	40-CB	40	None	20	----	0.06(DBE)
14	Davis-Besse	Ohio	PWR	40R-CB	30	None	0	0.15	0.08
15	WNP-2	Washington	BWR	34-CB	557	9	60	0.25	0.125
16	South Texas, U-1,2	Texas	PWR	59-CB	4000-6000	None	5	0.10	0.05
17	San Onofre, U-2,3	California	PWR	44-CB	850	None	24	0.67	0.33
18	Fort St. Vrain	Denver	HTGR	55R-CB	55	None	23	0.10	0.05
19	Arnold	Iowa	BWR	49R-CB	49	None	17	0.12	0.06
20	Limerick, U-1,2	Pennsylvania	BWR	50R-CB	15	None	deep	0.15	0.075
21	Trojan	Oregon	PWR	33R-CB	<5	None	>33	0.25	0.15
22	Midland, U-1,2	Michigan	PWR	57-CB	300	None	39	0.12	0.06

Abbreviations: CB = Containment Building; TB = Turbine Building; R = Foundation on Rock; U = Unit

TABLE 2
COMPARISON OF SEISMIC MEMBER FORCES
TWO-DIMENSIONAL SSI ANALYSIS

SSI Case ID	Coupled Auxiliary & Shield Bldgs. Elevation 100 ft					Containment Internal Structure Elevation 82.5 ft				
	Z-Shaking	X-Shaking		Y-Shaking		Z-Shaking	X-Shaking		Y-Shaking	
	Axial (10^3K)	Shear (10^3K)	Moment (10^3K-ft)	Shear (10^3K)	Moment (10^3K-ft)	Axial (10^3K)	Shear (10^3K)	Moment (10^3K-ft)	Shear (10^3K)	Moment (10^3K-ft)
R1	37.2	41.4	4433	46.8	4954	15.3	20.0	630.7	18.2	619.7
A2	31.7	40.3	4366	46.6	4963	14.8	15.3	432.3	15.8	448.3
B2	24.1	31.4	3642	36.6	3541	13.2	11.3	306.0	12.1	385.6
C2	25.5	18.9	1960	23.8	2314	11.3	9.6	209.8	11.5	332.9

SSI Case ID	Steel Containment Vessel Elevation 82.5 ft				
	Z-Shaking	X-Shaking		Y-Shaking	
	Axial (10^3K)	Shear (10^3K)	Moment (10^3K-ft)	Shear (10^3K)	Moment (10^3K-ft)
R1	6.6	7.1	871.7	6.7	774.6
A2	5.4	4.5	994.0	7.5	971.7
B2	4.1	4.4	580.9	9.3	1127
C2	3.7	5.0	620.0	6.6	1048

R1: Hard Rock Site

A2: Soft Rock Site

B2: Soft to Medium Soil Site

C2: Soft Soil Site

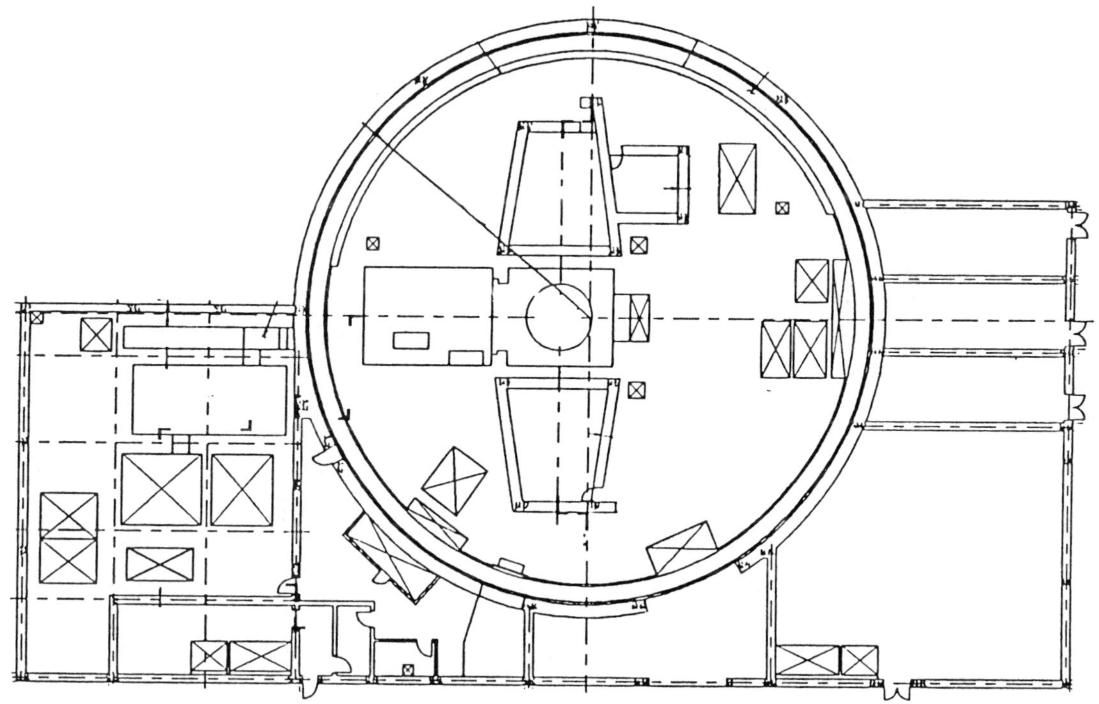

FIGURE 1A PLAN VIEW OF THE AP600 NUCLEAR ISLAND

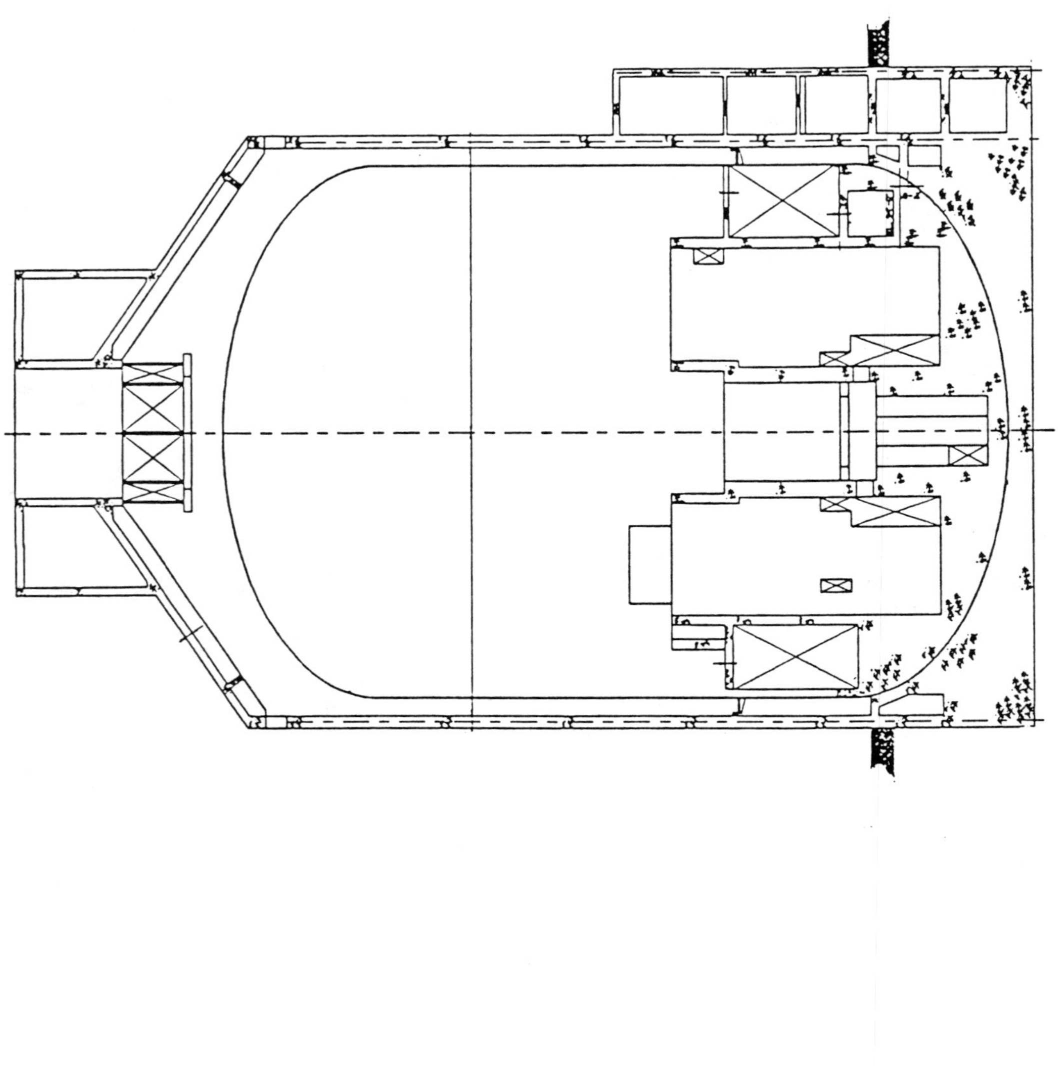

FIGURE 1B CROSS SECTIONAL VIEW OF THE AP600 NUCLEAR ISLAND

FIGURE 2 US PLANT SITES SURVEYED

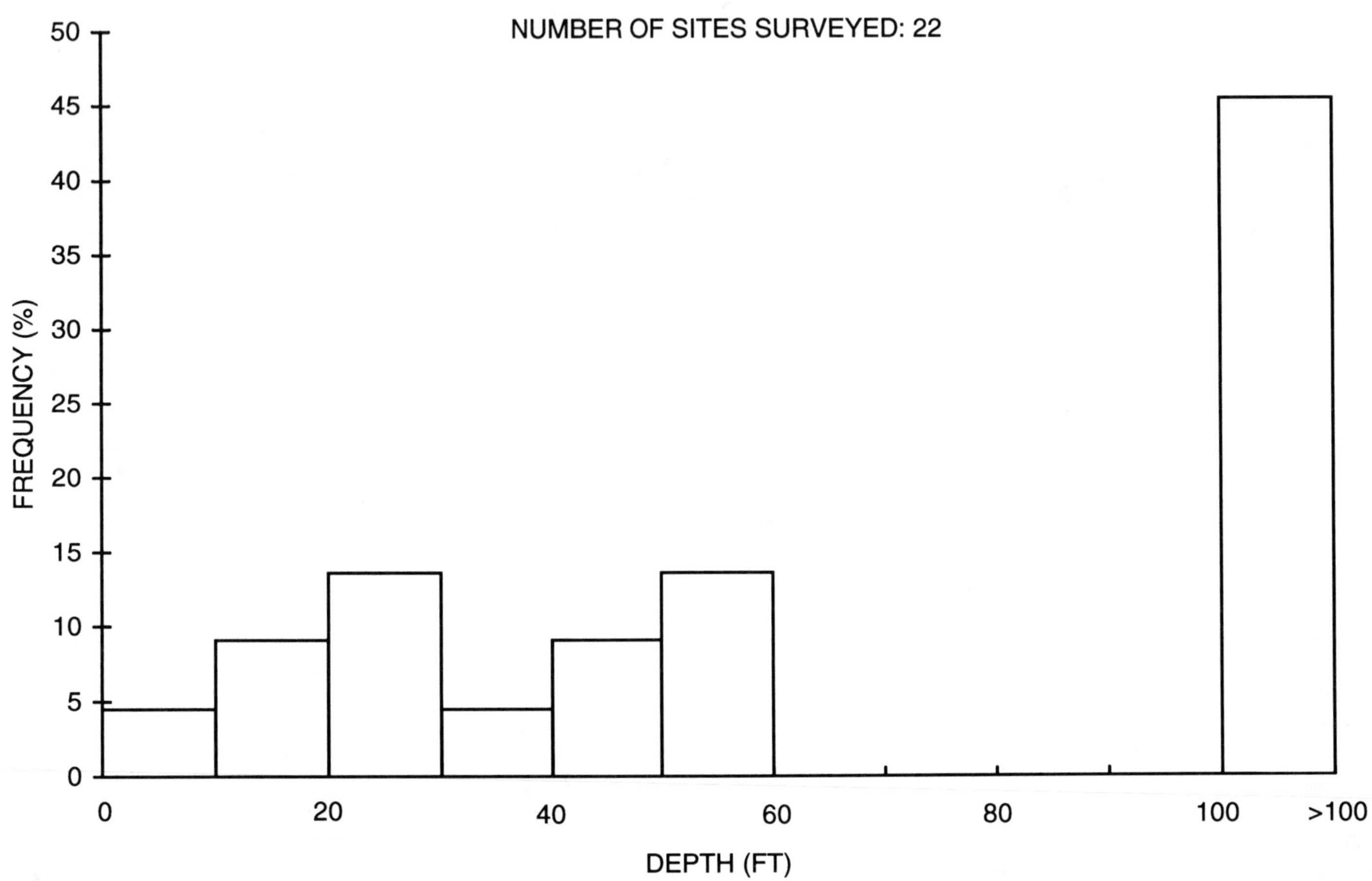

FIGURE 3 HISTOGRAM FOR DEPTH TO BASE ROCK

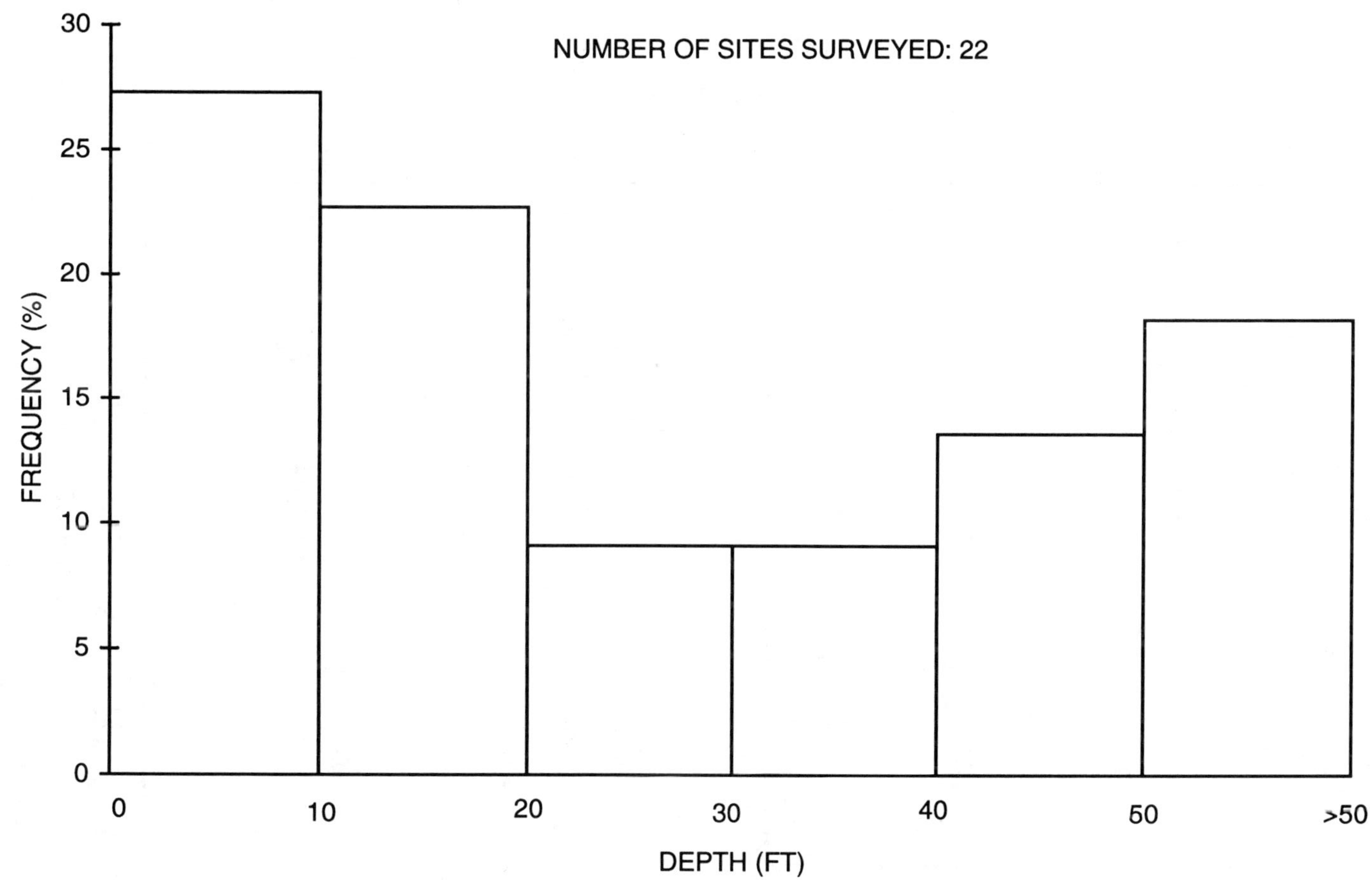

FIGURE 4 HISTOGRAM FOR DEPTH TO WATER TABLE

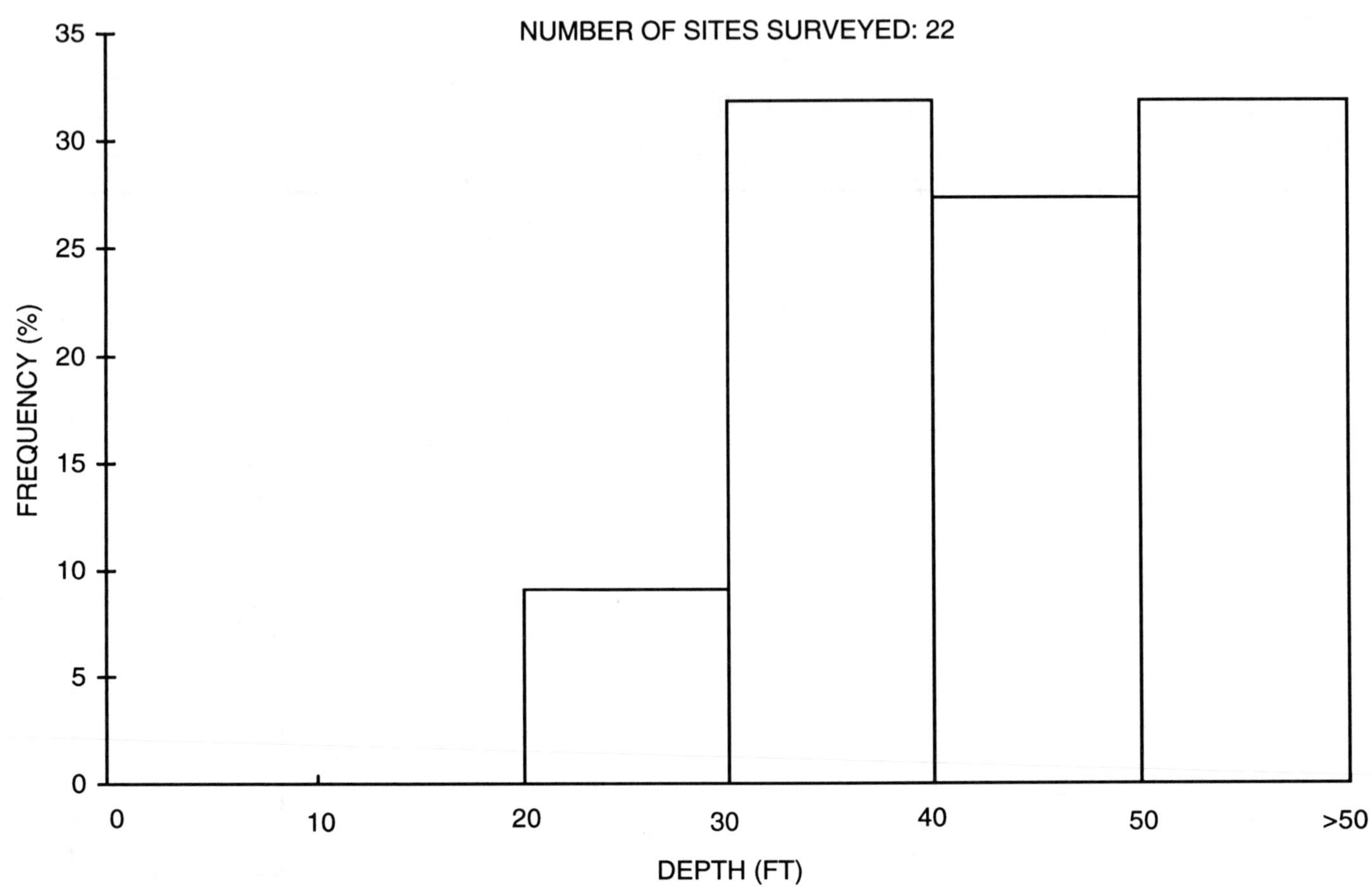

FIGURE 5 HISTOGRAM FOR EMBEDMENT DEPTH

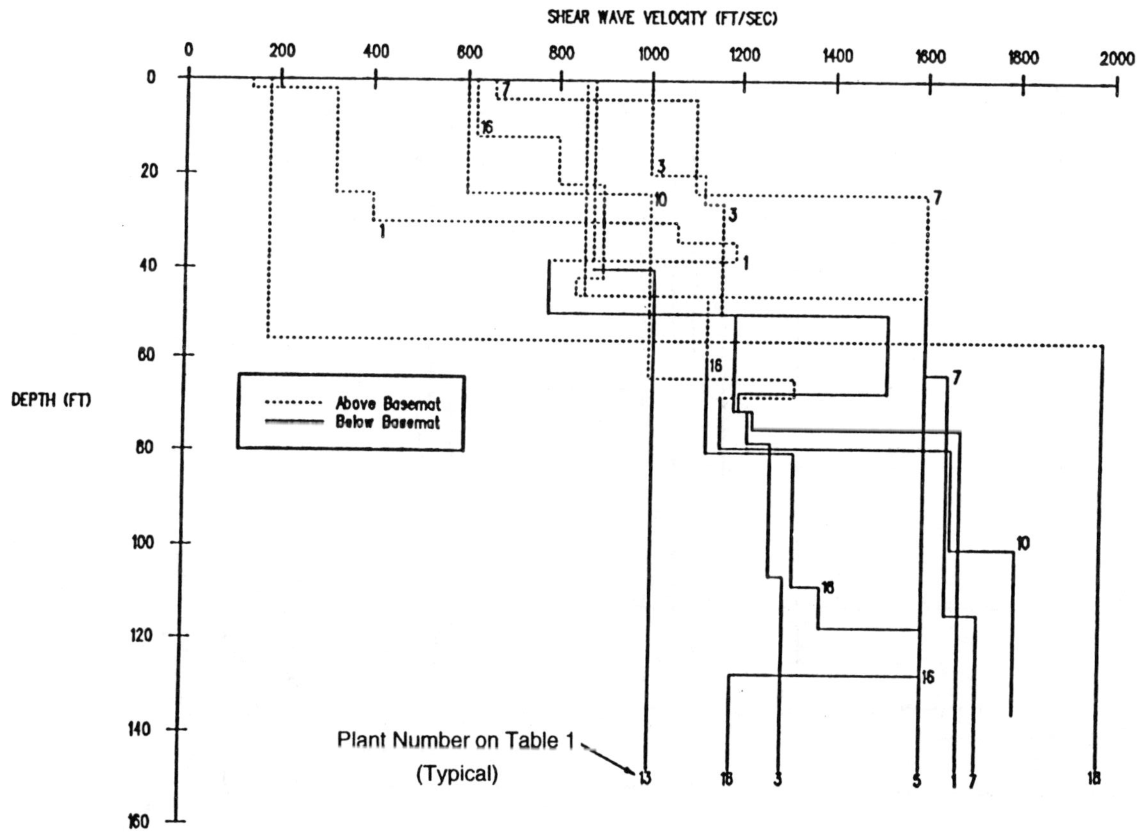

FIGURE 6A SOIL CONDITIONS FOR EXISTING NUCLEAR POWER PLANTS
(V_s LESS THAN 2,000 FT/SEC)

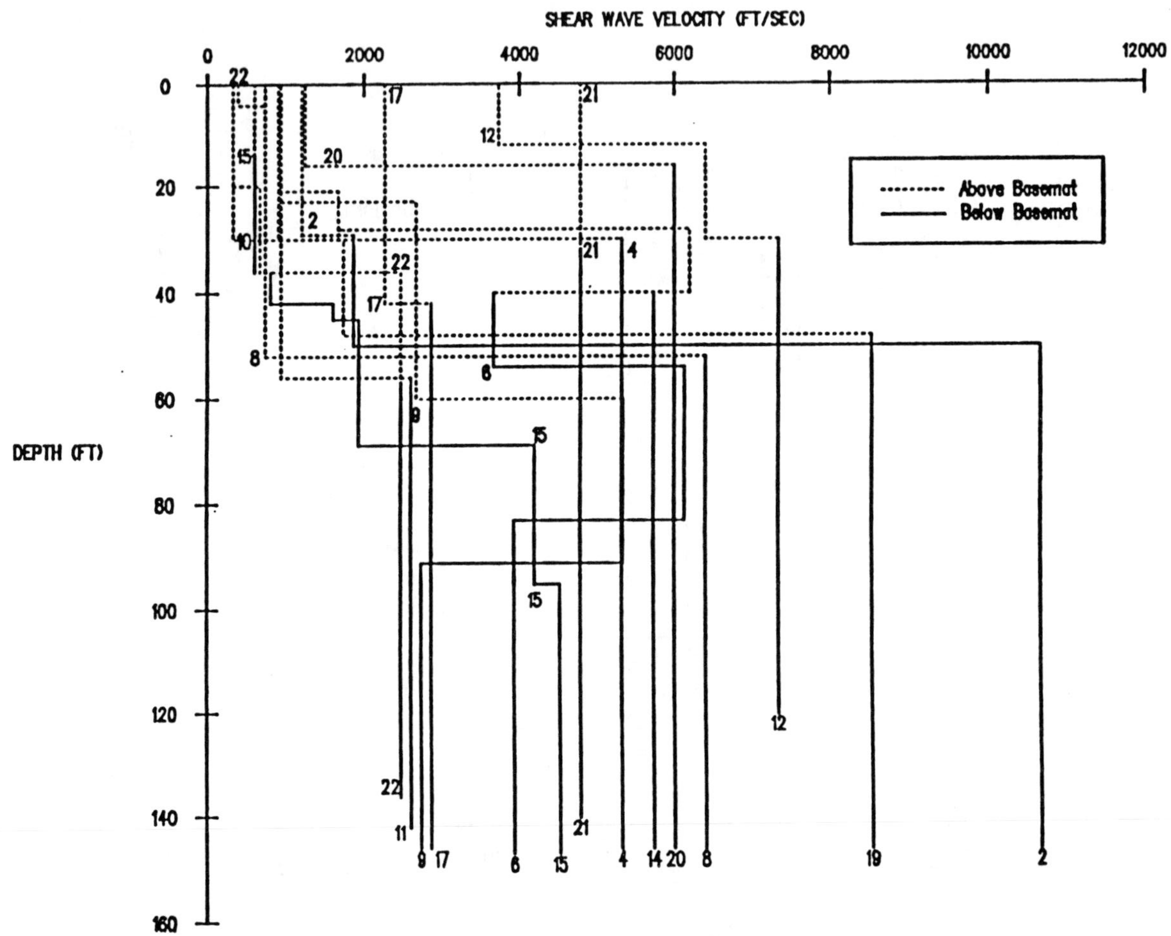

FIGURE 6B SOIL CONDITIONS FOR EXISTING NUCLEAR POWER PLANTS
(V_s GREATER THAN 2,000 FT/SEC)

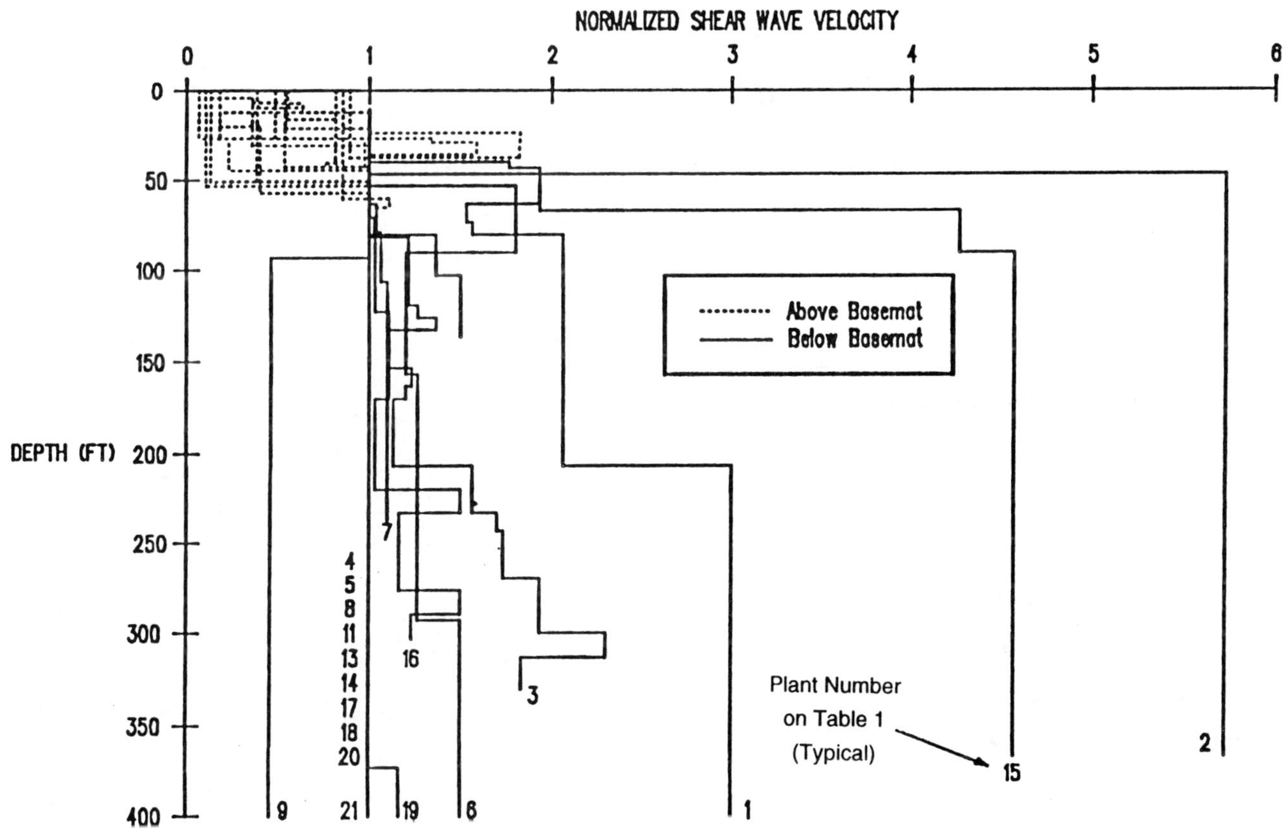

FIGURE 7 SOIL CONDITIONS FOR EXISTING NUCLEAR POWER PLANTS
(NORMALIZED SHEAR WAVE VELOCITY PROFILES)

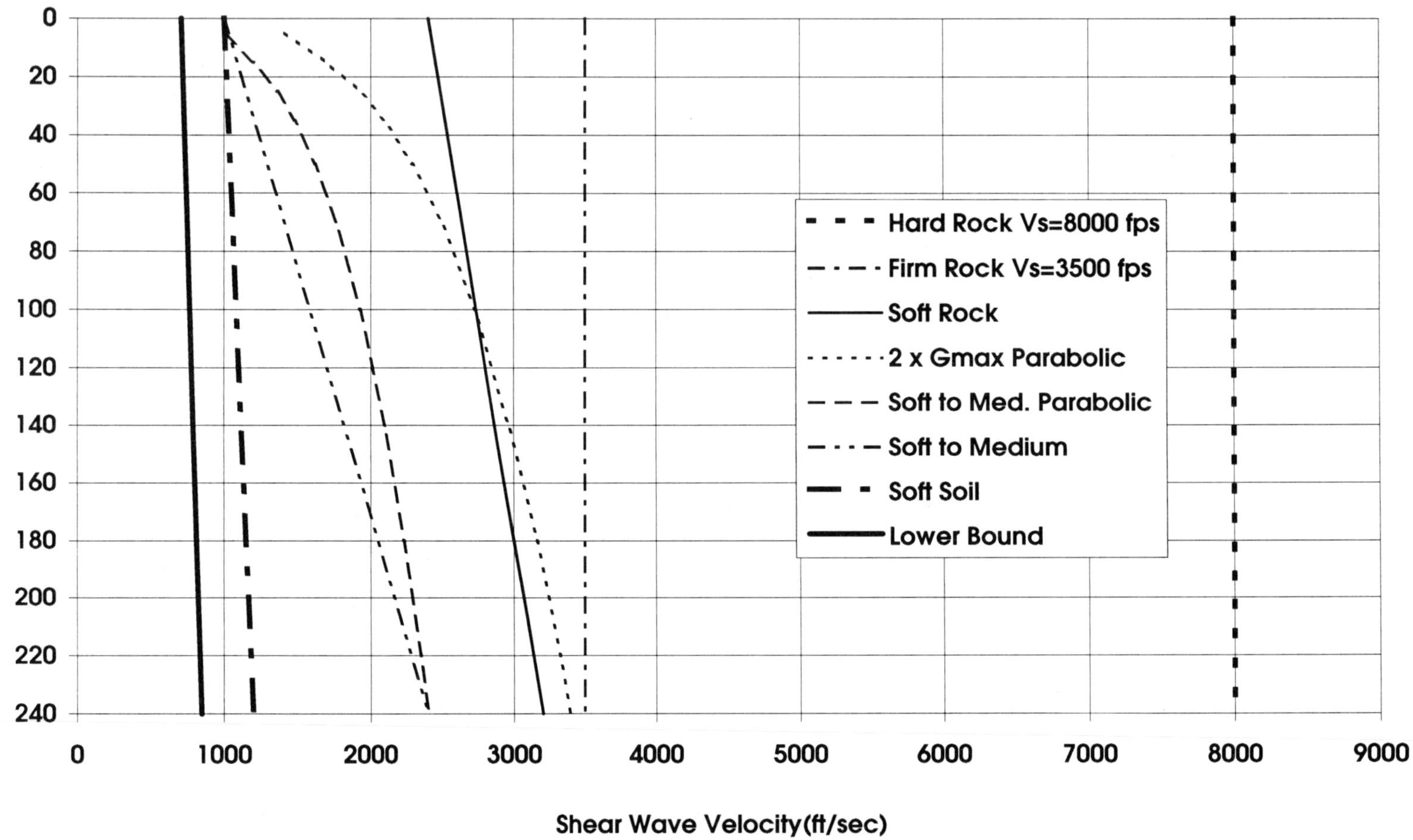

FIGURE 8 INITIAL SOIL PROFILES

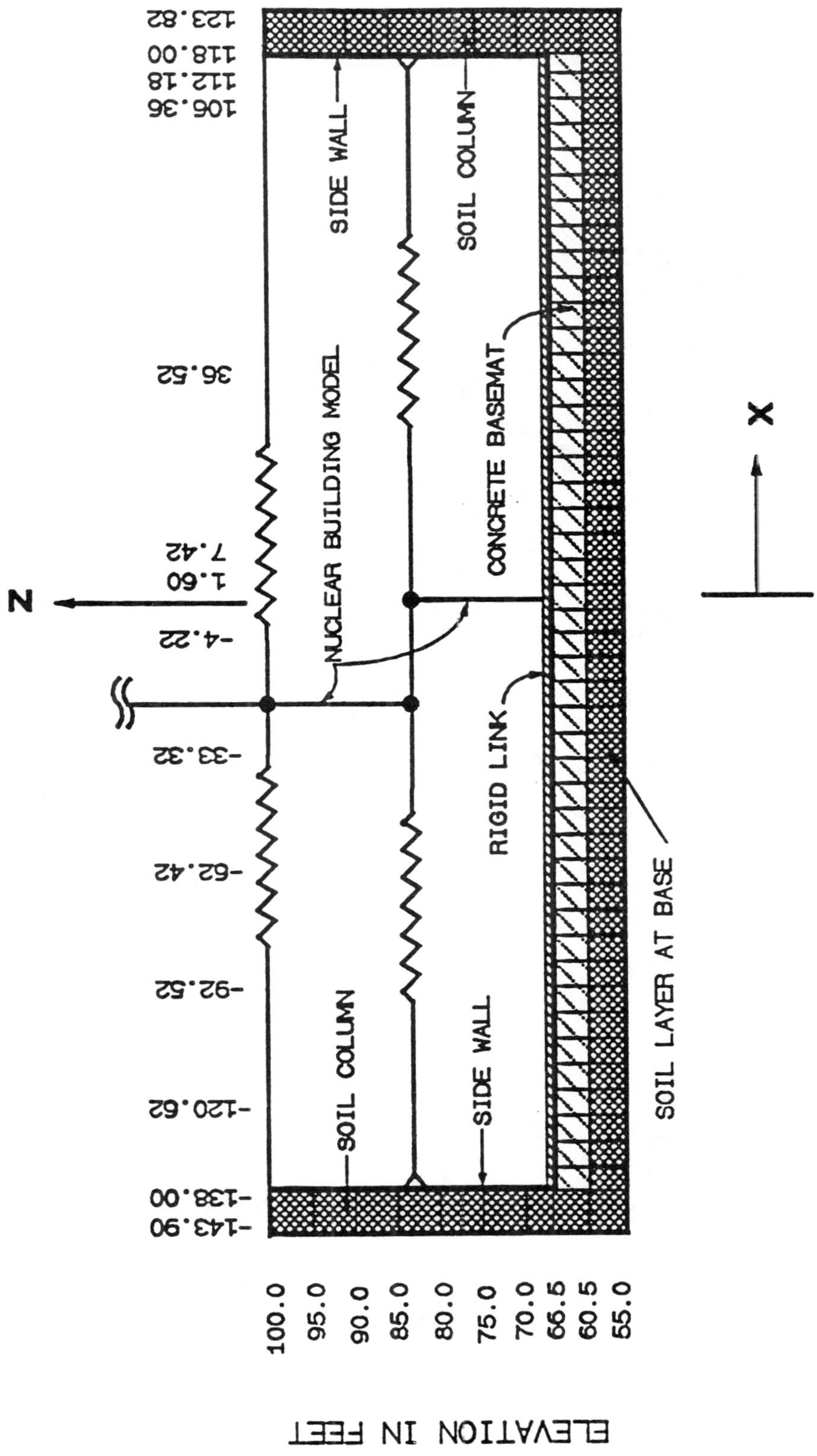

FIGURE 9 TYPICAL SASSI 2D MODEL OF THE AP600 NUCLEAR ISLAND

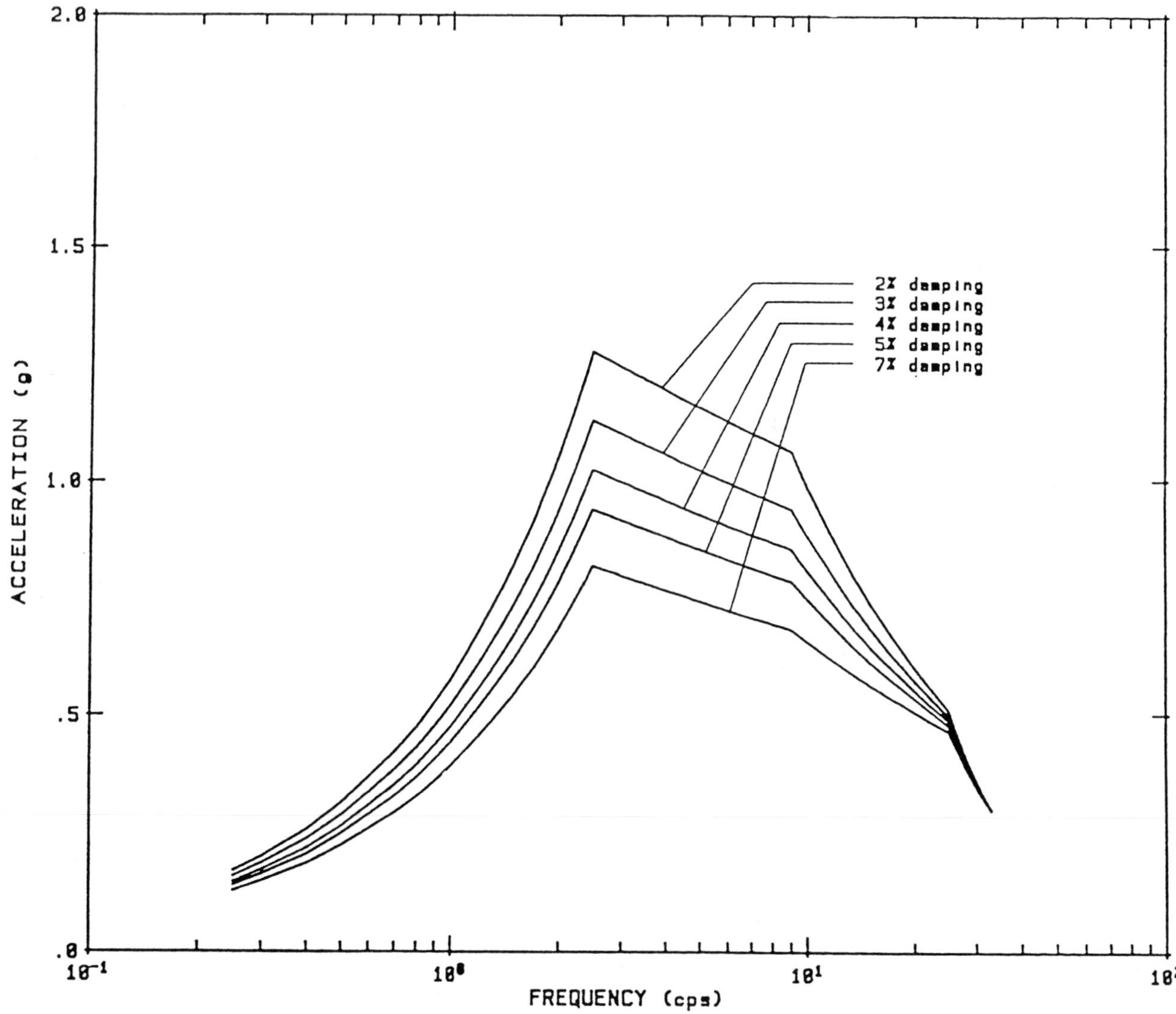

FIGURE 10 HORIZONTAL DESIGN RESPONSE SPECTRA, SAFE SHUTDOWN EARTHQUAKE

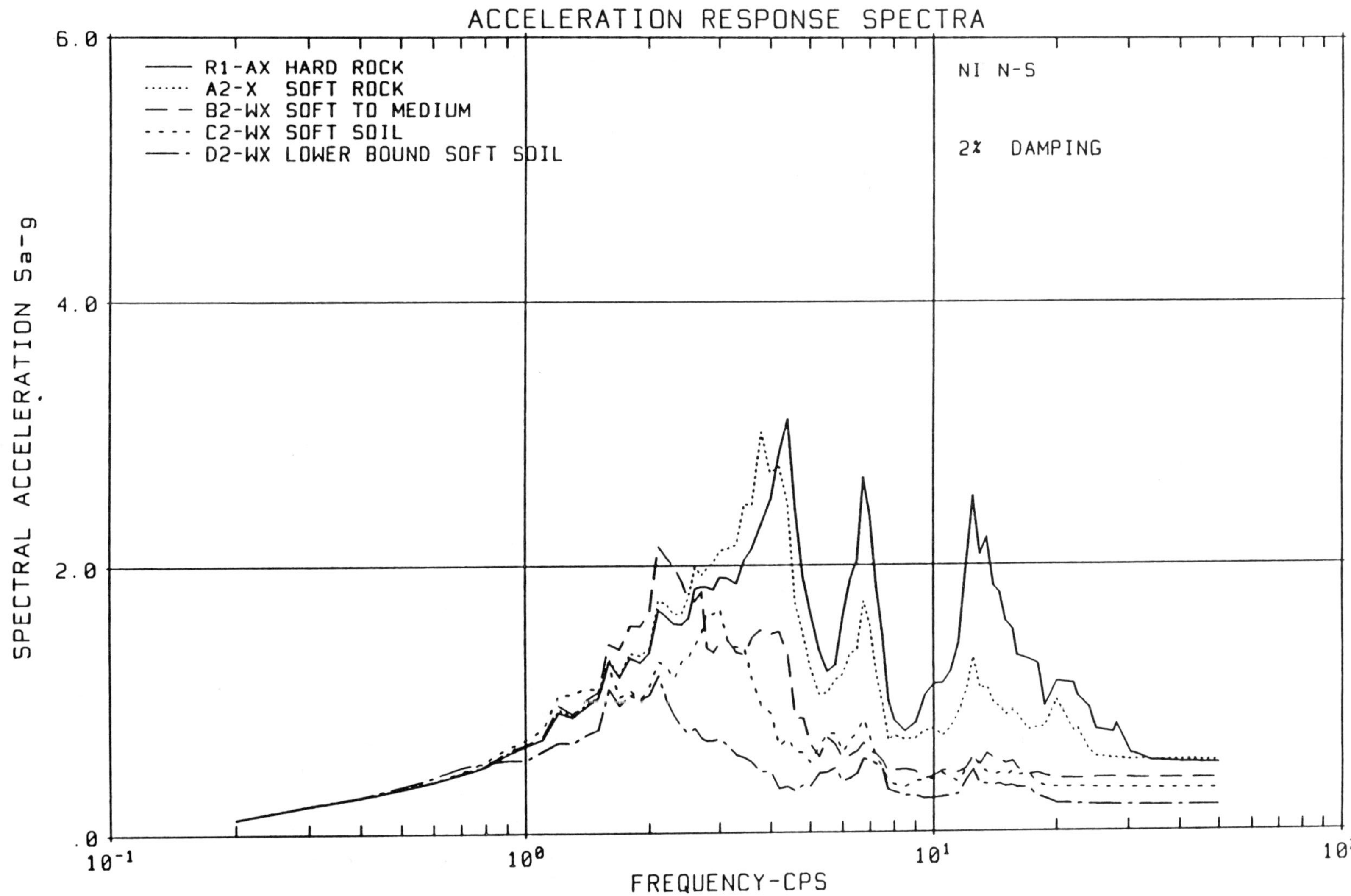

FIGURE 11 ACCELERATION RESPONSE SPECTRA AT ELEVATION 180 FT OF NUCLEAR ISLAND

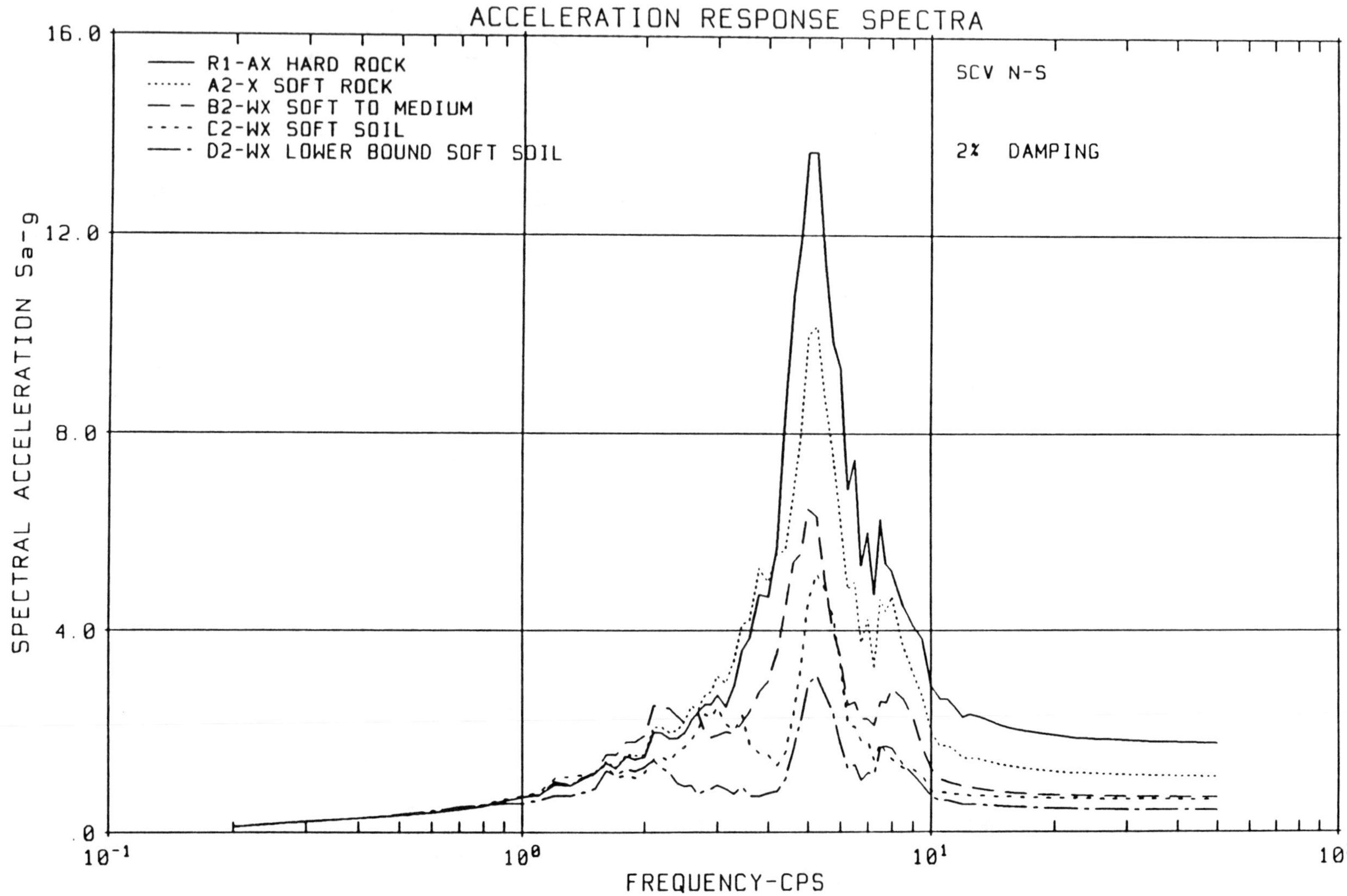

FIGURE 12 ACCELERATION RESPONSE SPECTRA AT ELEVATION 205 FT OF STEEL CONTAINMENT VESSEL

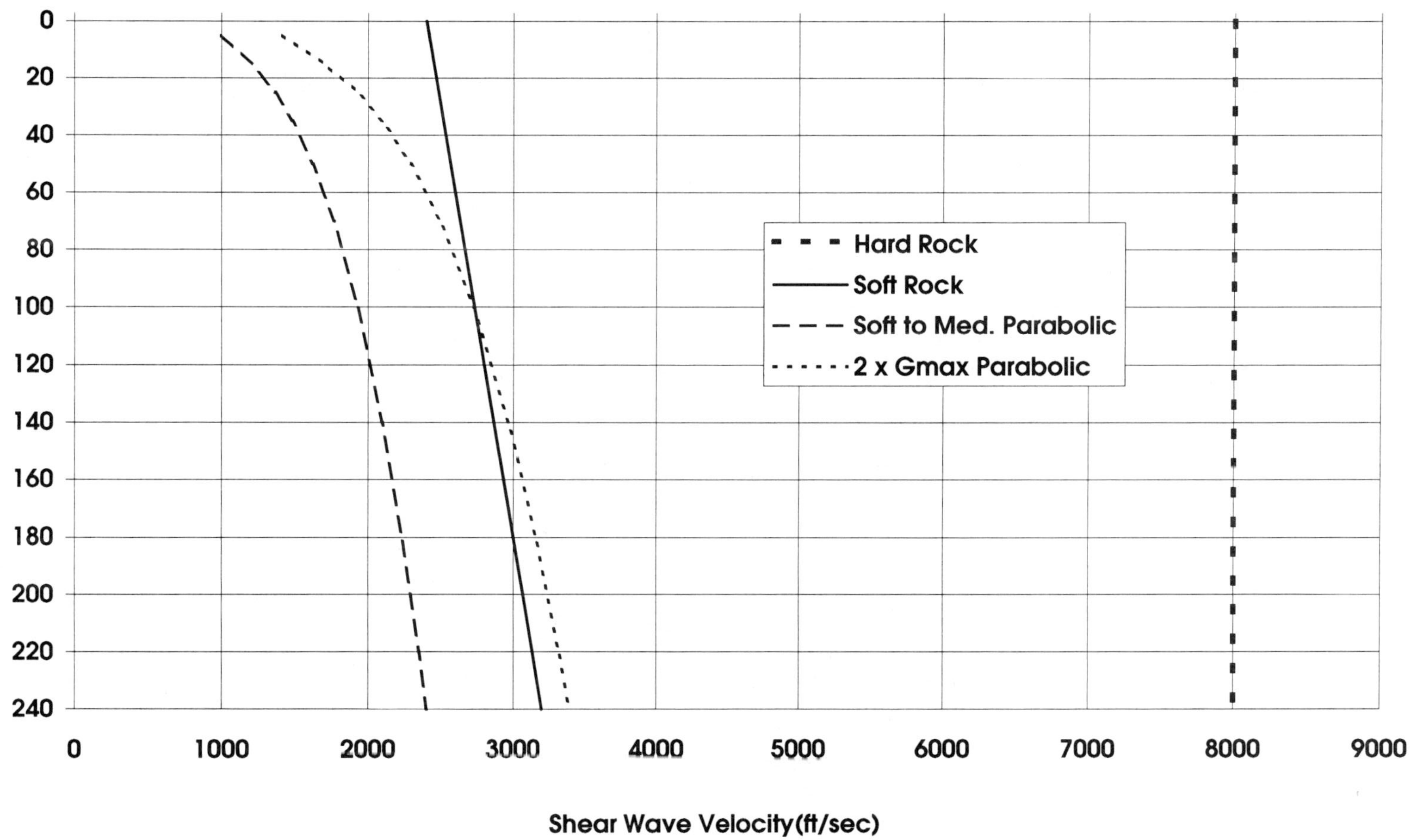

FIGURE 13 DESIGN SOIL PROFILE

PANEL DISCUSSION

Section III Special Working Group on Seismic Rules

Review of Current Issues

A panel discussion is planned for Wednesday July 24, 1996 to review issues that have been raised with the introduction of the new seismic design requirements for piping in the 1995 Edition of Section III of the ASME Boiler and Pressure Vessel Code. A Special Working Group on Seismic Rules was established in 1995 and given the responsibility for studying these issues. It reports to the Subgroup Design, Section III.

The panel will consist of a moderator and four panellists who will discuss the issues and answer questions put to them from the audience. The panellists are all active members of the Special Working Group (SWG) meaning that the experts involved in the work will be able to answer the questions directly. A list of the most significant issues identified is attached.

The panel will choose four issues out of these 11. Each panellist will take one issue and briefly summarize the main points in a five minute overview. The rest of the time will be available for questions and discussion. This will also provide an opportunity for feedback to the Codes and Standards Committee from the technical community attending the session.

Panel Participants

Moderator:	Richard W. Barnes	Chair, Subgroup Design SC-III ANRIC Enterprises
Panellists:	Ernest B. Branch	Chair, SWG on Seismic Rules Consultant
	John C. Minichiello	Secretary, SWG on Seismic Rules Commonwealth Edison Company
	Edward A. Wais	Member, SWG on Seismic Rules Wais and Associates Inc.
	Timothy M. Adams	Member, SWG on Seismic Rules Stevenson & Associates Inc.

List of Significant Issues

1. There is insufficient justification for the detailed strain limit rules in NB-3200

2. Do the higher stress allowables lead to unacceptable error in the linear elastic analysis for piping response.

3. Should the primary stress equations be used to control fatigue failure.

4. There are several issues relating to the analytical extrapolation of component tests.

5. Margin calculation questions such as actual thickness versus nominal thickness, temperature effects etc., need to be addressed further.

6. The combination of inertia or thermal expansion combined with seismic anchor motion needs to be addressed.

7. The effect of higher allowables on the acceptable flaw size in Section XI needs to be addressed.

8. The number of seismic load cycles assumed, 20 cycles, is not supported by the Japanese data.

9. Seismic events combined with large ground displacements could cause collapse failure which does not conform to the assumed failure mechanism in the new rules.

10. Margin definition, the method of calculating and setting acceptable margin level needs further review.

11. Higher stress limits should apply only tp building filtered loads and not to water hammer loads.

ADVANCED SEISMIC TECHNOLOGY IN ASIAN COUNTRIES

Introduction

K. Suzuki
Tokyo Metropolitan University
Hachioji, Tokyo, JAPAN

N. Shimizu
Iwaki Meisei University
Iwaki, JAPAN

This chapter consists of eight papers on advanced seismic technology and represents contributions from various asian countries. These have been gathered in one chapter consistent with the theme of the conference "Innovations In Pressure Vessels and Piping on an International Level". The topics cover analytical studies as well as large scale test results and applications to post -earthquake rehabilitation issues.

Reddy, Kushwaha, Mahajan and Suzuki propose a decoupling criteria for multi-connected equipments such as nuclear piping systems. Fujita, Nakamura, Ito and Iwamoto investigate the buckling of steel piers of elevated expressway observed at the Great Hanshin Earthquake by static and dynamic tests. Aoki provides a new analytical method for getting statistical properties of seismic response of the secondary systems. Hara and Iwasaki discuss an optimum design method for the pipe-support allocation , particularly, when seismic loading was provided. Shimizu, Suzuki, Watanabi, Ogawa and Kobayashi present various results obtained through large scale shaking test for 3-dimensional piping system having frictional motion between piping and supports. Liang describes two papers ; one is devoted to a decision support system for pre-earthquake planning of life lines and the other is related to a post-earthquake rehabilitation plan of lifeline networks. Chiba, Fujita and Tamayama present a genetic algorithm (GA) which can be applied to the support arrangement of piping systems.

Collection of these papers provide an insight to the diversified and advanced studies being carried out in various asian countries. Efforts of the authors and reviewers are deeply appreciated by the Seismic Engineering subcommittee.

PVP-Vol. 340, Seismic Engineering
ASME 1996

DECOUPLING CRITERIA FOR MULTI-CONNECTED EQUIPMENTS

G.R.Reddy, H.S.Kushwaha, S.C.Mahajan

Reactor Engineering Division
Bhabha Atomic Research Center
Bombay-400085, India.

Kohei Suzuki

Tokyo Metropolitan University
Mechanical Engineering Department
1-1, Minami Osawa, Hachioji-Shi,
Tokyo 192-03, Japan.

ABSTRACT

Nuclear Power Plant structures generally support equipments, piping systems, computing/controlling systems etc. These equipments and structures may interact with each other during an earthquake. This results in variation in the uncoupled response wherein uncoupled response is calculated for equipment and structure separately. The best way of accounting for interaction effects of structure and equipment is by coupling together and analysing for a given earthquake load. However, it may not be practicable to couple all the equipments to the structure because of three main reasons. These are: firstly, at the initial design stage of power plant structure all the details of equipments may not be available, secondly, coupling all the equipments may lead to numerical problems in computation because of large difference in stiffness and thirdly, the cost of computation may become very high and time consuming because by coupling all the equipments, size (e.g. number of numerical equations) of the problem becomes very large.

The above practical difficulties led the investigators to develop a criteria which can be used to identify those equipments which affect the structure/equipment frequencies beyond the acceptable limits. The reason for considering the frequency parameter is that in dynamics it has direct impact on the response of the structure/equipment. However, interaction effects on the frequency of the structure and equipment due to coupling is identified by decoupling criteria (USNRC,1990). United States Nuclear Regulatory commission developed this criteria after careful study of various works of different investigators. This criteria is best suited for a single structure and equipment and it has limitations in application to a complex structure-equipment system where a number of substructures are resting on a common foundation and which intern supports equipments.(e.g. Pressurised Heavy Water Building). This limitation is due to the fact that two or more substructures founded on common foundation base interact each other in some modes only. To understand this interaction effects,extensive work was carried out (Reddy at al.,1994) and modification in the decoupling criteria was evolved.

The above procedure is solely meant for equipments connected to the structure at a single location only. There is no criteria available in the literature/codes for equipments connected at multi locations to the structure.

In this paper the authors endeavor to present a decoupling model for multi-connected structure-equipment. The applicability of the model will be demonstrated with practical problems like reactor building and equipments. Here onwards equipment is referred as Secondary System (S.S) and structure as Primary System (P.S).

NOMENCLATURE

[K] = Stiffness matrix
[M] = Mass matrix
[I] = Unit matrix
f_{pj} = j th modal frequency of the primary system
f_{sj} = j th modal frequency of the secondary system
m_{psij} = j th modal mass of the i th primary subsystems including secondary system mass)
R_f = Ratio of frequency or modal frequency of the uncoupled secondary system to the uncoupled primary system
R_m = Ratio of mass or modal mass of the uncoupled secondary system to the uncoupled primary system
v_s = Shear wave velocity
$[\phi]$ = Modal vector matrix
λ = Eigen values
ω_s = Single degree of freedom secondary system frequency in rad/sec

DECOUPLIG CRITERIA:

Decoupling criteria (USNRC, 1990) is defined as follows.

i) Decupling can be done for any R_f, if $R_m < 0.01$

The roots of the equation (2) can be obtained easily using explicit procedure given in the literature. Having obtained the roots which are nothing but eigen values of the three dof system (Fig.4), the mode shapes can be calculated using following formulae.

$$\phi_s = \frac{-((k_s \times k_2/m_s \times m_2 + ((k_2+k_s/m_2) - \lambda)\, k_s/m_s)}{(k_s \times k_s/m_s \times m_2) - ((k_2+k_s/m_2) - \lambda)\,(\omega_s^2 - \lambda)}$$

$$\phi_2 = ((k_2/m_2) + (k_s/m_2) \times \phi_s)/\,((k_2+k_s)/m_2 - \lambda)$$

$$\phi_1 = 1.0$$

Development of a decoupling model:

The bottom spring of the model (Fig.4) represents soil under foundation raft. For this spring stiffness calculations, shear wave velocity of 2900 m/sec (table.1) has been considered. Shear wave velocity directly relates the condition of the foundation soil. Generally speaking, higher the value of the shear wave velocity higher is the stiffness. The practical assumption is that if the shear wave velocity is more than 1100 m/sec it is considered to be stiff rock. Using shear wave velocity and dimensions of the foundation as given in table.1, the stiffness of foundation soil is obtained using seismic code (ASCE-4, 1986).

Initially for a combined P.S. and foundation frequency of 3.5 Hz , the frequency of the S.S. of constant mass is varied from 0.25 rad/sec to 200 rad/sec with 0.25 rad/sec frequency interval and then for each value of S.S. frequency, coupled frequency is calculated using the procedure explained above. The variation of coupled frequency with respect to S.S. frequency is shown in Fig.5. From this figure, it can be seen that dotted lines which represents ±10% variation of uncoupled frequency of P.S. cuts the coupled frequency(1) curve at two points. These two points are at two different frequencies of the S.S.. This means that for a given mass ratio of S.S. to P.S. there are two frequency ratios of uncoupled S.S. to the uncoupled P.S. frequency at which there will be 10% variation of coupled frequency with respect to uncoupled frequency of P.S./S.S..

Therefore, for various mass ratios of S.S. to P.S. the above two frequency ratios have been identified by keeping P.S. frequency constant. Fig.6 shows mass ratio (R_m) v.s frequency ratio (R_f) for stiff soil base condition (v_s =2900 m/sec). In this figure it can be seen that there are two upper curves and two bottom curves . These are obtained for two different uncoupled fundamental frequencies of the P.S. (ie. 1 Hz and 10 Hz). This parametric study was performed to understand the effect of uncoupled frequency of the P.S. on the frequency and mass ratios. However, for checking the S.S. decoupling requirement , conservatively the curve obtained from stiff soil condition with 1 Hz P.S. frequency can be adopted. This also found (Reddy et al., 1996) conservative for different foundation soil (v_s = 580 m/sec) condition. The various regions where coupling required/not required have been identified and showed in Fig.6. The horizontal lines at frequency ratio 1.20 and 0.80 are drawn conservatively to account for tuning and are based on various case studies (Reddy et al.,1996). Here onwards this pictorial representation of decoupling criteria for multi- connected primary and secondary system will be referred as decoupling model. The applicability of this model is illustrated by considering typical reactor Internal Structure (IS) and Steam Generator (SG) for different fixidity conditions of IS.

Case studies:

Case.1: In this case, IS and SG of typical reactor building shown in Fig.2 have been considered. Coupled lumped mass beam model of reactor internal structure and SG is shown in Fig.7 which is not drawn to scale. In this,in addition to the IS considered in Fig.2(b), raft and springs representing soil have also been considered. In uncoupled condition both IS and SG will be separated and SG is fixed at support.1 and support.2. The numerals given in the Fig.7 are elevations of IS in meters. Frequencies and modal masses for both coupled and uncoupled condition of IS and SG have been calculated using standard finite element package COSMOS/M. Uncoupled modal frequencies and masses are listed in table.2. Figure.8, Fig.9 and Fig.10 respectively shows modeshapes for uncoupled SG, uncoupled IS, coupled IS and SG and are not drawn to the scale. Frequency and mass ratios of SG to IS have been calculated and listed in table.3. This data points have been indicated in decoupling model (Fig.11). The numbers indicated in the figure identifies ratios of i th modal frequency or mass of S.S (SG) to the j th modal frequency or mass of P.S. (IS). For example '21' indicates the ratio of 2nd modal frequency or mass of S.S. to the 1st modal frequency or mass of P.S.. It can be seen that the number of points falling in the 'coupling required ' region are four and they are 13,21,24,25. Therefore, it is obvious to expect 10% or more variation in first and second frequency of S.S. or first, third, fourth and fifth frequency of P.S. But about 4% and 5% in first and fourth mode of P.S. and 7% in second mode of S.S. and not much variation in other frequencies could be seen from table.4 which gives coupled system frequencies, modal masses, dof masses and change of uncoupled frequency to the coupled frequency of P.S/S.S.. It can be observed that the sum of (column 4 &5) dof mass is more than the modal mass (column.3). This is happening due to coupling effects. In fact, sign of modal mass is omitted for brevity and modal mass is conserved with sign. This effect can also be treated as a added mass in the case of fluid-structure interaction. In this table ik refers to the ith uncoupled frequency of S.S. and k^{th} frequency of the coupled system and jk refers j^{th} uncoupled frequency of P.S. to the kth frequency of the coupled system and these are referred in percentage frequency variation.

The reason for changing frequency less than 10 % as expected may be due to the decoupling model for this system is conservative as said in earlier discussion. This means that the decoupling model used was developed with rock base and similar condition may not be prevailing in IS below the SG support. This type of conservatism is also built in (USNRC, 1990) decoupling criteria for equipments connected at single location to the structure. From mode.1 of coupled condition (Fig.10) it can be said that the 4% variation in IS is due to inertia effect and not due to stiffness. Stiffness does not play any role in this mode ,because the SG is relatively more rigid than the IS. In this situation for 10% variation, more mass ratio is required. Unlike mode.1 in mode.5 and mode.6 interaction effects are clearly visible. This modes correspond to the 4 th mode of IS (Fig.9) and second mode of SG (Fig.8). Another reason for conservatism could be that the SG is not as simple as considered in developing decoupling model. However, for design point of view this conservatism is acceptable. In other modes the frequency have not changed significantly as indicated by the frequency and mass ratios. It may be due to the low modal mass ratios and frequencies are not tuned perfectly.

Case.2: In this case, to study the effect of flexibility of IS, IS is fixed at 109.00 m level. As like in case.1 uncoupled frequencies and modal masses of internal structure have been calculated and listed in table.5. This table also gives uncoupled frequencies of the SG. Similar to above case frequency and mass ratios have been calculated and are given in table.6. This data is marked in decoupling model (Fig. 12). It can be seen from this figure that the number of points falling in ' coupling required region ' are two and these are 21 and 23.With reference to this data, second frequency of S.S. or first and third frequency of IS should be changed by 10% or more. Tabel..7 shows a variation of 15% in second mode of SG and 6%,13% in first and third mode of IS.

Case.3: In this case, the IS is fixed at 115.00 m level . Uncoupled frequencies and modal masses are calculated and listed in table.8. This table also gives uncoupled frequencies of SG. Similar to above cases frequency and mass ratios have been calculated and given in table.9. This data is marked in decoupling model (Fig.13). It can be seen from this figure that the number of points falling in ' coupling required region' are two and these are 21 and 22. The mass ratio with respect to point 22 is 0.795 which is more than the mass ratio limit in the decoupling model. Due to this, point 22 is not shown in the figure, but it can be extrapolated that this also falls in the coupling required region. With reference to this data first and second frequencies of IS and second frequency of SG should be changed by 10% or more. Table.10 shows a variation of 6%,11% in first and second mode of IS and 8% in second mode of S.S..

DISCUSSIONS AND CONCLUSIONS

1. The decoupling model established herein for 10% variation of frequencies of P.S./S.S. is conservative for IS and SG of typical nuclear power plant. This is due to the fact that decoupling model established is based on stiff rock base and similar stiffness may not be existing in internal structure below the SG support. The conservatism may be eliminated if the SG rests on the IS very close to the foundation raft and not much overhang is present in SG. Nonetheless this model should be very good for design purpose.

2. In the decoupling model the area covered by two horizontal lines around frequency ratio 1.0 basically accounts for tuning of P.S. and S.S. From above studies it was observed that in some cases this range is conservative but from a design point of view it may be acceptable.

3. It can be carefully observed from different cases that for the mass and frequency ratios which are falling in the " coupling required " region, in most of the cases either IS or SG frequency is changing and both are changing in some cases. Therefore, it may not be correct to expect always a change in frequency of both IS and SG for a frequency and mass ratio falling in the " coupling required " region of decoupling model.

4. In the case of coupled model, it can be noticed from the tables 4,7,10 (column.5), that the total mass of SG participated is very much higher than its actual mass (99.5 tons). This is due to the interaction effects. In most of the modes it can be treated as added mass like in fluid-structure interaction. This mass in most of the cases is causing inertial effects on over all system and not significantly affecting the potential energy of the total system. Detailed studies in the view of energy contribution of the P.S. and S.S. for overall system may be required before final conclusion.

5. The decoupling model developed is for the case where the S.S. is supported at two locations. However, similar observations (Reddy et al.,1996) were made for a S.S. (e.g.: coupled SG and pump) having more than two supports.

REFERENCES.

1. ASCE standard, ASCE4, 1986, "seismic analysis of safety related nuclear structures and commentary on standard of seismic analysis of safety related structures".

2. G.R.Reddy et al., 1994, "Procedure of applying decoupling for complex primary and secondary systems". ASME seismic engineering conference proceedings. PVP-Vol.275-1,PP 129-139.

3. G.R.Reddy, Kohei Suzuki, S.C.Mahajan., 1996, "Decoupling criteria for multi- connected equipments". Under preparation.

4. United States Nuclear Regulatory Commission (USNRC), 1990, standard Review Plan (SRP), section 3.7.2, seismic system analysis.

Table.1 Foundation parameters

Description	Rock
Shear wave velocity (m/sec)	2900
Poisson's ratio	0.2
Weight density (t/m^3)	2.4
Foundation raft dimensions	5.5 m thick 62.0 m diameter

Table.2 case.1: Frequencies and modal masses

Mode	Frequency (Hz) IS	Frequency (Hz) SG	Modal mass (tons) IS	Modal mass (tons) SG
1	2.82	12.70	2435.40	14.31
2	6.95	28.43	1732.90	66.00
3	15.34	–	1469.10	--
4	27.80	–	997.01	--
5	31.41	–	673.30	--

SG- Steam Generator
IS - Internal Structure

For better understanding of the terms R_f and R_m Fig.1 may be referred.
ii) If $0.01 \le R_m \le 0.1$ decoupling can be done provided $0.8 \ge R_f \ge 1.25$.
iii) If $R_m > 0.1$ an approximate model of the S.S. should be included in the P.S..
iv) For rigid secondary system (natural frequency is more than 33 Hz) only mass of the S.S. can be included.

This criteria is based upon the fact that the difference in the uncoupled and coupled frequency of the P.S/S.S is in the range of 15%. Seismic code (ASCE-4 ,1986), also has defined similar criteria and indicated that the number of dominant modes can be identified as those frequencies in which modal mass is more than 20 % of the total mass of the S.S. or P.S.. This is very important aspect in the case of multi degree of freedom P.S. or S.S. to select important modes for which interaction is very important otherwise, virtually speaking , there will be a large number of modes which will not contribute significantly to the response of the P.S. or S.S..

The above criteria is best suited for single P.S. and S.S. and it has limitations in its application to a complex primary-secondary system where more than one primary subsystem are resting on common foundation. The limitations can be understood explicitly by considering a typical nuclear power plant reactor building as shown in Fig.2.

A typical reactor building is a complex structural system comprising of subsystems like inner and outer containments, internal structure and calandria vault, resting on common foundation. The internal structure which forms one primary subsystem supports S.S. such as steam generators, pumps, fueling machines, pressurisers etc. The calandria vault forming another primary subsystem supports calandria endshield assembly , control rod mechanisms etc. These subsystems interact with each other as they are resting on a common foundation raft. When a S.S is attached to one of the P.S. (Fig.2b), application of decoupling criteria on the basis of modal mass greater than 20 % of the total mass as described above is not straightforward. This is due to the fact that the two primary subsystems are founded on a common base and coupling may be observed between the two subsystems in some of the modes only. To understand this interaction effects, extensive work (Reddy et al.,1994) was carried out and a modification in the existing decoupling criteria was evolved, which is explained here under.

Modified decoupling procedure:

Consider the system shown in Fig.3 which consists of two primary subsystems. The S.S. is assumed to be of 2 degree of freedom (dof) which is supported on primary subsystem.2. With reference to this system, the procedure is explained as follows.

Calculate:
Frequencies of the structure including two primary subsystems and foundation: f_{p1}, f_{p2}, f_{p3}
Uncoupled frequencies of the S.S. : f_{s1} , f_{s2}
j^{th} modal mass of ith primary subsystem (including s.s. mass): m_{psij}
Subsystem modal mass of primary subsystem-2: m_{ps21}, m_{ps22}, m_{ps23}
Frequencies in which subsystem mass is more than 20% of its own mass : f_{p1}, f_{p3}

Then calculate:
Frequency ratios: f_{s1}/f_{p1}, f_{s1}/f_{p3}, f_{s2}/f_{p1}, f_{s2}/f_{p3}
Mass ratios : m_{s1}/m_{ps21}, m_{s1}/m_{ps23}, m_{s2}/m_{ps21}, m_{s2}/m_{ps23}

After evaluating the mass ratios and the frequency ratios, the coupling criteria explained earlier can be applied. In principle, the use of primary subsystem modal mass is recommended, unlike total mass of the combined primary system as mentioned in decoupling criteria.

The above two procedures are solely meant for equipment connected to the structure at one location only. There is no criteria available in the literature /codes for equipments connected at more than one location to the structure. In the following sections the latter issue has been discussed and new decoupling model for multi-connected equipment-structure has been presented.

INTERACTION OF MULTI-CONNECTED PRIMARY-SECONDARY SYSTEMS

An example for multi-connected primary-secondary system could be a Steam Generator (SG) and reactor building as shown in Fig.2. There are many such systems in the power plant but SG is one of the heavy equipments and may interact with structure significantly. Initially, a simple system is considered to understand the interaction effects and to develop decoupling model and then it is extended to the actual real life systems (Fig.2).

In the simple model shown in Fig.4, m_1,m_2,m_s represents the masses and k_1,k_2,k_s represents the stiffness of foundation, P.S., S.S, respectively. The mass parameters shown in figure closely represent typical reactor building internal structure and foundation. The stiffness parameters are explained later. For the sake of simplicity the S.S. is assumed to be connected to the P.S.at two locations as shown in the figure.

Characteristic equation for this model can be written as:

$$[\,[M]^{-1}\,[K] - \lambda\ [I]]\,[\phi] = 0 \qquad (1)$$

Where $[M] = \begin{bmatrix} m_1 & 0 & 0 \\ 0 & m_2 & 0 \\ 0 & 0 & m_s \end{bmatrix}$

and

$$[K] = \begin{bmatrix} k_1+k_2+k_s & -k_2 & -k_s \\ -k_2 & k_2+k_s & -k_s \\ -k_s & -k_s & k_s+k_s \end{bmatrix}$$

On expanding equation (1), the following cubic equation can be obtained.

$$\lambda^3 + p\lambda^2 + q\lambda + r = 0 \qquad (2)$$

Where

$$p = -(k_1+k_2+k_s)/m_1 - ((k_2+k_s)/m_2 + \omega_s^2)$$

$$q = (k_2+k_s)\,\omega^2 / m_2 + (k_1+k_2+k_s)/m_1\,((k_2+k_s)/m_2 + \omega_s^2) - k_s \times k_s/m_s * m_2 - k_2 \times k_2/m_1 \times m_2 - k_s \times k_s/m_1 \times m_s$$

$$r = -(k_1+k_2+k_s)/m_1\,((k_2+k_s/m_2)\,\omega_s^2 - k_s \times k_s/m_s \times m_2)) + (\omega_s^2 / 2 \times m_1 \times m_2)\,(2 \times k_2 \times k_2 + 3 \times k_2 \times k_s + k_s \times k_s)$$

Table.3 case.1: Frequency and mass ratios

j	i	Frequency ratio (R_f) 1	2	mass ratio (R_m) 1	2
1		4.50	10.08	0.006	0.027
2		1.83	4.09	0.008	0.037
3		0.83	1.85	0.010	0.045
4		0.46	1.02	0.014	0.066
5		0.40	0.91	0.021	0.098

i stands for Secondary System (SG) mode number
j stands for Primary System (IS) mode number

Table.4 case.1: coupled system modal parameters

Mode (1)	Frequency Hz (2)	modal mass tons (3)	dof mass tons IS (4)	SG (5)	jk (6)	ik (7)	% change in frequency (8)
1	2.71	2512.90	2376.05	136.88	11	--	4
2	6.78	1812.90	1855.35	42.45	22	--	2
3	12.90	142.67	108.89	33.77	--	13	2
4	15.41	1360.10	1404.35	44.28	34	--	1
5	26.54	401.47	294.49	106.99	45	25	5,7*
6	29.03	758.31	907.37	149.06	46	26	4,2
7	31.43	503.24	446.08	57.17	57	--	0

* indicates % change in frequency of SG

$$\% \text{ change in frequency} = \frac{\text{uncoupled frequency - coupled frequency}}{\text{uncoupled frequency}}$$

Table.5 case.2: Frequencies and modal masses

Mode	Frequency (Hz) IS	SG	Modal mass (tons) IS	SG
1	4.49	12.70	827.67	14.31
2	22.58	28.43	881.89	66.00
3	33.06	--	113.81	---

Table.6 case.2: Frequency and mass ratios

j	i	Frequency ratio (R_f) 1	2	mass ratio (R_m) 1	2
1		2.83	6.33	0.017	0.080
2		0.56	1.26	0.016	0.075
3		0.38	0.86	0.126	0.580

Table.7 case.2: coupled system modal parameters

Mode (1)	Frequency Hz (2)	modal mass tons (3)	dof mass tons IS (4)	SG (5)	jk (6)	ik (7)	%change in frequency (8)
1	4.24	894.33	808.32	86.00	11	--	6
2	12.91	37.65	55.10	17.45	--	12	2
3	21.90	796.07	711.29	84.79	23	--	3
4	28.65	86.16	154.87	68.72	34	24	13,1
5	32.73	108.31	93.60	14.71	35	25	1,15

Table.8 case.3: Frequencies and modal masses

Mode	Frequency (Hz) IS	SG	Modal mass (tons) IS	SG
1	4.75	12.70	741.39	14.31
2	30.02	28.43	83.05	66.00

Table.9 case.3: Frequency and mass ratios

j \ i	Frequency ratio (R_f) 1	2	mass ratio (R_m) 1	2
1	2.67	5.99	0.019	0.089
2	0.42	0.95	0.172	0.795

Table.10 case.3: coupled system modal parameters

Mode (1)	Frequency Hz (2)	modal mass tons (3)	dof mass tons IS (4)	SG (5)	jk (6)	ik (7)	% change in frequency (8)
1	4.48	806.90	725.68	81.22	11	--	6
2	12.98	19.05	31.71	12.66	--	12	2
3	26.71	68.19	12.08	56.11	23	23	11,6
4	30.77	31.35	56.07	24.71	24	24	3,8

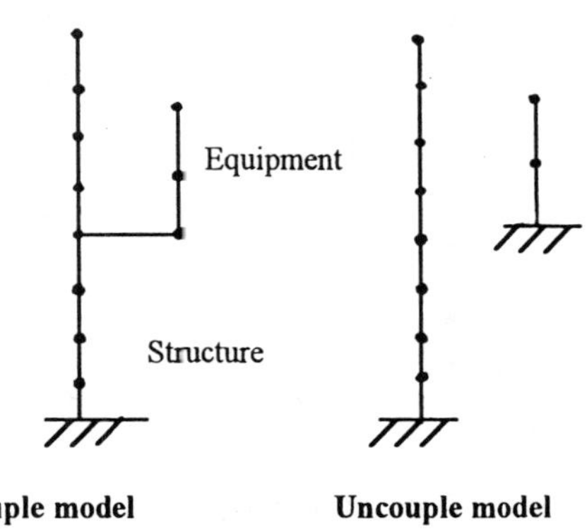

Fig.1 Beam model of typical cantilever structure supporting equipment at single location

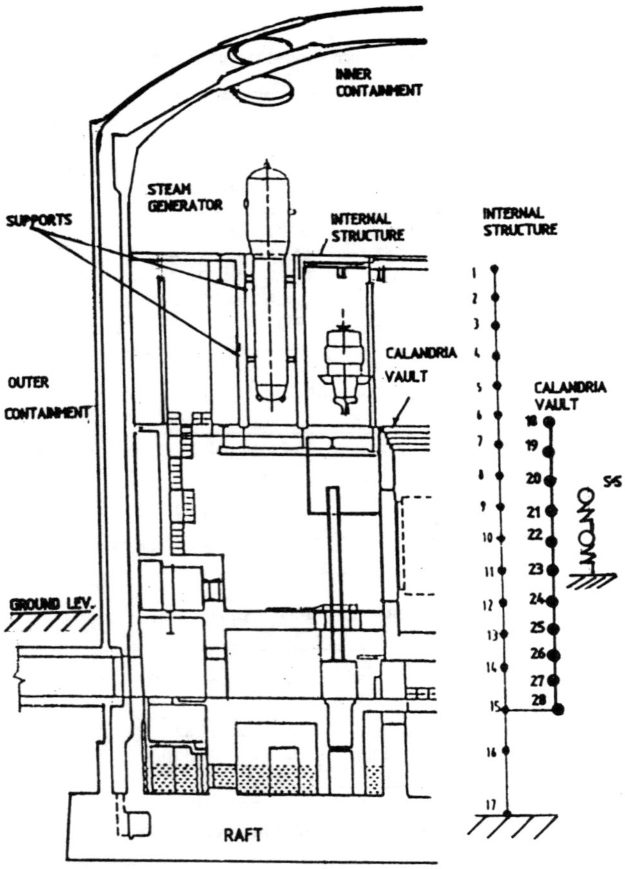

a. H alf section

b. Lumped mass beam model

Fig.2 Typical reactor building

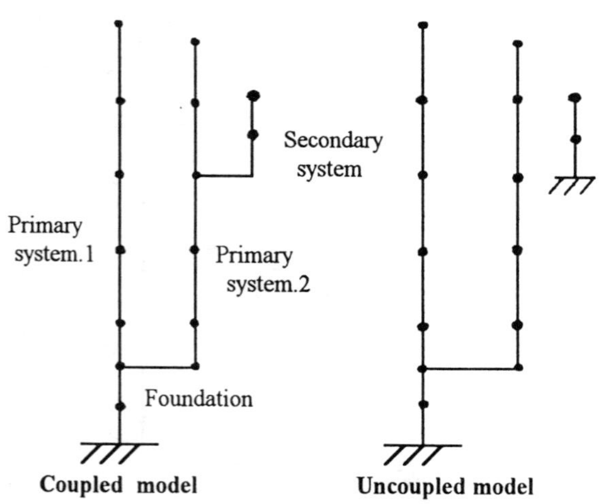

Fig.3 Beam model of two cantilever structures resting on common foundation

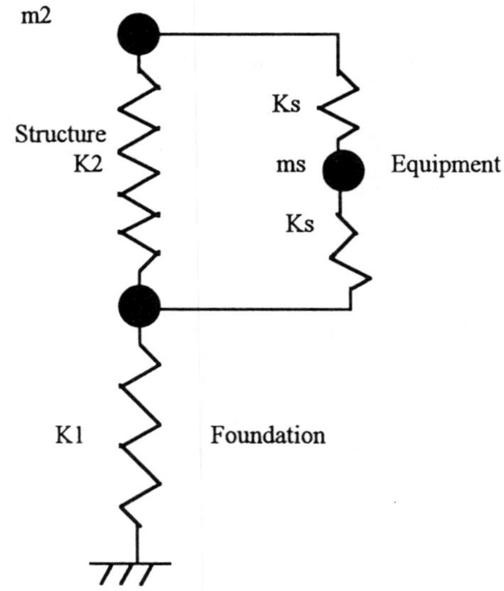

Fig.4 Spring-mass model of multi-connected primary-secondarysystem

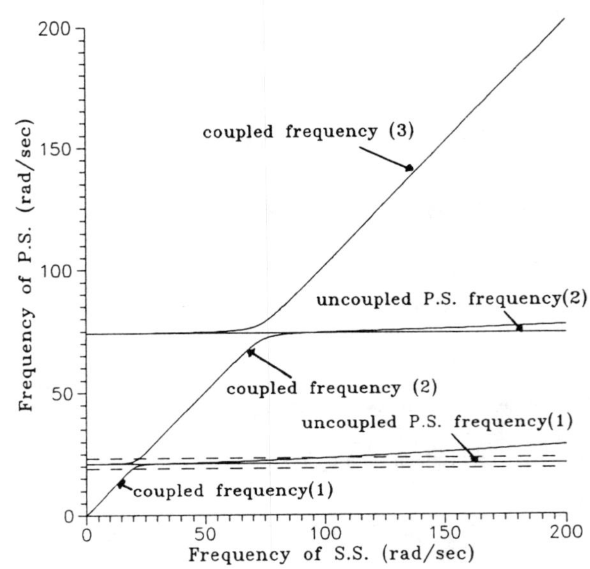

Fig.5 Secondary system frequency v.s primary system frequency

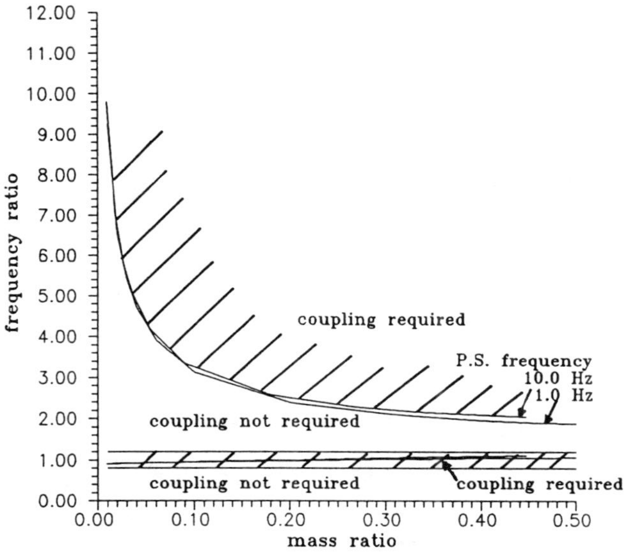

Fig.6 Frequency v.s mass ratio of secondary system to primary system

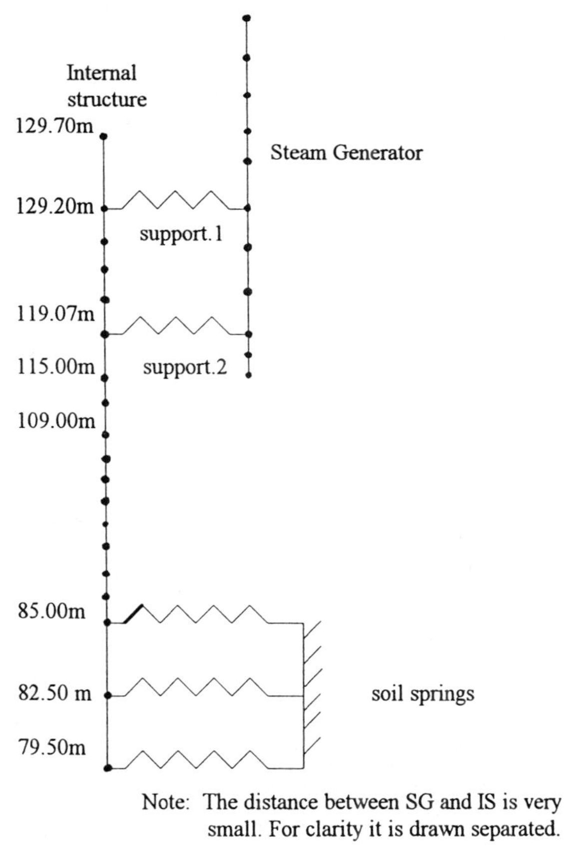

Fig. 7 Coupled lumped mass beam model of Reactor Internal Structure and Steam Generator

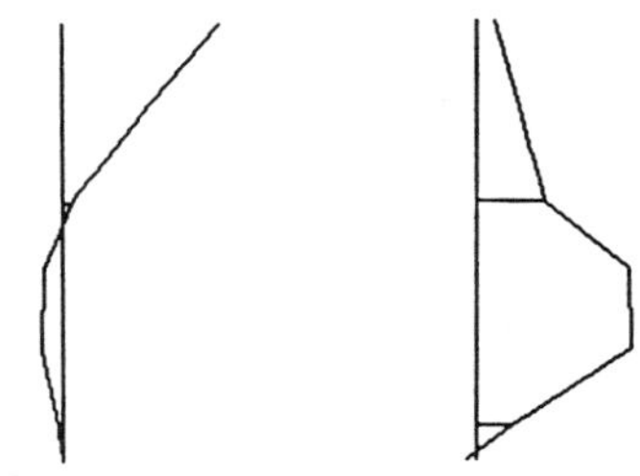

Mode.1 Frequency= 12.70 Hz Mode.2 Frequency = 28.43 Hz

Fig.8 Uncoupled modes and frequencies of SG

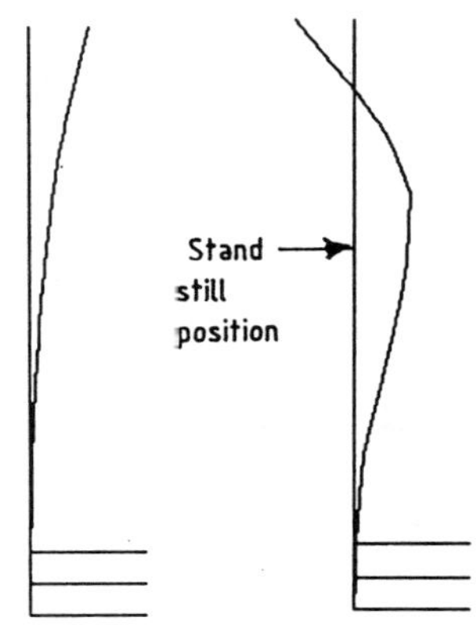

Mode.1 Frequency= 2.82 Hz Mode.2 Frequency= 6.95 Hz

Fig.9(a) Uncoupled modes and frequencies of IS (case.1)

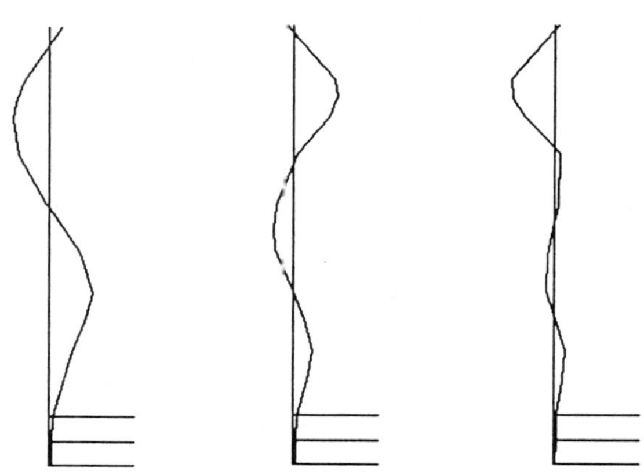

Mode.3 Frequency= 15.34 Hz Mode.4 Frequency= 27.80 Hz Mode.4 Frequency=31.41 Hz

Fig.9(b) Uncoupled modes and frequencies of IS (case.1

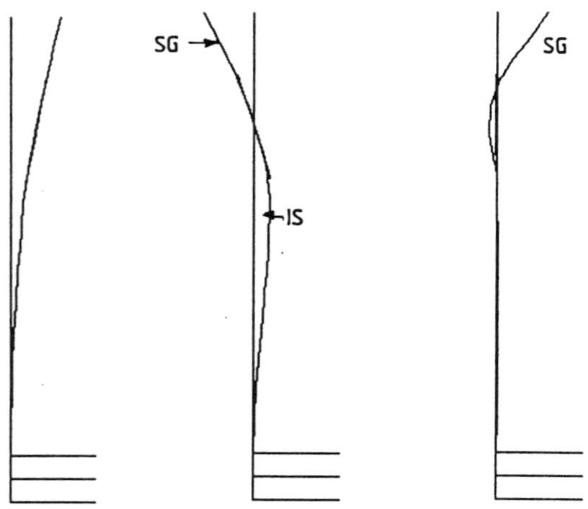

Mode.1 Frequency= 2.71 Hz Mode.2 Frequency=6.78 Hz Mode.3 Frequency= 12.90 Hz

Fig.10(a) Coupled modes and frequencies of IS and SG (case.1)

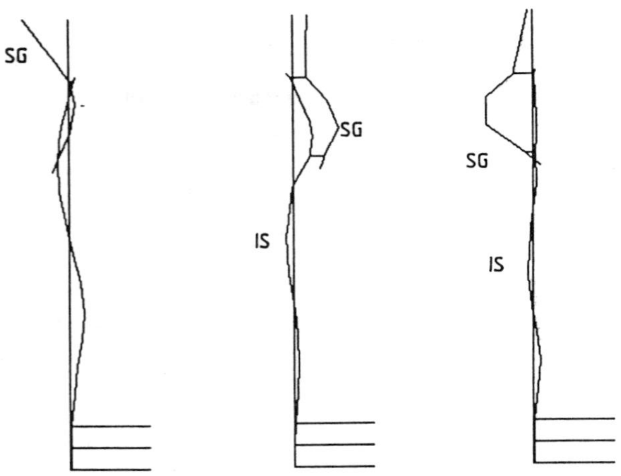

Mode.4 Frequency=15.41 Hz Mode.5 Frequency=26.54 Hz Mode.6 Frequency=29.03 Hz

Fig.10(b) Coupled modes and frequencies of IS and SG (case.1)

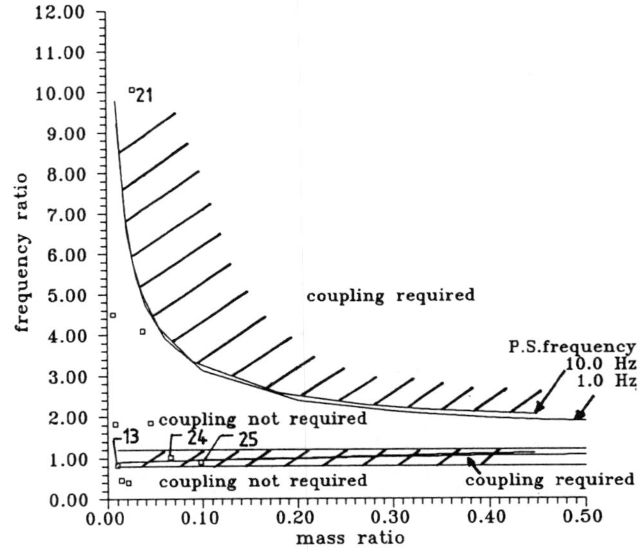

Fig.11 Frequency v.s mass ratio of secondary system to the primary system (case.1)

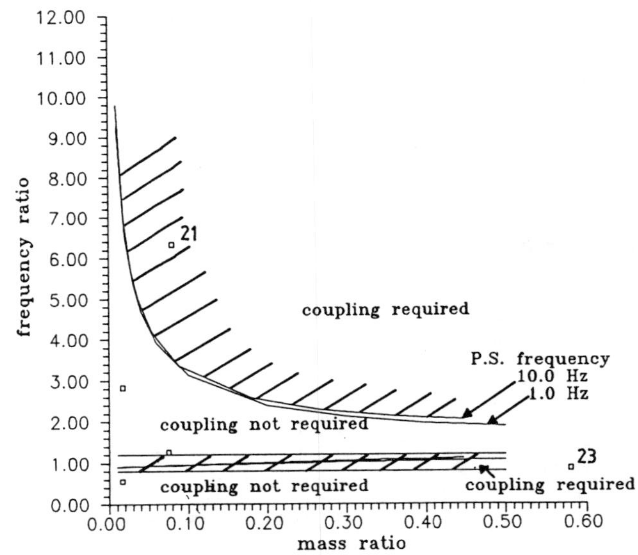

Fig.12 Frequency v.s mass ratio of secondary system to the primary system (case.2)

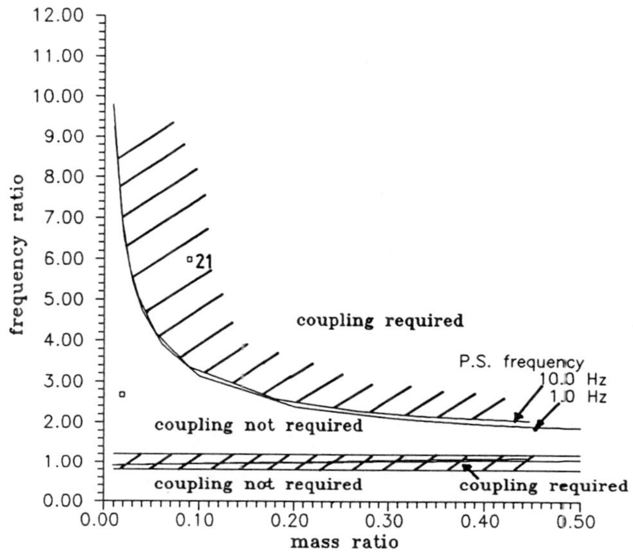

Fig.13 Frequency v.s mass ratio of secondary system to the primary system (case.3)

PVP-Vol. 340, Seismic Engineering
ASME 1996

INVESTIGATION ON THE BUCKLING OF STEEL PIERS OF ELEVATED EXPRESSWAY DUE TO THE GREAT HANSHIN EARTHQUAKE

Katsuhisa Fujita, Kazuhiro Nakamura, Tomohiro Ito and Yoichi Iwamoto
Takasago R&D Center, Mitsubishi Heavy Industries, Ltd.,
2-1-1 Shinhama Araicho, Takasago City, Hyogo Pref., 676 Japan

ABSTRACT

In the Great Hanshin Earthquake occurred in January 1995, extremely serious damages of the elevated expressways were found. Some of those piers are made of steel and many of them buckled and revealed plastic deformations.

One of them revealed the circumferencially uniform outward bulge at the middle portion in the axial direction, though the buckling of a long circular cylinder will occur at it's bottom end accompanied by more complicated deformation patterns typical to the bending buckling.

In this study, simple reduced scale models of the pier were fabricated and the dynamic and static buckling tests were performed. The effect of the girder weight or vertical excitation on the buckling phenomena were investigated. In addition to the tests, static buckling analysis using FEM was also performed in order to clarify the causes for this specified buckling phenomena.

NOMENCLATURE

A :	Cross sectional area
I :	Second moment of inertia of the section
M :	Moment
R :	Radius of the pier
W :	Self weight
Z :	Section modulus
ℓ :	Length of the pier
t :	Thickness of the pier
$\ddot{Z}$:	Input acceleration
α_H :	Horizontal response acceleration
α_v :	Vertical response acceleration
σ_b :	Bending stress
σ_{bcr} :	Bending buckling critical stress
σ_c :	Axial compressive stress
σ_{ccr} :	Axial compression buckling critical stress
σ_y :	Yield stress
θ :	Inclined angle
η :	= R

INTRODUCTION

The Great Hanshin Earthquake in January 1995 caused catastrophic damages on many buildings, elevated expressways, apartment houses and wooden houses. Especially, the elevated expressways supported by re-inforced concrete piers suffered extremely serious damages such as the collapse of the piers and shoes, falling and overturning of the bridge girder consequently occurred. On the other hand, some piers made of steel, buckled and revealed plastic deformations.

Several kinds of buckling deformation patterns were found, some of which could not be described according to the conventional buckling criteria (NASA 1968). One of them revealed a circumferencially uniform outward bulge at the middle portion in the axial direction, though the buckling of a long circular cylinder will occur at it's bottom end accompanied by more complicated deformation patterns typical to the bending buckling, in general. (Fujita, Ito and Wada 1991)

It was estimated that this pier suffered very large vertical seismic excitation in addition to the horizontal excitation, i.e., very large 3 dimensional seismic excitation. Also, it suffered very large steady axial compression due to the bridge girder weight. It was considered that these factors caused very large influences on the buckling deformation patterns of this pier.

In this study, simple reduced scale models of the pier were fabricated and the dynamic and static buckling tests were performed. The effects of the girder weight was simulated using a tension bar installed at the center of the pier model. In the static load tests, several cyclic forced displacement were applied to the models. In the dynamic buckling tests, the test model was installed on the shaking table and shaked by the earthquake wave measured at the Great Hanshin Earthquake.

The affects of the girder weight and vertical excitation on the buckling deformation were investigated. In addition to the tests, static buckling analysis using FEM was also performed in order to clarify the causes for this specified buckling phenomena.

BUCKLING PHENOMENA OF A STEEL PIER

Dimensions of a Steel Pier

Figure 1 illustrates the locations of the buckled piers (p4 and p5) and buckled portions. The length of the pier is around 15m and the diameter is around 2.2m. The thickness of the lower 8m of the pier is 25mm and the upper 7m of the pier has 19mm thickness. The material was SS400 for the upper part and SM400A for the lower part. In the lower 4m of the pier, concrete is filled to increase the strength of the pier.

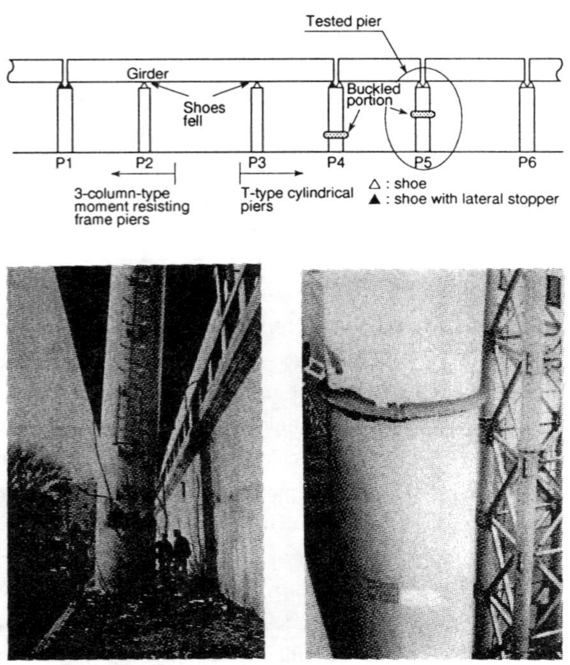

(a) p4 (b) p5

Fig.1 Buckling deformation for p4 and p5 pier

Buckling Phenomena

Buckling occurred at the bottom of p4 pier, and at the middle just above the 25mm-thick pier of p5 pier. Figure 1 also shows the photos of buckling deformation for p4 and p5. For p5, circumferentially uniform outward bulge was observed.

DYNAMIC BUCKLING TEST

Test Method

Reduced scale models for p5 pier were fabricated and were installed horizontally on the 1-dimensional shaking table.

After measuring the natural frequency of the model, dynamic buckling tests were performed using an earthquake wave measured at the Kobe marine meteorological observatory of which time scale was shifted shorter according to the model frequency. In one test, sine beat wave is used. Input acceleration level was increased step by step untill collapse occurred. For each step, the response displacement and acceleration were measured. Fig.2 shows the input acceleration wave and it's response spectrum.

Test Models

Figure 3 shows the configuration of the dynamic buckling test model. In this test, the qualitative evaluation of the buckling deformation pattern was the most important subject, thus the reduced scale models were employed satisfying only R/t and l/R (R : radius, t : thickness, l : length). The scale factor was settled around 24

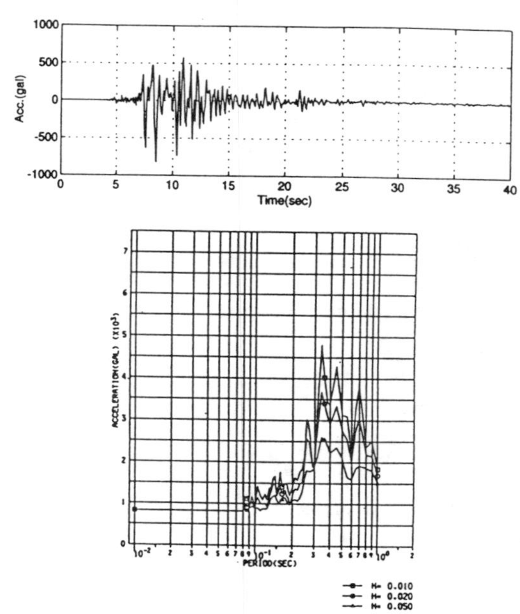

Fig.2 Input acceleration wave and response spectrum

considering the capacity of the shaking table etc.. To investigate the effect of the material, aluminum (Al) piers were also tested in addition to the steel piers. The actual pier was composed of two kinds of circular cylinders with different thickness, i. e., t=19mm and t=25mm. In this tests, short cylinders with 0.8mm-thickness were used to study the effect of the axial compressive stress and vertical excitation, and long cylinders composed of 0.8mm-thickness cylinder and 1.0mm-thickness cylinder, which simulated the actual pier, were used to study the axial location of buckling. Long cylinders are called double stage piers and short cylinders are called single stage piers.

To simulate the axial compressive weight, a tension bolt were inserted at the center of the piers. Here, the ratio between the yield stress and the axial compressive stress were settled the same value as the actual pier.

The ratio of the vertical input acceleration to the horizontal input acceleration was adjusted by inclining the pier model against the exciting direction.

At the top of the pier, a steel weight was attached simulating a girder weight. This steel weight was suspended by a fine wire to avoid bending due to this self weight because the pier models were installed horizontally on the shaking table.

The measured points are illustrated in Fig.3. Acceleration and the displacement of the top weight, and the strain of the cylinder were measured.

Table 1 Test Items

No.	Material	Stage Nos.	Radious R(mm)	Thickness t (mm)	Length ℓ (mm)	Weight (N)	Axial Compressive Stress (MPa)	Inclined Angle θ	Remarks
1	Al	Single	46.4	0.8	306	588	0	0°	Effect of the axial compressive stress (Al)
2			↑	↑	↑	↑	24.5	↑	
3			↑	↑	↑	↑	↑	45°	Effect of vertical excitation
4	Steel	Single	↑	↑	↑	2352	0	0°	Effect of the axial compressive stress (Steel)
5			↑	↑	↑	↑	39.2	↑	
6			↑	↑	↑	↑	↑	45°	Effect of vertical excitation
7			↑	↑	↑	↑	↑	0°	Sine beat wave
8	Steel	Double	↑ (46.6)	↑ (1.0)	456	↑	↑	↑	Double stage

cf. Values in parenthisis are for thicker cylinder.

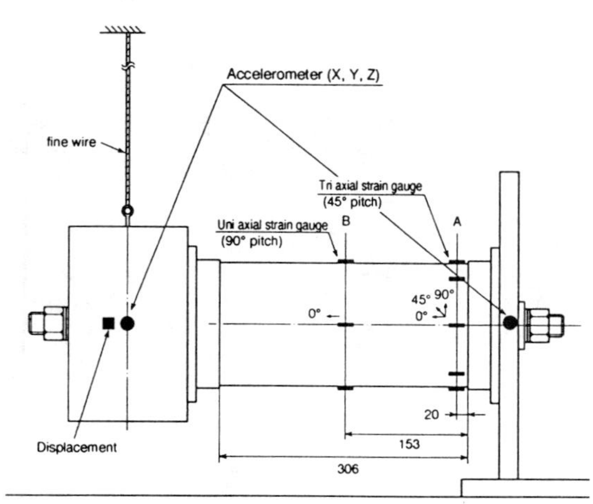

Fig.3 Dynamic buckling test model

Test Items

Table 1 shows the test items. The material of the pier, stage numbers, axial compressive stress and the inclined angle were varied as parameters.

Test Results

Natural frequency Before the buckling tests, natural frequencies of the pier models were measured. Table 2 shows the natural frequencies.

Effect of the axial compressive stress The actural compressive stress was around 39.2 MPa by the girder weight 4.9MN. Thus, for the steel pier models, two kinds of the axial compressive stress, 0MPa and 39.2MPa were applied.

Table 2 Natural frequencies of the pier models

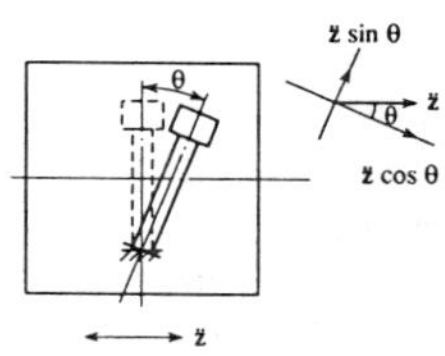

Material	Stage Nos.	Length ℓ (mm)	Weight (N)	Natural Freq. (Hz)
Aℓ	Single	306	588	15.0
Steel	Single	306	2352	10.2
	Double	456	2352	7.6

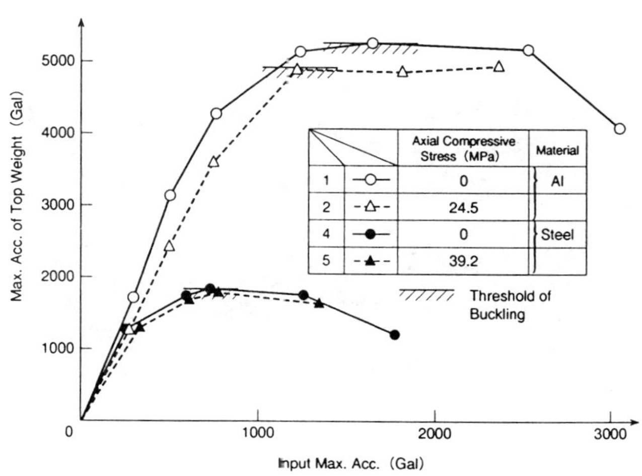

Fig.4 Effect of the axial compressive stress on the response acceleration

For the aluminum (Al) pier models, two kinds of axial compressive stress, 0MPa and 24.5MPa were applied.

The relationship between the response acceleration of the top weight and the input acceleration for these 4 cases were illustrated in Fig.4. The buckling deformation patterns for these cases are shown in Fig.5. When a circular cylinder buckles, the response acceleration will become maximum in general (Fujita et at al., 1991). Thus, from Fig.4, it is noted that buckling critical response acceleration reduced by 7% for aluminum pier model and by 4% for steel pier model. However, no effect of the axial compressive stress was observed for the buckling deformation patterns, i.e., uniform outward bulge was observed on the 3/4 in the circumferential direction at the bottom of the piers. For the aluminum pier models, slight inward deformations were also observed just above the outward bulge.

Effect of vertical excitation When the input acceleration is defined as $\ddot{Z}$, vertical input acceleration will become $\ddot{Z}\sin\theta$ and horizontal input acceleration $\ddot{Z}\cos\theta$. In this case, there is no phase difference between the horizontal and vertical input acceleration.

Fig.6 shows the response acceleration of the top weight and Fig.7 shows the buckling deformation patterns for the cases 5 and 6. The buckling critical response acceleration for the inclined pier made of aluminum was slightly lower than that of θ=0° pier, while that for the inclined pier made of steel was slightly higher than that of θ=0° pier. However, no effect of the vertical excitation was observed for the buckling deformation patterns in these cases.

Effect of input earthquake wave form Sine beat wave has the same acceleration magnitude in both + and - side. For the case No.7, sine beat wave was employed as an input earthquake wave and the results were compared to those obtained by the actual earthquake wave.

Figure 8 shows the buckling deformation pattern obtained by sine beat wave. These pattern was almost the same as those obtained by the actual earthquake wave, indicating that there is no effect of the input earthquake wave form on the buckling deformation pattern.

Effect of material As shown in Fig.5, for the aluminum piers, outward bulge was accompanied by slight inward deformation. This

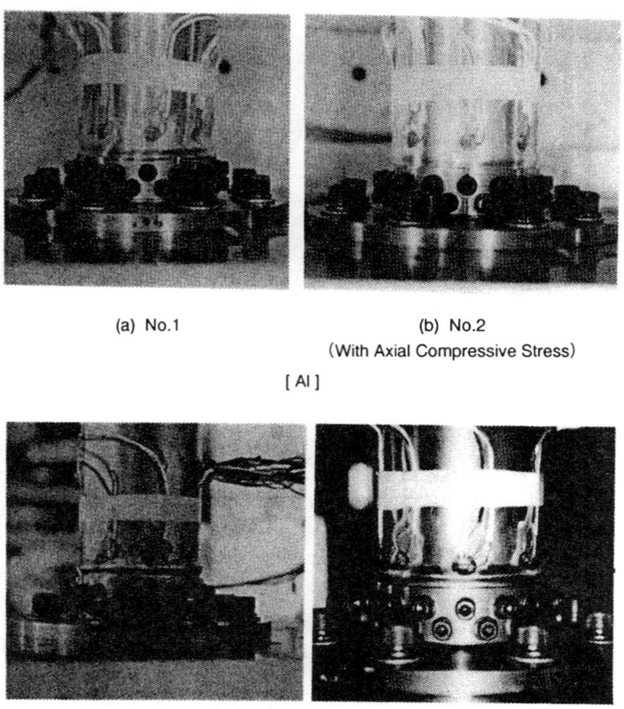

(a) No.1 (b) No.2 (With Axial Compressive Stress)

[Al]

(c) No.4 (d) No.5 (With Axial Compressive Stress)

[Steel]

Fig.5 Effect of the axial compressive stress on the buckling deformation pattern

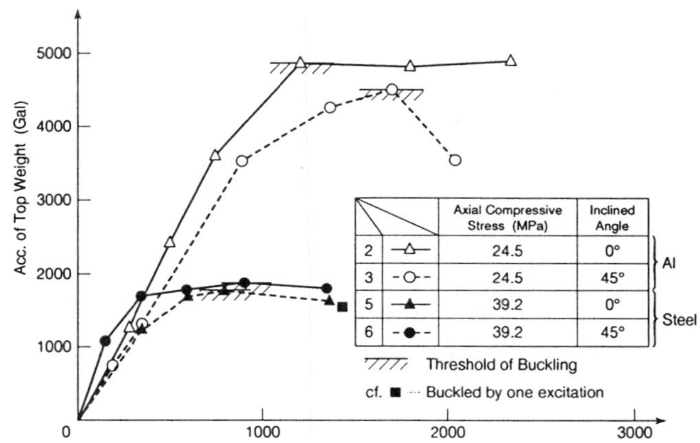

Fig.6 Effect of vertical input acceleration on the response acceleration

(a) No.5 (b) No.6 (With Vertical Excitation)

[Steel]

Fig.7 Effect of the vertical input acceleration on the buckling deformation pattern

means that the buckling phenomena for aluminum piers will be relatively closer to the elastic buckling. Young's modulus of aluminum is around 1/3 of that of steel, while yield point of aluminum is around 5/8 of that of steel, thus the elastic region of aluminum pier becomes relatively larger than that of steel pier. This will be one of the main reasons for the difference in the buckling deformation pattern between aluminum piers and steel piers.

Buckling location for the double stage pier Figure 9 shows the comparison of buckling critical response acceleration between the single stage pier and the double stage pier. Figure 10 illustrates the buckling deformation patterns for these cases. Buckling occurred at the bottom portion of the thinner cylinder of the double stage pier, and it's buckling deformation pattern is almost the same as that of the single stage pier. The buckling critical response acceleration for these piers were almost the same to each other, and this fact well describes the above mentioned buckling location.

For the actual pier, the buckling critical response acceleration was evaluated for both the thinner pier and thicker pier using the following conventional static buckling formulation. The results are shown in Fig.11. In the following equation, the effects of bending and axial compression are taken into considerations.

$$\left(\frac{\sigma_b}{\sigma_{bcr}}\right)+\left(\frac{\sigma_c}{\sigma_{ccr}}\right)=1.0 \quad \cdots\cdots (1)$$

$$\sigma_b=\frac{M}{I}\eta=\frac{M}{Z}=\frac{W\alpha_H \ell}{Z} \quad \cdots\cdots (2)$$

$$\sigma_c=\frac{W+W\alpha_v}{A} \quad \cdots\cdots (3)$$

where, α_H, α_V : Response acceleration in horizontal and vertical direction

W : Self weight

ℓ : Length

Z : Section modulus

(a) Front View (b) Side View

Fig.8 Buckling deformation pattern by sine best wave

(a) No.5 (Single Stage) (b) No.8 (Double Stage)

Fig.10 Buckling deformation patterns for single stage and double stage piers

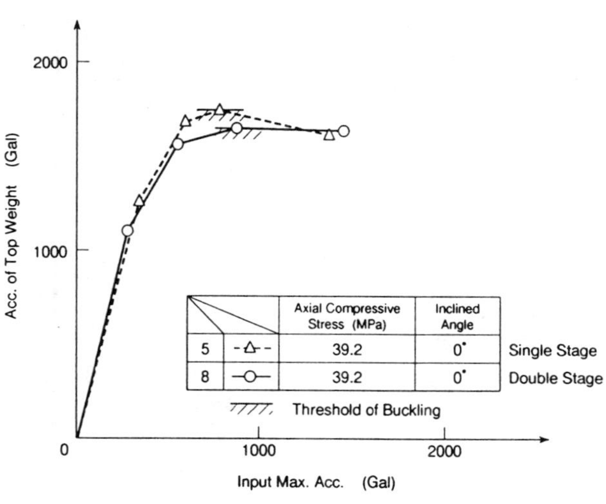

Fig.9 Comparison of buckling response acceleration between single stage pier and double stage pier

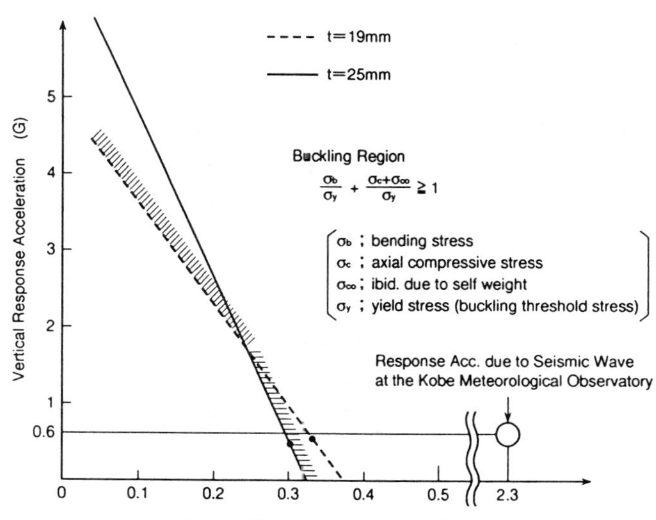

Fig.11 Buckling region for 19mm-thick pier and 25mm thick pier

A : Cross section area

σ_{bcr}, σ_{ccr} : Critical buckling stress for bending and axial compression (Here, both are 235MPa)

These rough estimations show that buckling for thinner pier and that for thicker pier will occur at almost the same response acceleration. Taking into consideration that the buckling phenomena will have very large deviations, it can be said that this result can also describe the test results shown in Fig.9 and Fig.10.

STATIC BUCKLING TESTS

Test Method

Tests have been made on scale models of the actual highway pier that showed the nearly uniform elephant-foot bulge. An axial compression load and a lateral load were applied statically on the models. Fig.12(1) shows the actual pier and Fig.12(2) shows the test set-up. The axial load was kept constant in each test and the deflection-controlled lateral load was applied cyclically in both directions.

Table 3 Static test cases

test condition \ case	No.1	No.2	No.3	No.4
axial compression load N (N)	0	6.3×10^3	6.3×10^3	12.7×10^3
axial compression stress				
σ_a = N/A (MPa)	0	41.0	41.0	82.7
σ_a/σ_y (-)	0	0.15	0.15	0.31
lateral displacement δ	±18mm 2.5 cycles	±16mm 1.5 cycles plus ±18mm 1 cycle	±9mm 10 cycles	±18mm ≈ 2 cycles

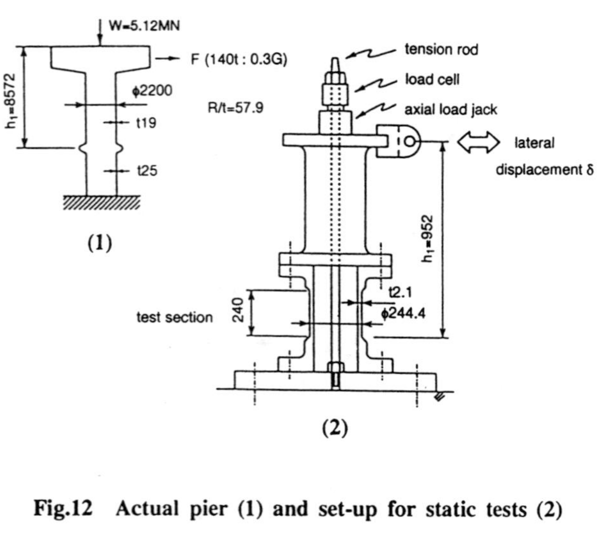

Fig.12 Actual pier (1) and set-up for static tests (2)

Test Models

Fig.13 shows the configuration of the test model. The test section of the model was machined out of carbon steel pipe (STPG410). The average thicknesses of the test section was 2.01mm. The radius to thickness ratio was 60.25. Test pieces were taken from the model to obtain the yield stress (0.2% proof stress) of the material, which was found to be 266MPa.

Test Cases

Four tests (Nos.1 to 4) shown in Table 3 were carried out. Models No.1 to 4 were used in the corresponding tests. Test No.1 was the case where the lateral displacement alone was applied. The applied lateral displacement was twice that corresponds to the maximum load in the first cycle. The same magnitude of the displacement was applied

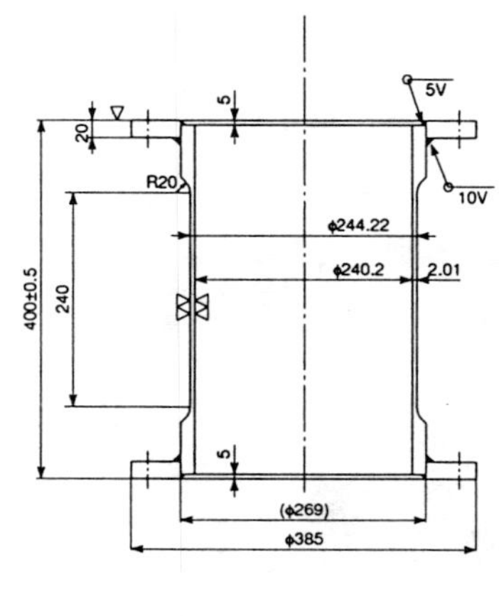

Fig.13 Static test models

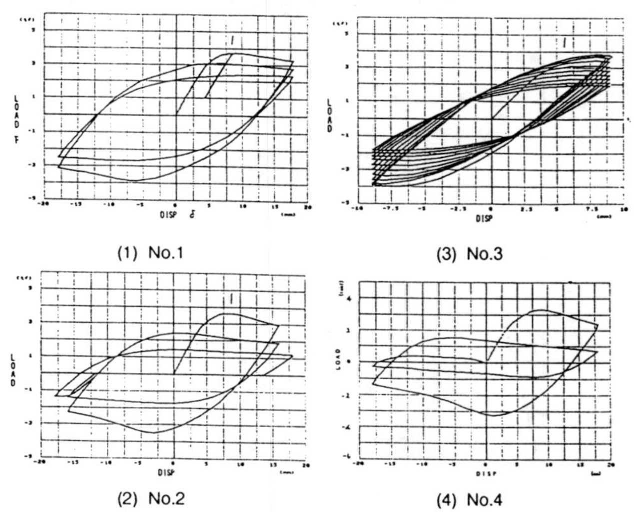

Fig.14 Lateral load F vs lateral displacement δ in tests No.1 to 4

in the opposite direction. A total of 2.5 cycles were applied. The models No.2 and 3 were tested with the axial compression load corresponding to the weight of the upper structure in the actual pier i.e. 41.0MPa in terms of the stress. No.2 was tested with the displacement amplitude similar to No.1, while No.3 was tested with the displacement amplitude about half that of No.2. No.4 was the case where the axial compression was twice those of cases 2 and 3.

Test Results

Fig.14(1) to (4) shows the load-displacement relations in tests No.1 to 4. The deformations of the models after the test are shown in Fig.15(1) to (4). It is noted that in Test No.1, although there was no axial load applied in this case, the deformation is almost uniform along the circumference—similar to those of other cases where axial loads were applied. Test No.4 was terminated just before the end of the second cycle, when the load carrying capacity had decreased to the limit and the model collapsed.

The maximum load of the first cycle of each case is plotted in Fig.16 with the collapse estimation line based on the elastic-perfectly plastic material behavior with the yield stress assumed equal to the 0.2% proof stress of the material. The experimental collapse loads are generally greater than the estimation line, but the overall tendency agrees well.

The maximum load of each half cycle of models No.2, 3 and 4 is plotted in Fig.17 against the axial shortening at the corresponding cycle. It is noted that for Nos.2 and 3, where the axial loads were the same, the relationships between the lateral load carrying capacity and the shortening are the same. In case No.4, the lateral load carrying capacity at the same shortening is less than Nos.2 and 3, i.e. the lateral load capacity is determined by the shortening (a measure of the amount of the bulge) and the axial load.

Fig.18 shows the increment of the shortening of the model in each cycle. Nos.2 and 4, whose lateral displacement ranges are practically the same, show the same behaoiors, while the model No.3, though the axial loads are the same as No.2, the increment of the shortening is smaller. The increment of the shortening grows proportionally with the number of the load cycle.

From the static buckling test, the following items were comfirmed;

(1) The circumferentially uniform deformation observed in some cylindrical steel piers after the Hanshin earthquake can be attributed to the cyclic bending, despite the resemblance to that caused by an axial compression. The mechanism can be understood that when a

(1) (3)

(2) (4)

Fig.15 Models after test Nos.1 to 4

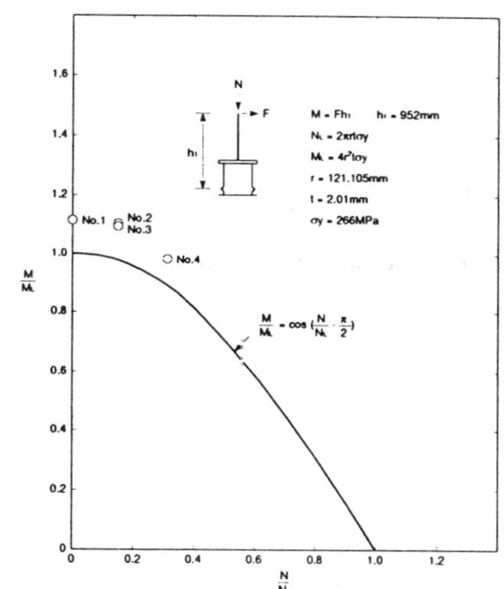

Fig.16 Initial load carrying capacities of static test models

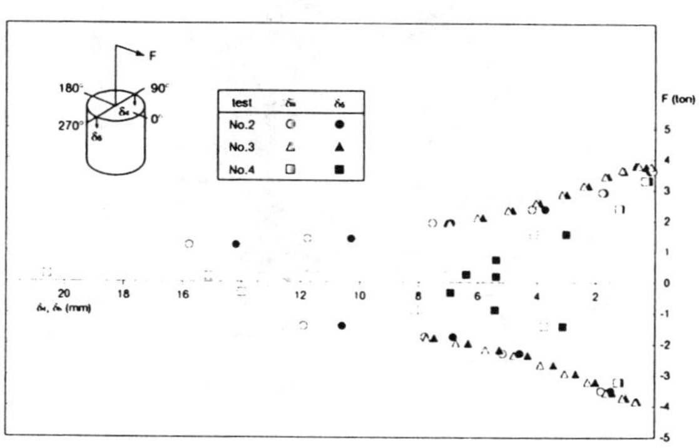

Fig.17 Decrease of lateral load carrying capacity in accordance with the shortening displacements

certain part of the circumference undergoes the bulging deformation due to the application of a bending moment, the rigidity of the part decreases and, in the next application of the bending moment, the adjacent part of the circumference should carry a greater portion of the load and is more likely to undergo the deformation than before, thus the bulging propagates circumferentially. The weight of the upper structure and the vertical seismic load may cause the increase of the deformation and the decrease of the ultimate strength, but they are not essential in the generation of the uniform deformation itself.

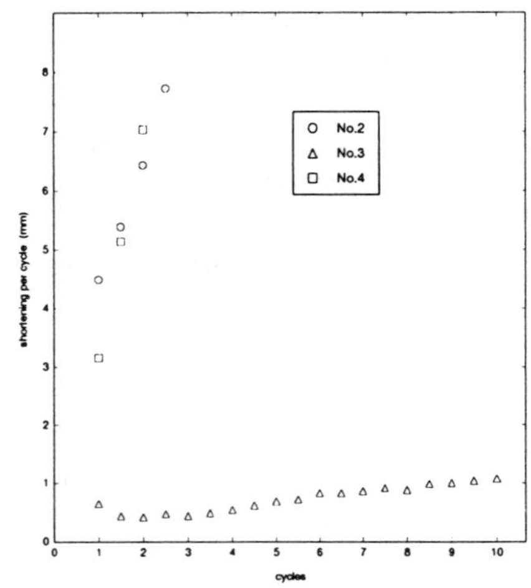

Fig.18 Increment of shortening per cycle

(2) The magnitude of the bulging deformation depends upon the range of the lateral displacement and the number of cycles. If the lateral displacement range is kept constant, the increment of bulging per cycle grows linearly with the load cycle. The increase rate is determined by the range of the lateral displacement. The influence of the axial load is not large.

(3) The lateral and vertical load carrying capacities decreases with the bulging deformation and when certain limit amount of bulging are reached, the pier collapses.

(4) The interaction of bending and axial compression loads can be expressed by the collapse of elastic-perfectly plastic cylinder as far as the first cycle is concerned.

DISCUSSIONS

Buckling Deformation Pattern

In the static buckling test, the circumferentially uniform outward bulge was reproduced, while in the dynamic buckling test, outward bulge was not observed for around 90° circumferential portion. These facts means that uniform outward bulge can easily occur for the forced cyclic displacement and for the cyclic inertia force, and the input energy will tend to accumulate in a direction of initial buckling. However,

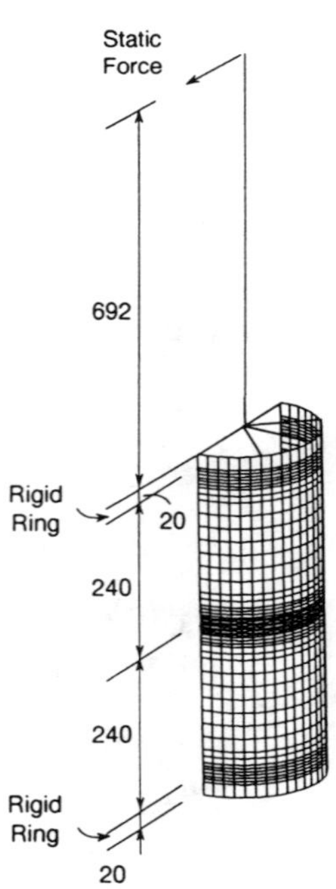

Fig.19 Analytical model for FEM

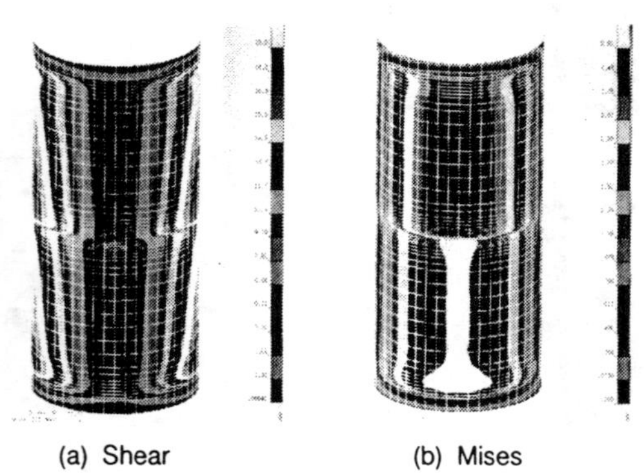

Fig.20 Stress distribution by static horizontal load

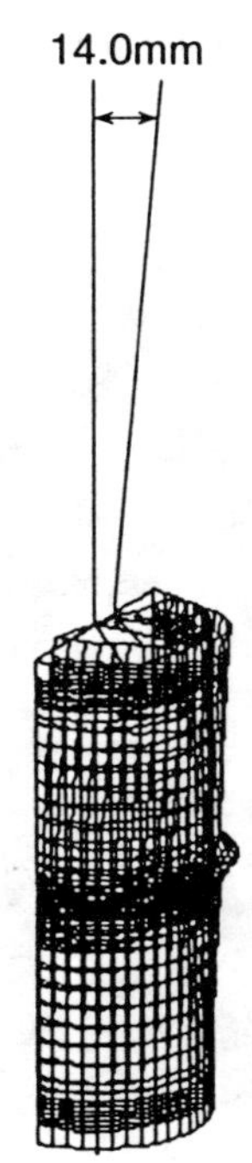

Fig.21 Buckling deformation due to static buckling analysis

the actual elevated expressways are multispan structures and also suffered 2 dimensional seismic load horizontally. Therefore, it is considered that just after the initial buckling occurred at one side of the pier, large forced displacement or very large inertia force acted in the another direction due to whirling at the next instance and the outward bulge was produced in that direction. Refraining this manner, the circumferentially uniform outward bulge was produced. The outward bulge itself is considered to possibly occur for the thick steel pier. These are the causes for the circumferentially uniform outward bulge.

Buckled Portion Even for a double stage pier model, buckling occurred at the bottom of the upper thinner pier in the dynamic buckling test just as the actual pier. When the inertia force is applied at the top of the pier, shear stress in the upper thinner pier will become about 32% larger than that in the lower thicker pier. Moreover, the maximum bending stress in the upper thinner pier will become about 5% larger than that in the lower thicker pier. It is thought that these stress distribution caused buckling at the middle portion in the axial direction of the pier p5.

To confirm this, static buckling analysis was performed using FEM. The analytical model is shown in Fig.19. FEM code used here was MARC. Figures 20, 21 show the buckling deformation and the stress distribution. This shows that the maximum stress and buckling will occur at the bottom end of the upper thinner pier.

CONCLUSIONS

This study was performec to clarify the causes of buckling (circumferencially uniform outward bulge) observed at the middle portion of the pier p5 of the elevated expressway due to the Great Hanshin Earthquake in 1995. The dynamic buckling tests, static buckling tests and the static buckling analysis were performed. The following conclusions were obtained.

(1) The circumferencially uniform outward bulge was caused by the very large cyclic horizontal load and whirling in addition to the fact that the pier was thick steel cylinder.

(2) The maximum stress of the pier occurred at the bottom of the upper thinner pier. This caused the middle portion buckling.

(3) The large girder weight and the vertical excitation seem to have promoted this buckling phenomena, but not essential.

REFERENCES

Fujita, K., Ito, T. and Wada, H., 1991, "Experimental Study on the Dynamic Buckling of a Cylindrical Shell due to Seismic Excitation (Plastic Buckling due to Bending Load), JSME Int. Journal, Series III, Vol.34, No.3, pp.339-344.

NASA, 1968, "Buckling of Thin-Walled Circular Cylinders," NASA Space Vehicle Design Criteria, SP-8007.

Niwa, A. and Clough, R. W., 1982, "Buckling of Cylindrical Liquid-Storage Tanks under Earthquake Loading," Earthquake Engineering and Structural Dynamics, Vol.10, pp.107-122.

PVP-Vol. 340, Seismic Engineering
ASME 1996

APPROXIMATE ANALYSIS METHOD FOR STATISTICAL PROPERTIES OF SEISMIC RESPONSE OF SECONDARY SYSTEM

Shigeru AOKI
Department of Mechanical Engineering
Tokyo Metropolitan College of Technology,
Shinagawa-ku, Tokyo
Japan

ABSTRACT

In this paper, effectiveness of a stationary approximation is examined. The mean square response and the first excursion probability of the secondary system such as pipings and mechanical equipment installed in the primary system such as building subjected to nonstationary random excitation are obtained. Results obtained by stationary approximation are compared with those obtained by nonstationary analysis for various values of damping ratio, natural period and mass ratio of the secondary system to the primary system.

1. INTRODUCTION

When structural system is subjected to nonstationary random excitation or the system is suddenly subjected to stationary random excitation, the response of the system is nonstationary random process. For the analysis of such response, the nonstationary random vibration analysis method is used. However, nonstationary analysis is generally complicated and time cosuming(Lin, 1967; Iyengar et al., 1970). Considering above mentioned problem, stationary approximation is sometimes used for simplicity(Hasselman, 1972). In the case of nonstationary random vibration of the secondary system, analysis is more complicated.

In this study, effectiveness of stationary approximation is examined. Stationary approximation is defined as follows: mean square response is obtained by multiplying stationary mean square response by square of the envelope function. Mean square responses obtained by the stationary approximation are compared with those obtained by the nonstationary analysis method. In the previous paper(Aoki, 1995), stationary approximation for the single-degree-of-freedom system is examined. In this paper, stationary approximation for the secondary system is considered. The following four items are examined for various values of damping ratio, natural period and mass ratio of the secondary system to the primary system. (1)the maximum value, (2)occurence time of the maximum value, (3)integral of mean square response with respect to time from 0 to infinity, (4)first excursion probability.

2. ANALYTICAL METHOD

2.1 Analytical Model and Excitation Model

As an analytical model of the secondary system, a two-degree-of-freedom system shown in Fig.1 is used. The upper system is the secondary system and the lower is the primary system. The equations of motion with respect to relative displacement $z_s = x_s - x_p$ and $z_p = x_p - y$ are

$$\ddot{z}_s + 2\zeta_s\omega_s(1+\gamma)\dot{z}_s + \omega_s^2(1+\gamma)z_s - 2\zeta_p\omega_p\dot{z}_p - \omega_p^2 z_p = 0$$

$$\left.\begin{array}{l}\ddot{z}_p+2\zeta_p\omega_p\dot{z}_p+\omega_p^2 z_p \\ \qquad -2\zeta_s\omega_s\gamma\dot{z}_s-\omega_s^2\gamma z_s=-\ddot{y}\end{array}\right\} \quad (1)$$

where ζ_s and ζ_p are the damping ratio of the secondary and the primary system, respectively and ω_s and ω_p are the natural circular frequency of the secondary and the primary system, respectively. γ is mass ratio of the secondary system to the primary system. As input excitation $\ddot{y}(t)$, nonstationary white noise which is given by multiplying envelope function I(t) representing amplitude nonstationary characteristic by stationary white noise $s_y(t)$, that is,

$$\ddot{y}(t)=I(t)s_y(t) \quad (2)$$

In this study, the following envelope function I(t)is used.

$$I(t)=(e^{-at}-e^{-bt})/|(e^{-at}-e^{-bt})|_{max} \quad (3)$$

Values of a and b are selected as 0.125 and 0.25, respectively. Fig.2 shows the envelope function.

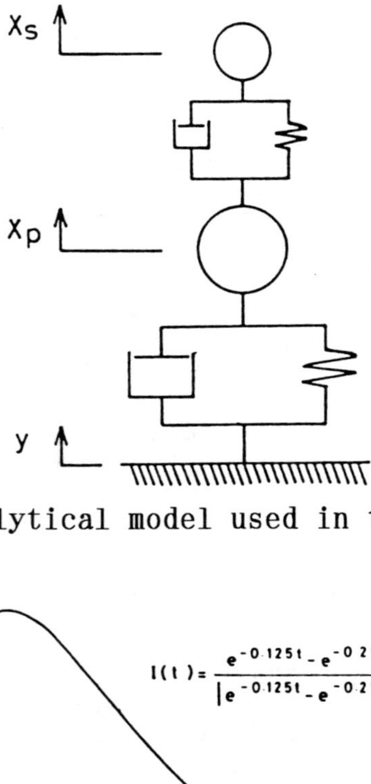

Fig.1 Analytical model used in this study

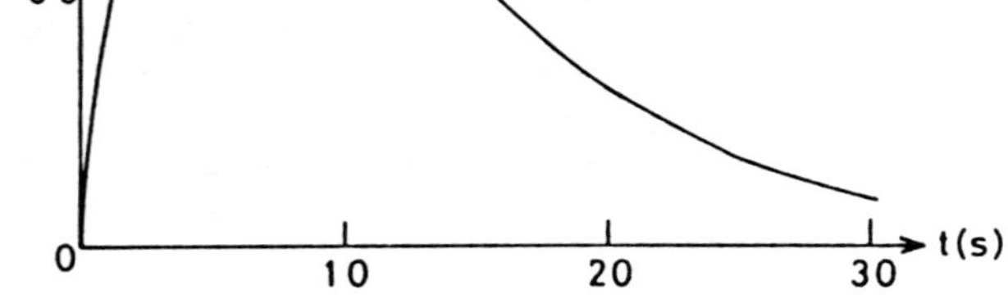

Fig.2 Envelope function

2.2 Mean Square Response

The mean square response of z_s is given by the autocorrelation function as follows.

$$R_z(t_1,t_2)=\int_{-\infty}^{\infty}G(\omega,t_1)G^*(\omega,t_2)S_0\,d\omega \quad (4)$$

where

$$G(\omega,t)=\int_0^t h(t-\xi)I(\xi)e^{i\omega\xi}d\xi \quad (5)$$

$G^*(\omega,t)$ is the complex conjugate of $G(\omega,t)$. h(t) is the unit impulse response function of the secondary system . S_0 is power spectral density function of white noise and $i=\sqrt{-1}$. The mean square response of displacement σ_{zs}^2 is given as,

$$\sigma_{zs}^2(t)=R_{zs}(t,t) \quad (6)$$

Eq.(1) is written as:

$$\dot{\mathbf{z}}=\mathbf{Gz}+\mathbf{f} \quad (7)$$

where

$$\mathbf{z}^T=\{z_s,z_p,\dot{z}_s,\dot{z}_p\} \quad (8)$$

$$\mathbf{G}=\begin{bmatrix} 0 & 0 & 1 & 0 \\ 0 & 0 & 0 & 1 \\ -\omega_p^2 & \omega_s^2\gamma & -2\zeta_p\omega_p & 2\zeta_s\omega_s\gamma \\ \omega_p^2 & -\omega_s^2(1+\gamma) & 2\zeta_p\omega_p & -2\zeta_s\omega_s(1+\gamma)\end{bmatrix} \quad (9)$$

$$\mathbf{f}^T=\{0,0,-y,0\} \quad (10)$$

The second moments of response are expressed as (Roberts and Spanos, 1990):

$$\mathbf{V}=\mathbf{GV}^T+\mathbf{VG}^T+\mathbf{D} \quad (11)$$

$$\mathbf{V}=\begin{bmatrix} \sigma_{zp}^2 & \kappa_{zpzs} & \kappa_{zp\dot{z}p} & \kappa_{zp\dot{z}s} \\ \kappa_{zpzs} & \sigma_{zs}^2 & \kappa_{zs\dot{z}p} & \kappa_{zs\dot{z}s} \\ \kappa_{zp\dot{z}p} & \kappa_{zs\dot{z}p} & \sigma_{\dot{z}p}^2 & \kappa_{\dot{z}p\dot{z}s} \\ \kappa_{zp\dot{z}s} & \kappa_{zs\dot{z}s} & \kappa_{\dot{z}p\dot{z}s} & \sigma_{\dot{z}s}^2 \end{bmatrix} \quad (12)$$

$$\mathbf{D}=\begin{bmatrix} 0 & 0 & 0 & 0 \\ 0 & 0 & 0 & 0 \\ 0 & 0 & 2\pi S_0\{I(t)\}^2 & 0 \\ 0 & 0 & 0 & 0 \end{bmatrix} \quad (13)$$

where κ is covariance.

The second moments of response are obtained from the moment equations as:

$$\frac{d\sigma_{zp}^2}{dt}=2\kappa_{zp\dot{z}p}$$

$$\frac{d\kappa_{zpzs}}{dt}=\kappa_{zs\dot{z}p}+\kappa_{zp\dot{z}s}$$

$$\frac{d\kappa_{zp\dot{z}p}}{dt}=\sigma_{\dot{z}p}^2-\omega_p^2\sigma_{zp}^2+\omega_s^2\gamma\kappa_{zpzs} -2\zeta_p\omega_p\kappa_{zp\dot{z}p}+2\zeta_s\omega_s\gamma\kappa_{zp\dot{z}s}$$

$$\frac{d\kappa_{zp\dot{z}s}}{dt}=\kappa_{\dot{z}p\dot{z}s}+\omega_p^2\sigma_{zp}^2-\omega_s^2(1+\gamma)\kappa_{zpzs} +2\zeta_p\omega_p\kappa_{zp\dot{z}p}-2\zeta_s\omega_s(1+\gamma)\kappa_{zp\dot{z}s}$$

$$
\left.\begin{aligned}
\frac{d\sigma_{zs}^2}{dt} &= 2\kappa_{zs\dot{z}s} \\
\frac{d\kappa_{zs\dot{z}p}}{dt} &= \kappa_{\dot{z}p\dot{z}s} - \omega_p^2\kappa_{zpzs} + \omega_s^2\gamma\sigma_{zs}^2 \\
&\quad -2\zeta_p\omega_p\kappa_{zs\dot{z}p} + 2\zeta_s\omega_s\gamma\kappa_{zs\dot{z}s} \\
\frac{d\kappa_{zs\dot{z}s}}{dt} &= \sigma_{\dot{z}s}^2 + \omega_p^2\kappa_{zpzs} - \omega_s^2(1+\gamma)\sigma_{zs}^2 \\
&\quad +2\zeta_p\omega_p\kappa_{zs\dot{z}p} - 2\zeta_s\omega_s(1+\gamma)\kappa_{zs\dot{z}s} \\
\frac{d\sigma_{\dot{z}p}^2}{dt} &= 2(-\omega_p^2\kappa_{zp\dot{z}p} + \omega_s^2\gamma\kappa_{zs\dot{z}p} \\
&\quad -2\zeta_p\omega_p\sigma_{\dot{z}p}^2 + 2\zeta_s\omega_s\gamma\kappa_{\dot{z}p\dot{z}s}) \\
&\quad +2\pi S_0\{I(t)\}^2 \\
\frac{d\kappa_{\dot{z}p\dot{z}s}}{dt} &= -\omega_p^2\kappa_{zp\dot{z}s} + \omega_s^2\gamma\kappa_{zs\dot{z}s} \\
&\quad -2\zeta_p\omega_p\kappa_{\dot{z}p\dot{z}s} + 2\zeta_s\omega_s\gamma\sigma_{\dot{z}s}^2 \\
&\quad +\omega_p^2\kappa_{zp\dot{z}p} - \omega_s 2(1+\gamma)\kappa_{zs\dot{z}p} \\
&\quad +2\zeta_p\omega_p\sigma_{\dot{z}p}^2 - 2\zeta_s\omega_s(1+\gamma)\kappa_{\dot{z}p\dot{z}s} \\
\frac{d\sigma_{\dot{z}s}^2}{dt} &= 2\{\omega_p^2\kappa_{zp\dot{z}s} - \omega_s^2(1+\gamma)\kappa_{zs\dot{z}s} \\
&\quad +2\zeta_p\omega_p\kappa_{\dot{z}p\dot{z}s} - 2\zeta_s\omega_s(1+\gamma)\sigma_{\dot{z}s}^2
\end{aligned}\right\} \quad (14)
$$

2.3 Stationary Approximation

The integral of Eq.(4) is complicated. For simplicity, it is assumed that I(t) is independent of the integral, then Eq.(4) is expressed as follows.

$$\sigma_{zs}^2(t) = \{I(t)\}^2 S_0 \int_{-\infty}^{\infty} |H_s(\omega)|^2 d\omega \quad (15)$$

where $H_s(\omega)$ is the frequency response function for displacement response of the secondary system to input acceleration. In this case, $H_s(\omega)$ is expressed as:

$$H_s(\omega) = \frac{R_n(\omega)}{R_d(\omega)} \quad (16)$$

where

$$R_n(\omega) = -(2\zeta_p\omega_p i\omega + \omega_p^2) \quad (17)$$

$$
\begin{aligned}
R_d(\omega) &= \omega^4 - \{2\zeta_p\omega_p + 2\zeta_s\omega_s(1+\gamma)\}i\omega^3 \\
&\quad -\{\omega_p^2 + \omega_s^2(1+\gamma) + 4\zeta_p\zeta_s\omega_p\omega_s\}\omega^2 \\
&\quad +(2\zeta_p\omega_p\omega_s^2 + 2\zeta_s\omega_s\omega_p^2)i\omega + \omega_p^2\omega_s^2
\end{aligned} \quad (18)
$$

The integral of Eq.(15) is that for stationary random process. This integral is obtained as(Roberts and Spanos, 1990):

$$
I_4 = \frac{\pi}{\lambda_4} \frac{\begin{vmatrix} \xi_3 & \xi_2 & \xi_1 & \xi_0 \\ -\lambda_4 & \lambda_2 & -\lambda_0 & 0 \\ 0 & -\lambda_3 & \lambda_1 & 0 \\ 0 & \lambda_4 & -\lambda_2 & \lambda_0 \end{vmatrix}}{\begin{vmatrix} \lambda_3 & -\lambda_1 & 0 & 0 \\ -\lambda_4 & \lambda_2 & -\lambda_0 & 0 \\ 0 & -\lambda_3 & \lambda_1 & 0 \\ 0 & \lambda_4 & -\lambda_2 & \lambda_0 \end{vmatrix}} \quad (19)
$$

where $\lambda_4 = 1$, $\lambda_3 = 2\zeta_p\omega_p + 2\zeta_s\omega_s(1+\gamma)$, $\lambda_2 = \omega_p^2 + \omega_s^2(1+\gamma) + 4\zeta_p\zeta_s\omega_p\omega_s$, $\lambda_1 = 2\zeta_p\omega_p\omega_s^2 + 2\zeta_s\omega_s\omega_p^2$, $\lambda_0 = \omega_p^2\omega_s^2$, $\xi_3 = 0$, $\xi_2 = 0$, $\xi_1 = (2\zeta_p\omega_p)^2$, $\xi_0 = \omega_p^4$ and

$$\sigma_{zs}^2(t) = \{I(t)\}^2 I_4 S_0 \quad (20)$$

The results using Eq.(19) are obtained when left-hand side of Eq.(14) is equal to 0. Thus, Eq.(15) is simpler than Eq.(6) or Eq.(14). If approximation of Eq.(15) is appropriate, nonstationary random vibration analysis becomes simple. In this paper, when mean square response is obtained by Eq.(15), it is referred as approximate value. When mean square response is obtained by Eq.(14), it is referred as exact value.

Integral of mean square response with respect to time from 0 to infinity is expressed as

$$I_z = \int_0^{\infty} \sigma_{zs}^2(t)dt \quad (21)$$

For stationary approximation, I_z is given as,

$$I_z = \frac{(a-b)^2}{2ab(a+b)} I_4 S_0 \quad (22)$$

2.3 First Excursion Probability

First excursion probability is obtained by applying above mentioned equations. It is assumed that failure occurs at the time when $|z_s(t)|$ first crosses the tolerance level. In this case, the first excursion probability is obtained by the following equation.

$$P_f(t) = 1 - \exp\{-2\int_0^t \nu(t)dt\} \quad (23)$$

$\nu(t)$ is the expected number of crossings of the tolerance level B_D per unit time. When probability density function of input is the normal distribution, $\nu(t)$ is approximately given as.

$$\nu(t) = \frac{1}{\pi}\frac{\sigma_{\dot{z}s}}{\sigma_{zs}}\exp\left(-\frac{B_D^2}{2\sigma_{zs}^2}\right) \quad (24)$$

In the case where $z_s(t)$ is stationary normal random process, Eq.(24) is written as,

$$\nu(t) = \frac{1}{\pi}\omega_s \exp\left(-\frac{B_D^2}{2\sigma_{zs}^2}\right) \quad (25)$$

The tolearance level is determined using the standard deviation of stationary response σ_{zss} as following equation.

$$B_D = n\sigma_{zss} \quad (26)$$

where n is a parameter. For the secondary system, σ_{zss} is given as

$$\sigma_{zss} = \sqrt{I_4 S_0} \quad (27)$$

In this paper, when first excursion

probability is obtained by Eq.(25), it is referred as approximate value. When first excursion probability is obtained by Eq.(24) using Eq.(14), it is referred as exact value.

3.RESULTS

3.1 Mean Square Response

(1)the maximum value of mean square response, (2)occurence time of the maximum value of mean square response, (3)integral of mean square response with respect to time from 0 to infinity are obtained. In this paper, ζ_p is fixed as 0.05. S_0 is selected as $1cm^2/s^4/rad/s$.

Table 1 shows results for differnt values of damping ratio of the secondary system ζ_s. Table 2 shows results for different values of natural period of the secondary system T_s. Table 3 shows results for different values of mass ratio of the secondary system to the primary system γ. Table 4 shows results for different values of natural period under the condition where $T_s = T_p$. From these tables, it is found that approximate value for the maximum value of mean square response is greater than exact value. Approximate value of occurence time of the maximum value is ealier than exact value. Approximate value of the maximum value and occurence time of the maximum value approach exact value as damping ratio becomes greater and natural period becomes shorter. Difference between approximate value and exact value becomes smaller as mass ratio becomes greater. Approximate value of integral of mean square response is equal to exact value independent of damping ratio, natual period and mass ratio.

Exact values of the maximum value $(\sigma_{zse}^2)_{max}$ can be obtained by the following equation using approximate

Table 1 Comparison of approximate solution with exact solution for mean square response (γ=0, ζ_p=0.05, $T_s = T_p$=0.5s)

ζ_s	Maximum Value(cm^2)		Time(s)		Integral($cm^2 \cdot s$)	
	Approximate	Exact	Approximate	Exact	Approximate	Exact
0.01	6.68	4.79		11.49	7.12X10	7.12x10
0.02	2.87	2.43	5.55	9.79	3.06x10	3.06x10
0.05	$8.07x10^{-1}$	$7.57x10^{-1}$		8.18	8.61	8.61

Table 2 Comparison of approximate solution with exact solution for mean square response (γ=0, ζ_s=0.01, ζ_p=0.05, T_p=0.5s)

T_s(s)	Maximum Value(cm^2)		Time(s)		Integral($cm^2 \cdot s$)	
	Approximate	Exact	Approximate	Exact	Approximate	Exact
0.25	$2.91x10^{-3}$	$2.71x10^{-3}$		6.90	$3.10X10^{-2}$	$3.10x10^{-2}$
0.5	6.68	4.79	5.55	11.49	7.12x10	7.12x10
1.0	1.15	$6.52x10^{-1}$		10.90	1.23x10	1.23x10

Table 3 Comparison of approximate solution with exact solution for mean square response (ζ_s=0.01, ζ_p=0.05, $T_s = T_p$=0.5s)

γ	Maximum Value(cm^2)		Time(s)		Integral($cm^2 \cdot s$)	
	Approximate	Exact	Approximate	Exact	Approximate	Exact
0	6.68	4.79		11.49	7.12X10	7.12x10
0.01	1.13	1.08	5.55	7.78	1.20X10	1.20x10
0.05	$2.72x10^{-1}$	$2.56x10^{-1}$		7.28	2.90	2.90

Table 4 Comparison of approximate solution with exact solution for mean square response (γ=0, ζ_s=0.01, ζ_p=0.05, $T_s = T_p$)

$T_s = T_p$	Maximum Value(cm^2)		Time(s)		Integral($cm^2 \cdot s$)	
	Approximate	Exact	Approximate	Exact	Approximate	Exact
0.25	$8.35x10^{-1}$	$7.32x10^{-1}$		8.84	8.90	8.90
0.5	6.68	4.79	5.55	11.49	7.12x10	7.12x10
1.0	5.34x10	2.73x10		15.72	$5.70x10^2$	$5.69x10^2$

values $(\sigma_{zsa}{}^2)_{max}$.

$$(\sigma_{zse}{}^2)_{max}=C(\sigma_{zsa}{}^2)_{max} \quad (28)$$

For the case where $T_s=T_p$ and $T_s=T_p\geqq0.1s$, C is approximately given as

$$C=\frac{256\zeta_s}{1+256\zeta_s}\ \frac{11}{10(1+T_s)}D \quad (29)$$

where

$$D=\frac{256(1+700\gamma)}{356+179200\gamma}\ :\gamma\leqq0.01,\quad 0.95:\gamma>0.01 \quad (30)$$

$(\sigma_{zsa}{}^2)_{max}$ is equal to I_4S_0 which is given by Eq.(19) or formulae given in the textbook (Crandall and Mark, 1963). Thus, $(\sigma_{zse}{}^2)_{max}$ can be easily obtained obtained by Eq.(28). The integral of the mean square response is total energy of the response. Approximate value is easily obtained by Eq.(22). Thus, stationary approximation is effective for evaluation of total energy.

3.2 First Excursion Probability

Fig.3 shows first excursion probability P_f for $\gamma=0$, $\zeta_s=0.01$, $\zeta_p=0.05$, $T_s=T_p=0.5s$. Approximate value is greater than exact value. When enuogh time passes, P_f becomes constant. Attention is focused on P_f which is obtained at the time when enough time passes.

Table 5 shows results for differnt values of ζ_s. Table 6 shows results for different values of T_s. Table 7 shows results for different values of γ. Table 8 shows results for different values of natural period under the condition where

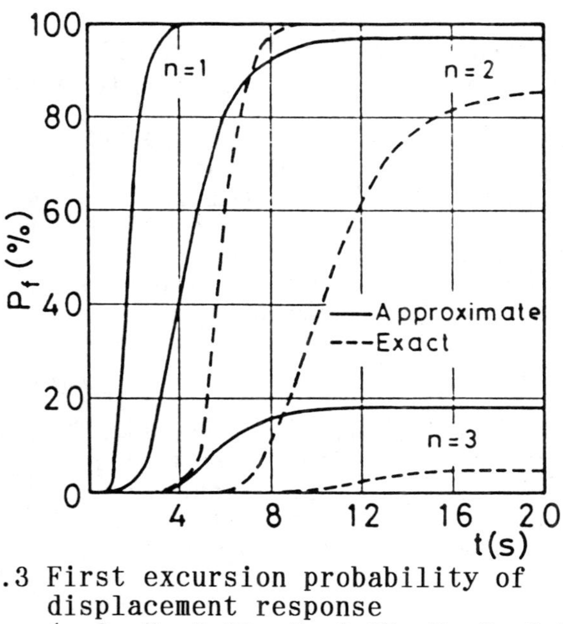

Fig.3 First excursion probability of displacement response ($\gamma=0$, $\zeta_s=0.01$, $\zeta_p=0.05$, $T_s=T_p=0.5s$)

$T_s=T_p$. From these tables, it is found that approximate value of first excursion probability approach exact value as damping ratio becomes greater and natural period becomes shorter. For relatively small vlaue of mass ratio($\gamma\leqq0.01$), approximate value approaches as mass ratio becomes greater. For relatively large value of mass ratio($\gamma>0.01$), difference between approximate value and exact value becomes larger as mass ratio becomes greater.

These results are same as those of the maximum value of mean squre response. Thus, when first excursion probability is obtained by stationary approximation method, results for the maximum response of mean square response should be considered.

4. CONCLUSIONS

Mean square response and first excursion probability of the secondary system subjected to nonstationary white noise is obtained by stationary approximation. Obtained results are summarized as follows.

(1) Approximate value for the maximum value of mean square response is greater than exact value. Approximate value of occurence time of the maximum value is earlier than exact value.

(2) Approximate value of the maximum value and occurence time of the maximum value approach exact value as damping ratio becomes greater and natural period becomes shorter. Difference between approximate value and exact value becomes smaller as mass ratio becomes greater.

(3) Exact value of the maximum value is obtained by Eq.(29) and mean square response of stationary random process.

(4) Approximate value of integral of mean square response is equal to exact value independent of damping ratio, natual period and mass ratio. The integral is total energy of the response. Thus, stationary approximation is effective for evaluation of total energy of the reapose.

(5) Approximate value of first excursion probability is greater than exact value.

(6) Approximate value of first excursion probability approach exact value as damping ratio becomes greater and natural period becomes shorter. For relatively small vlaue of mass ratio($\gamma\leqq0.01$), approximate value approaches as mass

Table 5 Comparison of approximate solution with exact solution for first excursion probability (γ=0, ζ_s=0.01, ζ_p=0.05, T_s=T_p)

ζ_s	n=2		n=3		n=4	
	Approximate	Exact	Aooroximate	Exact	Approximate	Exact
0.01	$9.68x10^{-1}$	$8.54x10^{-1}$	$1.80x10^{-1}$	$4.04x10^{-2}$	$4.58x10^{-3}$	$2.40x10^{-4}$
0.02	$9.68x10^{-1}$	$9.31x10^{-1}$	$1.80x10^{-1}$	$9.36x10^{-2}$	$4.58x10^{-3}$	$1.22x10^{-3}$
0.05	$9.68x10^{-1}$	$9.55x10^{-1}$	$1.80x10^{-1}$	$1.41x10^{-1}$	$4.58x10^{-3}$	$2.78x10^{-3}$

Table 6 Comparison of approximate solution with exact solution for first excursion probability (γ=0, ζ_s=0.01, ζ_p=0.05, T_p=0.5s)

T_s (s)	n=2		n=3		n=4	
	Approximate	Exact	Approximate	Exact	Approximate	Exact
0.25	$9.99x10^{-1}$	$9.89x10^{-1}$	$3.27x10^{-1}$	$1.97x10^{-1}$	$9.15X10^{-3}$	$3.97x10^{-3}$
0.5	$9.68x10^{-1}$	$8.54x10^{-1}$	$1.80x10^{-1}$	$4.04x10^{-2}$	$4.58x10^{-3}$	$2.40x10^{-4}$
1.0	$8.21x10^{-1}$	$4.08x10^{-1}$	$9.44x10^{-2}$	$4.36x10^{-3}$	$2.29x10^{-3}$	$6.85x10^{-6}$

Table 7 Comparison of approximate solution with exact solution for first excursion probability (ζ_s=0.01, ζ_p=0.05, T_s=T_p=0.5s)

γ	n=2		n=3		n=4	
	Approximate	Exact	Approximate	Exact	Approximate	Exact
0	$9.68x10^{-1}$	$8.54x10^{-1}$	$1.80x10^{-1}$	$4.04x10^{-2}$	$4.58x10^{-3}$	$2.40x10^{-4}$
0.01	$9.68x10^{-1}$	$9.59x10^{-1}$	$1.80x10^{-1}$	$1.52x10^{-1}$	$4.58X10^{-3}$	$3.25x10^{-3}$
0.05	$9.68x10^{-1}$	$9.53x10^{-1}$	$1.80x10^{-1}$	$1.40x10^{-1}$	$4.58x10^{-3}$	$2.79x10^{-3}$

Table 8 Comparison of approximate solution with exact solution for first excursion probability (γ=0, ζ_s=0.01, ζ_p=0.05, T_s=T_p)

T_s=T_p	n=2		n=3		n=4	
	Approximate	Exact	Approximate	Exact	Approximate	Exact
0.25	$9.99x10^{-1}$	$9.97x10^{-1}$	$3.27x10^{-1}$	$2.05x10^{-1}$	$9.15x10^{-3}$	$3.25x10^{-3}$
0.5	$9.68x10^{-1}$	$8.54x10^{-1}$	$1.80x10^{-1}$	$4.04x10^{-2}$	$4.58x10^{-3}$	$2.40x10^{-4}$
1.0	$8.21x10^{-1}$	$3.08x10^{-1}$	$9.44x10^{-2}$	$1.91x10^{-3}$	$2.29x10^{-3}$	$1.49x10^{-6}$

ratio becomes greater. For relatively large value of mass ratio(γ>0.01), difference between approximate value and exact value becomes larger as mass ratio becomes greater.

There are some assumptions in Eq.(24) and Eq.(25) for first excursion probabaility. In the future work, the author would like to compare results obtained in this study with first excursion probability obtained by simulation method and present a simple method for obtaining first excursion probability using approximate value.

REFERENCES

Aoki, S., 1995, "Approximate Analysis Method for Statistical Properties of Seismic Response", Proceedings of 1995 PVP Conference : Seismic Engineering-1995, PVP-Vol.312, pp.215-220

Crandall,S.H. and Mark,W.D., 1963, Random Vibration in Mechanical Systems, Academic Press

Hasselman,T., 1972, "Probabilistic Displacement Time History of a Single Degree of Freedom System", Proc. of ASCE Engineering Mechanics Division, Vol.98 June, pp.519-530.

Iyengar,R.N. and Iyengar,K.T.S.R., 1970, "Probabilistic Response Analysis to Earthquakes", Proc.of ASCE Engineering Mechanics Division, Vol.96 June, pp.207-225.

Lin,Y.K., 1967, Probabilistic Theory of Structural Dynamics, McGraw-Hill.

Roberts,J.B. and Spanos,P.D., 1990, "Random Vibration and Statistical Linearization, John Wiley & Sons

PVP-Vol. 340, Seismic Engineering
ASME 1996

Parametric Study of the Penalty Function in a Genetic Algorithm applied to the Support Arrangement for Piping Systems

Toshio Chiba, Ichiro Fujii, Shigeyuki Tamayama
Ishikawajima-Harima Heavy Industries Co., Ltd. , Yokohama, JAPAN
and
Fumio Hara
Science University of Tokyo, Tokyo, JAPAN

Abstract

A genetic algorithm (GA) is applied to the support arrangement of the piping systems in the nuclear power plant. In a GA, the form of the fitness function is the most important theme. It has a large influence on the result of support arrangement. There are many factors to solve for this problem; for example, stress of the piping system, the number of supports, strength of the supports, and so on. We attempt to select a more reasonable fitness function in trail and error. We illustrate the results of numerical experiments using an actual piping system. We found that the sigmoid function is good for the stress penalty function and that the balance of fitness function terms is important. The design results using the GA applied compare favorably with the results of the designer.

1 Introduction

The support arrangement of piping is an important factor when designing piping systems. The designer will spend much time in trial and error processes to design the support arrangement. It is therefore desirable that the support arrangement be performed automatically. Computers have been used to design piping systems. The support arrangement of piping systems is considered to be a combination optimal problem. It is known that there are many approaches for solving optimum problems; for example, a neural network, dynamic programming, and so on. The use of genetic algorithms has recently being applied to various problems. For example, they have been applied to determine the support arrangement of piping systems (Hara,1992), the optimization of ship structure designing (Okada,1992) and the parametric designing of airplanes (Davis,1991).

It is well known that the next three points are important to solve a optimal problem using a GA:

- How to model the problem into a GA,
- How to use genetic operators, and
- How to construct a fitness function.

In a previous report (Chiba,1995), we focused on the modeling and genetic operations. We presented our first application for the support arrangement problem of piping systems using a GA. It was confirmed that the use of a GA is feasible for determining the support arrangement of a piping system. After that, we studied the form of fitness functions and found they are a most important factor in the optimization process. A fitness function describes the know-how of the designer in the optimization process. It has a large influence on the result of the support arrangement. In this paper, we present the results of our parametric studies in making a better fitness function.

2 Nomenclature

σ_i	stresses of the element of piping system
F_j	reaction forces of the supports
N_{sup}	number of supports
N_{pos}	number of position to support
N_{elm}	number of elements of the piping system
K_l	spring constant of support
σ_{alw}	an allowable stress
F_{alw}	the capacity of a support
e	an incidence vector
α	weight of the term of support number in fitness function
β	weight of the term of spring constants in fitness function
γ	weight of the penalty term of stress in fitness function

δ	weight of the penalty term of reaction force in fitness function
ϵ_s	weight of the exponential term of stress penalty
ϵ_r	weight of the exponential term of reaction force penalty

3 Methodlogy

A parametric study was conducted to find a better fitness function. It was found that the stress of piping and the number of supports are the most important factors in the optimum support problem. So we focused on the stress of piping and the number of supports in the fitness function.

3.1 Gene coding

In a GA, each solution of the actual problem is encoded into a gene string. How the actual problem is encoded into the gene code is therefore important. The following describes the gene coding to determine the optimum support arrangement of a piping system.

First, the nodes which comprise the piping system must be numbered. This series number is the "node number". There are six degrees of freedom, the axial, vertical and horizontal directions and its rotations. Second, let a "freedom number" be assigned between the node numbers. Each freedom number characterizes the position and direction of a support. In addition, make several "dummy freedom numbers". If a support point is selected at a dummy freedom number, this support is treated as a dummy. This is shown in Table 1.

Table 1: Numbering System for Degrees of Freedom and Dummy Degrees

	Degree of freedom					
Node number	x	y	z	θ_x	θ_y	θ_z
1	0	0	0	0	0	0
2	0	1	0	2	0	3
3	0	4	0	5	0	6
4	0	7	0	8	0	9
5	0	10	0	11	0	12
6	0	13	0	14	0	15
7	0	0	0	0	0	0
Dummy degree	16	17	18	19	20	

3.2 Fitness function

It is considered that the support arrangement of a piping system is an optimal problem. It is important that the design of the objective function for solving the optimal problem applied the heuristic method. In a GA, the objective function called a fitness function. There are some independent variables in the optimum support arrangement of a piping system, for which the following conditions are considered:

1. It is better that the stresses of a piping system be minimal if the number of supports is equal.
2. If the stress level of the piping system is equal, the number of supports must be smaller.
3. Stress of the piping cannot be over the allowable value.
4. The reaction forces of a support cannot exceed the load capacity of the support.

We designed an objective function taking into consideration designer's experience and the design criteria. It is shown in Eq. 1.

$$\begin{aligned} Fitness(\sigma, N_{sup}, K, F) &= \max_{0<i<N_{elm}} \mathrm{RMS}(\sigma_i) \\ &+ \alpha n_{sup} \\ &+ \beta \sum_{j=0}^{n_{sup}} K_j \\ &+ \gamma P_{stress}(\sigma) \\ &+ \delta P_{reaction}(F) \qquad (1) \end{aligned}$$

4 A fitness function for solving optimal problems

4.1 Penalty functions

There are many restrictions in the design of support arrangements. Allowance of maximum stress is one of such condition. We added a penalty function to the fitness function to consider the allowable stress. The solution in which the stress is over the allowable stress is not acceptable, and must be rejected using the penalty function. In a GA, the progress is optimized by reducing the fitness in the population. The role of the penalty function is to add a penalty value to the fitness if maximum stress is over the allowable stress. Using the penalty function, the fitness solutions over the allowable value is increased and the selection probability is therefore decreased.

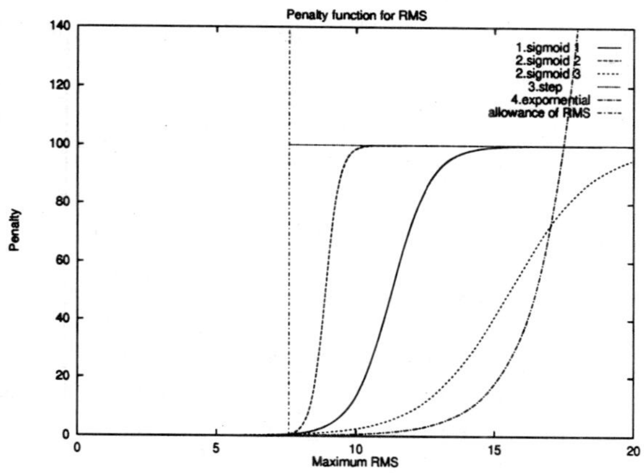

Figure 1: Penalty Functions

We applied three types of penalty functions as shown in Fig. 1. In the step function, the focus was on the stress near the allowable stress. In the exponential and sigmoid functions, the focus was on the stress over the allowable value.

- step function

$$P_{stress}(\sigma) = \begin{cases} 1 & : \ if \max\limits_{1 \le i \le N_{elm}} \mathrm{RMS}(\sigma_i) > \sigma_{alw} \\ 0 & : \ \text{others} \end{cases} \tag{2}$$

We considered the step function as a penalty function. Using the step function, a constant value was added to the fitness for stress over the allowable value. In some examinations, using the step function in the process of optimization does not work well. The solutions a small amount over allowable value are rejected because the step function is a discontinuous function.

- exponential function

$$P_{stress}(\sigma) = \begin{cases} \gamma(\exp(\epsilon_s(\max_{1 \le i \le}(RMS\sigma_i) \\ \qquad -\sigma_{alw})) - 1) \\ \qquad \text{if} \ \max\limits_{1 \le i \le}(RMS\sigma_i) > \sigma_{alw} \\ 0 \qquad \text{otherwise} \end{cases} \tag{3}$$

We applied the exponential function as a penalty function. We thought a larger penalty must be added to the fitness for those cases in which stress was over the allowable value. Fig. 1. shows the value of the increasing rate when the penalty value is 100 and stress is over 10 kgf/mm^2.

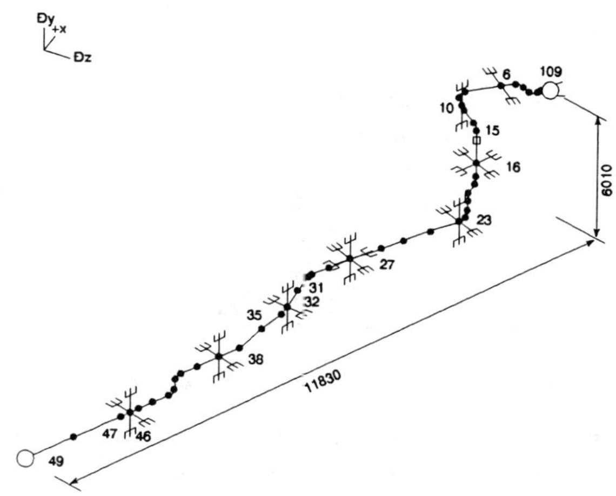

Figure 2: Model

- sigmoid function

$$P_{stress}(\sigma_{rms}) = \frac{100}{1 + \exp(-\frac{\sigma_{rms} - \sigma_{alw} - bl}{b})} \tag{4}$$

σ_{rms}	:	maximum stress
σ_{alw}	:	allowance stress
b	:	parameter for sigmoid form
$bl = b\log(-1.0 + \frac{1}{0.005})$	:	offset of sigmoid function

We applied the sigmoid function as a penalty function. The sigmoid function has the following advantages:

- A suitable penalty is added to fitness value if maximum stress is over the allowable value, and stress.
- It is a continuous function that does not continue to infinity.

It was supposed that the sigmoid function was more effective than the other two functions.

5 Numerical results

In this section , we investigated the effect of the penalty function and the term in the fitness function on the calculation. Reducing the number of supports is an important matter for reducing the cost. We tuned the fitness functions and examined them.

5.1 Analytical model

A piping system inside the reactor containment vessel of a nuclear power plant was chosen as the test model.

Table 2: Parameters used in GA Analysis

Parameters	Value
Generations	100
Population	100
Crossover	50 %
Mutation	5 %
α	1.0
β	1.0e-5
σ_{alw}	$7.6kgf/mm^2$

It was supported by restraints which were attached to a supporting structure and anchored at both ends. The model is shown in Fig. 2. The diameter of pipe was 25A and the allowable stress 7.6 kgf/mm^2. This model was of the high stress level, so it was difficult for designers to satisfy the allowable stress. We expected that the GA method would produce a better arrangement than designers for this model. We applied our program five times in each function. In each case the program started from the different initial populations. Table 2 shows the parameters used in the analysis.

5.2 Penalty functions

Fig. 1 shows the shapes of the penalty functions. Table 3 shows the results of the support arrangements using the above-mentioned three functions as penalty functions. The first column shows the function used in the examinations. The second shows the fitness value. The third lists the position of support arrangements. Each term means a node number and the direction of a support. The fourth column shows the maximum RMS stress of the piping. The fifth shows the number of supports. The last column shows whether or not the stress of the piping system is within the allowable value. We applied three penalty functions and compared the results of support arrangements.

5.3 Parameters in the penalty function

The sigmoid function was adopted as a penalty function in order to study the parameters of the penalty function and determine better parameters for optimization. We changed parameter$b$ (0.25, 0.7 ,and 1.5) in Eq. 4. The results were listed in Table 4.

5.4 Maximum number of supports

It is severe that seven supports be used for piping systems with a high stress level. Seven supports were de-

Table 3: Influence of Penalty Function on the Support Number

Func.	Fitness	Allocation of supports	Stress	n	
Step	116.0	19z 10y 39y 49y 51z 19y	9.84	6	×
	115.0	14y 51y 35y 27y 39z 50z	8.92	6	×
	13.2	35y 27z 51y 19y 10y 35z 50z	6.05	7	o
	115.5	51y 28z 20y 51z 39y 14y	9.43	6	×
	115.8	36y 10y 48y 29y 51z 43z 19y	8.66	7	×
Exp.	16.45	51z 29y 28z 14y 50y 36y 51y	9.09	7	×
	12.51	27z 36z 19y 51y 36y 51z 10y	5.37	7	o
	14.66	48z 19y 29z 11y 32z 36y 49y	7.52	7	o
	17.41	51z 39y 32z 20z 29z 14y 48y	9.79	7	×
	17.14	35y 48y 11y 28y 50z 52z 14y	9.60	7	×
Sig.	14.44	50y 36y 32y 51z 27y 19y 14y	7.06	7	o
	13.98	11y 19y 28y 51z 3z 49y 36y	6.70	7	o
	17.82	20y 52z 36y 10y 39y 49y 29y	8.61	7	×
	14.63	51y 20y 35y 27y 14y 51z	7.82	6	×
	26.54	51y 39y 10z 14y 19y 50z 29z	9.74	7	×

cided by the designer, and seem to have a critical point with an allowable solution. We set the maximum number of supports to seven. As stated above, we used 10 as the maximum number of supports, and examined abilities in both cases. The value of allowable stress had an actual value of $7.6kgf/mm^2$. The results of this examination are shown in Table 5. Fitness curves are shown in Fig. 5 and 6.

5.5 Applying exponential and quadratic function

The number of supports has a large influence on the fitness value. We tried to change the coefficient of a term of support and examine it. Until now, a linear function was used for the term of supports in the fitness function. We tried to use exponential and quadratic functions to emphasize the difference in the number of supports. For

Table 4: Influence of Coefficient b on the Support Number

b	Fitness	Allocation of supports	Stress	n	
0.25	11.32	51z 29z 32z 20y 51y 36y	9.81	6	×
	13.94	51y 49z 32z 28y 36y 19y 11y	6.78	7	o
	11.82	48z 52y 39y 5y 11y 20y	12.15	6	×
	11.76	52y 39y 48z 19y 10y	12.52	5	×
	11.81	51z 36z 19y 35y 49y 28z 35y	11.00	7	×
0.7	14.44	50y 36y 32y 51z 27y 19y 14y	7.06	7	o
	13.98	11y 19y 28y 51z 3z 49y 36y	6.70	7	o
	17.82	20y 52z 36y 10y 39y 49y 29y	8.61	7	×
	14.63	51y 20y 35y 27y 14y 51z	7.82	6	×
	26.54	51y 39y 10z 14y 19y 50z 29z	9.74	7	×
1.5	13.91	28y 49z 49y 10y 35y 36z 20y	6.52	7	o
	16.09	36y 27y 50z 49z 51y 14y 19y	8.20	7	×
	24.25	48y 52y 39y 28z 19y 52z 4y	11.35	7	×
	15.76	52y 48y 11y 27z 36y 19y 51z	7.97	7	×
	23.75	14y 36z 42z 29z 50z 50y 11y	11.23	7	×

this purpose , it was decided that coefficients would be 1 when the number of supports was 7. Consequently, this improvement was ineffective for our purpose. The results of this examination are shown in Table 6.

5.6 Increasing dummy points

In the previous section, we set the maximum number of supports at 10, and arrived at the allowable solution in many cases. Although we tried to find allowable solutions with less supports, the number of support were almost nine or ten. This algorithm used dummy points to select virtual points that can not be fitted with supports. This is equivalent to reducing the number of supports, so we increased the dummy points and examined them. We compared three cases as follows. This

Table 5: Influence of the Maximum Support Number on the Support Number

max	Fitness	Allocation of supports	stress	n	
7	15.01	20z 47y 31y 47z 11y 30y	7.47	6	o
	101.09	44y 14y 42z 20z 19z 35y	11.01	6	×
	25.52	10y 46y 42z 11y 32y 29y 20y	8.68	7	×
	16.98	31z 47y 904z 11y 20z 35y 44z	7.72	7	×
	14.63	35z 20z 45z 45y 35y 11y 19z	6.92	7	o
10	15.69	32y 903z 35y 47y 20z 45z 32z 11y 14y	6.39	9	o
	18.42	47z 20z 29y 19z 44y 36z 10y 35y 19y 43y	7.25	10	o
	16.21	29z 48y 29y 14y 10y 20z 31y 47z 46y	6.66	9	o
	18.26	31y 20z 45z 30y 47y 29y 11y 30z 31y	7.53	9	o
	17.22	11y 14z 35z 45z 44y 10y 39y 20y 32y 43y	6.65	10	o

piping model has 53 nodes that can be supported and each point has six degrees of freedom. That is to say, it has 318 freedom numbers. In a previous case, 106 dummy freedom numbers were used as dummy points. The ratio of freedom numbers to dummy freedom numbers was 3 : 1. The probability that a dummy freedom number was selected in a single GA operation was one in four. If the dummy freedom number is increased, the probability also rises and it is possible genes having less supports can be found. The results of this experimentation are shown in Table 7. Graphs of the fitness are shown in Fig. 3 and 4.

5.7 Coefficients of the fitness function

Taking a look at the value of each fitness function term in early generation, if supports were few, the penalty function had a strong influence on a gene having less supports. In such cases, the fitness function behaved practically as a single term function. We thought that since coefficient γ (a term of penalty function on maximum stress) was high in comparison to other coefficients, it should be applied. In this section, we reconsidered the balance of coefficients and examined it. To be effected by each term, we set the coefficient of 100 to 10. The improvement brought best results. The results of this analysis are shown in Table 8.

Table 6: Influence of $f(\alpha, n)$ of Support Number

f	Fitness	Allocation of supports	Stress	n	
Lin.	15.69	32y 903z 35y 47y 20z 45z 32z 11y 14y	6.39	9	o
	18.42	47z 20z 29y 19z 44y 36z 10y 35y 19y 43y	7.25	10	o
	16.21	29z 48y 29y 14y 10y 20z 31y 47z 46y	6.66	9	o
	18.26	31y 20z 45z 30y 47y 29y 11y 30z 31y	7.53	9	o
	17.22	11y 14z 35z 45z 44y 10y 39y 20y 32y 43y	6.65	10	o
exp.	24.78	42y 10y 35y 20z 45z 32z 14y 31z 1y	8.55	9	×
	28.34	35y 45y 901y 44z 14y 11y 35z 902y 20z 32y	6.00	10	o
	31.81	46z 11y 20z 45y 30z 30y 19y 41z 35y 902y	7.66	10	×
	15.82	14y 46z 10y 901y 19z 45y 35y 38y 29z	6.94	9	o
	18.17	20y 46y 37y 2y 11y 35y 45z 19z 10z	7.73	9	×
quad.	118.3	43z 14y 29y 43y 28z	14.7	5	×
	28.44	35y 904z 20z 11y 43y 43y 29z 47z	8.72	8	×
	21.40	35z 14y 11y 32z 20y 904z 28y 35y 45y 44z	6.58	10	o
	24.62	44z 35y 47y 35z 11y 19z 27z 3y 27z 20z	7.78	10	×
	20.42	45y 39z 10y 10z 44z 32y 20y 11y 11z	7.40	9	o

Table 7: Influence of Dummy Points on Support Number

Dummy	Fitness	Allocation of Supports	Stress	n	
106	15.69	32y 903z 35y 47y 20z 45z 32z 11y 14y	6.39	9	o
	18.42	47z 20z 29y 19z 44y 36z 10y 35y 19y 43y	7.25	10	o
	16.21	29z 48y 29y 14y 10y 20z 31y 47z 46y	6.66	9	o
	18.26	31y 20z 45z 30y 47y 29y 11y 30z 31y	7.53	9	o
	17.22	11y 14z 35z 45z 44y 10y 39y 20y 32y 43y	6.65	9	o
318	16.77	29z 44z 41y 46z 20y 44y 10y 19z 32y 14y	6.35	10	o
	14.26	31y 20z 32z 47z 19y 40z 47y 14y 10y	5.05	9	o
	13.01	11y 14y 32z 46z 20y 32y 46y 20z	4.84	8	o
	16.90	4z 47y 39z 20z 32z 1y 11y 19y 35y 46z	6.44	10	o
	19.47	11y 1y 46z 19y 30y 904y 46y 20z 1z 35y	7.58	10	o
636	15.84	19y 46y 32y 20y 32z 2z 3y 45z 10y	6.41	9	o
	18.46	2y 38y 11y 903y 31z 47z 10y 46y 20y 32y	7.26	10	o
	16.29	46y 5z 37z 20y 10y 46z 32y 29z 14y	6.71	9	o
	17.43	47z 14y 31z 38y 10y 4y 45z 901y 19y 32y	6.78	10	o
	16.61	45y 31z 46y 19y 44z 11y 20z 10y 35y	6.89	9	o

6 Discussion

6.1 Penalty functions

Using the step function, a good solution was obtained only one time in five trials. But applying the exponential or sigmoid functions, a good solution was achieved twice. It should be understood that it is difficult to design the support arrangement of this piping system because the system has a high stress level. Even in this case, a GA found solutions. It is especially efficient to use the exponential and sigmoid penalties is efficiently. Therefore we adopted the sigmoid function to our fitness function. The reason we did not adopt the exponential function is because it depends on a model. The value of the exponential function rises to infinity when large values are input. The value of a penalty function is the maximum RMS stresses value in a piping system. So it is difficult to decide the coefficient of the exponential function independent of the piping system. In contrast to the exponential function, the sigmoid function does not continue to infinity. So it can cope with any stress to piping systems.

6.2 Parameters in the penalty function

We investigated the effect of coefficient b in Eq.1 on the fitness solution. In the case using a coefficient of 0.25 in the penalty function, there are fewer supports than when using coefficients 0.7 and 1.5. Therefore, there is a tendency for maximum stress of the supports over the allowable value. The other way, the solution using a coefficient of 1.5 had more supports than that of coefficients 0.25 and 0.7. Therefore, we adopt 0.7 as the coefficient of the sigmoid function.

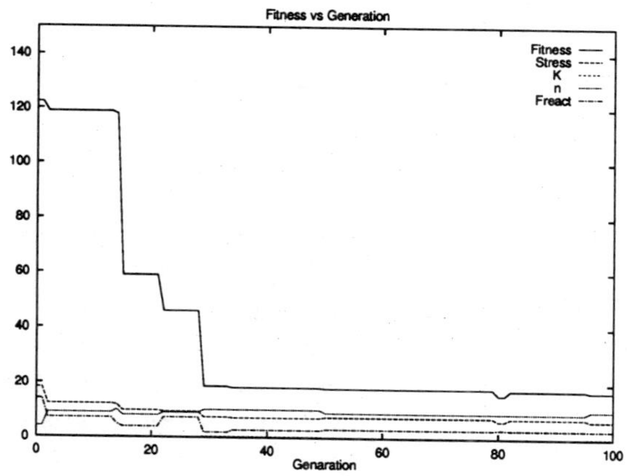

Figure 3: Number of dummy points = 106

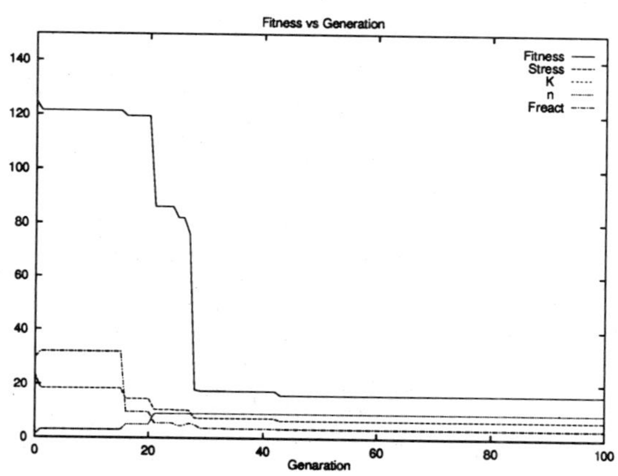

Figure 4: Number of dummy points = 636

6.3 Maximum number of supports

We investigated the effect of a maximum number of support on the fitness function. It is hard to reduce the number of supports used in a piping system with a high level of stress. We tried to increase the maximum number of supports to 10 and obtained more reasonable solutions than the case with a maximum number of seven supports. Because of the reduction in the maximum stress at each support, it is easy to get an allowable solution. Although the number of allowable solutions was increased surprisingly, the number of supports did not decrease. Fitness curves are shown in Fig. 5 and 6. It was found that the fitness in the case of 10 supports decreased rapidly.

Table 8: Influence of Coefficient γ on the Support Number

	Fitness	Allocation of supports	Stress	n	
10	16.69	27y 45y 39z 45z 30y 14y 11y 32y 20z	7.39	9	○
	16.44	35y 44z 47y 1y 11y 20y 19z 36y	7.96	8	×
	18.61	47y 20y 10y 40z 31y 48z 20z 36y 28y	8.58	9	×
	14.88	20y 19y 11y 4z 32y 45z 47y	7.57	7	○
	14.95	35y 20z 45z 31z 14y 11y 20y 36y 45y 904y	4.75	10	○
100	15.69	32y 903z 35y 47y 20z 45z 32z 11y 14y	6.39	9	○
	18.42	47z 20z 29y 19z 44y 36z 10y 35y 19y 43y	7.25	10	○
	16.21	29z 48y 29y 14y 10y 20z 31y 47z 46y	6.66	9	○
	18.26	31y 20z 45z 30y 47y 29y 11y 30z 31y	7.53	9	○
	17.22	11y 14z 35z 45z 44y 10y 39y 20y 32y 43y	6.65	10	○

6.4 Exponential and quadratic functions

To decrease the number of supports, we considered that the term of supports in the fitness function should have a large influence on the fitness function when the number of supports is increased. We tried using exponential and quadratic functions to emphasize the difference in increasing the number of supports. In addition to the severe stress conditions to get an allowable solution under 7 supports, the fitness value became larger using these functions, and such genes were not selected.

6.5 Dummy points

Taking the GA into account, dummy points were used to generate initial genes, crossover and mutation. We assumed that if dummy points were increased, a gene with less supports would be made. We selected three sets of dummy points (106, 318 and 636) and compared the results. But this improvement did not bring good results. The number of supports were always nine or ten. It was found that initial generations were influenced by many dummy points, almost gene had one or two supports in the case of 636. It was found that this was not an effective means for crossover or mutation.

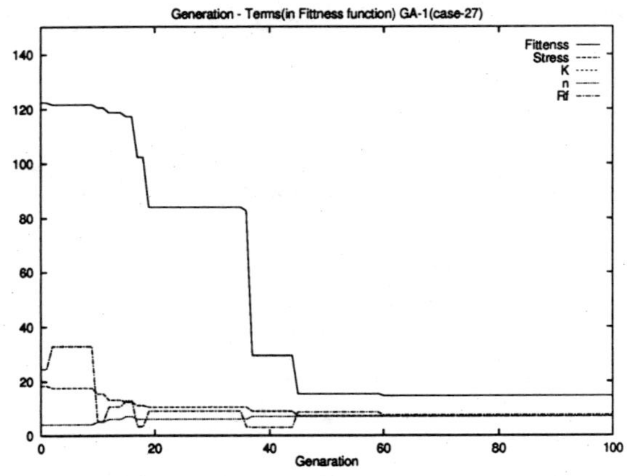

Figure 5: Maximum Number of Supports = 7

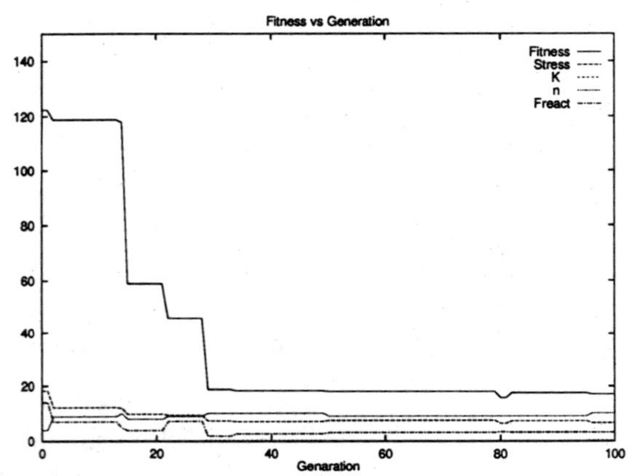

Figure 6: Maximum Number of Supports = 10

We could not find a clear difference between Fig. 3 and 4, which means that dummy points are not so effective on the fitness function.

6.6 Coefficients in the fitness function

In the previous fitness function, if some genes had less supports in early generation, the stress on each support tended to be high. A penalty function for the maximum stress penalized on them, such a genes, was not selected as a result. It was reconsidered balancing the terms of fitness during early generation did not have much of an influence. We changed coefficient γ from 100 to 10, so as to effect fitness both in terms of supports and penalty function for maximum stress. This proved to be a good way to decrease the number of supports, and we were able to get solutions that had seven supports and were allowable.

7 Comparison of designer's results with GA

We improved the fitness function to get better solutions and compared these results with those of designer. It was found that the stresses generated using GA were under allowable stress values, as well as under the values of the designer. The stress was more near the allowable stress than that of designer as shown in Table 9. This result shows that the design by GA is satisfactory. Results of this comparison are shown in Fig. 7.

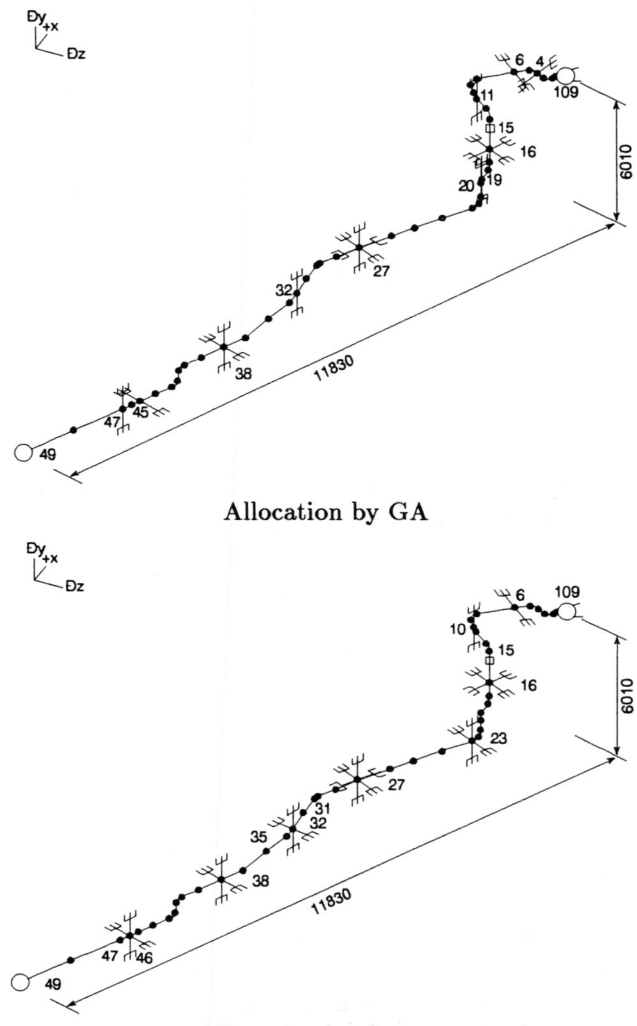

Figure 7: Comparison of Support Allocation by GA and Designer

Table 9: Comparaison of Results by GA and Designer

	Fitness	Number of Supports	Stress (kgf/mm^2)
GA	14.88	7	7.57
Designer	12.9	7	4.43

8 Conclusion

This paper presents a genetic algorithm applied to the support arrangement problem of piping systems. The results are as follows:

1. It is feasible to adopt the sigmoid function as a stress penalty function to optimize the support arrangement problem for a piping system with a high level of stress.

2. It is important to balance the terms of the fitness function, especially the terms of the number of supports and the penalty function for the maximum stress. By coordinating the term of the fitness function, reasonable solutions can be obtained.

References

[1] Chiba T. , et al., 1995 , "Optimum Support Arrangement of Piping Systems using Genetic Algorithm," ASME PVP-Vol.312, pp.447-454

[2] Davis L., 1991, "Handbook of Genetic Algorithms, " Van Nostrand Reinhold, New York

[3] Grefenstette J. J. ,1986, "Optimization of Control Parameters for Genetic Algorithms," IEEE Transactions on Systems, Man, and Cybernetics, Vol. SMC-16, No.1, pp.122-128

[4] Hara F., 1992, "Optimization of Pipe Support Allocation under Seismic Loading by Neural Network Algorithm," ASME PVP-Vol.237-2, pp.215-220

[5] Wang B. P. and Chen L., 1996, "Application of Genetic Algorithm for the Support Location Optimization of Beams," Computers and Structures, Vol.58, No.4, pp. 797-800

PVP-Vol. 340, Seismic Engineering
ASME 1996

OPTIMUM DESIGN FOR PIPE-SUPPORT ALLOCATION AGAINST SEISMIC LOADING

Fumio HARA and Akira IWASAKI

Department of Mechanical Engineering, Science University of Tokyo
1-3 Kagurazaka, Shinjuku-ku, Tokyo 162, Japan

ABSTRACT

This paper deals with the optimum design methodology of a piping system subjected to a seismic design loading to reduce its dynamic response by selecting the location of pipe supports and whereby reducing the number of pipe supports to be used. We will employ the Genetic Algorithm for obtaining a reasonably optimum solution of the pipe support location, support capacity and number of supports. The design condition specified by the support location, support capacity and the number of supports to be used is encored by an integer number string for each of the support allocation candidates and we prepare many strings for expressing various kinds of pipe-support allocation state. Corresponding to each string, we evaluate the seismic response of the piping system to the design seismic excitation and apply the Genetic Algorithm to select the next generation candidates of support allocation to improve the seismic design performance specified by a weighted linear combination of seismic response magnitude, support capacity and the number of supports needed. Continuing this selection process, we will find a reasonably optimum solution to the seismic design problem. We examine the feasibility of this optimum design method by investigation the optimum solution for 5, 7 and 10 degree-of-freedom models of piping system, and find that this method can offer us a theoretically reasonable solution to the problem. We will be, thus, liberated from the severe uncertainty of damping value when the pipe support guaranties the design capacity of damping. Finally we will discuss the usefulness of the Genetic Algorithm for the seismic design problem of piping systems and some sensitive points when it will be applied to actual design problems.

INTRODUCTION

Damping value is one of the most influential dynamics parameters on seismic response of piping system when the floor response spectrum(FRS) methods are used in seismic qualification analysis. Disregard of the importance of damping value in seismic response analysis, the damping value contains a large uncertainty due to unspecified conditions at design stage such as friction, impact, and so forth. To overcome the uncertainty in damping value, we have employed a rather conservative design value of damping which has led to over-stiff piping system with excessive number of supports ill-suited to resisting a thermal transient as well as to in-service surveillance on piping system.

In Japan the damping ratio of 0.5% had been used in design analysis of nuclear piping systems and this value was considered to generate over-conservative seismic responses in piping system. To improve the conservatism in the damping design value, a study project was conducted to investigate the mechanisms in generating damping in nuclear piping systems and to establish more rational design values of damping [1][2][3][4]. Based on the results obtained in the project, the design values of damping in unclear piping systems were determined as follows[5]: Nuclear piping system were categorized into three

Category I includes piping systems supported mainly by snubbers and guide-framed restraints,

Category II is for those supported mainly by rod constraints and hangers,
and Category III for other pipings. Corresponding to each of the categories, the following design value of damping is used: 2.5%(2%) for the piping supported mainly by 4 or more snubbers and guide-framed restraints, 1.5%(1%) for the piping having 4 or more supports such as snubbers, restraints, or hangers, and 1.0%(0.5%) for the piping not included in categories I and II. Note that values in parenthesis are for pipings without thermal insulator.

These design values were determined with a certain safety margin from the values obtained experimentally to account for uncertainty included in the damping values. The uncertainty seems to come from, on one hand, the complex mechanism of damping generation resulting in its nonlinear characteristics with respect to the dynamic parameters of piping system and from, on the other hand, unspecified dynamic conditions at design stage. To overcome this uncertainty, we have employed the conservative design damping values as shown above. However if we wish to use rather high damping values such as average ones, we would be confronted with the uncertainty problem in damping values again. To eliminate the uncertainty in the damping value of actual piping system, it may need a costly design analysis as well as an expensive in-service maintenance to keep the damping value within a design range. This seems not a proper not a proper way to tackle the problem.

A different approach has been taken in Japan for trying to solve the uncertainty problem in damping values of piping systems: Development of seismic energy absorbing devices utilizing the plasticity of materials or friction. Concerning these devices several reports have appeared in the proceeding of ASME PVP Conferences [6][7][8][9] as well as in SMiRT Conference proceedings [10]. The devices can give us an idea that the damping values in the piping system installing these ones would guarantee a design value of damping which may be much higher than those conventionally used as the design values and they may be controlled as a specified value by proper design and manufacture of the devices. This evokes a new design concept of damping control support which could guarantee the high-damping value to the piping system in question.

When we will use these damping control supports, we will encounter a couple of design problems about how to solve the nonlinear dynamic characteristics of the damping control supports and where the supports should be installed to generate the most effective performance in seismic reduction. Thus, we are interested in the optimum design problem of the allocation of the damping control supports in the given piping system where not only the location of the supports to be installed but also the capacity of the supports as well as the number of the supports to be installed are very important for the rational design of the piping system. When the conventional design of pipe support allocation is considered, the performance of the seismic design of piping system is mostly dependent on the designer's engineering sense or experience since the concept of the optimum design of piping systems has not been perceived among the designers. The conservatism in seismic design of piping system seems to have pushed the concept of optimum design away outside our consideration. However as it is pointed out above, the optimum selection of the damping control supports as well as optimum allocation of these on the piping system are very important to achieve the qualified performance of seismic response reduction against the design seismic loading. Thus we will need a feasible design methodology to achieve the optimum solution to damping control support allocation in actual system.

From the view of optimum design of piping support allocation in piping systems, Hara reported a neural network application to the selection of the pipe supported allocation in nuclear piping systems[11], but the methodology showed some difficulty in tuning some parameters used in searching for the optimum solution. Chiba et al applied the Genetic Algorithm for finding the optimum allocation of the pipe supports in practical piping systems [12] and showed a certain credit for the use of the GA method in practical cases. They employed FRS method to evaluate the seismic response of the piping system, which seems to be not proper when nonlinear damping control supported are used since the FRS is a priori calculated for an assumed damping ratio for each vibration mode. This assumption is not always guarantied by the selected damping control supports. This paper deals with the optimum methodology for the seismic response reduction of a piping system a design seismic loading where the capacity of damping control supports and their number as well as their location on the given piping system are considered. We will employ the Genetic Algorithm in searching for the optimum solution to the problem and examine the feasibility of the method by employing 5, 7 and 10 degree-of-freedom piping models. We use the direct numerical integral method to evaluate the seismic response of the piping system equipped with the candidate damping control supports to ensure the accuracy of the design performance of the piping system. Then we will discuss the capability of the Genetic Algorithm with respect to calculation capability in finding the optimum one as well as some sensitive points when we will apply the method to the design of actual piping system.

OPTIMUM DESIGN METHOD

Design Performance Index

When we consider the optimum design of the piping system subjected to seismic external loading, we need the index to measure the optimally of the piping system. In practical view point of piping system design the most important factor is stress induced by the seismic loading and then the reaction force generated in the damping devices which must be reasonably small and must not exceed the allowable design values. When we use many supports or newly developed damping control supported in a piping system, we can reduce these and reaction force as small as possible. However the use of many damping control supports causes the cost of the piping system to be higher. Then we must take into account the number of damping control supports to be installed in the piping system. We also must take into account the capacity of the damping control supports since the large capacity damping control supports are usually expensive which causes the piping system cost to be become high. The safety consideration is deeply involved in determining the value of allowable stress or reaction force, and then we do not here pay much attention to how to determine these values. Not here that the stress induced in the piping system has a definite relationship to the response magnitude of the piping system to the seismic loading. then we can employ the dynamic response magnitude of the piping system, instead of the actual stress, induced by the seismic loading, for the optimization calculation.

The index to measure the goodness of the seismic design is then considered as a weighted sum of these factors stress of seismic response magnitude, reaction force, capacity and number of damping control supports. If you would like to stress more on the structural integrity of the piping system against the seismic loading, the weight for the pipe stress must be large than those for other factors. Consideration to the weighting on design factors offers a different character in the quality of the piping system.

The constraints imposed upon the stress and reaction force are usually taken into consideration by employing Lagrangian parameters and these are added to the index mentioned above. Thus the optimality index can take a unified form. Another consideration to the constraints is done by simply eliminating those piping system candidates violating the constraints in the optimalization calculation. Seismic Response Analysis.

There are several methods of evaluating the seismic response magnitude of a piping system, and the most typical and conventionally used one in the practical design analysis is floor response spectrum(FRS) method. This method is based on linear dynamic response analysis and, thus, if the piping system includes nonlinear damping control supports, the method is considered to give us only an approximation. Because the damping ratios used for generating the FRS is assumed beforehand and the nonlinear damping control supports can not guarantee the damping ratios used in FRS. Thus if the FRS is used in the optimization calculation, the accuracy of the seismic response magnitude must be examined and then the careful judgment must be taken to the result obtained from the optimization calculation

The most accurate method for obtaining the seismic response magnitude is of course a direct numerical integration, but this takes a huge computation time for a large piping system supported by nonlinear damping control supports. For academic investigation of the optimum design characteristics, the direct numerical integration may be reasonable. But for practical optimum design, we need to develop a faster numerical method to evaluate the seismic response magnitude with a certain accuracy compatible to the optimality evaluation, since we need to undertake a huge number of response evaluation in the optimization calculation.

Optimization Algorithm

The pipe support allocation or damping control support allocation is one of the typical combinatorial optimization problems and this problem can be formulated as a well known "Traveling Salesman Problem" [11]. Thus we need a special algorithm for finding the optimal solution to the problem. Hara showed a possibility of the Boltzmann Machine algorithm for this pipe support allocation problem [11] but he revealed some difficulty in tuning the algorithmic parameters in the Boltzmann Machine. Chiba et al also applied the Genetic Algorithm(GA) to the similar problem of pipe support arrangement [12] and found that GA may be a powerful algorithm for finding a reasonable optimal solution of the pipe support allocation problem. While it needed a huge number of the seismic response calculation in the optimization process, at this time the GA may be one of the most suitable algorithm for the combinatorial optimization problems [13].

Consideration to Designer Interface

When we undertake an optimization in the damping control support allocation, it is noted here that any kinds of optimization algorithm expect theoretical one gives us an approximate solution to the problem unless the calculation would continue indefinitely. Then we need to cease the calculation in searching for the optimal solution at a certain stage of the calculation, which means that the one obtained at the final stage of the calculation must be one of the approximate solution close to the theoretically optimal one. Thus the solution must be considered as a quasioptimum solution and it is important for the designer to know how close the solution is to the optimum or how robust the solution is since the approximate solution can be adopted as the piping system to be installed in an actual plant.

Secondarily it is worth pointing out that it is rather difficult to include every kind of design factor in the index mentioned above since the designer has some kind of sensitivity to the 1 topological configuration of the piping system in the environment of the plant facility layout. Those factors may be related to an artistic feature of piping system and we do not know at this moment how to express this kind of artistic feature in the form of numerical index. Thus the solution to the optimum design problem must be considered as one candidate for the designer to decide the final arrangement of damping control supports for the given piping system.

Other considerations should be taken for the designer to make a good decision in the seismic design of piping system, i.e., 3 dimensional drawing of the piping system together with environment, vibration modes and natural frequencies, stress distribution along the piping, reaction force distribution over the damping control supports, the robustness of the piping system against the other seismic loading, actual piping cost and so forth.

When we discuss the optimum design of piping systems, it is better for the optimization calculation to be linked to a CAD system since the considerations mentioned above would be easily undertaken by using the CAD system.

OPTIMUM DESIGN FORMULATION

Piping System Model and Response Analysis

For demonstrating the optimum design procedure in the damping control support allocation, a dynamic model of the piping system is essentially needed to evaluate the seismic response magnitude. However the model is not necessarily three dimensional when we do not focus on the numerical analysis method of seismic responses of piping systems in the optimization calculation but on the optimization characteristics of the optimum design procedure, since we have already developed numerical analysis techniques for evaluating the seismic response of three dimensional piping systems. Thus here we use a 1 dimensional piping system and it is modeled as multi-mass-and-spring model as shown in Fig.1, where, for instance A, B, $\cdots$ D denote mass and the symbol -VVV- represents spring. Note here that we do not include pipe-structural damping into the model since the damping is assumed to be generated by the damping control supports to be installed.

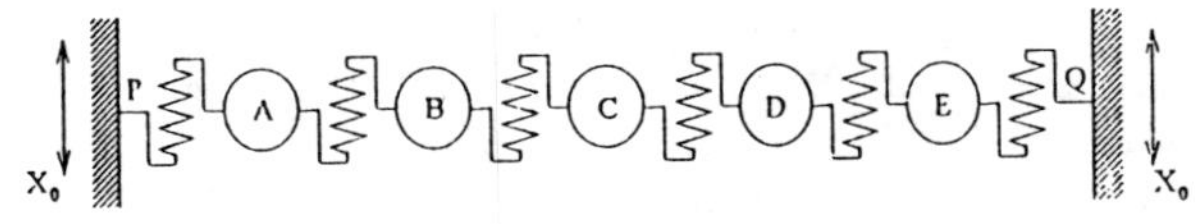

Fig.1 Multi-mass-and-spring model of a piping system

When the piping system depicted in Fig.1 is subjected to a 1 dimensional seismic excitation at both ends P and Q, the equation of motion of each mass is expressed as

$$M\ddot{Z}_i + K(Z_i - X_0) = 0 \tag{1}$$

where M is mass, K spring constant, Z_i is absolute displacement of the mass i and X_0 is seismic excitation in terms of displacement. When we introduce the relative displacement X_i to the floor displacement X_0 as $X_i = Z_i - X_0$, Eq.(1) is rewritten as

$$[M]\{\ddot{X}\} + [K]\{X\} = -[M]\ddot{X}_0[I] \tag{2}$$

Where [I] is unit matrix.

When introducing a damping control support S_i, the support is assumed to generate a specified damping capacity described in terms of the viscous damping coefficient C_i. It is also natural that the damping control support has a certain stiffness K_i for supporting the piping system. Thus we say the damping control support can work as a pipe support with a specified damping capacity. This model of the damping control support is based on the energy absorbing devices developed already [6][7][8][9], when we employ an equivalent linear model of the nonlinear energy absorption characteristic. The damping control support is thus modeled as shown in Fig.2. The damping control support is installed at a mass point of the piping system model, where one end of the damping control support is attached at the mass, and the other end is on the floor. The force needed to generate the relative displacement X_i and relative velocity $\dot{X}_i$ is easily obtained as

$$F_i = K_i X_i + C_i \dot{X}_i \tag{3}$$

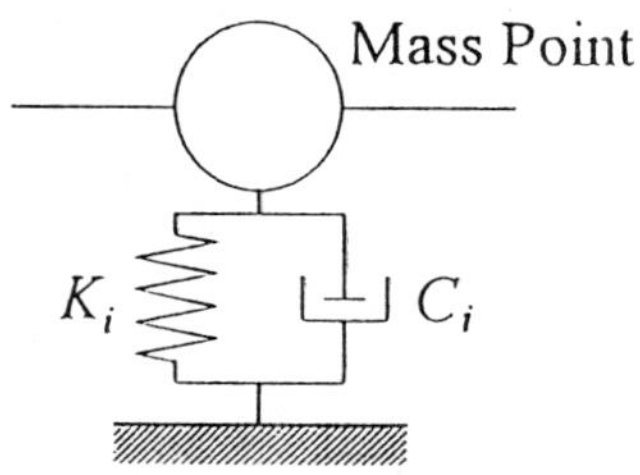

Fig.2 Damping control support model

then the equation of motion of the piping system including the effect of the damping control support generating force F_i is written as

$$[M]\{\ddot{X}\} + [K]\{X\} = -[M]\ddot{X}_0[I] - \{F\} \qquad (4)$$

where variables in [] is matrix form, ones in {} is vector form. Note the vector $\{F\}$ is composed of the force F_i, i indicating the mass at which a damping control support is attached.

The seismic excitation is here simply generated as artificial earthquake acceleration in the following; A uniformly distributed random acceleration R, $(-1 \leq R \leq 1)$ is applied to a single degree-of-freedom vibration model with the natural frequency f_n and damping ratio ζ and the acceleration response of the vibration system is calculated by a numerical method Runge-Kutter-Gill method. THe time step is 0.02 sec and the 2000 digits of the acceleration are obtained, An envelope function defined as

$$E(t) = \begin{cases} t^2/9 & (0 \leq t < 3) \\ 1 & (0 \leq t < 10) \\ \exp\{-0.2092(t-10)\} & (10 \leq t \leq 20) \end{cases} \qquad (5)$$

is multiplied with the obtained stationary acceleration and the artificial earthquake acceleration is obtained as

$$\ddot{X}_0(t) = E(t)\ddot{y}(t) \qquad (6)$$

Fig.3 shows a time history of the artificial earthquake acceleration wave form for the duration of 20 seconds.

Applying the artificial earthquake acceleration to Eq.(4) and the relative acceleration of the piping system is numerically calculated by Runge-Kutter-Gill method and the absolute acceleration of the piping system is evaluated by simply adding $\ddot{X}_i(t)$ and $\ddot{X}_0(t)$ time-wisely. The magnitude of the seismic response of the piping system is evaluated as the sum of the root mean squares of the absolute acceleration $\ddot{Z}_i$, $i = 1, 2, \cdots$ the number of mass in the vibration model of the piping system.

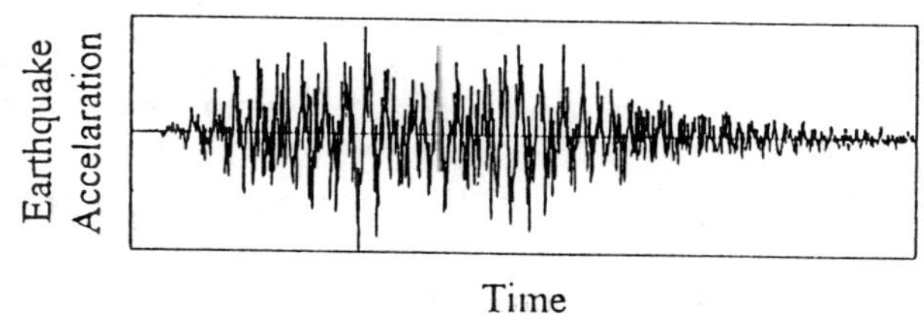

Fig.3 Time history of artificial earthquake acceleration

Design Index

When we would assumedly install several damping control supports, say, S_i $i = 1, 2, \cdots n$ smaller than the number of mass points in piping system model, the equation of motion of the piping system described as in Eq.(4) has a different structure owing to the selection of damping control support capacity and the install location. Thus we can easily expect that the seismic response magnitude evaluated by the sum of root mean squares of the response acceleration at each mass would vary with the allocation of damping control supports along the piping system model. When the seismic design of piping system is interested from the view of the response reduction as well as from the piping system cost, we can use the following design index

$$Fitness = RMS + \alpha \sum K_i + \beta n \qquad (7)$$

where RMS is the sum of the root mean squares of the absolute acceleration of each mass, K_i is the stiffness of the ith damping control support which represents the cost of the support, and n is the number of damping control supports to be installed. When the index is small, the piping system becomes a good-designed one from the view of the high-seismic response reduction together with low cost and a small number of damping control supported.

Note here that we simply neglect the constraints to be imposed on the seismic response magnitude and the reaction force affected on the damping control supports. If it should be included in the optimization process, we can eliminate those candidates violating the constraints from the candidate family of the damping control support allocation.

Genetic Algorithm Formulation

As an algorithm to search for the optimum solution to the damping control support allocation problem, in which the capacity, location and number of damping control supports minimizes the index defined in Eq.(7), the Genetic Algorithm(GA) may be one of the suitable ones since the problem is transform to the Traveling Salesman Problem[11] and the GA is found that it is powerful to the optimal solution to the Traveling Salesman Problem [14][15].

We define the set of damping control support consisting of S_i, $i = 1, 2, \cdots, m$ and also the allocation state by an integer number string, say, (0, 0, 3, 1, 0) expressing the damping control support allocation, where the location of an integer in the string indicates the position of mass in the multi-degree-of-freedom model of the piping system, 0 indicates no damping control support installation and the nonzero integer k means that the damping control support S_k is installed at the mass position corresponding to the location in the string. For this string we construct the equation of motion of the piping system and evaluation the seismic response magnitude from the numerical method. Also K_i and the number of damping control supports used in the piping system are directly evaluated, and thus the design index is definitely determined for this string.

Initial population of the candidate family of the damping control support allocation state is determined as follows: For each mass point, 0 is selected at 50% probability and nonzero k_s, $(k = 1, 2, \cdots m)$, are selected at random under probability of 50% and the population of these strings is predetermined, say, 10 to 20 depending on the complexity of the design index consisting of design factors. If the population size is large, the GA is almost equivalent to a random search technique.

Selection of damping control support allocation candidates is done on the string domain in the following manner: the strings performed better fitness to the problem or corresponding to the index value large than the average are selected and other string are eliminated from the optimization calculation. A pair of the string randomly selected from the survived strings undertake a crossover operation by simply exchanging the two integers located at the same mass position on each string which is randomly selected. This crossover process is undertaken at a certain probability.

Mutation is undertaken at a certain probability, in which, at a mass position randomly chosen, a randomly selected damping control support is installed or the damping device allocation in the substring is simply reversed.

When these genetic operations, selection, crossover and mutation are undertaken on the given string population, we will obtain the new damping control support candidate family which are used to evaluate the design index. And then the similar procedure will be taken until the prespecified number of iteration cycle. Then the string corresponding to the smallest design index may be the most suitable damping control support allocation among the many allocation candidates experienced the artificial selection through the criterion of the design index.

OPTIMIZATION PROGRAM

The optimization calculation based on the GA is shown in Fig.4, where g is the number of iteration, and GENERATION is the prespecified number of iteration when the iteration cycle reaches the number, the calculation will cease.

The concrete parameter values for the piping model are as follows: Spring constant K=200,000[N/m], damping ratio is 0.1 which is calculated by the stiffness and damping coefficient of the damping control support together with the mass of the pipe model, and the 1st natural frequency of the piping system model is 2.0[Hz].

The GA parameters are as follows:

Population size is 18, crossover probability is 50%, mutation probability is 5%, iteration cycle number is 100, and coefficient α and β are 1.0×10^{-5} and 1.0.

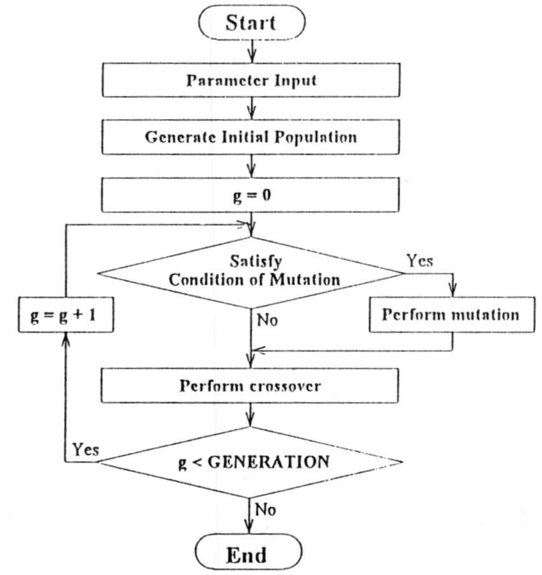

Fig.4 Flow chart of the GA optimization procedure

RESULTS

We have examined the capability of the GA for finding the best solution of the damping control support allocation in 5, 7 and 10 degree-of-freedom piping models.

5 degree-of-freedom model

Table 1 shows the specification of damping control supports to be installed on the piping system modelled as a 5 degree-of-freedom vibration system where the stiffness K and damping coefficient are given for each damping control support $S_1, S_2, \cdots, S_5$. When we count the number of allocation by selecting damping control supports among S_1 to S_5, it is 3,996 for the 5 degree-of-freedom system: and for these cases, we have examined the seismic response magnitude in terms of RMS and selected the top five for a single damping control installation, the two damping control supports and the three damping control supports to be used. The result is shown in Table 2. The best solution obtained by the GA in terms of the defined design index is shown in Table 3 where we have performed the optimization calculation 100 times and , for 99 times out of 100, we have obtained the best one. In Table 3, Fitness means the value of the design index.

Fig.5 shows the time history of the number of the best solution and the second best one obtained among 100 trial cases at each iteration stage. When the iteration in the optimization calculation reachs the 100th, 99 cases among 100 give the best solution.

Table 1 Damping control support specification for 5 degree-of-freedom vibration model

Support	k [N/m]	c [Ns/m]
S_1	10000	260.52
S_2	20000	285.39
S_3	30000	308.26
S_4	50000	349.53
S_5	100000	435.94

7 degree-of-freedom model

Table 4 shows the specification of 5 damping control supports $S_1, S_2, \cdots, S_5$ to be installed on the 7 degree-of-freedom piping model. For the 140,616 cases of allocation of damping control supports on the 7 degree-of-freedom model we evaluated the seismic response magnitude in terms of RMS value and selected the top five allocation for a single, two, three damping control supports to be installed.

Table 2 Top five allocation of damping control supports with respect to RMS seismic response in 5 degree-of-freedom model

(Number of Mass Point : 5)

	Number of Suport = 1		Number of Suport = 2	
Rank	Solution	RMS	Solution	RMS
1	0 0 4 0 0	5.3051	0 0 2 1 0	4.4835
2	0 0 2 0 0	5.7169	0 0 2 2 0	4.5723
3	0 0 3 0 0	5.9221	0 0 3 1 0	4.5750
4	0 0 1 0 0	6.4141	0 0 1 1 0	4.6012
5	0 0 0 1 0	7.9203	0 1 0 3 0	4.6311

Number of Suport = 3	
Solution	RMS
0 1 1 1 0	3.5120
0 1 1 2 0	3.5899
0 1 2 1 0	3.6079
5 0 1 1 0	3.6372
4 0 1 1 0	3.6485

Table 3 The best solution of damping control support allocation obtained by the GA (5 degree-of-freedom model)

Number per 100	Solution	RMS	Fitness
99	0 0 2 1 0	4.4835	6.7835
1	0 0 4 0 0	5.3051	6.8051

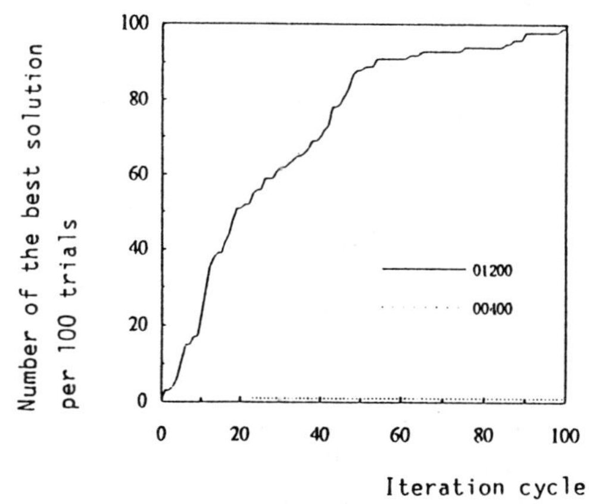

Fig.5 Trend in the number of the best solution obtained by GA for the GA for 5 degree-of-freedom model

The result is shown in Table 5. The best solution obtained by the GA with respect to the design index is shown in Table 6 where we have we have carried out the optimization of 100 times and 99 cases out of 100 give the best solution when the GA is employed for optimization calculation.

Fig.6 is the time history of the number of the best solution and the second best one obtained among 100 trial cases at each iteration stage. This figure shows that for 99 cases out of 100 we can obtain the best solution by the GA.

Table 4 Damping control support specification for 7 degree-of-freedom vibration model

Support	k [N/m]	c [Ns/m]
S_1	10000	196.37
S_2	20000	215.12
S_3	30000	232.35
S_4	50000	263.46
S_5	100000	328.60

Table 5 Top five allocation of damping control supports with respect to RMS seismic response in 7 degree-of-freedom model

(Number of Mass Point : 7)

	Number of Suport = 1		Number of Suport = 2	
Rank	Solution	RMS	Solution	RMS
1	0 0 0 3 0 0 0	7.8825	0 0 1 0 2 0 0	6.0281
2	0 0 0 2 0 0 0	8.8610	0 0 1 0 1 0 0	6.0889
3	0 0 0 1 0 0 0	9.2344	0 1 0 0 2 0 0	6.1482
4	0 0 0 0 2 0 0	10.5203	0 3 0 0 1 0 0	6.2785
5	0 0 0 4 0 0 0	11.5670	0 1 0 0 3 0 0	6.3467

Number of Suport = 3	
Solution	RMS
0 1 0 1 1 0 0	5.1713
0 1 0 1 0 1 0	5.1999
0 1 0 5 0 1 0	5.2919
0 0 1 1 1 0 0	5.3248
0 1 0 5 0 2 0	5.3582

Table 6 The best solution of damping control support allocation obtained by the GA (7 degree-of-freedom model)

Number per 100	Solution	RMS	Fitness
99	0 0 1 0 1 0 0	6.0889	8.2889
1	0 0 1 0 2 0 0	6.0281	8.3281

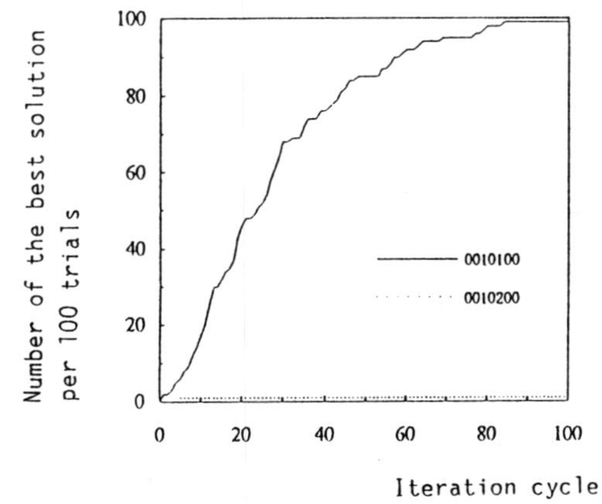

Fig.6 Trend in the number of the best solution obtained by the GA for 7 degree-of-freedom model

10 degree-of-freedom model

Table 7 shows the specification of 5 damping control supports $S_1, S2, \cdots, S_5$ to be installed on the 10 degree-of-freedom piping model. For the 30,236,976 cases of allocation of damping control supports on the 10 degree-of-freedom model we evaluated the seismic response magnitude in terms of RMS value. The top five allocations for a single, two, and three damping control supports were selected and shown in Table 8. The best solution obtained by the GA with response to the design index is shown in Table 9, where we have carried out the optimization calculation of 100 times. The 91 cases out of 100 can give the best solution when the GA is utilized in the optimization calculation.

Fig.7 shows the convergence process of the optimization calculation in terms of the number of the best solution and the second best one obtained among the 100 trial cases at each iteration stage.

Table 7 Damping control support specification for 10 degree-of-freedom vibration model

Support	k [N/m]	c [Ns/m]
S_1	10000	143.25
S_2	20000	156.92
S_3	30000	169.50
S_4	50000	192.19
S_5	100000	239.71

Table 8 Top five allocation of damping control supports with respect to RMS seismic response in 10 degree-of-freedom model

(Number of Mass Point : 10)

Rank	Number of Suport = 1 Solution	RMS	Number of Suport = 2 Solution	RMS
1	0 0 0 0 0 1 0 0 0 0	13.0726	0 0 0 0 1 1 0 0 0 0	8.7494
2	0 0 0 0 0 0 2 0 0 0	15.3094	0 0 0 1 0 1 0 0 0 0	8.7546
3	0 0 0 0 0 3 0 0 0 0	16.2620	0 0 1 0 0 1 0 0 0 0	8.8101
4	0 0 0 0 0 2 0 0 0 0	16.3617	0 0 0 1 0 0 1 0 0 0	8.8859
5	0 0 0 0 0 0 0 0 1 0	18.4764	0 0 2 0 0 1 0 0 0 0	9.1545

Number of Suport = 3 Solution	RMS
0 0 1 0 0 5 0 1 0 0	7.8538
0 0 0 1 0 1 0 1 0 0	8.0360
0 0 2 0 0 5 0 1 0 0	8.0388
0 0 1 0 0 5 0 0 1 0	8.0834
0 0 0 3 0 1 0 1 0 0	8.1116

Table 9 The best solution of damping control support allocation obtained by the GA (10 degree-of-freedom model)

Number per 100	Solution	RMS	Fitness
91	0 0 0 0 1 1 0 0 0 0	8.7494	10.9494
6	0 0 0 0 1 0 1 0 0 0	8.7546	10.9546
1	0 0 0 0 1 0 0 1 0 0	8.8101	11.0101
1	0 0 1 0 0 1 0 1 0 0	8.1117	11.4117
1	0 0 1 0 2 0 1 1 0 0	7.0365	11.5365

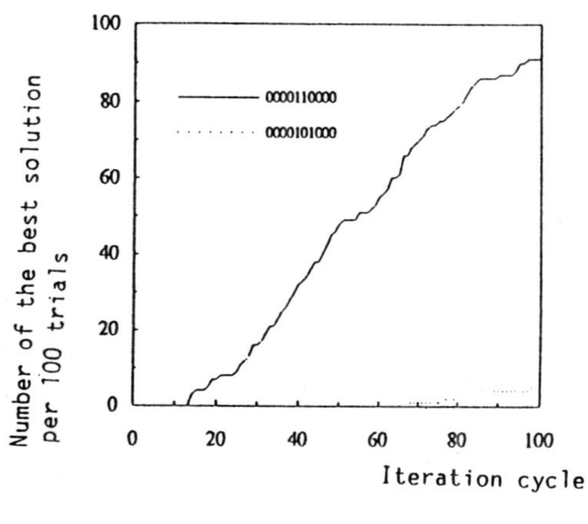

Fig.7 Trend in the number of the best solution obtained by the GA for 10 degree-of-freedom model

DISCUSSION

Optimality of the GA Solution

We think the practical designers are interested in the optimality of the solution obtained by the GA optimization, and thus it is worthy to compare the damping control support allocation obtained by the GA and the one from the examination of the seismic response magnitude for the total cases of the damping control support allocation. For the three cases of the piping model, we can say that the best solution obtained by the GA at the 100th iteration is surely the optimum solution among the total cases of the damping control support allocation when we compare the Table 2 and 3.

Table 5 and 6, and Table 8 and 9. Figs. 5, 6, and 7, the GA optimization calculation is said to give almost always, or at about 90% probability, the optimum solution to the damping control support allocation in the piping system.

Rationality of the GA Solution

For one dimensional piping system modeled as 5, 7 or 10 degree-of-freedom mass-spring system, the best solution of the damping control support allocation has particular feature such as 1) the damping control support(s) is located in the middle of the piping system and 2) the multiple, not single, damping control support allocation is better than the single, large damping capacity support allocation. The first feature is so called the vibration antinode allocation, and the second one may be the multiple allocation.

Vibration antinode allocation is easily understood from the vibration analysis of a string beam with a single damping control support installed at the position of a from one end subjected to the sinusoidal displacement excitation. The damping control support generates the force

$$P(t) - ky(a,t) - c\frac{\partial}{\partial t}y(a,t) \tag{8}$$

when the vibrational displacement of the beam is denoted as y where k and c are the stiffness and damping coefficient of the damping control support. Applying the modal expansion method to the equation of motion of the beam, we will obtain the modal damping for ith vibration mode. If we simply adopt the simply supported boundary condition at the both ends of the beam, the model damping is obtained as

$$C = 2c\sin^2\frac{n\pi a}{L} \tag{9}$$

Eq.(9) means that we will generate the largest damping at the central position of the pipe when the damping control support is installed for a symmetric vibration mode.

When we look at the Table 1, 4, or 7, the damping coefficient for the damping control support S_1 is more than the half of the one for the S_4 or S_5. Then it is easily understood that when we use two damping control supports with a small capacity, we can perform better seismic response reduction.

Selection of GA Parameters

The most influential parameters in the GA optimization calculation are population size, crossover probability and mutation probability and these parameters were selected in trial and error manner in this paper. However, the parameter values adopted in this paper are clearly seen, and they give a good result in the optimization calculation for the damping control support allocation problem. The discussion on the reason why these parameters give a good result seems to be beyond the scope of the paper which stresses the rather practical point of the seismic optimum design of piping system.

Some Remarks on the Optimum Design of Support Allocation. When we will use the nonlinear damping control supports for reducing the seismic response magnitude in the actual large-scale piping system, the response evaluation is a big issue with respect to computation time even if present computers have a high speed computation capability, since we need a huge number of the response calculation. The solution to this problem may be 1) to develop a suitable equi-linear model of the damping control support, 2) to develop a numerical algorithm much faster than conventional numerical integration methods, or to develop a technique to utilized a parallel computing system.

As pointed out by Chiba et al in their paper [12], the GA has certain sensitive parameters such as penalty functions to include the constraints imposed upon the stress and reaction force into the design index. The most probable way to avoid a high sensitivity of the index to support allocation states is to balance each design factor in the index. However there is another way to cope with this problem: eliminating the support allocation candidates violating the constraints from the family of the candidate population in the GA procedure.

To make these problem clearer, we need more systematic investigation to the feasibility of the GA procedure in searching for a reasonably optimum design of the nonlinear damping control support allocation in actual 3 dimensional piping system.

CONCLUSIONS

We have discussed the practical importance of the damping control supports that work for generation a prespecified capacity of damping and the optimum design of damping control support allocation on piping system. To search for the best solution to the allocation of the damping control supports on the piping system, we employed the Genetic Algorithm which simulates artificially the natural selection in biological system and showed its powerfulness in finding the best solution. The best solution obtained from a limited among of iteration was shown to converge to the optimum solution at more than 90% possibility.

Further studies must be done to apply the GA procedure to actual 3 dimensional piping systems under more practical, technical constraints, especially the development of faster numerical integration of the equation of motion of piping system is urgently needed for costless optimization calculation.

REFERENCES

[1] Shibata, H. et al., Investigation on the Design Damping Values for Seismic Analysis of Nuclear Plant Piping Systems, Paper K11/3, 5th SMiRT Transaction, 1979

[2] Shibata, H. et al., A Study on Damping Characteristics of Equipments and Piping Systems for Nuclear Power Facilities, paper K13/4, 6th SMiRT Transaction, 1981

[3] Shibata, H. et al., Test and Evaluation about Damping Characteristics of Hanger Supports for Nuclear Power Plant Piping Systems, Paper K6/4, 6th SMiRT Transaction, 1981

[4] Shibata, H. et al., An Experimental Study of Damping Characteristics with Emphasis of Insulation for Nuclear Power Plant Piping Systems, Paper K6/5, 6th SMiRT Transaction, 1981

[5] Cloud, R.L., Hara,F., and Tomkins, B., LWR Piping: Advances in Design, Analysis and Performance, Transaction of 10th SMiRT, Vol.F, pp 185-217, 1989

[6] Aiken,I.D. and Kelly,J.M., Experimental Study of Friction Damping for Steel Frame Structure, ASME PVP Damping - 1988, PVP-Vol 133, pp. 95-100, 1988

[7] Namita,Y. et al., Development of Energy Absorbor and Its Application to Piping System in Nuclear Power Plants, ASME PVP Active and Passive Damping-1991, PVP Vol-211, pp.51-56, 1991

[8] Fujita,K. et al., Development of Friction Damping as a Seismic Support for the Piping System in Nuclear Power Plants, ASME PVP Active and Passive Damping-1991, PVP-Vol. pp. 57-62, 1991

[9] Shibata, H. et al., Development of the Elasto-Plastic Damper as a Seismic Support for Piping System in Nuclear Power Plants, ASME PVP Active and Passive Damping-1991, PVP Vol.211, pp. 63-69, 1991

[10] Nomura, T., et al., Study of Lead Extrusion Damper as a Seismic Support, Transaction of 10th SMiRT, Vol. K2, pp. 733-737, 1989

[11] Hara, F., Optimization of Pipe Support Allocation under Seismic Loading by Neural Network Algorithm, ASME PVP Ssismic Engineering, PVP Vol. 237-2, pp. 215-220, 1992

[12] Chiba, T et al, Optimum Support Arrangement of Piping Systems Using Genetic Algorithm, ASME PVP Seismic Engineering-1995, PVP Vol. 95-SEI, pp., 1995

[13] Agui, T. and Nagao, T., Genetic Algorithm, Shonkoudo Pub. p. 185, 1993

[14] Kitano,H., Genetic Algorithm, Sangyo Tosho Pub. p. 328, 1993

[15] Yomezawa, Y., Genetic Algorithm, Morikita Publishing Co. p. 184, 1993

PVP-Vol. 340, Seismic Engineering
ASME 1996

LARGE SCALE SHAKING TABLE TEST ON MODAL RESPONSES OF 3-D PIPING SYSTEM WITH FRICTION SUPPORT

Nobuyuki SHIMIZU
Department of Mechanical Engineering,
Iwaki Meisei University
Iwaki, Fukushima, Japan

Kohei SUZUKI , Tetsuya WATANABE
Department of Mechanical Engineering,
Tokyo Metropolitan University
Hachioji,Tokyo, Japan

Nobuyuki OGAWA
National Research Institute for Earth
Science and Disaster Prevention
Tsukuba, Ibaraki, Japan

Hiroe KOBAYASHI
Ishikawajima-Harima Heavy Industries Co.
Yokohama, Kanagawa, Japan

ABSTRACT

Friction between pipe and supporting structure is generally known to reduce seismic responses of the piping system. Vibration tests using large-scale piping model of three dimensional configuration with friction support were carried out to evaluate reduction effect of piping response. The piping responses were governed by the first and the second mode of vibration. The test data of load and velocity via displacement showed that the mathematical model of friction in plane motion was reasonably described by $f = -\mu N v / |v|$, and was applicable to evaluate the response of the piping with friction support. A seismic response analysis procedure of the piping system with friction was developed by an approximate modal analysis and confirmed to be adequate to evaluate the piping response by comparing simulated results and test results, and consequently, response reduction effect due to friction was evaluated.

1. Introduction

The dynamic characteristics of various kinds of piping and supporting structures installed in industrial facilities have to be recognized as one of the significant issues from economic and safety design consideration. Piping systems are generally connected to supporting structures at several points. When attention is directed to the dynamic interaction between the piping and its supporting structure, they frequently behave nonlinear frictional vibration due to contact motion between piping and supporting structures. Studies on response reduction of structures due to friction have been reported by several papers. Ferri (1992) have studies damping effect on beams with various types frictional support. Seismic analyses of structures with friction have been studied by Shah (1983) and Hundal (1979). A series of experiment has been conducted by the authors (Suzuki,1995, Shimizu,1995) for four years on a relatively simple U shaped cantilever piping system. The study was performed to elucidate the response reduction of piping system due to energy dissipation under the action of friction force. As the first mode of vibration mainly governed the responses of the piping system, response estimation was relatively simple.

To establish a response evaluation procedure for a more complex piping system with friction, a large-scale 3 dimensional mock-up piping model was set-up and tested in this study. Mathematical model of friction was proposed and confirmed to be applicable to formulate the equations of motion of piping system with friction. An approximate modal response analysis method by using the mode shapes of piping without friction was proposed.

2. Experiment

2.1 Experimental Set-up

Photo 1 and 2 show the experimental set-up and the contact portion between the pipe and the teflon plate with a load sensor. The schematic figure and the size of the test mock-up of the piping are illustrated in Fig.1. In the figure, the mark ⊖ is indicated anchor point. The piping model, which is made of carbon steel, has a total length of 24.75m, a diameter of 165.2mm, a thickness of 5mm and a total weight of 527kg. In order to induce the first and the second mode of vibration, a main axial direction of the piping was set in accordance with 45 degree with respect to the excitation direction. The friction support of the piping marked as Ⓕ in Fig.1 was prepared to give a frictional coupling between the piping and the supporting base (see Photo 2). The friction support was composed of a shoe welded to the piping and the teflon plate fixed to the load cell which clamped to the supporting base on the shaking table as shown in Fig.2. The short bar shaped shoe, which is shown in Fig.2, is made of carbon steel with a diameter of 45mm and a length of 20mm. The friction forces for X and Z directions and the normal force for Y direction acting on the plate were measured by the load cell.

2.2 Experimental Procedure

The experiment has been performed by using the large scale shaking table (15m × 15m) in the summer of 1995 at the National Research Institute for Earth Science and Disaster Prevention. Capacity and performance of the table is listed in Table 1. The 50 sensors were installed on the piping, including 34 accelerometers, 11 strain guages, 2 displacement sensors amd one load cell measuring three directions at a time. The location of the sensors is shown in Fig.3. All responses were sampled in 1000 Hz and stored in 1000 Hz if necessary, or 200 Hz in a digital recorder. The following two types of wave were used as an input acceleration for driving the shaking table.

1) To concurrently excite almost same amount of displacement amplitude of the first and the second mode of vibration of the piping with friction, a "combined sine wave" with an amplitude ratio of 1 : 2 and 0 dgree phase difference having frequency of 4.65 Hz and 7.05 Hz was prepared.

2) To excite randomly the first and the second mode of vibration of the piping with friction, a "narrow band random wave" with frequency band of 3Hz ~ 9 Hz as shown in Fig.4 which simulates an artificial earthquake ground motion was prepared.

The narrow band random wave was simulated by the following procedure. After summing up sufficient number of sinusoidal waves in time domain to form the white noise and then transforming it into frequency domain by FFT. The transformed signal was shaped as shown in Fig. 4 in frequency domain so that the maximum response of the first and the second mode of vibration at friction point could be approximately identical. After the signal being transformed again from frequency domain to time domain by IFFT, the envelope function proposed by Jennings (1968) was multiplied. Thus, the final "narrow band random wave" was simulated.

During the vibration testing, the supporting conditions of the piping and the types of input acceleration have been changed according to the intent of the test. Sensors were arranged so as to coincide the principal directions of the sensors with the main axial direction and two other perpendicular directions of the piping. The piping responses of the friction point and the corresponding friction forces induced by the first and the second mode of vibration are the main concern of this study.

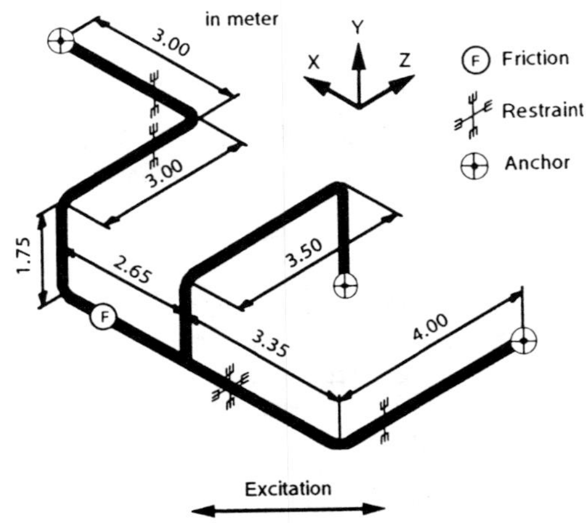

Fig. 1 Experimental ste-up of piping system

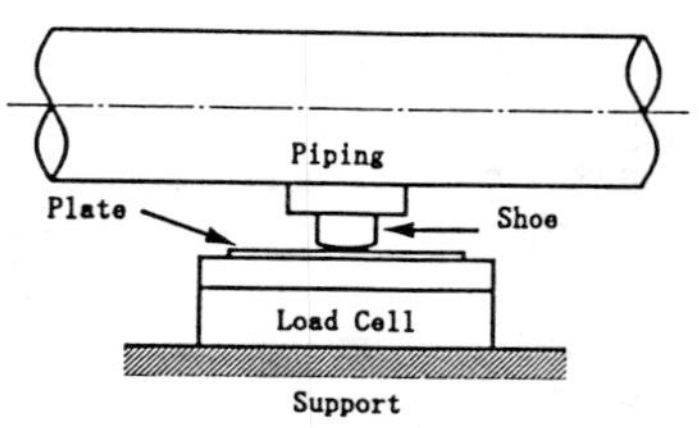

Fig. 2 Detail of friction support

Photo 1 Overview of experimental set-up of piping system

Photo 2 Contact portion of pipe and plate with load sensor

Table 1 Capacity and performance of shaking table

Size of table	15m × 14.5m grid-boxtype welded structure
Driving method	Electro-hydraulic servo type
Power of actuator	360ton (90 × 4units)
Maximum load	500ton
Maximum displacement	±220mm
Maximum vilocity	75cm/sec
Maximum acceleration	0.5G (500 ton hour) 0.94G (200 ton hour) 2.0G (shaking table only)
Frequency range	DC~50Hz

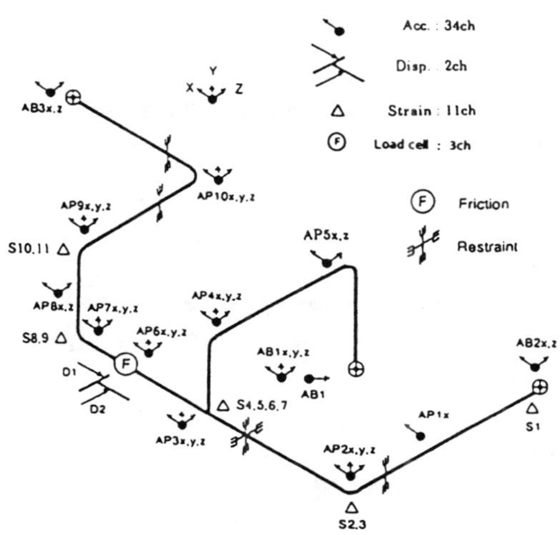

Fig. 3 Location of sensors

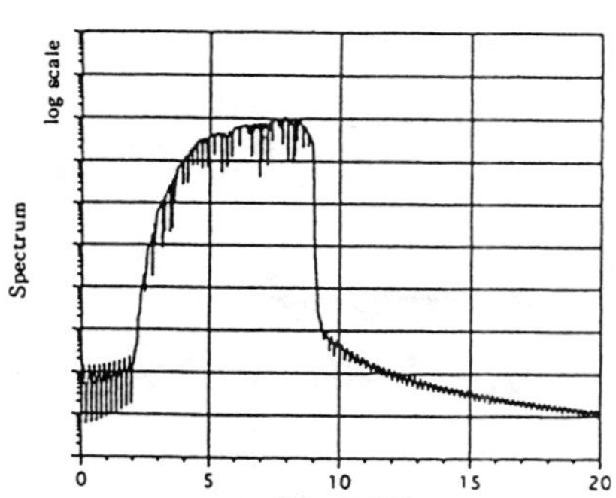

Fig. 4 Power spectrum of narrow band random wave

2.3 Experimental Results

Figure 5(a) and 5(b) are the typical results showing the input shaking table acceleration and the corresponding piping responses for the combined sine wave and for the narrow band random wave. From the top to the bottom, the input acceleration, the piping displacement relative to the shaking table at the friction point (F) for X and Z directions, 3 directional forces of the load cell acting from the piping in X, Z and Y directions and the piping accelerations at the point (F) for X and Z directions are shown in the figures. Other responses have been also

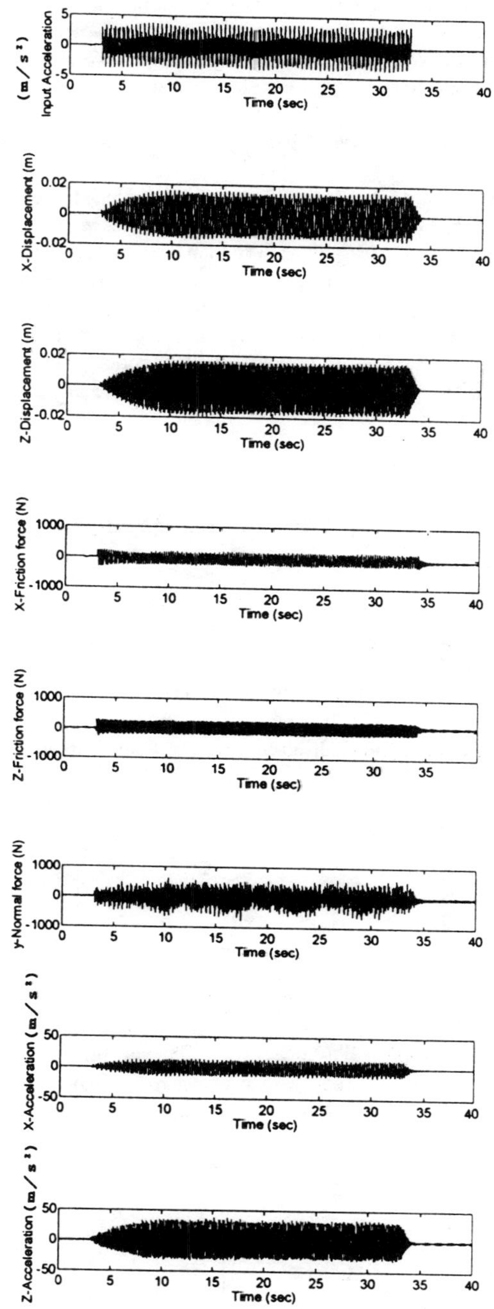

(a) Combined sine wave

Fig. 5 Input acceleration and corresponding piping responses

obtained for other test conditions. Figure 6 shows the maximum displacements at the point Ⓕ against the maximum input acceleration. From Fig.6, the maximum displacements of the piping with friction marked by ● and without friction marked by ✱ are both observed to increase with the increasing maximum input level of acceleration. By comparing the values with and without friction, a great amount of response reduction is observed. The value by the mark ○ will be explained in section 4.2.

Fig.7 shows histograms of friction force to the combined sine wave. We can observed that the friction force histogram varies with the input level of acceleration. The peak of the histogram occurs at 255[N] for the input level of 2[m/s^2] and 196[N] for 5.5[m/s^2]. Fig.8 shows the friction force giving the peak agaist the input level of acceleration. When the input level increases, friction force decreases to an almost constant value of 196[N]. As the normal force acting to the friction plate is about 1470[N], the friction coefficient is μ = 196/1470 = 0.13 in the region of higher input level.

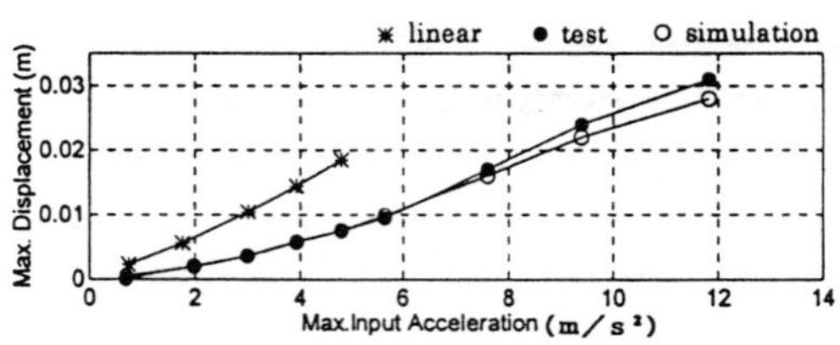

Fig. 6 SRSS of maximum displacements for X and Z direction

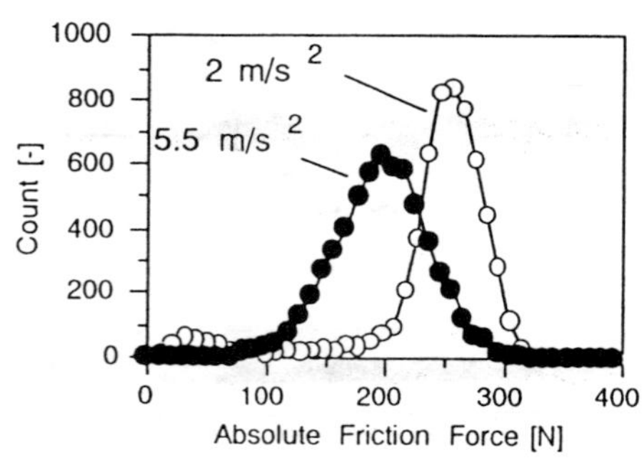

Fig. 7 Histogram of friction force

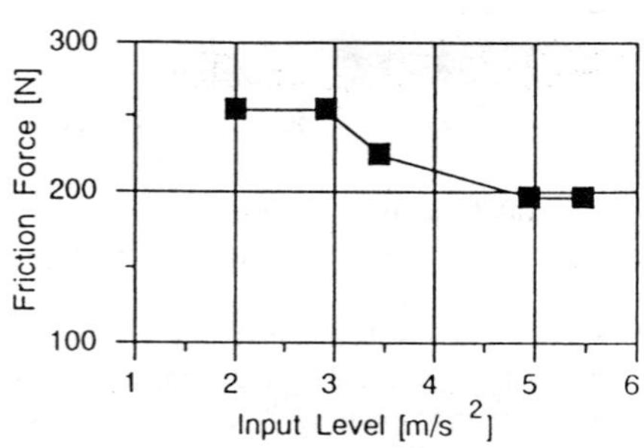

Fig. 8 Friction forces against input level

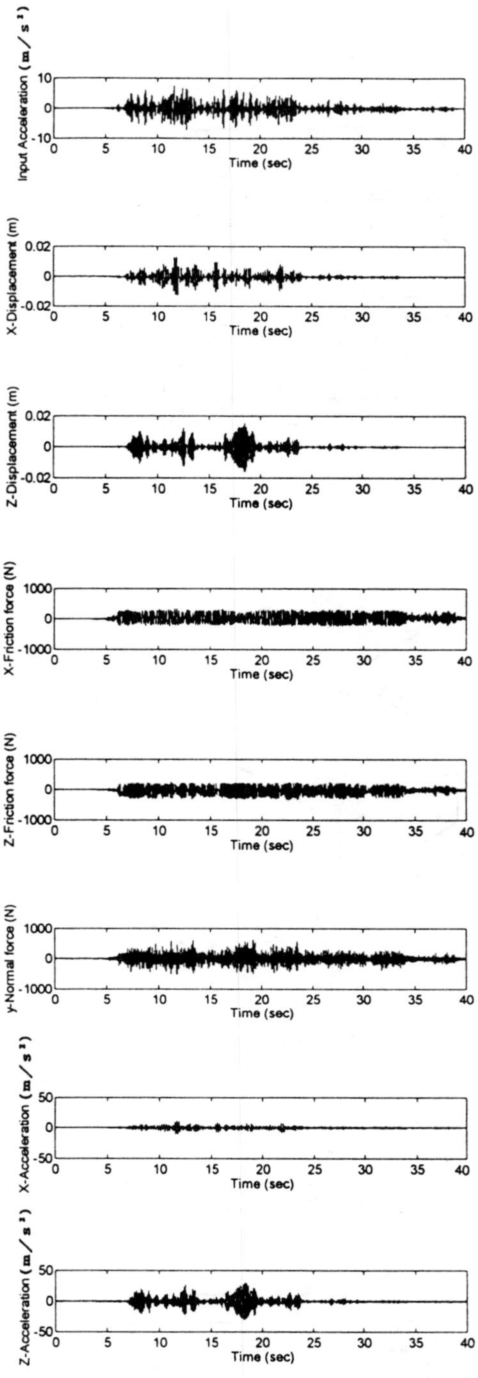

(b) Narrow band random wave

Fig. 5 Input acceleration and corresponding piping responses

3. Friction Model

3.1 Proposed Model

In a uni-directional motion, it is understood that friction force exerted on a body from a base shown in Fig.9(a) is written as

$$f = -\mu N \frac{\dot{x}}{|\dot{x}|} \tag{1}$$

where, μ is the friction coefficient and N the normal force perpendicular to direction of motion of the body. On the other hand, we propose the motion of the body on a plate in 2-dimensional X-Z plane in Fig.9(b) obeys the following rule

$$f = -\mu N \frac{v}{|v|} \tag{2}$$

from the natural extension of Eq.(1) to 2-dimensional plane motion. Here, the friction force vector f which acts on the body from the plate is assumed to be proportional to the velocity vector v of the body. Eq.(2) reduces to Eq.(1), when the motion is restricted to uni-direction.

3.2 Comparison with Experimental Results

Figure 10 and 11 show the piping displacements in X and Z direction and friction forces in X and Z direction of the piping exerted from the teflon bearing plate. As was known from the experiment of the combined sine wave input that the piping displacement is composed of the first and the second mode of vibration, we assume that the piping displacements in X and Z direction are written as

$$u_x = u_{x1}\cos(\omega_1 t - \phi_{x1}) + u_{x2}\cos(\omega_2 t - \phi_{x2}) \tag{3.a}$$

$$u_z = u_{z1}\cos(\omega_2 t) \quad + u_{z2}\cos(\omega_1 t - \phi_z) \tag{3.b}$$

Least square error fit of the experimental displacement shown in Fig.10 to Eq.(3) determined parameters as

$u_{x1} = -0.014[\mathrm{m}], u_{x2} = 0.0016[\mathrm{m}], \omega_1 = 29.14[\mathrm{rad/\,s}],$

$\omega_2 = 44.31[\mathrm{rad/\,s}], \phi_{x1} = 0.279[\mathrm{rad}], \phi_{x2} = -1.19[\mathrm{rad}],$

$u_{z1} = -0.0159[\mathrm{m}], u_{z2} = -0.0012[\mathrm{m}], \phi_z = 0.939[\mathrm{rad}]$

By differentiating Eq.(3.a) and (3.b) with respect to time, the velocities in X and Z direction of the piping at the contact point is obtained as

$$\dot{u}_x = -u_{x1}\omega_1 \sin(\omega_1 t - \phi_{x1}) - u_{x2}\omega_2 \sin(\omega_2 t - \phi_{x2}) \tag{4.a}$$

$$\dot{u}_z = -u_{z1}\omega_2 \sin(\omega_2 t) \quad - u_{z2}\omega_1 \sin(\omega_1 t - \phi_z) \tag{4.b}$$

From Eq.(2),

$$f = -\mu N \frac{v}{|v|} = f_x i + f_z k\,, \tag{5}$$

$$f_x = -\mu N \frac{\dot{u}_x}{\sqrt{\dot{u}_x^2 + \dot{u}_z^2}}, \quad f_z = -\mu N \frac{\dot{u}_z}{\sqrt{\dot{u}_x^2 + \dot{u}_z^2}}\,. \tag{6}$$

Friction force of the piping acting on the contacting plate f_x' and f_z' in X and Z direction is, $f_x' = -f_x$, $f_z' = -f_z$ and therefore,

$$f_x' = \mu N \frac{\dot{u}_x}{\sqrt{\dot{u}_x^2 + \dot{u}_z^2}}, \quad f_z' = \mu N \frac{\dot{u}_z}{\sqrt{\dot{u}_x^2 + \dot{u}_z^2}} \tag{7}$$

Figure 11 shows friction forces of X and Z direction by the experiment and by the simulation using the velocities of X and Z directions into Eq.(7). Where, in the simulation, the values $\mu = 0.13$ and $N = 150 \times 9.8$ N were used. Both are in good agreement. From these results, Eq.(7), consequently, Eq.(2) is satisfactory for describing friction force against velocity of a body.

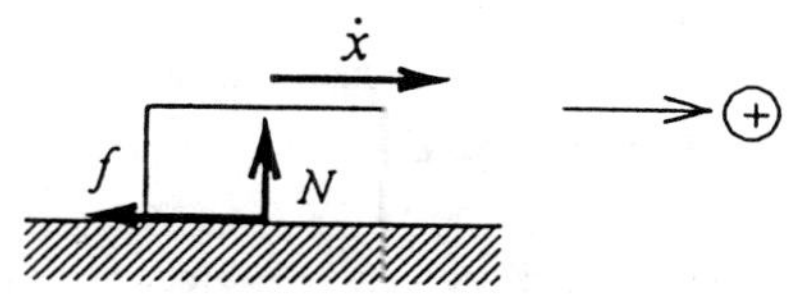

Fig. 9(a) Uni- directional motion of a body on plate

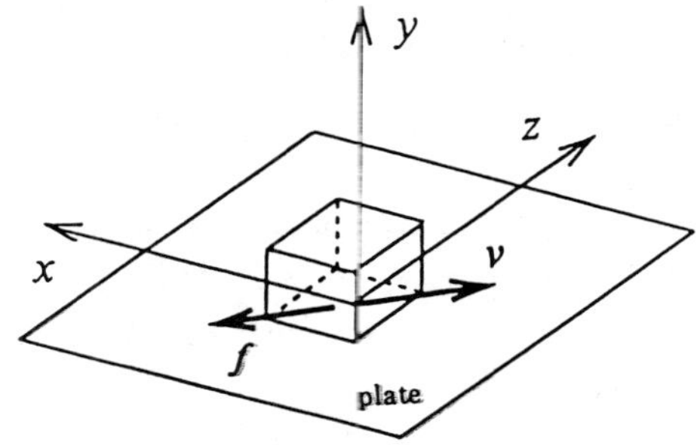

Fig. 9(b) In plane two dimensional motion of a body

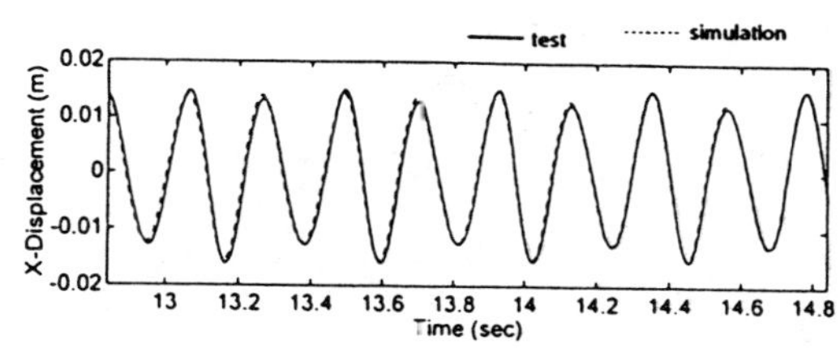

(a) Displacement of X direction

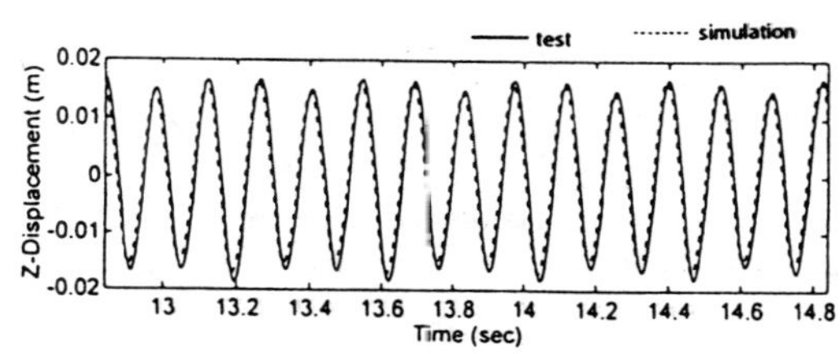

(b) Displacement of Z direction

Fig. 10 Piping displacement in X and Z direction

4. Theoretical Analysis

4.1 Equations of Motion

Equations of motion of the piping system same as used in the experiment are written as

$$[M_p]\{\ddot{u}_p\} + [K_p]\{u_p\} = -[M_p]\{e_p\}\ddot{u}_g + \{f_p\} \tag{8}$$

where, $[M_p]$ and $[K_p]$ are the mass and the stiffness matrices and $\{u_p\}$, $\{e_p\}$ and $\{f_p\}$ are the displacemet, the input excitation, and the friction force vector of the piping system. $\ddot{u}_g$ is the acceleration of the shaking table. Beam element which models the piping has 6 degrees of freedom at a node. Therefore, the total number of degrees of freedom of the piping is $6N$. The piping is in slip state under the action of friction force at the contact point is considered here. As, in the experiment, the first and the second mode of vibration have been observed to the combined sine wave and the narrow band random wave, we assume that the displacement of the piping is approximately written as

$$\{u\} \equiv \{u\}_1 q_1(t) + \{u\}_2 q_2(t) = [U]\{q\} \tag{9}$$

by using the first and the second natural mode of vibration of the piping without friction; $\{u\}_1$ and $\{u\}_2$, where, $[U] = [\{u\}_1, \{u\}_2]$; $(6N \times 2)$, $\{q\} = (q_1, q_2)^T$; (2×1). Substituting Eq.(9) into Eq.(8), and premultiplying both side of the result by $[U]^T$

$$[U]^T[M_p][U]\{\ddot{q}\} + [U]^T[K_p][U]\{q\} = -[U]^T[M_p]\{e_p\}\ddot{u}_g + [U]^T\{f_p\} \tag{10}$$

is obtained. Eq.(10) is rewritten as

$$\begin{bmatrix} M_1^* & 0 \\ 0 & M_2^* \end{bmatrix}\begin{Bmatrix} \ddot{q}_1 \\ \ddot{q}_2 \end{Bmatrix} + \begin{bmatrix} K_1^* & 0 \\ 0 & K_2^* \end{bmatrix}\begin{Bmatrix} q_1 \\ q_2 \end{Bmatrix} = -\begin{Bmatrix} m_1^* \\ m_2^* \end{Bmatrix}\ddot{u}_g(t) + \begin{Bmatrix} f_1^* \\ f_2^* \end{Bmatrix} \tag{11}$$

where, $M_i^* = \{u\}_i^T[M_p]\{u\}_i$, $K_i^* = \{u\}_i^T[K_p]\{u\}_i$, $m_i^* = \{u\}_i^T[M_p]\{e_p\}$, $f_i^* = \{u\}_i^T\{f_p\}$. The friction force vector is written as

$$\{f\} = \left(\cdots \quad \frac{-\mu N \dot{u}_x}{\sqrt{\dot{u}_x^2 + \dot{u}_z^2}} \quad N \quad \frac{-\mu N \dot{u}_z}{\sqrt{\dot{u}_x^2 + \dot{u}_z^2}} \quad 0\,0\,0 \quad \cdots \right)^T \tag{12}$$

where, the elements other than described above are all zero in Eq.(12). From Eq.(11) and (12), modal equations of motion are written as

$$\ddot{q}_1 + 2\zeta_1\omega_1\dot{q}_1 + \omega_1^2 q_1 = -\frac{m_1^*}{M_1^*}\ddot{u}_g - \frac{\mu F}{M_1^*}\frac{\dot{q}_1}{\sqrt{\dot{q}_1^2 + \dot{q}_2^2}} \tag{13.a}$$

$$\ddot{q}_2 + 2\zeta_2\omega_2\dot{q}_2 + \omega_2^2 q_2 = -\frac{m_2^*}{M_2^*}\ddot{u}_g - \frac{\mu F}{M_2^*}\frac{\dot{q}_2}{\sqrt{\dot{q}_1^2 + \dot{q}_2^2}} \tag{13.b}$$

in terms of the first and the second mode of vibration, where ζ_i, ω_i, m_i^*, M_i^*, μN and $q_i(t)$ are the modal damping ratio, natural frequency, mode participation mass, modal mass, friction force and modal displacement for mode i.

4.2 Simulation

Natural frequencies and mode shapes of the test piping system were calculated by the finite element method. Table 2 lists the modal

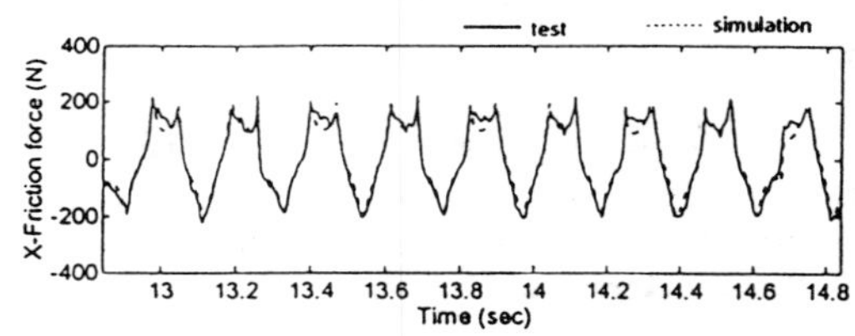

(a) Friction force of X direction

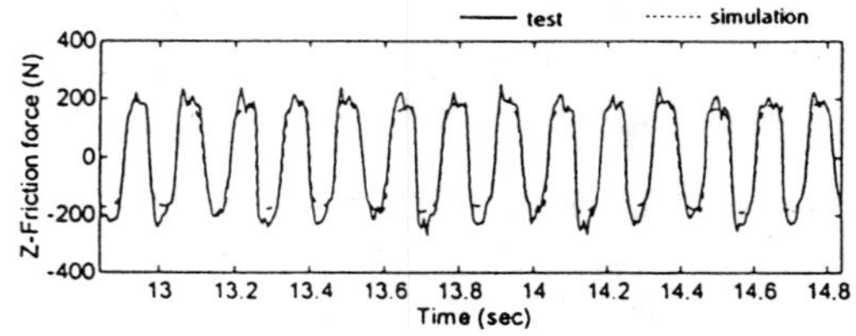

(b) Friction force of Z direction

Fig. 11 Friction force of piping exerted from teflon plate in X and Z direction

parameters and Fig.12(a) and 12(b) show the mode shapes of the piping system. In Table 2, the values of the natural frequencies listed are those obtained by the experiment and were used in the computer simulation. Figure 13 and 14 show the simulated and the tested displacements of the piping at the point (F) in Fig. 1 for the narrow band random wave. Both are in good agreement. Maximum responses of the computed displacement have been shown in Fig.6 by the mark ○ together with the experimental ones. Both are considerably in good agreement. In the figure, it is also shown that the responses without friction marked by * (indicated as linear) is remarkably reduced when the piping system has friction. When the maximum input acceleration is less than the value of about 1~2 m/s², the piping, at the contact point, does not move relative to the plate. When the input acceleration exceeds the above mentioned value, the inertia force of the piping exceeds the resistance of friction force, and consequently, the piping starts to move relative to the plate.

5. Concluding Remark

From the above study, the following conclusions were obtained,

(1) Friction exerted between the piping and the teflon plate in 2-dimensional plane motion can be approximately modeled as two dimensional Coulomb friction. This model was confirmed to be reasonable by comparing the experimental results and the simulated results.

(2) A procedure for obtaining modal equations of motion of a piping system is presented and is confirmed to be satisfactory for evaluating the modal responses of the piping system with friction by comparing the piping responses by the experiment and by the simulation.

(3) Because energy dissipation by friction is large enough to reduce the piping response, the large amount of response reduction was observed by the experiment and this was estimated by the simulation.

The project was carried out as a joint research among Iwaki Meisei University, Tokyo Metropolitan University, the National Research Institute for Earth Science and Disaster Prevention, and Ishikawajima-Harima Heavy Industries Co., Ltd. The authors express their sincere thanks to Mr. Chikahiro Minowa, the National Resarch Institute for Earth Science and Disaster Prevention and Mr. Tatsuya Fujiwaka, Ishikawajima-Harima Heavy Industries Co., Ltd. and Mr. Yoshiomi Hanawa, a student of Iwaki Meisei University and other members who supported their projects.

References

(1)Suzuki,K., Watanabe,T., Mitsumori,T., Shimizu,N., Kobayashi, H., and Ogawa N., 1995, "Experimental Study on Seismic Responses of Piping Systems With Friction-Part 1: Large Scale Shaking Table Vibration Test", ASME J. of Pressure Vessel Technology, Vol. 117, pp.245-249.

(2)Shimizu,N., Suzuki,K., Watanabe,T., and Mitsumori T., 1995, "Experimental Study on the Effect of Softness of Supporting Structures on Piping Responses", ASME, Proceedings of 1995 PVP Conference, Vol.312, Seismic Engineering, pp.211-214.

(3)Ferri,A.A. and Bindemann,A.C., 1992, "Damping and Vibration of Beams with Various Types of Frictional Support Conditions", ASME J. of Vibration and Acoustics, Vol. 114, pp.289-296.

(4)Shah,V.N. and Gilmore,C.B., 1983, "Seismic Analysis of Structures with Coulomb Friction", ASME J. of Pressure Vessel Technology, Vol. 105, pp.171-178.

(5)Hundal,M.S., 1979, "Response of a Base Excited System with Coulomb and Viscous Friction", J of Sound and Vibration, 64(3), pp.371-378.

(6)Jennings,P.C., Housner,G.W. and Tsai,N.C., 1968, "Simulated Earthquake Motions", Report of Earthquake Engineering Research Laboratory, California Institute of Technology, Pasadena, California.

Table 2 Modal parameters

	1st mode	2nd mode
Natural frequency	4.65 (Hz)	7.05 (Hz)
Mode participation mass	2.28×10^2 (kg)	-1.69×10^2 (kg)
Modal mass	2.74×10^2 (kg)	2.89×10^2 (kg)
Modal stiffness	2.34×10^5 (N/m)	5.67×10^5 (N/m)
Modal damping ratio	0.02	0.01
Dynamic friction coefficient	0.13	0.13
Normal reaction force at friction support	1470 (N)	1470 (N)

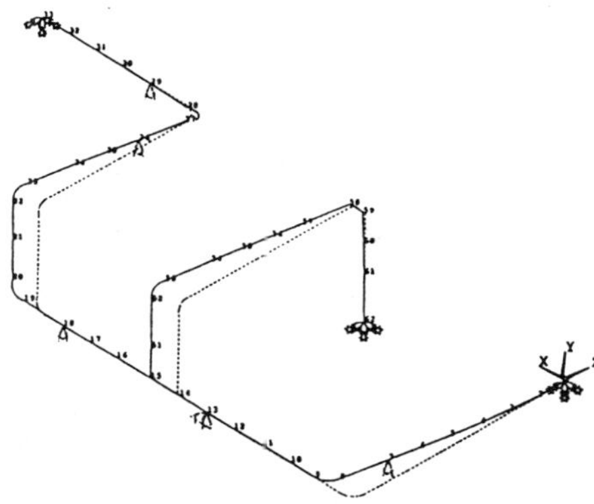

(a) first mode

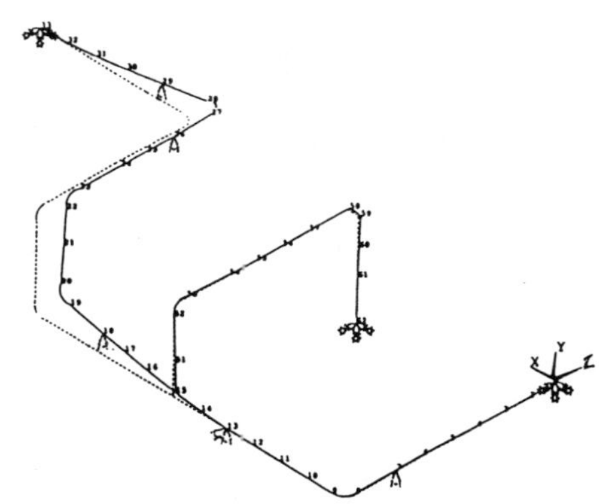

(b) second mode

Fig. 12 Mode shapes of piping system

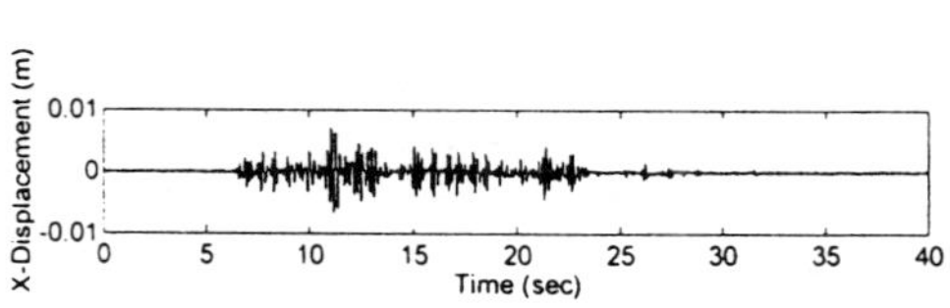

(a) Experiment

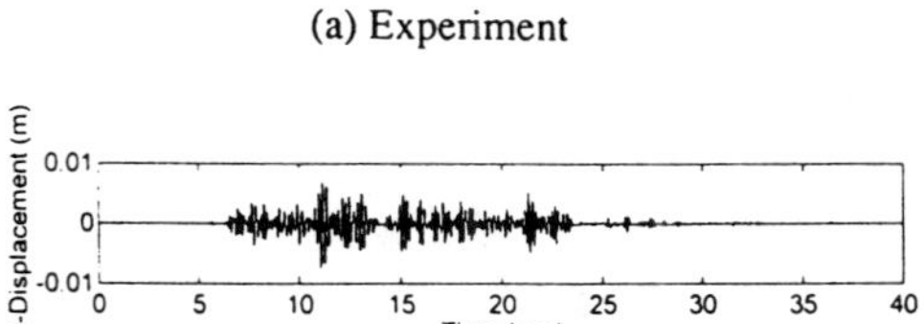

(b) Simulation

Fig. 13 Comparison of X directional displacement of piping at point(F)

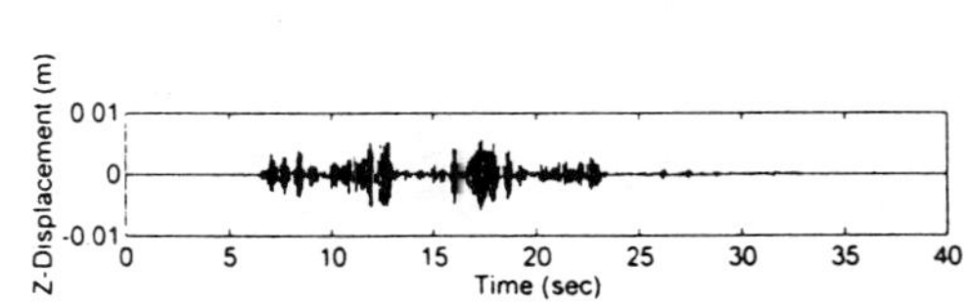

(a) Experiment

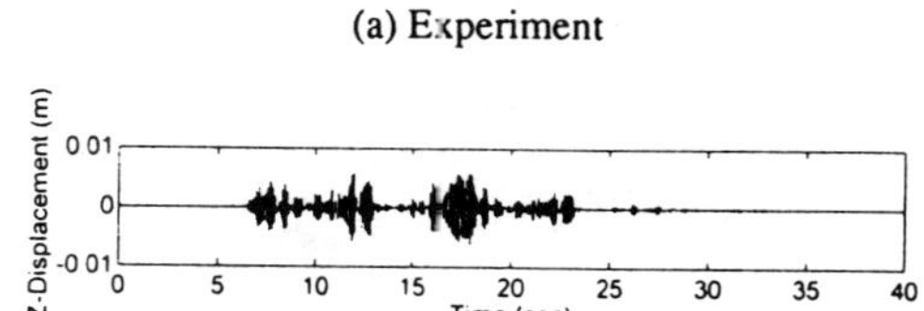

(b) Simulation

Fig. 14 Comparison of Z directional displacement of piping at point(F)

PVP-Vol. 340, Seismic Engineering
ASME 1996

A DECISION SUPPORT SYSTEM FOR PRE–EARTHQUAKE PLANNING OF LIFELINE NETWORKS

Jian–Wen Liang
Department of Civil Engineering
Tianjin University
Tianjin 300072
China

ABSTRACT

This paper describes the frame of a decision support system for pre–earthquake planning of gas and water networks. The system is mainly based on the earthquake experiences and lessons from the 1976 Tangshan earthquake. The objective of the system is to offer countermeasures and help make decisions for seismic strengthening, remaking, and upgrading of gas and water networks.

INTRODUCTION

Earthquake experiences (Sun, 1990; Lund, 1995; Takada and Ueno, 1995) have shown that lifeline networks such as gas and water networks have fatal effects on the ability of the disaster prevention and hazard mitigation of a city, for the disruption of lifeline networks due to an earthquake may cause serious secondary disasters. For example, in the 1906 San Francisco earthquake, the lack of water was mainly responsible for the great fire following the earthquake, and lack of potable water led to epidemics. Similarly, the 1923 Kanto earthquake in Tokyo resulted in destruction of 40% of the city by fire. In the 1976 Tangshan earthquake, the water supply of Tangshan City suspended completely, and two months later, the water supply was restored to 70% of that before the earthquake, which caused indescrible difficulties to the earthquake disaster relief. Even in the 1995 Kobe earthquake, it took two and an half months to finish restoration of water networks, and the reason for the delay was due to too many numbers damages to pipelines and a shortage of water pressure enough for locating the leakages.

Earthquake investigations have also indicated that pipe material, joint type, ground condition, etc. have great effects on the damages of pipelines, and some appropriate seismic countermeasures can prevent or mitigate the damages of pipelines. For example, during the 1976 Tangshan earthquake, there was no one damage to an asbestos cement pipeline in Tanggu district of Tianjin City due to adding rubber rings to joints, but the damage ratio (damage number per kilometer) was more than 20 / km for another same pipeline using asbestos cement as joint filler (Shen, 1986).

In view of the importance of lifelines to a city and their serious secondary disasters after an earthquake, the earthquake resistance and disaster prevention planning of lifeline networks has attracted a great attention (Chen, 1989). In China, the regulation for Compilation of Plan of Urban Earthquake Resistance and Disaster Prevention was promulgated in 1985 (Ministry of Construction, 1985). Lifeline networks such as gas network and water network, as a part of infrastructures of a city, have been the focal point of the plan compilation.

Tianjin City is the third largest city in China, which has a population of more than 9,000,000 and a area of 1100km^2. The urban district has a population of more than 4,800,000 but only a area of 160km^2, whose average population density amounts to 30,000 / km^2, and even 60,000 / km^{2}for the downtown district. There are very complex gas and water networks in the urban district, the malfunction of which due to an earthquake will result in inestimable disasters.

In history, Tianjin city and its nearby area suffered from 2 earthquakes of $M > 8$, 10 earthquakes of $M7$ to $M7.9$, and 30 earthquakes of $M6$ to $M6.9$ (Tianjin Earthquake Resistance Office, 1984).

Tianjin City, about 100km southwest of Tangshan City, was situated on the alluvial plain. In the urban district, there are 7 ancient river courses and 45 ancient puddles and swamps, which has formed a very complicated geological condition for lifeline networks. In the 1976 Tangshan earthquake, the seismic intensity of urban district is 8 degree (Figure 1), and severe damages due to liquefaction and subsidence occurred to the water networks (Sun, 1990), and the damage ratio in the urban district amounted to 0.18 for the transmission lines, and 1.13 / km for the distribution lines. There were 1300 leakages

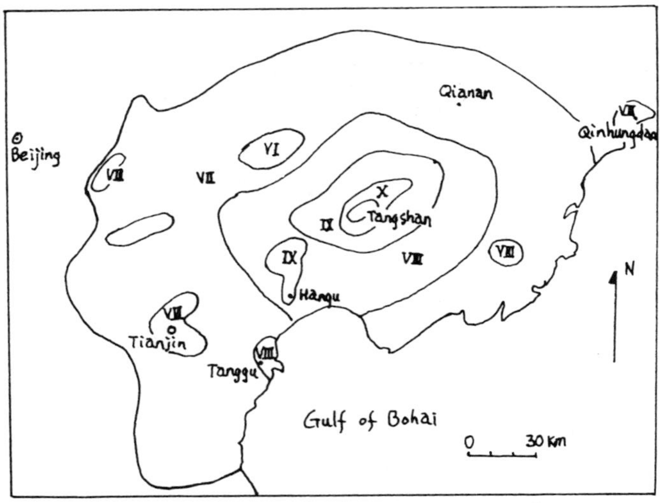

FIGURE 1. SEISMIC INTENSITIES OF TIANJIN IN THE TANGSHAN EARTHQUAKE

to the water networks in the urban district. It took one month for the water supply to be restored to normal state. In Tanggu district and Hangu district, the water networks suffered from more serious damages. The damage ratio of the Tanggu district was to the amount of 4.18 / km, and the water supply was restored to 50% of that of before earthquake two months later, and 100% six months later. In Hangu district, the damage ratio was 10.00 / km, and the water supply was restored to 30% three months later, and 100% one year later.

Based on the experiences and lessons from the 1976 Tangshan earthquake and other earthquakes in recent years, a decision support system (DSS) for pre–earthquake planning of lifeline networks has been developed by Tianjin University, with the collabration of Nankai University and National Center for Water Research. The objective of this system is to offer some countermeasures and help make decisions for seismic strengthening, remaking, and upgrading of gas and water networks, so to improve the ability of earthquake resistance and disaster prevention of Tianjin City. As a supplement to the pre–earthquake planning, the emergency plan which is being developed is briefly introduced in this paper.

SYSTEM COMPONENTS

The DSS is composed by three part: database system, knowledge base system, and decision making system.

Database System

This system generates the database for pipeline networks, ground condition, distribution of important consumers such as hospitals, refuges, etc. The system, with the aid of GIS technique, has the functions of inquiry, mapping, statistics, etc.

The data for pipeline networks includes: use, spatial position, diameter, thickness, length of segment, joint type, pipeline material, elastic modulus, maximum displacement and angle of joint; distribution of valves, water resources and reserviors, gas storage tanks, regulator stations, district regulators, etc.

The data for macro and micro topography includes: spatial position for alluvial plain, ancient river valley and basin, fault zone, reclaimed layer, backfilled layer, ground water level.

The data for soil condition includes: classification, shear wave velocity, thickness of surface layer, potential of liquefaction.

Knowledge Base System

The knowledge base is composed by four part: the national code of seismic design and evaluation, expert knowledge, opinions and suggestions, earthquake experiences and lessons from the 1976 Tangshan earthquake, and flexible decision mathematic model.

The national code of seismic design and evaluation (Ministry of Construction, 1978) acts as the most fundamental knowledge base. It is the guideline for initial seismic evaluation of pipelines and structures.

The author has summarized some research achievements published in proceedings and journals, including the achievements by the author and his co–authors. All these have become the second part of the knowledge base. This part includes the method for detailed seismic evluation of pipelines and seismic reliability assessment of networks. The references are not listed here for limited space.

The author has made a systematic study on a large amount of post–earthquake investigations of gas and water networks in the 1976 Tangshan earthquake (Hua, 1986; Shen and Yao, 1986; Sun, 1979; Sun; 1990; Tianjin Earthquake Resistance Office, 1984; Xu, 1986; Yao and Shen, 1986; Zhao and Du, 1986) and in Haicheng earthquake (Wang and Cui, 1979), including some damage analyses and lessons (Shen, 1986; Shen, 1988a; Shen, 1988b; Sun, 1983; Sun, 1985; Ye, 1988), and some experiences of disaster prevention (Zhang et al, 1986; Zhou and Chen, 1989; Sun and Feng, 1989; Fu, 1989; Hou and Shen, 1989). The summation of the experiences and lessons has become the third part of the knowledge base. It is one of the most important knowledge bases, which offers countermeasures for strengthening, remaking, and upgrading.

The flexible decision model (Shi, 1994) is used here for optimum of different schemes. This is due to that the information which the decision maker has is always not enough, and the ability of the decision maker is limited, and the decision maker usually expresses his desires by descriptive statements, and also the constraints of the decision problem and the objective the decision maker seeks are flexible. This model is needed when the decision to a few possible schemes should be made.

Decision Making System

The intelligent decision making technique is used here for identification of problems, analysis of problem, logical inference, and judgement.

BASIC POLICY FOR PLANNING

The pre–earthquake planning is divided into three levels. The first and second levels are preventive plan, and the third one is emergency plan.

The first level is seismic evaluation of existing pipelines and structures. In this level, we should find out the actual ability of earthquake resistance, then strengthen the pipelines and structures whose abilities are not enough.

The second level is seismic reliability assessment of networks, from which we could find out the weak links and key links, then remake or upgrade them.

The third level is an emergency plan which is based on the fact that the damages to pipelines and structures are unavoidable for a severe earthquake, and the emergency plan is needed to mitigate the secondary disasters, so to shorten the time for restoration of the citylife.

PREVENTIVE PLAN

The preventive plan, including the first and second levels of the basic policy for planning, is the most important component of the whole plan. It contains seismic risk analysis, seismic microzoning, seismic evaluation of existing pipelines and structures, seismic reliability assessment of networks, and seismic countermeasures for strengthening, remaking, and upgrading.

Seismic Risk

Seismic risk analysis is to estimate the magnitude, epicenter, frequency of future earthquakes for Tianjin City in 100 years; and to determine the possible largest earthquake intensity for Tianjin City.

The seismic intensity for Tianjin City is defined as 8 degree, but for important pipelines and structures in the networks the seismic intensity should be taken as 9 degree. This is one of the most important experiences and lessons from the Tangshan earthquake.

Seismic Microzoning

As is well known, local ground condition has a great effect on the distribution of seismic intensity, of course on the seismic behavior of buried pipelines. The seismic microzoning of Tianjin City had been carried out (Xie et al, 1988), the data of which becomes an important component of the database.

The seismic microzonation includes microzonation for ground motions and microzonation for large ground deformations such as liquefaction, subsidence, etc.

Seismic Evaluation of Pipelines

The objective of the seismic evaluation of pipelines and related structures is to find out the actual ability of earthquake resistance, and to strengthen the pipelines and structures whose ability of earthquake resistance is not enough.

The principles for the seismic evaluation are based on the importance consideration:

- The important pipelines and structures should have good aseismic ability, and keep undamaged when the earthquake intensity is 9 degree. The important pipelines and structures may include water intake, supply station, transmission line, reservior, etc. for water network; gas station, storage tank, high–pressure lines, regulator station, medium–pressure line, district regulator, etc. for gas networks;
- The ordinary structures should have enough aseismic ability, and allow damaged partly but not hard to repair;
- Some pipelines, which have a limited effect on the operation of whole networks and are hard to strengthen and easy to repair timely, allow damaged partly;
- Some pipelines, which have a limited effect on the operation and are very hard to strengthen or very high expense to repair, allow damaged.

The evaluation process can be divided into two steps. The first step is to do a initial evaluation to all the pipelines and structures using the code (Ministry of Contruction, 1978). The second step is to do further analysis to pipelines which have high risk in the evaluation of first step. The emphasis should be placed on pipelines buried in irregular sites such as faults, two different soils, three soils, alluvial valley, riverbed, cut–and–fill site, backfilled area, easily liquefied area, etc.

The strength, ductility, and buckling evaluations or one or two of the three can be conducted.

Seismic Strengthening

Seismic strengthening is the most essential measure to improve the aseismic ability of pipelines and structures, but the decision should be made based on economical considerations. Some important earthquake experiences from the 1976 Tangshan earthquake are listed here:

- The support of gas storage tank is the weak part which should be strengthened;
- For assembled water tank, the integrity of the roof and the connection between the roof and the wall should be reinforced;
- Utilize steel pipelines in the situation of transmission lines and crossing rivers, liquefied area, faults, etc. irregular ground conditions. But for other situation, it is better not to use steel pipe for its easily corroded property;
- Using rubber ring flexible joint for cast–iron, prestressed reinforced concrete, reinforced concrete, and asbestos cement pipelines is the most effective aseismic measure;
- Transmission lines should use prestressed reinforced concrete pipelines with rubber ring joint for that transmission lines can resist the effect of earthquake intensity 9 degree;
- The reinforced concrete pipeline with rubber ring joint can only used for earthquake intensity 8 degree;
- The cast–iron pipeline with rubber ring and cement joint can only used for earthquake intensity 8 degree, and the joint should be installed for whole lines;
- The asbestos cement pipeline with rubber ring joint can only be used for distribution lines;
- Install flexible joint at connection between pipes and equipments, and structures, and valves, and fire hydrants, etc;
- The interval of joints should not be too long, which depends on ground conditions;
- For transmission line, adding another line is not better than using aseismic material and flexible joint. The diameter for transmission lines should not be less than 250mm;
- The angle for inclination of pipelines through riverbed

should be less than 30 degree.

Reliability Assessment of Networks

The objective is to assess the seismic connectivity for gas networks and seismic serviceability for water networks, so to find the weak and key links, then remake or upgrade them.

The key link is defined as one with high importance, and the weak link as one with low reliability. The priority of remaking and upgrading should be the links with high importance and low reliability.

For gas networks, the connectivity assessment is divided into two levels. At the first level, we can regard regulator stations from high–pressure lines as resources, and district regulators as consumers; and at the second levels, we can take district regulators as resources, and the end of distribution lines as consumers.

For water networks, the seismic serviceability is also calculated at two levels. At the first level, the supply stations can be regarded as resources, and distribution reserviors as consumers; and the distribution reserviors as resources and the end of distribution lines as consumers at the second level.

Seismic Remaking and Upgrading

Seismic remaking and upgrading are the effective countermeasures to improve the seismic reliability of networks, since it is impossible and unnecessary to strengthen all the pipelines. Some countermeasures derived from earthquake experiences from the 1976 Tangshan earthquake are listed:

- Set up multiple water resources, and fully utilize ground water wells such as pressure supplement wells and local wells. Since 1983, Tianjin City has been supplied through long distance delivery from Luan River, 234km away from Tianjin City. There are 814 bridges built for the water delivery. It is difficult to continue supplying water from Luan River after a severe earthquake. Therefore, the ground water wells such as the former pressure wells and local wells should be jointed into the water networks, and the valves can be opened to supply water after an earthquake;
- Install two independent powers for supply stations and distribution reserviors. For supply stations the oil power–generators should also be used. The power supply for pumps of pressure supplement wells and local wells should have priority to be restored, for these wells can be recoverd easily and rapidly;
- Divide water networks into several blocks, and each block into several subblocks. Install valves between blocks and between subblocks to keep them in contact. The valves may be close or open at different situations. One distribution reservior should be built in each block, and the reservior capacity should be designed from the minimum water needs for drinking and firefighting in restoration process based on the acceptable inconvenience and risk;
- Set up redundant lines for the important transmission lines, and arrange distribution lines into ring shape. Install valves as many as possible for distribution lines, so to isolate the damaged portions to supply water through bypass ways;
- Add redundant distribution lines for important consumers such as hospitals and refuges, etc.;
- Remake the existing water networks, and add some redundant pipelines to improve the reliability;
- Use prestressed reinforced concrete pipes with rubber ring joint for water transmission lines;
- Use ductile cast–iron pipes with rubber ring joint for water distribution lines;
- Use steel pipelines for low–pressure gas lines;
- Avoid as much as possible water transmission lines and high–pressure gas lines crossing faults, easily liquefied areas, easily cleft areas, easily subsided aeras, backfilled areas, etc. If impossible, use steel pipes for water transmission lines, and take further measures.

EMERGENCY PLAN

The emergency plan, which is the third level of basic policy for planning, is a supplement to the preventive plan above. The objective of this plan is to mitigate the secondary disasters, so to shorten the time for restoration of the citylife.

Just as mentioned above, after an earthquake serious secondary disasters may cause a city standstill. For gas networks, fires and explosions may occur if gas pipelines are damaged. Unnecessary shutoff of gas networks may lead to more serious ramifications, and even exacerbate damages already caused by the earthquake. If water pipelines are damaged, the water supply for firefighting will be short, or waterheads may not be enough even there is water supply.

The best way is to shut off and isolate the damaged portions, so to maintain gas and water supply for important consumers such as hospitals, refuges, and firefighting. But this is often difficult soon after an earthquake. Therefore, the emergency plan is to be an earthquake monitoring system.

Blocking of Lifeline Networks

For gas networks, each network is divided into several blocks for medium–pressure lines, and each block is divided into several subblocks for low–pressure lines. One automatic shutoff valve is installed between blocks and between subblocks. Each block has one or a few regulator stations from high–pressure transmission lines, and each subblock has one or a few district regulators from medium–pressure lines.

Each water network is also divided into several blocks, and each block into several subblocks. Each block has one water reservior with automatic shutoff valves. One valve is installed between blocks and between subblocks.

Earthquake Monitoring

A few seismic sensors are laid in each subblock, and the measured PGA (Peak Ground Acceleration) and PGD (Peak Ground Displacement) values are transmitted by radio to the Disaster Relief Headquarter where damage estimations to pipelines and residential buildings are carried out.

Damage Estimation

The damage ratios of pipelines are estimated by the ground condition, the intensity of earthquake ground motion represented by PGA, and deformability of pipelines (Nishio, 1994).

The damage ratios of residential buildings are estimated by PGA, ground condition, structure type, age of building, and number of story of building (Yang et al, 1992).

Since the ground condition has a close relationship with pipeline damages, the microzonation of Tianjin City with mesh size of 500m × 500m is used here. The ground condition in each mesh (subzone) is classified into four types by combining soil type (hard soil, moderate hard soil, moderate soft soil, and soft soil) with the thickness of surface layer.

The damage ratios of pipelines and residential buildings are firstly calculated in each subzone where the PGA value is taken equal to that of the same ground condition in one subblock. Then, the damage ratios in each subblock are obtained by taking weighted averages of those of subzones.

The statistical relationships (Sun, 1985) between damage ratio and earthquake intensity, and pipe diameter, and pipe material, and the relationships (Tianjin Earthquake Resistance Office, 1984) between damage ratio and earthquake intensity, and age of building, and structure type, and number of story of building, obtained in the 1976 Tangshan earthquake, may be used to modify the damage estimation.

Decision Making for Gas Networks

The remote control system sends instructions to shut off the blocks and subblocks once the decisions are made:

- Cut off the valves between blocks to separate all the blocks independent;
- Shut off regulator stations in blocks or district regulators in subblocks.

The earthquake experiences of Tianjin City in the 1976 Tangshan are used here that the shutoff operation is carried out when earthquake intensity reaches 8 degree (PGA value 250gal).

Decision Making for Water Networks

Since the water for Tianjin City is supplied through long distance delivery, preserving water is an important task once an earthquake occurs. This is also because that the fire is the most fatal disaster due to the so high population density of Tianjin City.

Unlike the gas networks, the damage to the water networks is estimated at two level. At the first level, the damage ratios of pipelines and residential buildings are calculated based on monitored PGA and PGD values, in which the damage estimation method for gas networks are used here. At the second level, the damage positions of water networks in supply are further estimated, with the aid of GIS, based on monitored pressures and flowrates of the networks, the PGA and PGD distribution, and the experiences of site effects on seismic behavior of water pipes from the Tangshan earthquake.

When an earthquake occurs, first emergency operation is to close the valves among blocks to eliminate the interaction, and the valves among the subblocks, either. Simultaneously, shut off the reservior to preserve water, which depends on the intensity of the earthquake. If the PGA values of more than half of the seismic sensors in one block exceed 125gal, the half valves of the reservior in this block will be shut off to preserve water; and if the PGA values of more than half of the sensors exceed 250gal, the all valves will be shut off. Then, that when the reservior will resupply water and how many valves will be opened depends on the following parameters: the damage estimation of water networks and residential buildings, and whether there are fires, and the fires' distribution from aerial reconnaissance and residents reports by telephone.

The second emergency operation is to shut off the subblocks by hand based on the damage estimation to the pipelines and the residential buildings, and the distribution of fires.

If one subblock suffers from slight damages, maintain water supply; if serious damages, of course shut off the supply, then the water for firefighting will have to be supplied by sending ground water through temporary pumps and pipelines combining fire engines; if moderate damages, the water supply will depends on whether there are fires, how great the fires are, and the distribution of fires. Send ground water by temporary pumps and pipelines to fight fires if there are ground water wells nearby. If there is no ground water wells, the water has to be supplied through networks by isolating the damaged portions with the aid of damage position estimation report above. This needs an accurate estimation of damaged positions, and usually needs a long time for the closing of valves by hand.

The shutoff of the subblock also depends on the damage estimation to the residential buildings. If there are slight damages to buildings, maintain the water supply; if the damages are moderate or serious, the supply will be cut down for there is no water need in this subblock, even though there is no damage to the network.

CONCLUSIONS

A decision support system for pre–earthquake planning of lifeline networks has been developed. The system, mainly based on the experiences and lessons from the 1976 Tangshan earthquake, can offer some countermeasures and help make decisions for seismic strengthening, remaking, and upgrading of gas and water networks, so to improve the ability of earthquake resistance and disaster prevention of Tianjin City. As a supplement to the pre–earthquake planning, an emergency plan which is being developed is also introduced.

ACKNOWLEDGEMENTS

It is a great honor for the author to express his deep appreciation to Prof. Kohei Suzuki, Tokyo Metropolitan University of Japan for his kind help. The author also gratefully acknowledges the financial support by the National Science Foundation of China under grant No.59308082.

REFERENCES

Chen, S., 1989, "On Countermeasure for Reducing Earthquake Disasters," *Forty Years of Research on Earthquake Resistance Engineering in China 1949–1989*, Seismological Press, pp.161–165. (in Chinese)

Fu, S., 1988, "Urban Earthquake Risk Analysis and Microzoning," *Earthquake Resistance and Disaster Prevention Countermeasures*, Henan Scientific and Technical Publishing House, pp.237–240. (in Chinese)

Hou, Z., and Shen, S., 1989, "Seismic Evaluation and Measures of Existing Buried Pipeline Systems," *Forty Years of Research on Earthquake Resistance Engineering in China 1949–1989*, Seismological Press, pp.216–219. (in Chinese)

Hua, Y., 1986, "Earthquake Damages to Gas Engineering Facilities of Tianjin City, " *The Mammoth Tangshan Earthquake Damages*, Vol.3, Seismological Press, pp.490–508. (in Chinese)

Lund, L.V., 1995, "Lifeline Utilities Lessons, Northridge Earthquake," *Proceedings, 4th U.S. Conference on Lifeline Earthquake Engineering*, ASCE, San Francisco, pp.676–683.

Nishio, N., 1994, "Damage Ratio Prediction for Buried Pipelines Based on the Deformability of Pipelines and the Nonuniformity of Ground," *Journal of Pressure Vessel Technology*, ASME, Vol. 116, pp.459–466.

Shen, S., 1986, "Earthquake Resistance of Buried Pipelines," *Engineering Earthquake Resistance*, No.2, pp.34–37. (in Chinses)

Shen, E., 1988a, "Repair and Reconstruction of Water Supply System in Tangshan City," *Earthquake Resistance and Disaster Prevention Countermeasures*, Henan Scientific and Technical Publishing House, pp.539–543. (in Chinese)

Shen, S., 1988b, "Aseismic Strengthening of Water–Supply Works," *Earthquake Resistance and Disaster Prevention Countermeasures*, Henan Scientific and Technical Publishing House, pp.463–465. (in Chinese)

Shen, S., and Yao, C., 1986, "Earthquake Damages to Water Supply Engineering Facilities," *The Mammoth Tangshan Earthquake Damages*, Vol.3, Seismological Press, pp.434–446. (in Chinese)

Shi, Y., 1994, "Flexible Decision," *Acta Aeronautica et Astronautica Sinica*, Vol.15, pp.468–471. (in Chinese)

Sun, S., 1979, "Earthquake Damage to Pipelines," *Proceedings, 2nd U.S. National Conference on Earthquake Engineering*, EERI, Stanford, pp.61–68.

Sun,S., 1983, "Pipeline Damage and its Relationship with Joints," ASME, PVP–Vol.77, pp.374–377.

Sun, S., 1985, "Earthquake Resistance and Disaster Prevention of Buried Pipelines," *Research Report of Beijing Municipal Engineering Research Institute*, Beijing. (in Chinese)

Sun, S., 1990, "Earthquake Damages to Buried Pipelines in China," *Earthquake Resistance of Buried Pipelines*, Academic Publisher, Beijing, pp.1–11. (in Chinese)

Sun, S., and Feng, Q., 1989, "Seismic Analysis of Buried Pipelines," *Forty Years of Research on Earthquake Resistance Engineering in China 1949–1989*, Seismological Press, pp.193–198. (in Chinese)

Takada, S., and Ueno, J., 1995, "Performance and Restoration of Lifeline Systems in the Great Hanshin Earthquake," *Presented at 4th U.S. Conference on Lifeline Earthquake Engineering*, ASCE / TCLEE, San Francisco.

Tianjin Earthquake Resistance Office, 1984, *Engineering Damages of Tianjin City in Tangshan Earthquake*, Tianjin Scientific and Technical Publishing House. (in Chinese)

Ministry of Construction, 1978, *The Aseismic Design Code for Water Supply, Wastewater, Gas, and Heating Engineerings*, Beijing. (in Chinese)

Ministry of Construction, 1985, *Compilation of Plan of Urban Earthquake Resistance and Disaster Prevention*, Beijing. (in Chinese)

Wang, S., and Cui, C., 1979, "Water Pipeline Damages in Haicheng Earthquake and the Lessons," *The Haicheng Earthquake Damages*, Seismological Press, pp.283–299. (in Chinese)

Xie, Y., Gao, J., and Xu, J., 1988, "Earthquake Resistance and Disaster Prevention Countermeasures of Tianjin City," *Earthquake Resistance and Disaster Prevention Countermeasures*, Henan Scientific and Technical Publishing House, pp.103–108. (in Chinese)

Xu, D., 1986, Earthquake Damages to Buried Water Networks of Urban Area of Tangshan City," *The Mammoth Tangshan Earthquake Damages*, Vol.3, Seismological Press, pp.447–467. (in Chinese)

Yang, Y., Wang, Z., Yang, Y., Li D., and Yang L., 1992, "An Assistant Intelligent Decision System for Prediction Earthquake Damage to Urban Existing Buildings," *Earthquake Engineering and Engineering Vibration*, Vol.12, No.1, pp.77–88. (in Chinese)

Yao, C., and Shen, S., 1986, "Earthquake Damages to 4000T Assembled Prestressed Water Tank of Tangshan City," *The Mammoth Tangshan Earthquake Damages*, Vol.3, Seismological Press, pp.479–481. (in Chinese)

Ye, Y., 1988, "Aseismic Empirication of Lifeline System," *Earthquake Resistance and Disaster Prevention Countermeasures*, Henan Scientific and Technical Publishing House, pp.390–396. (in Chinese)

Zhang, S., Lou, B., and Peng, K., 1986, "Measures for Engineering Earthquake Resistance and Disaster Preparedness" *Earthquake Countermeasures*, Seismological Press, pp.97–133. (in Chinses)

Zhao, H., and Du, S., 1986, "Earthquake Damages to Gas Facilities," *The Mammoth Tangshan Earthquake Damages*, Vol.3, Seismological Press, pp.482–488. (in Chinese)

Zhou, X., and Chen, Y., 1989, "Research and Practice for Earthquake Resistance and Disaster Prevention Planning of City," *Forty Years of Research on Earthquake Resistance Engineering in China 1949–1989*, Seismological Press, pp.166–171. (in Chinese)

PVP-Vol. 340, Seismic Engineering
ASME 1996

A COMPREHENSIVE FRAMEWORK FOR POST–EARTHQUAKE REHABILITATION PLAN OF LIFELINE NETWORKS

Jian–Wen Liang
Department of Civil Engineering
Tianjin University
Tianjin 300072
China

ABSTRACT

Post–earthquake rehabilitation process of lifeline networks can be divided into three stages: emergency operation, short–term restoration, and long–term restoration, and different rehabilitation measures should be taken for different stages. This paper outlines the post–earthquake rehabilitation plan of lifeline networks which is being developed for Tianjin City. The objective of this plan is to shorten the time for restoration of lifeline networks and to reduce secondary disasters.

INTRODUCTION

Many earthquake experiences (Liu, 1986; Lund, 1995; Schiff et al, 1995; Takada and Ueno, 1995) have shown that the rehabilitation of lifeline networks has fatal effects on post–earthquake hazard mitigation of a city. The disruption of lifeline networks such as gas, water, and power networks will cause serious secondary disasters. For example, in the 1906 San Francisco earthquake, the lack of water was mainly responsible for the great fire following the earthquake, and lack of potable water led to epidemics. Similarly, the 1923 Kanto earthquake in Tokyo resulted in destruction of 40% of the city by fire. In the 1976 Tangshan earthquake, the water supply of Tangshan City suspended completely, and two months later, the water supply was restored to 70% of that before the earthquake, which caused indescrible difficulties to post–earthquake disaster relief (Liu, 1986). Even in the 1995 Kobe earthquake, more than 170 fires broke following the earthquake some of which spread out of control of fighters, and burned out a total area of 60 hectares with 7,500 houses and buildings (Kameda, 1995). It took two and an half months to finish restoration of water networks, and the reason for the delay was due to too many numbers of a pipeline damage and a shortage of water pressure enough for locating the leakages (Takada and Ueno, 1995).

Earthquake experiences have also indicated that effective rehabilitation plan of lifeline networks is playing a more and more important role in post–earthquake restoration process of lifeline networks, especially in emergency restoration stage (Chen, 1989; Lagorio, 1994). Now, this topic has been paid more and more attention (Liu and Singh, 1994). Some work have been done on this subject for large cities in the world, such as Tokyo (Katayama, 1983; Iwata and Saito, 1985; Iwata and Nakane, 1988; Kano et al, 1992; Yamazaki et al, 1994; Yoshikawa et al, 1995), Osaka (Takada and Ogawa, 1995), San Francisco (Aschheim and Lynn, 1995; Lau and Eidinger, 1995), and Wellington (Brunsdon et al, 1995).

Tianjin City is the third largest city in China, which has a population of more than 9,000,000 and a area of 1100km^2. The urban district has a population of more than 4,800,000 but only a area of 160km^2, whose average population density amounts to 30,000 / km^2, and even 60,000 / km^{2}for the downtown district. There are very complex power, gas, and water networks in the urban district (Xie et al, 1988), the malfunction of which due to an earthquake will result in inestimable disasters. It is very important to work out a comprehensive post–earthquake restoration plan of the lifeline networks for Tianjin City.

In history, Tianjin city and its nearby area suffered from 2 earthquakes of $M > 8$, 10 earthquakes of $M7$ to $M7.9$, and 30 earthquakes of $M6$ to $M6.9$ (Ma, 1988). The epicenters are shown in Figure 1, and the distribution of earthquake intensity of Tangshan earthquake is shown in Figure 2 (Liu, 1986).

Tianjin City was situated on the alluvial plain, in the urban district of which there were 7 ancient river courses, 45 ancient puddles and swamps (Tianjin Earthquake Resistance Office, 1984), and all these ancient areas were backfilled. This formed a very complicated geological condition for lifeline networks.

Tianjin City is about 100km southwest of Tangshan City. In the 1976 Tangshan Earthquake, the earthquake intensity of urban district is 8 degree (Figure 2), and severe damages due to

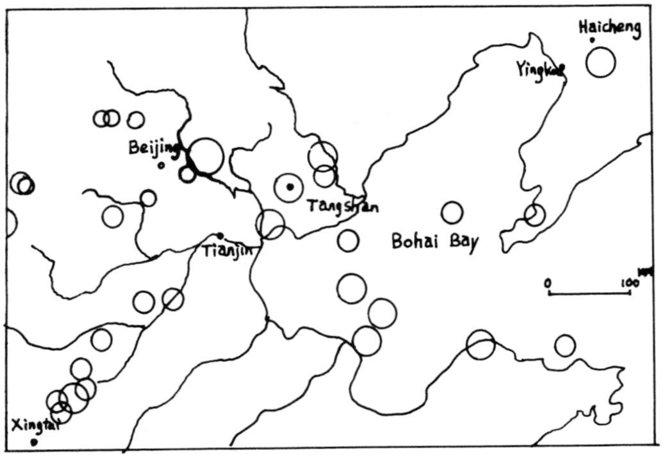

FIGURE 1. HISTORICAL EARTHQUAKES OF TIANJIN AND ITS NEARBY AREA

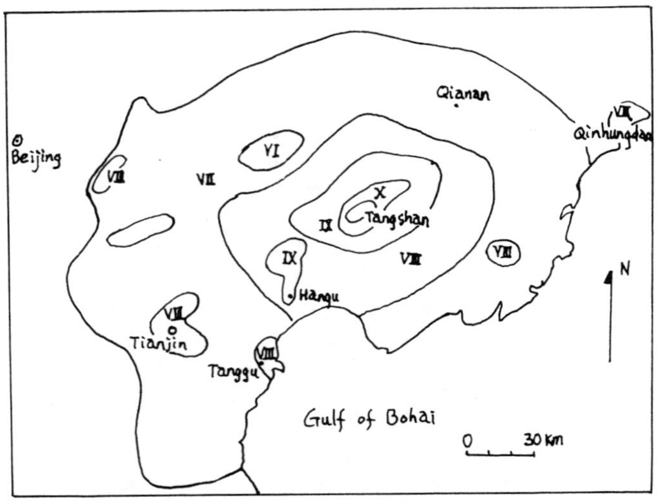

FIGURE 2. DISTRIBUTION OF EARTHQUAKE INTENSITY OF TANGSHAN EARTHQUAKE

liquefaction and subsidence occurred to the water networks, and the damage ratio in the urban district amounted to 0.18 (number of damage / km) for the transmission lines, and 1.13 / km for the distribution lines. There were 1300 leakages to the water networks in the urban district. It took one month for the water supply to be restored to normal. In Tanggu district and Hangu district, the water networks suffered from more serious damages. The damage ratio of the Tanggu district was to the amount of 4.18 / km, and the water supply was restored to 50% of that of before earthquake two months later, and 100% six months later. In Hangu district, the damage ratio was 10.00 / km, and the water supply was restored to 30% three months later, and 100% one year later.

Based on the lessons from the 1976 Tangshan earthquake and other earthquakes, a comprehensive plan for post–earthquake rehabilitation of lifeline networks for Tianjin City is being developed by Civil Engineering Department of Tianjin University, with the collabration of Nankai University and National Center for Water Research. The objective of this plan is to shorten the time for restoration of lifeline networks and to reduce the secondary disasters. It is hoped that this plan could minimize the impact of a severe earthquake on the city.

CLASSIFICATION OF REHABILITATION STAGES

Recent earthquake experiences have shown that post–earthquake restoration process of lifeline networks can be divided into a few different stages (Katayama, 1983; Kameda et al, 1984; Chen, 1989). Based on the behaviors of lifeline networks in restoration, the interactions among lifeline networks, and the earthquake experiences of lifeline networks, we divide the post–earthquake restoration process of lifeline networks into three stages: emergency operation, short–term restoration, and long–term restoration. Each stage has its own objective and plan.

GAS NETWORKS

Among all the lifeline networks, gas network needs the most urgent measures, for secondary disasters like fires and explosions may occur if gas pipelines are damaged. Unnecessary shutoff of gas networks may lead to more serious ramifications, and even exacerbate damages already caused by the earthquake, e.g., serious inconvenience to the customers for a rather long time. So, the best way is to shut off and isolate the damaged portions of the networks timely. But this is often difficult soon after an earthquake. Therefore, an earthquake monitoring system is being developed for the emergency operation.

EMERGENCY OPERATION

The emergency operation is the most important one of the three stages, and can win the most remarkable outcomes in this stage if appropriate countermeasures are taken in time.

The objective of the emergency operation is to shut off and isolate the damaged portions in one to two hours. This emergency operation plan is to be an intelligent control system, which includes four subsystems: earthquake monitoring system, damage estimation system, decision making system, and remote control system.

EARTHQUAKE MONITORING SYSTEM Earthquake monitoring is a effective way for damage estimation of gas networks, and a few systems are now being developed (Yamazaki et al, 1994; Yoshikawa et al, 1995; Takada and Ogawa, 1995; Ballantyne, 1995).

In Tianjin City, gas customers are supplied by two networks: one is natural gas network, and the natural gas is delivered by long distance pipelines from oilfield; and another is coal gas network.

Each network is divided into several blocks for medium–pressure lines, and each block is also divided into several subblocks for low–pressure lines. Each block has one or a few regulator stations from high–pressure transmission lines, and customers are supplied by low–pressure distribution lines through district regulators.

The first emergency operation is to cut off the valves between blocks to separate all the blocks independent. The second emergency operation is to shut off regulator stations in blocks or district regulators in subblocks based on damaged estimations. A few seismic sensors are laid in each subblock, and the measured PGA (Peak Ground Acceleration) values are transmitted by radio to the Disaster Relief Headquarter where damage estimations to pipelines and customers' buildings are carried out.

DAMAGE ESTIMATION SYSTEM The damage estimation to pipelines and customers' buildings are calculated with the aid of a Geographic Information System (GIS) (Liang, 1996).

The damage ratios of pipelines are estimated by method (Nishio, 1994) which takes into account three fundamental factors, namely, the ground condition, the intensity of earthquake ground motion represented by PGA, and deformability of pipelines.

The damage ratios of customers' buildings are estimated by PGA, ground condition, structure type, age of building, and number of story of building.

Since the ground condition has a close relationship with pipeline damages. The microzonation of Tianjin City with mesh size of 500m × 500m (Xie et al, 1988) is used here. The ground condition in each mesh (subzone) is classified into four types by combining soil type (hard soil, moderate hard soil, moderate soft soil, and soft soil) with the thickness of surface layer.

The damage ratios of pipelines and customers' buildings are firstly calculated in each subzone where the PGA value is taken equal to that of the same ground condition in one subblock. Then, the damage ratios in each subblock are obtained by taking weighted averages of those of subzones.

The damage of pipelines in some backfill areas are also monitored by PGD (Peak Ground Displacement) seismic sensors, since in the 1976 Tangshan earthquake pipelines suffered from serious damages due to liquefaction and subsidence.

The statistical relationships between damage ratio and earthquake intensity, and pipe diameter, and pipe material (Sun, 19909), and the relationships between damage ratio and earthquake intensity, and age of building, and structure type, and number of story of building (Jin and Zhang, 1988; Li, 1988; Wang and Yang, 1988; Yang et al, 1992), obtained in the 1976 Tangshan earthquake, are also used to modify the damage estimation.

DECISION MAKING SYSTEM The decision for shutoff of subblocks and blocks is made using the damage estimation above. The earthquake experiences of Tianjin City in the 1976 Tangshan and Ninghe earthquake are used here that the shutoff operation is carried out when earthquake intensity reaches 8 degree (PGA value 250gal).

REMOTE CONTROL SYSTEM The remote control system sends instructions to shut off the blocks and subblocks once the decisions are made.

SHORT–TERM PLAN

This stage has two objectives: one is further detailed damage investigation of networks, especially for the areas in supply, another is restoration of important customers such as hospitals and refuges. This stage may last for one to two weeks.

Although the emergency operation has been conducted based on damage estimation, detailed damages of gas networks should still be investigated, especially for areas in supply. This can be done by sending damage investigation team. The guide for post–earthquake investigation (Schiff, 1991) can be used as the specific investigation schemes. The reports from customers through telephone should also be considered.

The remedial measures should be taken for areas in supply to shut off the damaged distribution lines and service lines by hand, to repair the slight leakages, and to eliminate hidden disasters.

Two strategies for restoration may be chosen. The strategy one restores the most important customers such as hospitals and refuges. The strategy two emphasizes the restoration of slight damaged areas.

The priority for restoration should be:

- long distance gas pipelines from oilfield;
- key equipments such as storage stations and regulator stations;
- medium–pressure lines;
- district regulators;
- low–pressure lines.

The detailed restoration process of medium–pressure lines and low–pressure lines will not be illustrated here.

LONG–TERM PLAN

The objective of this stage is to restore the gas supply to the pre–earthquake level, which may need three to six months and will depend on the clearance of collapsed buildings.

This stage is also a opportunity to reform, upgrade, and strengthen the gas networks. We should learn quickly from the earthquake and apply the earthquake lessons into the long–term restoration. The author thinks this is one of the most important countermeasures.

Other countermeasures may be taken as follows:

- Divide the networks into blocks;
- Subdivide the middle block into subblocks;
- Install automatic shutoff valves for subblocks;
- Replace the damaged and corroded pipelines;
- Use mechanical joint for steel pipelines;
- Replace cast–iron medium–pressure lines with steel pipelines;
- Install flexible joints when crossing faults and at backfill interfaces;
- Replace or compact the easily liquefied soil, or add extra weight to pipelines.

WATER NETWORKS

The water pipeliness of Tianjin City are composed by cast–iron pipelines, prestressed concrete pipelines, asbestos cement pipelines, and reinforced concrete pipelines, and are more vulnerable than gas networks.

If water pipelines are damaged, the water supply for firefighting and to hospitals will be short, and waterheads may not be enough even there is water supply, and the water will also be wasted for nothing. This will lead to secondary disasters such as fires out of control, and the loss of city's own salvation ability. If whole water networks are shut off, though supply can be preserved, there will be more serious difficulties to the restoration of the city.

EMERGENCY OPERATION

The objective of this stage is to shut off and isolate the damaged portions, so to maintain water supply for firefighting, and to hospitals and refuges. This stage may need 24 to 48 hours.

Just like the gas networks, to shut off and isolate the damaged portions, the most important thing is still damage estimation for water networks (Eguchi et al, 1990; Liang, 1996). So, the emergency operation plan for water networks is to be an intelligent control system which includes earthquake monitoring system, damage estimation system, decision making system, and control system.

The earthquake monitoring system for the gas networks is used here also for the water networks. The measured PGA and PGD values are used for the damage estimation of water networks.

The water for Tianjin City is supplied through long distance delivery from Luan River, 234km away from Tianjin City. There are 814 bridges built for the water delivery. It is a very difficult thing to continue supplying water from Luan River after a severe earthquake (Liang, 1996). Therefore, preserving water is an important task once an earthquake occurs. This is also because that the fire is the most fatal disaster for Tianjin City due to the so high population density.

The water network is divided into several blocks, too, and each block into several subblocks, though the segmentation of blocks is different from that of the gas networks. One water reservior with automatic shutoff valves is built in each block (Liang, 1996).

Unlike the gas networks, the damage to the water networks is estimated at two level. At the first level, the damage ratios of pipelines and residential buildings are calculated based on monitored PGA and PGD values, in which the damage estimation methods for gas networks are used here. At the second level, the damage positions of water networks in supply are further estimated, with the aid of GIS, based on monitored pressures and flowrates of the networks, the PGA and PGD distribution, and the earthquake experiences of site effects on seismic behavior of water pipes from the Tangshan earthquake.

When an earthquake occurs, first emergency operation is to close the valves among blocks to eliminate the interaction, and the valves among the subblocks, either. Simultaneously, shut off the reservior to preserve water, which depends on the intensity of the earthquake. If the PGA values of more than half of the seismic sensors in one block exceed 125gal, the half valves of the reservior in this block will be shut off to preserve water; and if the PGA values of more than half of the sensors exceed 250gal, the all valves will be shut off. Then, that when the reservior will resupply water and how much valves will be opened depends on the following parameters: the damage estimation of water networks and residential buildings, and whether there are fires, and the fires' distribution from aerial reconnaissance and residents reports by telephone.

The second emergency operation is to shut off the subblocks based on the damage estimation to the pipelines and the residential buildings, and the distribution of fires.

If one subblock suffers from slight damages, maintain water supply; if serious damages, of course shut off the supply, then the water for firefighting will have to be supplied by sending underwater through temporary pumps and pipelines combining fire engines; if moderate damages, the water supply will depends on whether there are fires, how great the fires are, and the distribution of fires. Send underwater by temporary pumps and pipelines to fight fires if there are underwater wells. If there is no underwater wells, the water has to be supplied through networks by isolating the damaged portions with the aid of damage position estimation report above. This needs an accurate estimation of damaged positions, and sually needs a long time for closing the valves by hand.

The shutoff of the subblock also depends on the damage estimation to the residential buildings. If there are slight damages to buildings, maintain the water supply; if the damages are moderate or serious, the supply will be cut down for there is no water need in this subblock, even though there is no damage to the network.

SHORT–TERM PLAN

The objective of this stage is to investigate the damages to water networks in detail, and to repair water resources and transmission lines. This stage may last for one week.

The following restoration plans may be considered:

- Investigate the damages of the networks. Priority is placed to blocks and subblocks in supply. The emphasis should be place on water resources, equipments, transmission lines through liquefaction areas, crossing Hai River, through backfill areas, through ground cleaves, and permanent deformations.
- Close the customers' valves to preserve water.
- Shut off the damaged portions, and repair them if the damage is slight.
- Restore electricity to water plants and repair the equipments.
- Repair or replace the transmission lines based on the investigation above.
- Supply water in a certain pressure to transmission lines to find the damages by checking water leakages if there is not enough water, and repair or replace them.
- Supply water in a high pressure to distribution lines to find damages, and repair or replace them.
- Do rash repairs and replacements for pipelines using the following methods. For a cast–iron pipe, inlay slug if the pullout displacement at joint is small (Shen, 1988); or inlay rubber ring, and then add cement around joint. If the joint or body of a cast–iron pipe is damaged, cut off the damaged part, and add short pipe or insert a new pipe, then use quick–condence cement as joint filler

(Shen, 1988). For a concrete pipe, add quick–condence cement around joint and body.

LONG–TERM PLAN

This stage is to restore the whole water networks, which may last for two to three months.

This stage is also the opportunity to reform, upgrade, and strengthen the whole networks. We should learn quickly from the earthquake and apply the lessons to this restoration.

The following measures may be used:

- Block distribution lines and install valves;
- Install automatic shutoff valves to reservior;
- Introduce rubber rings to cast–iron and concrete pipelines;
- Replace old–type pipelines with new aseismic pipelines.

POWER NETWORKS

Power network is the most important one among all the lifeline networks, whose suspension will make a city at a standstill. After an earthquake, the power networks are the first priority to be restored, for electricity is essential to maintaining the public order, supplying water, driving machines, etc. The damage of power pipelines may also cause other secondary disasters such as fires.

EMERGENCY OPERATION

The objective of this stage is to restore power supply to Disaster Relief Headquarter, water plants, hospitals, refuges, etc. as quickly as possible. This period may need 12 hours.

The following operations may be considered:

- Introduce power from other electric networks;
- Generate electricity for Headquarter, hospitals, water plants, etc. by oil power generators before the supply of electricity ;
- Utilize aerial cables for supply where the power pipelines are damaged;
- Investigate damages of power stations, transformer substations, overhead transmission lines, underground transmission lines, and distribution lines;
- Shut off the power supply of residential areas based on the damage estimation to buildings.

SHORT–TERM PLAN

The objective is to restore power to the important consumers listed above.

Two strategies for power restoration are united for this stage. One strategy is that restores the most important facilities first. Another is that emphasizes restoration of maximum power.

The measures are listed as followed:

- Restore the important facilities above;
- Restore the power supply based on the damage estimation to residential buildings.

LONG–TERM PLAN

- Strengthen the buildings of power plants, steel towers, and other facilities in the restoration.
- Place the overhead transmission lines and distrbution lines under ground.

POWER–WATER–GAS INTERACTION

The interactions among the lifeline networks can affect the post–earthquake restoration process both in positive and negative ways (Kameda, 1994). Therefore, we should try our best to avoid the negative interactions. The most important and effective countermeasure is that the restoration of the gas, water, and power networks should be unified under the direction of the Disaster Relief Headquarter.

POWER–WATER INTERACTION

There are some interactions between power and water networks:

- Restoration work competition;
- Water supply suspension due to power shortage;
- Power generation stop due to water pipeline damage.

The following measures should be considered:

- Utilize additional power feeder lines to reduce the probability of power failure;
- Utilize backup power;
- Utilize oil power generators.

WATER–GAS INTERACTION

The main interactions between gas and water networks are:

- Restoration work competition;
- No water for fighting fires due to gas leakages.

The effective measures are listed as followed:

- Use other resources such as groundwater resources, pressure supplyment wells, local wells, etc.;
- Supply water from distribution reserviors;
- Supply water through bypass way by isolating damaged portions;
- Prepare enough pumps, temporary pipelines, fire engines, etc.

POWER–GAS INTERACTION

The interaction between power and gas networks can be summerized:

- Restoration work competition;
- Fire due to interaction between power and gas;
- Gas production stop due to power shortage.

Some measures can be used:

- Emergency shutoff of damaged power networks;
- Emergency shutoff of gas networks based on the damage estimation to pipelines and customers' buildings.

CONCLUSIONS

A comprehensive plan for post–earthquake rehabilitation of lifeline networks which is being developed for Tianjin City has been outlined. The emergency operation, short–term plan, and long–term plan for the restoration of gas, water, and power networks are described.

We should not only learn from past earthquakes but also learn quickly from an earthquake which has just happened, and apply the lessons to the post–earthquake restoration.

The rehabilitation plan should be modified and improved constantly.

It is believed that this comprehensive plan will reduce the secondary disasters, shorten the time for restoration of lifeline networks, and minimize the impact of a severe earthquake on the city.

ACKNOWLEDGEMENTS

It is a great honor for the author to express his deep appreciation to Prof. Kohei Suzuki, Tokyo Metropolitan University of Japan for his kind help. The author also gratefully acknowledges the financial support by the National Science Foundation of China under grant No.59308082.

REFERENCES

Aschheim, M., and Lynn, A., 1995, "Rapid Reconnaissance Response Plans for Post–Earthquake Investigations," *Proceedings, 4th U.S. Conference on Lifeline Earthquake Engineering*, ASCE, San Francisco, pp.447–454.

Ballantyne, D., 1995, "Monitoring and Control of Lifeline Systems to Enhance Post–Earthquake Operation," *Proceedings, 4th U.S. Conference on Lifeline Earthquake Engineering*, ASCE, San Francisco, pp.533–540.

Brunsdon, D.R., Hopkins, D.C., Norton, J.A., 1995, "A Comprehensive Framework for Response Planning," *Proceedings, 4th U.S. Conference on Lifeline Earthquake Engineering*, ASCE, San Francisco, pp.573–579.

Chen, S., 1989, "On Countermeasure for Reducing Earthquake Disasters," *Forty Years of Research on Earthquake Resistance Engineering in China 1949–1989*, Seismological Press, pp.161–165. (in Chinese)

Eguchi, R.T., Chrostowski, J.D., Ayala, A.G., and Lee, G.M., 1990, "Early Post–Earthquake Damage Detection for Lifeline Systems," *Proceedings, 4th U.S. National Conference on Earthquake Engineering*, EERI, Palm Springs, California, Vol.1, pp.875–884.

Iwata, T., and Nakane, H., 1988, "Restoration Planning System for Earthquake Damaged City Gas Pipelines," *Proceeeings, 9th World Conference on Earthquake Engineering*, Tokyo–Kyoto, Vol.7, pp.171–176.

Iwata, T., and Saito, K., 1983, "Restoration Planning for Earthquake Damaged Gas Pipeline Networks," ASME, PVP–Vol.77.

Jin, G., Zhang, J., 1988, "Application of Prediction Method for Building Damage," *Earthquake Resistance and Disaster Prevention Countermeasures*, Henan Scientific and Technical Publishing House, Zhengzhou, pp.247–258. (in Chinese)

Kameda, H., 1994, "Multi–Phase Evaluation of Lifeline System Performance Under Earthquake Environment," *Proceedings, 2nd China–Japan–US Trilateral Symposium on Lifeline Earthquake Engineering*, Xi'an, pp.1–16.

Kameda, H., 1995, "Engineering Overview of the Disaster Caused by the Hyogoken–Nanbu (Kobe) Earthquake of January 17, 1995," *Paper presented at 4th U.S. Conference on Lifeline Earthquake Engineering*, San Francisco.

Kameda, H., Goto, H., and Kasuga, T., 1984, "System reliability and Serviceability of Water Supply Pipelines Under Seismic Environment," *Proceedings, 8th World Conference on Earthquake Engineering*, San Francisco, Vol.6, pp.491–498.

Kano, H., Uragami, Y., and Iwata, T., 1992, "Restoration Work Control System for Earthquake Damaged City Gas Pipelines," *Proceedings, 10th World Conference on Earthquake Engineering*, Madrid, pp.5535–5539.

Katayama, T., 1983, "Emergency Operation of Rehabilitation of Water Supply System After Earthquake," ASME, PVP–Vol.77, pp.389–396.

Lagorio, H.J., 1994, "Importance of Lifeline Systems in Rapid Urban Post–Earthquake Recovery Strategies," *Proceedings, 2nd China–Japan–US Trilateral Symposium on Lifeline Earthquake Engineering*, Xi'an, China, pp.295–292.

Lau, B., and Eidinger, J.M., 1995, "Water–Power Lifeline Interaction," *Proceedings, 4th U.S. Conference on Lifeline Earthquake Engineering*, San Francisco, pp.541–548.

Li, H., 1988, "Review of Prediction Method for Earthquake Disaster," *Earthquake Resistance and Disaster Prevention Countermeasures*, Henan Scientific and Technical Publishing House, Zhengzhou, pp.241–246. (in Chinese)

Liang, J.W., 1996, "A Decision Support System for Pre–Earthquake Planning of Lifeline Networks," *Proceedings, 1996 ASME Pressure Vessels and Piping Conference*, Montreal.

Liu, H., (Eds.) 1986, *"The Mammoth Tangshan Earthquake Damages,"* Vol.3, Seismological Press, Beijing. (in Chinese)

Liu, S.C., and Singh M.P., 1994, "Earthquake Hazard Mitigation and Infrastructure System Research: Interface and Integration," *Proceedings, 5th U.S. National Conference on Earthquake Engineering*, EERI, Chicago, Vol.4, pp.911–918.

Lund, L.V., 1995, "Lifeline Utilities Lessons, Northridge Earthquake," *Proceedings, 4th U.S. Conference on Lifeline Earthquake Engineering*, ASCE, San Francisco, pp.676–683.

Ma, X. (Ed.), 1989, *Lithospheric Dynamics Atlas of China*, State Seismological Bureau, ICL Publication No.125, China Cartographic Publishing House, Beijing. (in Chinese)

Nishio, N., 1994, "Damage Ratio Prediction for Buried Pipelines Based on the Deformability of Pipelines and the Nonuniformity of Ground," *Journal of Presure Vessel Technology*, ASME, Vol. 116, pp.459–466.

Schiff, A., (Ed.), 1991, *Guide to Post–Earthquake Investigation of Lifelines*, Monograph No.3, TCLEE, ASCE.

Schiff, A., Tognazzini, R., and Ostrom, D., 1995, "Power Systems after the Northridge Earthquake: Emergency Operations and Changes in Seismic Equipment Specifications, Practice, and System Configuration," *Proceedings, 4th U.S. Conference on Lifeline Earthquake Engineering*, ASCE, San Francisco, pp.549–556.

Shen, E., 1988, "Repair and Reconstruction of Water Supply System in Tangshan City," *Earthquake Resistance and Disaster Prevention Countermeasures*, Henan Scientific and Technical Publishing House, Zhengzhou, pp.539–543. (in Chinese)

Sun, S., 1990, "Earthquake Damages to Buried Pipelines in China," *Earthquake Resistance of Buried Pipelines, Academic Publisher, Beijing, pp.1–11. (in Chinese)*

Takada, S. Ogawa, Y., 1995, "Seismic Monitoring and Real Time Damage Estimation for Lifelines," Proceedings, 4th U.S. Conference on Lifeline Earthquake Engineering, ASCE, San Francisco, pp.224–231.

Takada, S., and Ueno, J., 1995, "Performance and Restoration of Lifeline Systems in the Great Hanshin Earthquake,"

Presented at 4th U.S. Conference on Lifeline Earthquake Engineering, ASCE / TCLEE, San Francisco.

Tianjin Earthquake Resistance Office, 1984, *Engineering Damages of Tianjin City in Tangshan Earthquake*, Tianjin Scientific and Technical Publishing House, Tianjin. (in Chinese)

Wang, K., and Yang, Y., 1989, "Damage Prediction Method for Buildings," *Forty Years of Research on Earthquake Resistance Engineering in China 1949–1989*, Seismological Press, pp.186–192. (in Chinese)

Xie, Y., Gao, J., and Xu, J., 1988, "Earthquake Resistance and Disaster Prevention Countermeasures of Tianjin City," *Earthquake Resistance and Disaster Prevention Countermeasures*, Henan Scientific and Technical Publishing House, Zhengzhou, pp.103–108. (in Chinese)

Yamazaki, F., Katayama, T., and Yoshikawa, Y., 1994, "On–Line Damage Assessment of City Gas Networks Based on Dense Earthquake Monitoring," *Proceedings, 5th U.S. Conference on Earthquake Engineering*, Chicago, Vol.4, pp.829–837.

Yoshikawa, Y., Kano, H., Yamazaki, F., Katayama, T., and Akasaka, N., 1995, "Development of SIGNAL: An Early Warning System of City Gas Networks," *Proceedings, 4th U.S. Conference on Lifeline Earthquake Engineering*, ASCE, San Francisco, pp.160–167.

Yang, Y., Wang, Z., Yang, Y., Li D., and Yang L., 1992, "An Assistant Intelligent Decision System for Prediction Earthquake Damage to Urban Existing Buildings," *Earthquake Engineering and Engineering Vibration*, Vol.12, No.1, pp.77–88. (in Chinese)

DEVELOPMENTS IN SEISMIC CODES AND STANDARDS

Introduction

L. H. Geraets
Tractebel Electricity & Gas International
Brussels, BELGIUM

and

M. L. Aggarwal
Ontario Hydro
Toronto, Ont., CANADA

Interpretation of code rules and development of seismic qualification methods is a very important area. With increasing demand on design optimization and designing for sites with more and more severe seismicity, a continuous effort is being made to understand and interpret code rules, and implement new standards both for new sites as well as for retrofit designs being assessed for fitness for service. This chapter contains eight papers related to current developments in seismic codes and standards, and their relevent applications.

Seismic margin approach is gaining acceptance and popularity, and at the same time has its own advantages and disadvantages. Anoba, Laur, Heiba and Barak present practical insights and lessons learned from implementing a seismic margin approach to an IPEEE relay evaluation.

Usmani and Baughman present recommendation for damping values and treatment of modelling uncertainities in seismic analysis of CANDU reactor and fuelling machine. It is noted that use of more realistic damping values is becoming popular in contrast to the early seismic engineering days of using overly conservative assumptions.

In past few years, snubber replacement and snubber reduction programs are being implemented extensively in the nuclear industry. Baltus and Billington discuss their experience on snubber replacement program at Vandellos II, and Baltus and Palusamy present their experience on seismic monitoring system replacement at Temelin plant.

Now when extensive data base is available on performance of structures and equipment under real earthquake loadings, experience based seismic qualification is another approach that is gaining popularity. T. Adams presets a strategy for implementation of experience based seismic equipment qualification in IEEE and ASME industry standards.

Yong Zhao and Wilkowski preset a case study for assessment of consistency between ASME section III and XI rules, providing an insight on maximum stress ratios for flaw acceptance.

With the new trend on global perspectives and increasing buisness at international level, industry is faced with the problem of correlation and acceptance of various standards. Barnes, Cooper and Landers present a study on failures in piping manufactured to different standards.

Lu Ming-Wan and Li Jian-Guo discuss the concept of primary structure to distingwish primary stresses and give an insight on the code rules and code philosophy.

The papers in this chapter thus cover a wide range of seismic qualification methods and implementation of code rules and standards. Such a continuing will bring a better understanding of the code rules and will contribute to achieving better and optimized designs at relatively lesser cost.

The efforts of the authors and the reviewers that contributed to the quality of these papers are appreciated.

PVP-Vol. 340, Seismic Engineering
ASME 1996

PRACTICAL INSIGHTS AND LESSONS LEARNED FROM IMPLEMENTING A SEISMIC MARGINS APPROACH TO AN IPEEE RELAY EVALUATION

Richard Anoba, Michele Laur, Mahmoud Heiba, and Denis Barak
Science Applications International Corporation (SAIC)
615 Oberlin Road
Suite 300
Raleigh, North Carolina 27605
PH: 919-832-7242
FAX: 919-832-7243

ABSTRACT

The Nuclear Regulatory Commission's (NRC) guidance, NUREG-1407, on performing Individual Plant Examinations for External Events (IPEEE) allows a Seismic Margins Assessment (SMA) to address the IPEEE seismic issue regarding the survivability of plant equipment during a seismic event and the capability of the plant to safely shutdown following the seismic event. The NRC guidance further allows the use of the NRC SMA approach or the Electric Power Research (EPRI) approach to evaluate these issues. In addition, the NRC's Unresolved Safety Issue A-46 (USI A-46) can also be efficiently addressed with the EPRI SMA evaluation. These methodologies are useful in determining a minimum list of equipment required for safe shutdown (i.e. Safe Shutdown Equipment Lists). The resultant "Safe Shutdown Equipment List", SSEL, is used to identify the electrically powered equipment required for safe shutdown for which a relay evaluation is required. The evaluation of relays associated with "safe shutdown equipment" is necessary since equipment can fail to operate properly due to mechanical failure or improper operation/actuation due to relay failure. SAIC has performed relay evaluations on five nuclear power plants. This paper is intended to summarize the practical insights and lessons learned from use of the EPRI SMA approach.

1.0 INTRODUCTION

The Nuclear Regulatory Commission's (NRC) guidance, NUREG-1407, on performing Individual Plant Examinations for External Events (IPEEE) allows a Seismic Margins Assessment (SMA) to address the IPEEE seismic issue. SAIC has implemented this approach to evaluate the relays associated with safe shutdown equipment at five nuclear plants. The paper describes the work performed to prepare and evaluate an Essential Relay List (ERL). Basic technical guidance for this effort was obtained from the Generic Implementation Procedure (GIP) (Ref. 5.1) and from the Electric Power Research Institute (EPRI) report EPRI NP-7148-SL (Refs. 5.2 and 5.5). Finally, this paper is intended to summarize the practical insights and lessons learned from use of the EPRI SMA approach

2.0 PROCEDURE FOR PREPARING THE A-46/IPEEE ESSENTIAL RELAY LIST (ERL)

2.1 Assumptions and Preliminary Steps

In preparation of an ERL, it was necessary to develop a comprehensive associated relay list (ARL), which is a list of relays associated with SSEL equipment control circuits. To form the ARL, a list of the electrically powered equipment appearing on the SSEL was first derived. Next, the applicability of the list of electrical control circuit schematics for this equipment was verified. The ARL was then developed to include all relays and other contact devices in the control paths (given in both primary and cascading schematics) for the equipment in the SSEL requiring relay evaluation. Included in the ARL are the normal and desired (i.e., the state desired for safe shutdown) states for the SSEL equipment requiring relay evaluation (this information was confirmed during the SSEL development).

As EPRI NP-7148-SL (Ref. 5.2) indicates, similar circuits can be functionally evaluated by category/type. Hence, the electrical control circuit drawings were sorted by system, then by circuit type, according to the following categories: Motor-operated valves (MOVs), pumps, solenoid-operated valves (SOVs), switchgear (SWGR), engineered safeguards actuation system (ESAS) circuits, and other special cases. Within each circuit category, the schematics were further categorized as either "primary" or "cascading." A schematic is termed "primary" if it represents the direct control circuit for the particular SSEL component. If the schematic does not indicate corresponding coils for one or more associated relays, the coil(s) may be shown on one or more additional schematics, which are referred to as "cascading."

Following the development of the ARL, a relay walkdown was required by the GIP (Ref. 5.1) and EPRI NP-7148-SL (Ref. 5.2) to gather site-specific information for the evaluation. In accordance with the relay evaluation methodology outlined in the GIP (Ref. 5.1) and in EPRI NP-7148-SL (Refs. 5.2 and 5.5), the following assumptions were made in performing the functional analysis on the SSEL equipment control circuits:

1. Relays/contact devices will be exposed to a 30-second earthquake.

2. Relays/contact devices will not be permanently damaged, with the exception of two specific models: the GE IJD (non IE) and the English Electric YCG, as listed in Appendix E of EPRI NP-7148-SL (Ref. 5.2).

3. "Chatter" is the inadvertent opening or closing of a contact with a sustained output of 2 milliseconds (Ref. 5.2).

4. Relay/contact device failure modes are a) contact chatter causes inadvertent and undesired equipment actuation and b) contact chatter causes failure of equipment to actuate as desired.

2.2 Description of the Functional Analysis

The functional analysis was performed on the SSEL equipment control circuits on a system-by-system basis, as recommended in EPRI NP-7148-SL (Refs. 5.2 and 5.5). Table 1-1 is a list of typical systems that contained and/or controlled SSEL equipment for a Pressurized Water Reactor (PWR).

Table 1-1. Systems Containing/Controlling SSEL Equipment

- Engineered Safeguards Actuation System (ESAS)
- Reactor Protection System (RPS)
- AC Power (ACP)
- DC Power (DCP)
- Service Water (SW)
- Emergency Feedwater (EFW)
- High Pressure Injection (HPI)
- Heating, Ventilation and Air Conditioning (HVAC)
- Low Pressure Injection (LPI)
- Reactor Building Cooling (RBC)
- Reactor Building Isolation System (RBIS)
- Reactor Coolant System (RCS)

This list would vary somewhat from plant to plant based on reactor type (i.e. Pressurized Water Reactors vs. Boiling Water Reactors) and on plant specific differences.

Following the development of the list, a functional analysis was performed on the control circuits for the SSEL equipment in each of these systems.

Initial functional analyses were performed on the circuits containing devices on the ARL to eliminate devices that can be easily identified as "non-essential." A device was considered "non-essential" if, based on its general role in the control circuit, its malfunction as a result of a seismic event could not potentially prevent the safe shutdown equipment from accomplishing its function (e.g. devices with contact pairs only in alarm circuits and indication circuits). This judgment was made without considering the operational requirements of the associated SSEL equipment. Thus, relays/other devices were functionally screened out if all of their contact-pairs control annunciators and/or alarm devices, or if they control equipment other than that listed in the SSEL, as these devices have no effect on the proper operation of SSEL control circuits. In addition, any mechanical or solid state devices were eliminated from further consideration at this stage, as such devices are defined to be "inherently rugged" by the GIP (Ref. 5.1).

Next, using the safe shutdown equipment state information, a more detailed functional analysis was used to determine if chatter of certain relay contacts in the control path was functionally acceptable (Refs. 5.1, 5.2, and 5.5). Based on the normal and desired states of the SSEL equipment being controlled by a given circuit, the specific function of each device within that circuit was examined. When this information was considered for a particular item of equipment, the device was evaluated to determine whether chatter would be acceptable for a given contact pair. If acceptable, that contact pair was eliminated from further evaluation. If all of the contacts on a given relay or contact device could be so-screened, that device was considered "non-essential."

2.3 Relay Screening using Operator Action

In accordance with EPRI NP-7148, relays whose malfunction can be corrected by operator action by resetting of the circuits were also eliminated from further evaluation. These relays were considered "non-essential," as long as the following conditions were met:

a) The operator is provided with specific indications of system status.

b) There is sufficient time for the operator to perform the necessary diagnostic evaluations and the necessary resetting actions.

c) There is immediate access to the indications and controls necessary to perform the operator action(s).

d) The number of systems to be reset is not excessive (i.e., the operator could realistically perform all of the proposed reset actions in a reasonable amount of time).

e) Detailed procedures are available to guide in diagnosing the situation and in carrying out the necessary steps to reset the subject system(s).

In addition, the impact of the valve misplacement/misposition was evaluated with regard to overall functioning of the system. If misplacement of the valve was determined to be acceptable, all relays associated with the particular valve were designated as non-essential. Conversely, any relay that could not be screened out using the above screening methods was marked "essential," requiring seismic evaluation.

2.4 Results of the Functional Analysis

A comprehensive ARL was developed which included relays and other contact devices in the control paths (given in both primary and cascading schematics) for the equipment in the SSEL requiring relay evaluation. After performing Functional Analyses on the SSEL control circuitry containing relays found on the ARL, it was determined that:

a) devices were determined to be inherently rugged (mechanical devices), or Not Vulnerable (*NV*) devices; or

b) relays and other devices were determined to be Chatter Acceptable *CA*, or that their malfunction in the case of the seismic event could not prevent a safe shutdown equipment from accomplishing its function; or

c) relays and other devices were determined to be covered by the "switchgear rule of the box," by which, in order to qualify these relays, it is possible to automatically consider most switchgear devices as essential and to evaluate only the seismic adequacy of the cabinets housing these devices; or

d) relays and other devices were considered to be essential to the operation of the SSEL equipment, and, therefore, need to be seismically adequate.

For those relays and other devices that were determined to be essential, a seismic adequacy analysis was performed.

3.0 SEISMIC ADEQUACY ANALYSIS

In addition to following the guidance in the GIP (Ref. 5.1) and EPRI NP-7148-SL (Refs. 5.2 and 5.5), the Seismic Adequacy Analysis used here was based on the Nuclear Regulatory Commission's (NRC's) Supplement No. 1 to Generic Letter (GL) 87-02 (Ref. 5.8), EPRI NP-7147-SL (for seismic capacity data) (Refs. 5.3, 5.6, and 5.7), and EPRI NP-5223-SL (also for seismic capacity data) (Ref. 5.4).

The resultant ERL contained relays and other devices with contacts that are essential to the control of required safe shutdown equipment. These devices generally fell into one of three main categories (Ref. 5.1). The first category of devices were auxiliary relays. These devices included electromechanical and pneumatic timing relays used for a variety of general control functions. Other devices on the ERL were protective electromechanical relays whose function was to protect equipment from system faults and other abnormal (e.g. over-current and under-voltage) conditions. The third main category of devices on the ERL were contactors.

A seismic Adequacy Analysis was performed on each of the devices discussed above, to the extent that applicable data were currently available (Refs. 5.3, 5.4, 5.6, and 5.7).

3.1 Preparation for the Seismic Capacity Screening

Before performing the Seismic Capacity Screening (SCS), it was necessary to conduct a "bad actor review," or an organized search for low ruggedness relays, as described in EPRI NP-7148-SL (Ref. 5.2). "Bad actor" relays that appear on the ERL are exempt from the SCS process that uses the Generic Equipment Ruggedness Spectra (GERS) provided in EPRI NP-7147-SL (Refs. 5.3, 5.6, and 5.7). Following the conduct of the "bad actor" relay search, a walkdown was performed on the remaining relays identified on the ERL.

This relay walkdown was required prior to performing the SCS (Refs. 5.1, 5.2 and 5.5). The objectives of the relay walkdown included:

1) Obtaining information needed to determine cabinet, panel and/or rack types which house or support essential relays,

2) Obtaining the information necessary to qualify the switchgear and other cabinets evaluated with the "switchgear rule of the box,"

3) Conducting spot check of essential relay mountings [(the capacity data in EPRI NP-7147-SL (Refs. 5.3, 5.6, and 5.7) and EPRI NP-5223-SL (Ref. 5.4) was applicable only for relays/other devices mounted, adjusted, calibrated, and maintained in accordance with manufacturer's instructions)],

4) Conducting spot check of essential relay types and locations (to verify the accuracy of plant data base information), and

5) Verifying the adequacy of the anchorage of the cabinets and other enclosures which house the essential relays.

Specific information from this walkdown was used to determine the seismic demand values for the essential relays/other devices. The seismic adequacy of the cabinets/other enclosures housing essential relays was reviewed during the walkdown. A "spot check" of relay mountings and types was also performed. Finally, a selection of control room panels as well as local diesel generator panels were "spot-checked" for proper anchorage.

3.2 Determination of the Seismic Capacity

According to the GIP (Ref. 5.1), the seismic capacity of relays/other devices can be established using three different methods: 1) generic seismic test data, 2) earthquake experience data, and 3) specific seismic test data. The first option amounts to using the available test data on a variety of devices and equipment, which have been reduced to generic equipment ruggedness spectra (GERS) in the complementary reports EPRI NP-7147-SL (Refs. 5.3, 5.6, and 5.7) and EPRI NP-5223-SL (Ref. 5.4). The second method searches for seismically unacceptable relay types, or "bad actors," (Ref. 5.2), identified as such as a result of the data obtained on relay performance, specific failures, vulnerabilities, and from actual earthquake experience in power plants and other facilities. Finally, the third option is the use of plant-specific and relay-specific test data, generally maintained by the plant and/or relay suppliers. All three methods were used during the various projects when appropriate data were available. The methods used were directly dependent on the type of available data. Limitations in the magnitude and type/category of data provided the analyst with one of the important lessons learned during these analyses.

The application of GERS data is the most involved determination of seismic capacity and, hence, warrants some additional discussion. First of all, it is necessary to verify that it is permissible to apply this data to the devices on the ERL. In order for the GERS data to be acceptable, it is necessary that the essential relays/other devices be properly mounted and the essential relay/other device types and locations listed in the plant data base be correct. It is also required that the anchorage of cabinets and other enclosures housing these devices generally meet applicable acceptance standards (Refs. 5.3 and 5.4). Finally, a number of various caveats (in the form of a checklist for each relay GERS and notes on many of the individual model capacities) must be met in order for the relay GERS (Refs. 5.3, 5.4, 5.6. and 5.7) to apply. Certain of these caveats are considered confirmed by the relay walkdown "spot check," while others should be verified from the drawings or from plant records. In all cases where these criteria are met, it is considered permissible to use the GERS sources to determine the seismic capacity for each of the ERL devices covered by these sources.

By definition, a GERS is the spectral acceleration capacity (in g's) defined at the relay/other device mounting point; above this acceleration level, the device cannot be expected to function without chatter or damage. Chatter in excess of two milliseconds is considered a "failure." The GERS capacities are given in terms of 1) the Peak Spectral Acceleration (PSA) capacity occurring between frequencies of 4 to 16Hz and 2) the Zero Peak Acceleration (ZPA) value, used for frequencies higher than 33Hz. For relays, different spectra are provided with varying degrees of availability for different operating modes (i.e. operate/non-operate) and normally open or closed contacts, for which there can be significant differences in seismic ruggedness. EPRI NP-7147-SL (Refs. 5.3, 5.6, and 5.7) provides a large number of GERS for auxiliary and protective relays. Each of these GERS applies to a specific subclass of relays for certain years of manufacture (all prior to 1985). EPRI NP-5223-SL (Ref. 5.4) contains the MCC contactors GERS.

3.3 Determination of the Seismic Demand

For devices on the ERL that are not covered by the switchgear "rule of the box," EPRI NP-7148-SL (Refs. 5.2 and 5.5) and the GIP (Ref. 5.1) describe SCS as a four-level process. In "Level 1," a relay located below 40 feet above the grade of the plant can be screened out as sufficiently rugged if it is contained in a low- (MCC-type) or high-amplification (switchgear-type) cabinet and its GERS capacity value (or relay-specific test value) is greater than 5 or 8g, respectively.

"Level 2" estimates the seismic demand for a relay based on an In-Structure Response Spectrum (ISRS), multiplied by an appropriate Safety Factor (SF), and then multiplied by an in-cabinet Amplification Factor (AF). The ISRS can be obtained by multiplying the ground response spectrum by a factor of 1.5, where applicable, or they can be taken as the ISRS actually determined from the building's dynamic analysis. These "level 2" demands were used in the phased SCSs approach described in Section 3.44 of the GIP (Ref. 5.1). A relay was seismically screened out if its GERS capacity equals or exceeds the conservatively determined "Level 2" demand in both the low frequency range (4Hz to 16Hz) and in the high frequency range (33Hz and above) (Refs. 5.2 and 5.3).

If a relay did not pass the SCS using "Level 2" demands, Level 3 and 4 criteria were applied in the evaluation of the remaining relays.

3.4 Seismic Capacity Screening of the Essential Relay List

According to the guidelines for the SCS presented in EPRI NP-7148-SL (Refs. 5.2 and 5.5) and supplemental guidance from the NRC (Ref. 5.8), relays/other devices that are not "bad actors" (Section 3.1) are considered "acceptably rugged" if two conditions are met. First, the peak seismic demand spectrum must be bounded by the peak seismic capacity spectrum in the frequency range of 4 to 16Hz and second, for high frequencies, the capacity ZPA must exceed the demand ZPA. As discussed in Section 3.2, to determine the applicable GERS capacity value from EPRI NP-7147-SL (Refs. 5.3, 5.6, and 5.7), it is necessary to know the operating mode (i.e. operate/non-operate and contact position) of that relay during the earthquake (i.e. after offsite power is lost). For the first pass, the overall worst-case capacities were compared to the relay demands (the first time using the conservatively determined demands), with no consideration of operating mode. A relay may fail this first pass due to an inadequate PSA or an inadequate ZPA value, which could potentially be "upgraded" during the second pass. In the second pass, it was necessary to determine whether each relay would be in the operate or the non-operate mode (i.e. energized or deenergized for auxiliary relays and actuated or non-actuated for protective relays) upon loss of offsite power (Ref. 5.3). The capacity for each remaining relay was then upgraded to a higher value, provided the "worst-case" value for the contact positions under "operate" or "non-operate" was higher than the capacity used in the previous pass. Finally, the third and most specific screening pass required the knowledge of the normal positions of the unscreened relays' contacts and the normal operating state of the equipment in question. This entailed examining the control circuit drawings for the specific contacts, and determining if in the equipment's normal operating state these contacts are normally closed or normally open. Upon completion of this screening process, those relays that failed the evaluation were categorized as "outlier" relays requiring some form of resolution.

4.0 INSIGHTS AND LESSONS LEARNED

The lessons learned from conducting the IPEEE relay evaluation projects fall into two categories. First, the functional screening process was far more labor intensive than previously anticipated. Second, the seismic capacity screening process was less successful than expected. Both of these "problems" can be directly attributed to the complexity of issue and to the lack of available data on relays and associated devices in use at nuclear power plant facilities.

The problems associated with conducting the functional analysis can be directly attributed to the quality of the data sources maintained at facilities and to the depth of those sources. To conduct a functional screening of electric circuits for this analysis, it was necessary to review many primary and cascading electrical schematics. The quality of the individual schematics varied greatly. Drawings were on some occasions unreadable at best even after repeated attempts to produce a more readable schematic. This made it difficult to not only understand the circuit shown on the drawing of poor quality but it also made the task of tracing the circuit to additional drawings more difficult and time consuming. In addition, it was difficult to locate some drawings either because the cataloging and maintenance of the documents required procurement of the documents from several different locations or documentation of the location/source of the drawings did not exist. For example, some drawings were buried in technical manuals which are maintained in locations distant from the location of the majority of the electrical schematics. In addition, poor documentation of the relay operating states (normally closed vs. normally open and energized vs. deenergized) added to the time required for the analysis. Finally, some of the "architect/vendor" schematics took a form that did not lend itself to a functional analysis. When combined, these difficulties exponentially increased the required resources to conduct an accurate and complete functional analysis of the circuits associated with relays on the ERL.

With regard to the seismic capacity screening efforts, the process was less than successful and more labor intensive because of a lack of data on many relays currently in use in these facilities. By less than successful, it is meant that a number of relays could not be eliminated or categorized as an "outlier relay" because of a lack of seismic capacity data. In an effort to completely evaluate these relays, more exhaustive searches for seismic capacity information were conducted from a variety of sources. The lack of data for these "relays with unknown capacity" may be partly due to the fact that some of these relays are newer models than the relays addressed in the EPRI studies identified as references in this paper. In which case, they have probably been constructed to a higher standard than their older counterparts, presumably posing less of a threat to the safe shutdown of these facilities. The poor documentation of the relay states also impacted the seismic capacity screening process in that the seismic capacity is directly dependent on whether a relay is opened or closed, and energized or deenergized. In any event, the lack of information escalated the amount of resources required to accurately perform the analysis and in the end resulted in the categorization of a number of ERL devices as "relays with unknown capacity."

In conclusion, except for the devices identified as "relays with unknown capacity," the IPEEE relay evaluations were successfully concluded. However, the cost in manpower was greater than anticipated due to the degraded maintenance of the documents required to conduct the analysis and the lack of seismic capacity data.

5.0 REFERENCES

5.1 EPRI NP-5223-SL, Generic Seismic Ruggedness of Equipment, August 1991.

5.2 EPRI NP-7147-SL, Seismic Ruggedness of Relays, August 1991.

5.3 EPRI NP-7147-SL-Volume 2, Seismic Ruggedness of Relays, September 1993.

5.4 EPRI NP-7147-SL-V2-Addendum 2, Seismic Ruggedness of Relays, April 1995.

5.5 EPRI NP-7148-SL, Procedure for Evaluating Nuclear Power Plant Relay Seismic Functionality, December 1990.

5.6 EPRI NP-7148-SL-Volume 2, Procedure for Evaluating Nuclear Power Plant Relay Seismic Functionality, September 1993.

5.7 Generic Implementation Procedure (GIP) for Seismic Verification of Nuclear Plant Equipment, R/2 Corrected, February 1992.

5.8 Supplement No. 1 To Generic Letter (GL) 87-02 That Transmits Supplemental Safety Evaluation Report No. 2 (SEER No. 2) On SLUG Generic Implementation Procedure, Revision 2, As Corrected On February 14, 1992 (GIP-2), NRC, May 22, 1992.

PVP-Vol. 340, Seismic Engineering
ASME 1996

RECOMMENDATIONS FOR DAMPING AND TREATMENT OF MODELING UNCERTAINTY IN SEISMIC ANALYSIS OF CANDU NUCLEAR POWER PLANT

S. A. Usmani
Atomic Energy of Canada, Ltd.
Mississauga, Ontario
Canada

P.D. Baughman
EQE International
Stratham, N.H.

ABSTRACT

The seismic analysis of the CANDU nuclear power plant is governed by Canadian Standard series N289. However, the dynamic analysis of some equipment and systems such as the CANDU reactor and fuelling machine must treat unique components not directly covered by the broad recommendations of these standards. This paper looks at the damping values and treatment of modeling uncertainty recommended by CSA N289.3, the current state of knowledge and expert opinion as reflected in several current standards, testing results, and the unique aspects of the CANDU system. Damping values are recommended for the component parts of the CANDU reactor and fuelling machine system: reactor building, calandria vault, calandria, fuel channel, pressure tube, fuelling machine and support structure. Recommendations for treatment of modeling and other uncertainties are also presented.

INTRODUCTION

The reference CANDU 6 nuclear power plant is currently conservatively qualified for a design basis earthquake with peak ground acceleration of 0.2g. AECL has been evaluating options for placing the standard reference design at higher seismic sites. Part of the assessment involves recommending updated seismic analysis assumptions and inputs with respect to damping and treatment of uncertainty.

Several seismic analysis standards were consulted for input to the recommendations including CSA N289.3 [1], USNRC Standard Review Plan [2] and Regulatory Guides [3,4,5,6], ASCE Standard 4 [7], NUREG/CR-0098 [8] and EPRI NP-6041 [9].

DAMPING

Damping is specified in a model to take into account the various energy dissipation mechanisms in a structure. These normally occur due to nonlinearities in the structure (gaps, friction, etc.) or inelastic cycling. Taking these effects into account as viscous damping is a gross approximation. Therefore, damping is specified conservatively.

Damping is specified as a percent of critical damping for a single degree of freedom system and increases with the level of stress and distortion in the structure. The damping factors can be incorporated in a modal analysis in a straightforward manner if the damping is constant throughout the structure. For structures made up of parts having different dampings, methods have been developed to assign modal damping factors based on relative participation of the different parts in each mode. Seismic analysis standards usually recommend using strain energy as the weighting factor, although other methods are also acceptable. Since the specified damping factors are conservative, the weighting method will not produce an unconservative result.

Table 1 gives modal damping factors from the reference CANDU 6 documents and various standards and research papers. Where a range of dampings is given, the lower range is to be used for structures and equipment which are lightly stressed; i.e., all elements are stressed less than about one-half yield. The upper range is for structures and equipment with elements which are stressed greater than about one-half yield.

Note that ASME Code Case N-411 allows higher damping (2 to 5% depending on frequency) for piping. This has been accepted by USNRC provided the design basis ground response spectrum has an 84% nonexceedence probability shape. The CSA N289.3 ground response spectrum is a 90% nonexceedence probability shape. EPRI NP-6041 recommends 5% damping for piping regardless of frequency. Hadjian and Tang [11] recommend higher damping values for piping over 150 mm diameter, but Malcher, et. al., [14] recommend 4% for all sizes and stress levels.

The dampings specified in the CSA standard are seen to be lower than dampings specified in the US standards. Prior to the issuance of Regulatory Guide 1.61, plants in the US were designed using dampings which were lower than the Regulatory Guide 1.61 dampings. They also used design ground response spectra which were based on Housner's recommendations. Housner's recommended spectrum shape was derived from the average spectral ordinates of several recorded earthquakes. Later, Newmark, Blume and Kapor provided a recommended spectrum shape based on the 84th percentile spectral ordinates of a larger group of earthquake records. The USNRC used this as the basis for Regulatory Guide 1.60. It issued Regulatory Guide

1.61 at the same time. Regulatory Guide 1.61 specified higher dampings than were previously being used because the use of the Regulatory Guide 1.60 spectra introduced more conservative input (84% nonexceedence probability spectral shape) than the Housner spectra. Therefore, use of more realistic (median) damping values were justified. ASCE 4 has carried this forward with clarification. It differentiates between a seismic analysis of a structure for its own design and a seismic analysis to generate floor response spectra for subsystem analysis. For design, it recommends using the high stress damping values. For spectra generation, it recommends using damping values corresponding with the actual stress levels in the structure due to the earthquake level which is being analyzed for.

Newmark and Hall in NUREG/CR-0098 recommended ground response spectra for use in analysis of nuclear plant structures. These have since become the standard for use in the U.S.A. and elsewhere for evaluating existing nuclear power plant structures and equipment, and have been accepted by the USNRC in many applications, such as seismic margin analysis. Newmark and Hall also recommended dampings. For structures, the damping levels are higher than Regulatory Guide 1.61 or ASCE 4. For piping they are the same as Regulatory Guide 1.61; however, this was before ASME Code Case N-411. EPRI NP-6041 is the current standard for evaluating existing nuclear power plants. It recommends damping similar to those of NUREG/CR-0098 (it also uses the NUREG/CR-0098 84% spectral shape) for structures, and 5% damping for piping.

Hashimoto, et. al., [12] examined data on structural damping and concluded that Regulatory Guide 1.61 values were conservative except for bolted steel structures for which it recommended a damping of 6% rather than 7% for a structure stressed above 1/2 yield. It also indicated that friction bolted structures should be considered the same as welded structures. Kennedy, et. al., [13] recommends median damping values for structures using data from several references including [8] and [12]. The upper level of these values are higher than Regulatory Guide 1.61 and ASCE 4; they reflect the structure stressed beyond or just below yield. Kennedy also recommends -1σ values, which are seen to be the same as the Regulatory Guide 1.61 values for SSE. Kennedy also recommends median values for piping (5%), equipment (5%) and cable trays (15%).

The CSA Standard and the CANDU 6 reference plant documents specify that the design basis earthquake (DBE) loads be considered as ASME Service Level C per the Canadian practice. The current ASME Service Level C stress limits for pressure vessels are 1.2Sm or 1.0Sy for primary membrane, and 1.8Sm or 1.5Sy for primary membrane plus bending; for piping the stress limits are higher. Thus, components may be stressed beyond yield and certainly are stressed beyond the zero to one-half yield regime.

It is clear that the damping values of CSA N289.3 can be justifiably increased, for the following reasons:

- Elements of the model are being stressed beyond the 1/2 yield threshold, and use of damping values corresponding to low stress levels is not warranted.
- The specified ground response spectrum shape is a 90% nonexceedence probability shape in which case the median values of damping are appropriate.
- The consensus of expert opinion, as discussed above, is that median damping values are higher than those specified in CSA N289.3.

The following generic damping values, which are in the range of median values of Table 1, are presented for consideration as being justified:

<u>Structure or Component</u>	<u>(1)</u>	<u>(2)</u>
Pressure vessel or piping	3	5
Welded or friction bolted steel structures	3	5
Rigid components	2	4
Bearing bolted steel structures	4	6
Prestressed concrete structures	3	5
Reinforced concrete structures	4	7

The values in column (1) would be used if the structure or system stresses do not exceed about one-half yield. The values in column (2) would be used if portions of the structure or system are stressed greater than one-half yield. The higher values would be used for design of the structure or system for DBE, while for generation of floor response spectra the values corresponding to the actual stresses should be used. For combined structure-system models, such as a combined reactor building-reactor structure-HT system model, the actual stresses in the structure should be considered in choosing structure damping values. The damping values for soil and rock appear to be reasonable, median values and are not recommended to be changed.

The analysis model of the fuel channel and fuelling machine (Figure 1) was reviewed in order to recommend damping values for use in the analysis for these specific components. The reference plant seismic analysis of the fuel channel and fuelling machine is a time history, direct integration analysis using a partly non-linear model. The nonlinear components are three bilinear springs: the pitch and yaw controls on the fuelling machine, which are quite soft, and the fuelling machine z-direction restraint, which is initially soft and then becomes abruptly quite stiff when the stop is engaged. The model is assembled from several linear submodels: a representative simplified model of the calandria vault, calandria vessel and end shields; a model of the fuel channel; and a model of the fuelling machine and bridge , columns and supports. There are a number of cases analyzed in which the model characteristics are changed to envelope all possible configurations: several fuel channel positions, the fuelling machine disengaged, etc. Using the conventional assumptions and inputs the seismic analysis yields high seismic loads. The most significant one is the seismic loading of the fuel channel from the connected fuelling machine which gives high stresses in the fuel channel component. There are also calculated highly stressed areas on the bridge and columns.

Consistent with the discussion above ,the following damping values are presented for consideration as being justified for the specific components in the seismic analysis:

Fuelling machine columns	4%
Fuelling machine bridge	6%
Fuelling machine	5%
Z-motion spring	7%
Pitch spring:	6%
Yaw spring	15%
Carriage	6%
Cradle	6%
Fuel channel	6%
Calandria	5%
Vault	7%

These damping values are derived from the recommended generic values recommended earlier as explained below.

The fuelling machine columns (Figure 2) are very stiff steel columns which are rigidly braced to the reactor building structure. The damping for rigid components would be appropriate. Since the columns are much stiffer than the bridge, they will not contribute significantly to the fuel channel/fuelling machine response so the damping being higher or lower will not affect the response.

The fuelling machine bridge (Figure 2) is a stiff welded frame structure. However, the elevators, ball screws and roundways all have small gaps which are not modeled. These will dissipate energy during seismic motion. Therefore, the bridge will behave more as a bearing bolted steel structure than as a rigid welded steel structure, and 6% damping is appropriate. The bridge response could be significant because of the flexibility at the elevators.

The fuelling machine itself (Figure 3) is modeled in several parts. The snout, magazine, ram housing and gear box are modeled with beam elements. These components are akin to pipes, and piping damping (5%) is therefore considered appropriate. The fuelling machine axial motion is controlled by the z-direction spring. The axial movement would cause flexing of the catenary system z-motion loop, which will be highly damped similar to cables in cable trays. It is noted that cable tray systems exhibit damping of 15% or higher [7,15] due to rubbing of the cables. Because of this, It is judged appropriate to specify at least 7% damping for the Z-direction spring (note that this will be significant for the uncoupled case). Rotation in the lateral (x and y) directions, when the machine is uncoupled from the fuel channel, is controlled by the pitch and yaw springs. The pitch and yaw spring dampings were determined experimentally and should be as currently specified.

The carriage and cradle assemblies (Figures 3) are very stiff and would not be highly damped. However, the drive unit assembly, latch assembly, gimbals, etc. will dissipate energy by friction during strong motion shaking, and a higher damping is warranted. It is judged that the damping for a bearing bolted structure (6%) is appropriately justified.

The fuel channel (Figure 4) is again similar to a pipe section and piping damping (5%) would be appropriate. The channels are surrounded by water which will add damping. ASCE 4 recommends that added damping due to submergence be limited to 0.5% unless detailed analysis or testing is done to determine higher values. The feeder pipes (Figure 9, Item 11), which are attached to the ends of the channels outside of the calandria end plates but are not included in the model, have been determined experimentally to have 8% or higher damping. During a strong motion earthquake these highly damped feeder arrays will add damping to the connected fuel channels. Inside the fuel channel, any slight motion of the numerous (twelve per channel) fuel bundles against each other and against the retaining fuel channel, will dissipate dynamic energy and thus adding some damping. The relative motion of channels in its bearing, friction between spacers and pressure tube and spacer and calendria tube will also dissipate some energy. Considering these extra energy dissipation mechanisms it is judged appropriate and conservative to add 1% to the generic value for piping. Therefore, the recommended damping for the fuel channel is 6% (5% piping damping plus 1% for submergence and additional dissipation mechanisms discussed). Note that this will most affect the lateral response; the axial response is mostly controlled by the calandria end shields.

The calandria (Figures 5 and 6) is a steel pressure vessel which is quite large in diameter and relatively (in comparison to PWR Reactor Vessel) thin wall. Additionally, the calandria contains heavy water fluid and is also submerged in the reactor -vault water. It also has numerous pipes and pipelike items as attachments or penetrants. Application of the conventional damping values which are even lower than specified for a thick and rigid PWR reactor vessel is overly conservative. A more realistic damping of 6% is recommended. The end shields are not submerged; however, they move as rigid bodies, so specifying the same damping will not affect the system response. It is noted that the end shields are filled with steel balls which could potentially dissipate energy caused by relative motion. Experiments on a reactor core filled with pebbles or balls [15] showed a wide spread in various damping behavior. The review of the reference plant analysis showed the end shields mostly move as rigid bodies so the steel balls would not add significant damping, while in [15] the core structure containing the balls was flexing and had a significant contribution However, the relative movement of the end shields and numerous channels in the free end bearings and effect from numerous feeders (which are akin to dampers) at the fixed end would result in energy dissipation adding to the damping. This effect cannot be quantified except experimentally, but it makes the addition of 1% over piping damping, further justified.

The vault (Figure 6) is a reinforced concrete structure so damping is specified at 7%. The walls of the vault are very thick for shielding. If the stress levels in the concrete is determined to be very low, a lower damping should be used. However, if this is so then the vault will behave as a rigid element, and the damping actually specified will have little effect on the response of the fuelling machine or fuel channel.

TREATMENT OF UNCERTAINTY

Variation of Soil Properties

CSA N289.3, Section 4.2.7 specifies that for evaluating soil-structure interaction effects, the averaged measured dynamic soil properties shall be varied by ±50% (a factor of 1.5 and 0.5) unless a smaller variation can be justified. The reference plant analysis justified and used a variation of ±33%.

The USNRC criteria specifies that the soil properties be varied by a factor of 2.0 for the upper bound and 0.5 for the lower bound, unless otherwise justified by site data. ASCE 4 specifies a factor, Cv, to account for uncertainties in soil-structure interaction (SSI) analysis and soil properties. The soil shear modulus is multiplied by 1+Cv to obtain the upper bound, and is divided by 1+Cv to obtain the lower bound. Cv is determined statistically to cover the mean plus or minus one standard deviation, but is not to be less than 0.5. If insufficient data are available to address uncertainties in soil properties, Cv is taken as unity. This means that the upper bound shear modulus is twice the average, while the lower bound is one half the average.

EPRI NP-6041, Appendix E discusses that uncertainty in modeling the soil/rock at the site consists of two parts: (1) the soil configuration (layering or stratiography) and (2) the dynamic stress-strain behavior. Citing various studies, it concludes that dynamic stress-strain uncertainty could range from 0.5 to 1.0, and considering only uncertainty in shear modulus itself and the equivalent linear estimates to be used in SSI analysis, a coefficient of variation of 0.35 to 0.50 is warranted. It should be noted that for a rock site, the soil configuration uncertainty is small, and uncertainties in stress-strain behavior other than the shear modulus itself would be small. Therefore, for a rock site a coefficient of variation, Cv, of 0.5 could be used, making the upper and lower bounds +50%, -33%.

ASCE 4, Section 3.3.1.7 addresses uncertainties in the soil-structure interaction analysis by either a probabilistic analysis (discussed at length in the commentary) or by varying the soil shear modulus between the best estimate value times (1+Cv) and the best estimate value divided by (1+Cv), with the minimum value of Cv being 0.5.

The above requirements were developed in the mid-1970s and early 1980s as analytical treatment of soil-structure interaction developed. The focus was on soft soils where the soil response dominated the overall structural response. Prior to this, buildings on rock or hard soil were modeled as fixed base, while buildings on medium to soft soils were modeled with soil springs. The soil spring constants were calculated using elastic half-space theory. Only best estimate values were used. Parameter variation was accounted for by broadening the peaks of in-structure response spectra by 10% for rock sites and 15% for soil sites.

Peak Broadening

The USNRC peak broadening criteria [5] is intended to account for uncertainties in the structural frequencies owing to uncertainties in such parameters as the material properties of the structure and soil, damping values, soil-structure interaction techniques and the approximations in the modeling techniques used in the seismic analysis. The computed floor response spectra from the floor time history motions are smoothed and the peaks associated with each of the structural frequencies broadened by a frequency, ±Δfj, where:

$$\Delta fj = \left[(0.05fj)^2 + \sum_{n=1}^{P}(\Delta fjn)^2\right]^{1/2} \geq 0.1fj.$$

Δfjn denotes the variations in the jth mode frequency, fj, due to variation in parameter n and P is the number of significant parameters considered. A value of 0.1fj should be used if the computed value of Δfj is less than 0.1fj. If the above procedure is not used, Δfj should be taken as 0.15fj.

The USNRC regulatory position does not state explicitly what parameters would be included or how one determines the variation. This is left to the analyst. Presumably, one would examine the data on the various parameters, find the mean and standard deviation, then use the range between plus and minus one standard deviation to compute Δfjn. In practice, for a model with lumped parameter soil springs the seismic analysis is conducted with best estimate properties, including soil properties, and Δfj is taken as 0.15fj. When a finite element soil-structure interaction analysis is conducted, the three soil cases corresponding to mean, upper bound and lower bound are enveloped and the result broadened by ±0.15fj. In a number of cases the mean was broadened by ±0.15fj, while the upper and lower bound soil case results were broadened by ±0.10fj.

ASCE 4 gives the following criteria. It recommends using a peak spreading criterion of ±0.15fj when the seismic analysis uses median soil properties. This is primarily because the maximum permissible reduction in peak spreading is to ±0.10fj, which is not significant enough to warrant the extra effort of parameter variation (the effort is better spent on reducing other, more significant conservatisms.) Where soil parameters, by themselves, may produce variation greater than ±0.15fj in the frequency range associated with soil-structure interaction effects, the recommended procedure is to analyze the three soil cases but apply no further broadening to account for uncertainties in other parameters such as structural modeling or material properties. This is because, for those frequencies which are substantially affected by soil parameter uncertainties, the effects of other parameters variations are, in comparison, small.

Further, ASCE 4 recommends that in conjunction with peak broadening, the peak amplitude be reduced by 15% provided that the damping specified for the subsystem analysis is less than 10%. The reason is that the broadening criteria only consider the effects of uncertainties with support structure response characteristics which produce increased subsystem responses. However, the same uncertainties which lead to broadening of in-structure response spectrum peaks also lead to reduction in the peak spectral amplitudes with a given probability of exceedence. ASCE 4 cites studies which compare equal probability of exceedence spectra with deterministic in-structure spectra and show that, for 2% damping, the deterministic spectrum peaks may be more than a factor of two greater than the equal probability of exceedence spectrum peaks.

ASCE 4 states that the studies which have been performed comparing equal probability of exceedence in-structure response spectra with deterministic in-structure spectra reveal that the equal probability of exceedence spectra have much broader spectrum peaks and much lower peak amplitudes than do the deterministic spectra. Therefore, if deterministic peaks are broadened to account for the effects of uncertainties, considerable conservatism is introduced within this broadened peak region unless a corresponding reduction in peak amplitudes are allowed. A reduction of spectrum peak amplitudes such that the probability of nonexceedence for the resulting spectra is not less than 90% represent a rational seismic design basis for subsystem design. In lieu of performing a probabilistic evaluation, a 15% reduction in peak amplitude of deterministic spectra is deemed reasonable and conservative.

ASCE 4 cautions that the 15% reduction is applicable for in-structure floor response spectra with low damping. Generally, the level of conservatism is increased at very low damping levels (like 0.5% damping), and decreases for higher damping levels. In cases where very high damping levels are adopted (e.g., 15% as for cable trays) the analyst should confirm that the 15% margin to the 90% nonexceedence probability is maintained.

For time history analysis of subsystems, it is also necessary to consider the effects of uncertainties. This is done by varying the time interval of the time history on the assumption that the frequency content of the ground time history has been filtered so that frequencies close to the structure frequencies dominate the in-structure response. The time interval variation is similar to varying the structure frequency (or period) and the overall effect is similar to frequency shifting of floor response spectra as opposed to peak broadening. It is recommended that only the median soil properties be used in the reactor building analysis with the soil spring model. The subsystem analysis should consider the resulting time history at the original digitization time interval and at a 15% longer and 15% shorter time interval. Because this does not exactly correspond to peak broadening an amplitude reduction is probably not justified.

Recommendations

Based on the above discussion , the following is proposed as recommendation for consideration :

- Re. CSA N 289.3 Section 4.2.7, Dynamic Soil Properties, it is appropriate to require property variation only for soft to medium stiffness soils. A cutoff shear wave velocity could be established above which only the median or best estimate properties would be used.
- Re. CSA N289.3 Section 5.5.2.2, Design Floor Response Spectra Input, consistent with ASCE 4-86, Section 3.4.2.3 and commentary, it is appropriate to clarify that the variation in structural frequency includes both the soil and structure, that minimum broadening is 15% unless justified, that if the soil variation produces more than 15% variation no additional broadening is done and if soil variation produces less than 15% variation the broadening is increased to 15%, and that in conjunction with broadening the peak amplitudes may be reduced by an equal percentage or to the 90% nonexceedence probability level.
- Re. CSA N289.3 Section 5.5.2.1, Time-History Input, it is appropriate to consider that the time scale variation applies only to the mean soil case.

The subgrade material uncertainty is not a dominating factor in the overall uncertainty. A CANDU reactor building analysis that varied the stiffness of the soil springs by ±50% saw approximately 10% variation in the building frequency. Therefore, it would be appropriate to use only the best estimate soil properties and spreading the peaks by ±15%. Further, it is justifiable to also reduce the peak amplitudes by 15% for equipment dampings up to 10%. It would be prudent to incorporate above recommendations in CANDU practice to be consistent with ASCE 4.

For a rock site, we would not expect a significant reduction in response to result from a finite element soil-structure interaction analysis. Therefore, performing the analysis with a soil-spring model along with the CSA specified damping for the soil spring may be adequate. In this case there is generally no reason to do a finite element analysis unless it could be shown to be beneficial [16] in specific cases . Therefore a case study should be used to determine application of appropriate methodology.

It is noted that the lower range of the proposed shear wave velocities, 600 m/sec is below the rock range and could be characterized as a stiff soil. A SASSI soil-structure interaction could result in a significant reduction in the peak of in-structure response spectra. On a US nuclear power plant with a shear wave velocity of 731m/sec (2400 ft/sec), significant reductions were obtained compared to the original analysis which used lumped parameter soil springs.

For treatment of uncertainty, it is recommended that only the best estimate rock properties be used in the model. The resulting floor response spectra should be broadened by ±15%. Furthermore, it will be appropriate to recommend reducing the peak amplitudes by 15% for equipment dampings up to 10%. For time history analysis, likewise it is recommend to use only one soil case and vary the time interval of the resulting time history by ±15%. However, any reduction in amplitude should be justified by means such as a probabilistic analysis.

ACKNOWLEDGMENT

The contribution from various AECL and EQE experts to the study on which this paper is based is duly acknowledged. In particular assistance and input received from M. Elgohary, R. Ricciuti, L. Lee, A.S. Banwatt, S.A. Soliman, B. Ly, T.S. Aziz and J.K. Biswas of AECL is greatly appreciated.

REFERENCES

1. Canadian Standards Association, Design Procedures for Seismic Qualification of CANDU Nuclear Power Plants, CAN-N289.3-M81, September, 1981.

2. US Nuclear Regulatory Commission, Standard Review Plan, NUREG-0800, Sections 3.7.1 and 3.7.2, Revision 2, 1989.

3. USNRC, Design Response Spectra for Seismic Design of Nuclear Power Plants, Regulatory Guide 1.60, Revision 1, December 1973.

4. USNRC, Damping Values for Seismic Design of Nuclear Power Plants, Regulatory Guide 1.61, October 1973.

5. USNRC, Development of Floor Design Response Spectra for Seismic Design of Floor-Supported Equipment or Components, Regulatory Guide 1.122, Revision 1, February 1978.

6. USNRC, Combining Modal Responses and Spatial Components in Seismic Response Analysis, Regulatory Guide 1.92, Revision 1, February, 1976.

7. American Society of Civil Engineers, Seismic Analysis of Safety Related Nuclear Structures and Commentary, ASCE Standard 4-86 (and Draft Revision, November 1994).

8. Newmark, N.M., and Hall, W.J., Development of Criteria for Seismic Review of Selected Nuclear Power Plants, NUREG/CR-0098, May 1978.

9. Campbell, R.D., et. al., A Methodology for Assessment of Nuclear Power Plant Seismic Margin, EPRI NP-6041, October 1988.

10. WSN PSAR Chapter 3, "Design of Structures and Systems," 1992.

11. Hadjian, A.H., and Tang, H.T., "Light Water Reactor Piping System Damping," PVP-Vol. 197, 1990.

12. Hashimoto, P.S., et. al., "Summary of Research into Structure Damping," Fourth Symposium on Current Issues Related to Nuclear Power Plant Structure, Equipment and Piping, December, 1992.

13. Kennedy, R.P., et. al., Methodology for Developing Seismic Fragilities, EPRI TR-103959, June, 1994.

14. Malcher, L., et. al., Seismic Investigations of the HDR Safety Program, NUREG/CR-4236.

15. Kleine-Tebbe, A., et. al., "Seismic Safety of the Core Structure in a HTR Plant with Pebble Bed," PVP-Vol. 144, 1988.

16. Ricciuti, R., et.al., "Application of the SASSI Soil Structure Interaction Method to CANDU 6 NPP Seismic Analysis" PVP Conference 1996, Montreal (to be published).

Table 1

Comparison of CANDU 6 (reference) Damping Factors (%) with Other Standards

(1) CANDU 6 Reference Plant
(2) CSA N289.3
(3) USNRC Regulatory Guide 1.61
(4) NUREG/CR-0098
(5) ASCE 4-86
(6) EPRI NP-6041
(7) Hadjian and Tang [12]
(8) Hashimoto, et. Al. [13]
(9) EPRI TR-103959 [14]

Structure or Component	(1)	(2)	(3)	(4)	(5)	(6)	(7)	(8)	(9)
Pressure vessel or piping 300 mm (12 in) diameter or greater	4	3	2-3	2-3*		5	8		5
Piping between 150 mm (6 in) and 300 mm (12 in) diameter	2	2	1-2	1-2*		5	6		5
Small piping under 150 mm (6 in) diameter	1-2	1	1-2	1-2*		5	5		5
Welded steel structures	3	3	2-4	3-7	2-4	5		3-4	3-7
Rigid components	2	2		2-5					7-10
Bolted steel structures	5	5	4-7	7-15	4-7	7		4-6	3-7
Prestressed concrete structures	3	3	2-5	3-10	2-5	5		3-5	3-10
Reinforced concrete structures	5	5	4-7	5-10	4-7	7		4-7	
Soft soil	10	5							
Competent rock (granite)	4**	2							

* USNRC has accepted ASME Code N-411 specifying 5% for piping at frequencies under 10 Hz, 2% damping for piping over 20 Hz, and a linear variation between these frequencies.

** The reference CANDU reactor building analysis used 10% soil damping for rocking mode and 30% soil damping for translational mode.

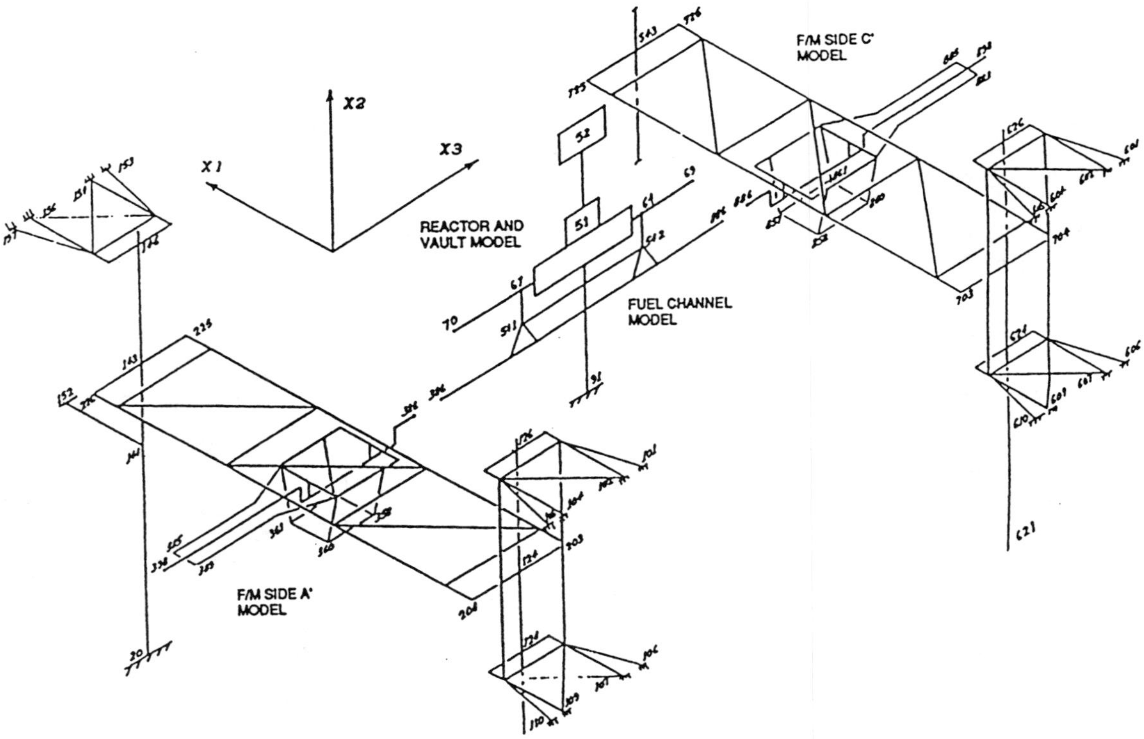

Figure 1. ATTACHED FUELLING MACHINE AND REACTOR

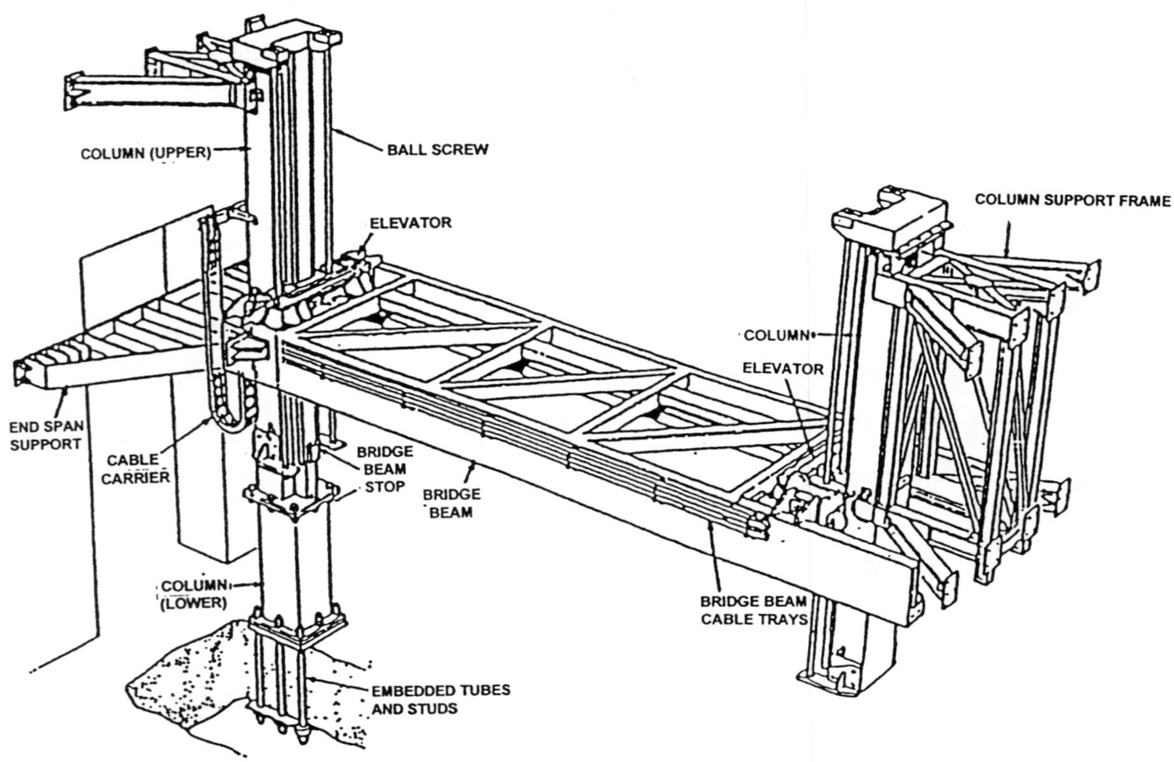

Figure 2. FUELLING MACHINE BRIDGE AND SUPPORT COLUMN ASSEMBLY

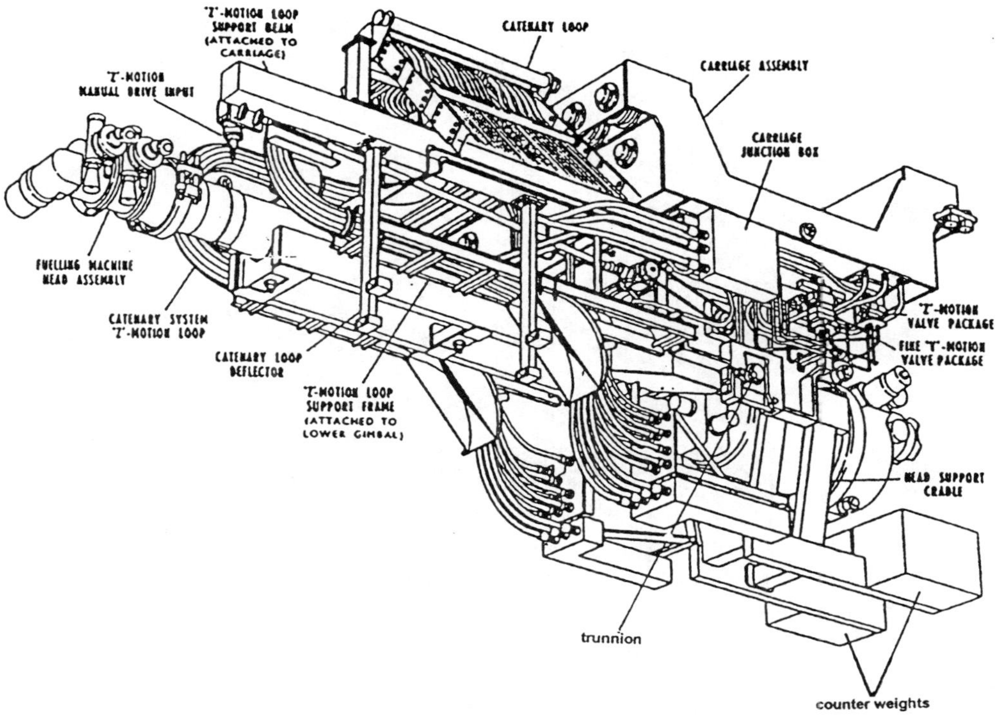

Figure 3. FUELLING MACHINE CARRIAGE AND HEAD ASSEMBLY

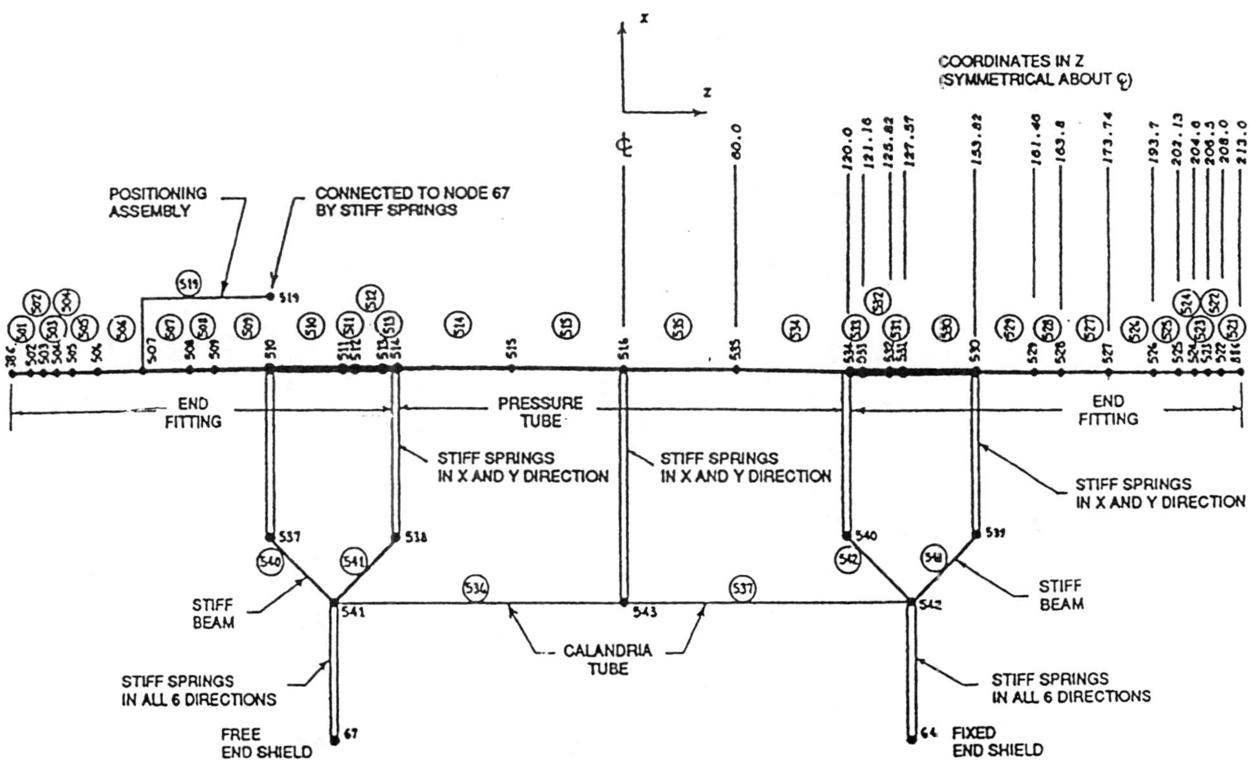

Figure 4. SCHEMATICS OF THE SEISMIC MODEL OF THE FUEL CHANNEL

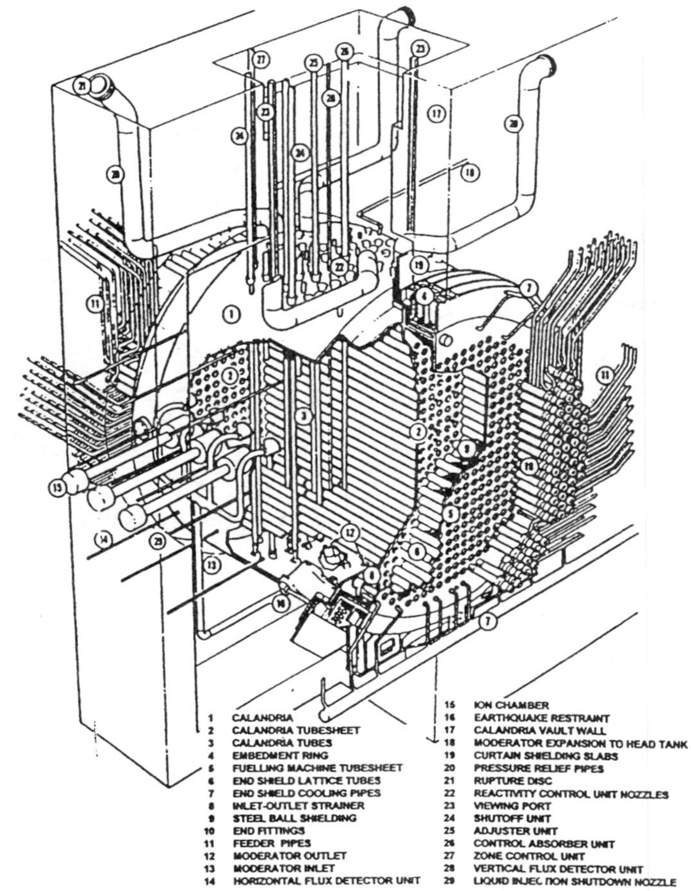

Figure 5. REACTOR ASSEMBLY

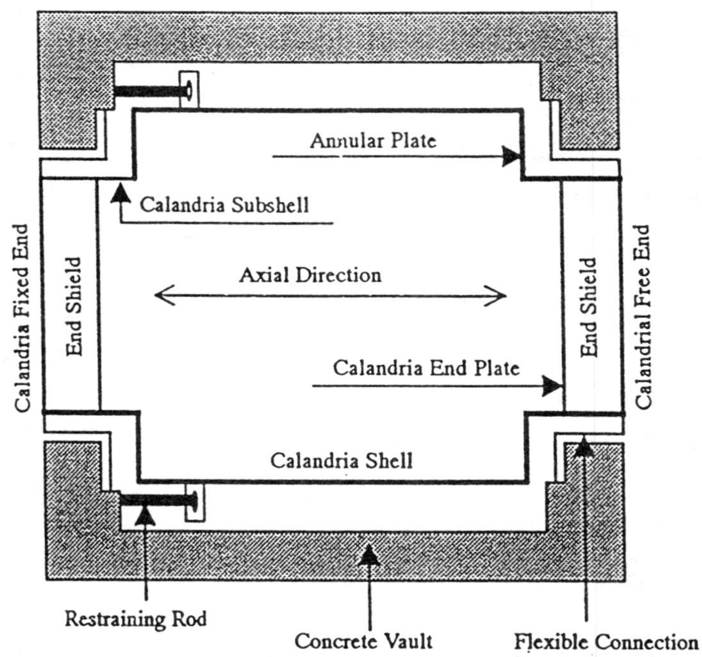

Figure 6. SCHEMATICS OF THE CONNECTION BETWEEN THE REACTOR ASSEMBLY AND THE CONCRETE VAULT

PVP-Vol. 340, Seismic Engineering
ASME 1996

SNUBBER REPLACEMENT WITH LIMIT STOPS AT VANDELLOS II

R. Baltus and **A. Billington**
Westinghouse Energy Systems Europe
Brussels (Belgium)

ABSTRACT

The paper describes the particularities and reports the results of the single largest project with Limit Stops to–date, i.e. a plant–wide snubber replacement at Vandellos II plant in Spain. This modern PWR plant had already undergone a snubber reduction program at the end of the construction stage, taking benefit from relaxation of dynamic loading criteria such as ASME Code Case N411 damping and Leak Before Break. The original analysis criteria were used, except for an innovative modal combination method named Generalized Algebraic Combination.

A 100% elimination was achieved for the first package of 19 lines, i.e. removal of all 115 snubbers during the 1995 outage.

INTRODUCTION

The Limit Stop is a maintenance free piping support which provides an alternative to unreliable snubbers. It limits the dynamic response of the piping during an earthquake or other dynamic event like hydrodynamic transient, but allows for totally free thermal expansion. The non–linear elastic behaviour of the Limit Stop may be analysed by the use of the time–history analysis method. An alternative approach is an equivalent linear methodology that linearizes the problem and permits piping analysis in the conventional manner [5]. This methodology has been automated in the GAPPIPE® computer program for seismic analysis with response spectrum method. The program has been independently verified and approved by the US Nuclear Regulatory Commission [4], thus facilitating the acceptance by the spanish authority, the Consejo de la Seguridad Nuclear (CSN).

Vandellos II is a modern PWR plant located in Spain, that was initially designed with about 500 snubbers on primary equipment and auxiliary piping. Mid–way through the design phase of the plant, dynamic loadings on piping and equipment had been alleviated through the introduction of ASME Code Case N411 damping, for seismic analysis, and of Leak Before Break methodology, which removed the requirement to postulate breaks in primary loop piping. Therefore the demand for snubbers diminished and a snubber reduction program was carried out for most of the auxiliary piping systems. It took advantage of existing thermal stress margins in certain lines, by replacing some snubbers with rigid supports. Vandellos II started operation with an "optimised" population of 305 snubbers and reduced it further to 288 snubbers, located in 68 auxiliary piping lines, after elimination of the RTD bypass lines. These snubbers have a mechanical type of design from Anchor–Darling for the small sizes and from Vibrachoc for all the others.

However, owing to snubber degradation, CNV was obliged to increase rapidly the size of the sample for in–service functional testing. Since the 4th refueling outage, they had to inspect 100% of 220 Vibrachoc snubbers and between 25 and 40% of the 68 Anchor–Darling models. Faced with the prospect of having to refurbish or replace many snubbers, Vandellos utility decided to embark on a plant–wide snubber elimination and replacement program employing Limit Stop technology. The 3 year program was instigated in the autumn of 1994.

DESCRIPTION OF VANDELLOS PROGRAM

The primary objective of the Vandellos program is to eliminate the majority, if not all, of the 288 snubbers installed in 68 auxiliary piping lines. The lines were split into 3 packages defined in Table 1.The 19 lines of Package 1 are listed in Table 2. They cover the complete spectrum of safety class, pipe size, interface equipment, and thermal conditions, and are required to withstand a combination of dynamic loading conditions including earthquake, LOCA induced displacements and steam hammer hydraulic forces.

Piping Analysis Strategy

The introduction of gapped supports causes certain variations in pipe stresses and support loads with respect to the existing situation with snubbers. In some cases the loads increase due to impact loads at gap closure. In other cases the loads decrease be-

TABLE 1 – PIPING LINE PACKAGES

Package Number	Piping Lines	Snubbers	Implementation Year
1	19	115	1995
2	47	136	1996
3	2	37	1997

cause the equivalent stiffness decreases and produces a beneficial shift in modal frequencies. Therefore it is necessary to verify that the piping and support systems have enough margins to cover these variations.

In older plants, designed to the initial US standards, these margins are available from relaxation in certain design criteria, mainly the use of higher damping ratios for seismic analysis as per ASME code case N411, and reduction of pipe break postulation requirements as per Leak–before–break methodology. These margins have permitted for example to launch plant wide replacement of snubbers with Limit Stops on a one to one basis for some US plant [1], based on a very limited piping analysis effort for a small sample of piping lines.

TABLE 2 – CHARACTERISTICS OF PIPING LINES IN PACKAGE 1

Line Description	Class	Number of Snubbers	Nominal (a) Diameter (inch)	Design Dynamic Loadings(b)		
				Seismic	LOCA	Steam Hammer
Accumulator Discharge A	1	6	12	RC	TH	–
Accumulator Discharge B	1	2	12	RC	TH	–
Accumulator Discharge C	1	6	12	RC	TH	–
RHR Suction A	1	4	12	RC	TH	–
RHR Suction B	1	6	12	RC	TH	–
Safety Injection to Hot Leg A	1	2	6	RC	TH	–
Safety Injection to Hot Leg B	1	2	6	RC	TH	–
RHR Return/Safety Inj.	1	9	6	RC	TH	–
Main Steam A	2	12	32	RC	SE	TH
Main Steam B	2	4	32	RC	SE	TH
Main Steam C	2	13	32	RC	SE	TH
RHR Outside Containment	2	5	14	RS	–	–
RHR Outside Containment	2	4	14	RS	–	–
Steam Generator Blowdown B	2	7	3	RS	SE	–
Steam Generator Blowdown C	2	7	3	RS	SE	–
Steam Generator Drain A	2	6	2	RS	–	–
Steam Generator Drain B	2	4	2	RS	–	–
Steam Generator Drain C	2	3	2	RS	–	–
Main Steam to Turbine	NNS	13	28	–	–	TH

Notes: (a) The most representative size is indicated for lines with mixed sizes
(b) Dynamic analysis methods are as follows:
– (RC) Response Spectrum with Coupled primary equipment model (RCL or SG)
– (RS) Response Spectrum with Seismic anchor motion
– (TH) Time History
– (SE) Static Equivalent

tion requirements as per Leak–before–break methodology. These margins have permitted for example to launch plant wide replacement of snubbers with Limit Stops on a one to one basis for some US plant [1], based on a very limited piping analysis effort for a small sample of piping lines.

In the case of Vandellos, these margins have been used already in previous snubber reduction programs, and it was decided to reanalyse systematically each piping line in order to verify the acceptability of the new piping stresses and displacements, as well as loads on supports and equipment. All original design requirements were verified, including break postulation criteria. The only new methodology improvement is the use of an innovative formula for Generalised Algebraic Combination (GAC) of the closely spaced modes, which guaranties more stable and realistic results for seismic analysis of piping systems subject to limited support changes. This method is further described in the next section.

Whenever possible, the new support configuration was further optimized by reducing the number of Limit Stops.

Modal combination with GAC

The response spectrum method assumes basically that the natural vibration modes are excited randomly by a seismic event, and combines them by SRSS (Square Root of the Sum of the Squares). However, some correction must be made for the modes with closely spaced frequencies, to account for a potential resonance and phase correlation of modal responses.

The formula provided by the US NRC in the Regulatory Guide 1.92 [3] leads to reasonably conservative results in most of the cases, but leads to totally distorted results in some specific cases of coupled models. This is typically the case when small mass auxiliary piping models are coupled to the large mass model of the Reactor Coolant Loop, like in Class 1 line analysis, or in the case of a small branch lines attached to large auxiliary piping. The scientific community and the nuclear industry started to develop algebraic modal combination methods as alternatives to the absolute sum method. P.G Mertens (1994) developed a consistent method called Generalized Algebraic Combination (GAC) [2]. It has been independently verified and approved for the first time by the Swiss nuclear licensing authority for its application to a major piping analysis project at Beznau NPP [6]. It has been used systematically for all class 1 lines at Vandellos. The following section illustrates its application to an accumulator line.

GAC Application to Vandellos

To compare the GAC(A) method with the RG 1.92 10% method, six different SSE response spectrum analyses were performed on the accumulator discharge line BH01 (non–optimised) with different support (snubbers or limit stops) and coupled or not to the reactor coolant loop (RCL), using each of the two modal combination techniques in turn.

BH01 is a 12 inch line, connected to the RCL on one side and to the accumulator tank on the other side (Figure 1). The mass of the tank is about 70 tonnes. The total mass of the line, including 3 valves, is about 15 tonnes, i.e. about 20% of the tank mass. The total mass of the RCL model (including one loop and the reactor vessel) is about 1000 tonnes, i.e. about 70 times bigger than the line, and about 15 times bigger than the tank.

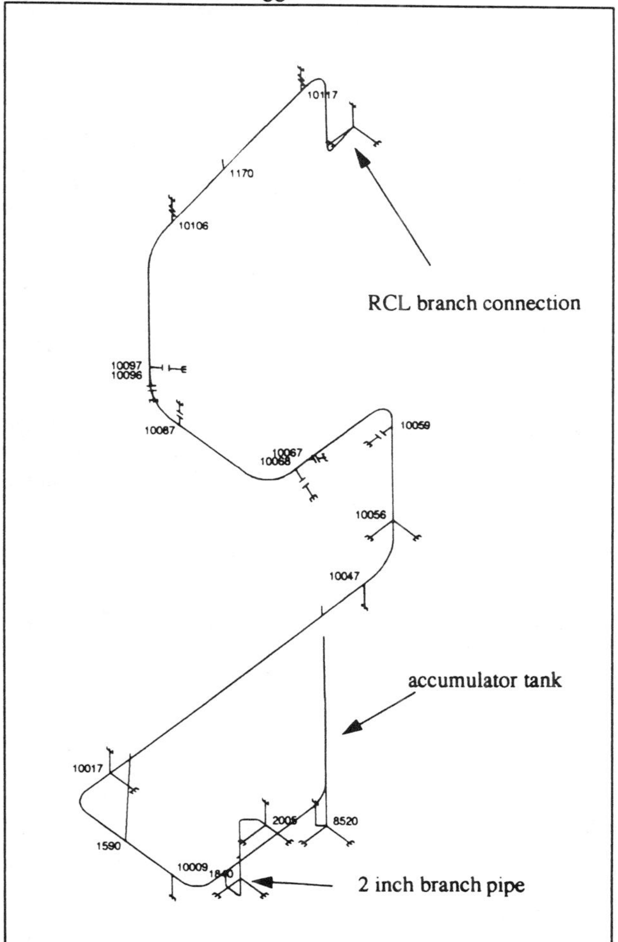

FIGURE 1 – ACCUMULATOR LINE BH01 AND TANK

The first comparison to be made is of the reactions at the base of the accumulator tank skirt, which should not be influenced by support changes in the line nor by the coupling to RCL. It can be seen on Figure 2 that in all the cases the GAC method results are in excellent agreement with the theoretical values, whereas RG 1.92 method yields unstable results, with sometimes asymmetrical distortion exceeding by a factor of 2.5 the theoretical reaction. The shear force increase represents more than 30 times the maximum force transmitted by the pipe to the tank nozzle, which does not make sense.

Figure 3 shows the bending moments in the accumulator piping for the four analysis cases with Limit Stops. It can be seen that there is good agreement between GAC and R.G.1.92 in the uncoupled analyses. When the RCL is coupled, for the GAC case, we see, as expected an increase in response in the zone closest to the loop and a good agreement with the uncoupled analyses further from the loop. However, for the corresponding R.G. 1.92 case the picture is very different, not only is there a huge amplification close to the loop but also the remotest zone of the line sees a large increase in response.

In conclusion it has been shown that the RG 1.92 modal combination method is both unstable and leads to grossly over estimated results, while the GAC method is stable and leads to realistic results for both snubber and Limit Stop configurations.

ANALYSIS RESULT SUMMARY

The package 1 analyses were successfully completed with the following results:

- all 115 snubbers eliminated
- 99 snubbers replaced by Limits Stops
- 3 snubbers replaced by rigid supports
- 13 snubbers removed
- 2 rigid supports removed

All piping stresses and fatigue usage factors were reconciled to the original code requirements. The variations were small enough not to change at all the break locations postulated in high energy lines.

The support loads were reconciled for 379 different supports. Beside the 115 snubber supports, there were 22 other supports which required some structural adjustments, i.e less than 6 percent.

There were 73 vent and drain lines connected to the 19 lines, supported by 49 small supports. All of them were reconciled for new loads or displacements without any modification to their design.

LIMIT STOP DESIGN AND SIZES

The Limit Stops are manufactured to ASME NF standards by a US manufacturer. The catalog sizes range from 1 to 6 with load capacities of 650 lbs to 120 Kips in level A/B. The design of their connection pieces was adapted to fit the metric dimensions of Vibrachoc clamps and rear brackets. A matrix of equivalent sizes was established in order to give a choice between two or three different Limit Stop sizes for replacement of any european snubber size. For replacement of large snubbers with smaller Limit Stops, special pin conversion kits were provided to fit the small diameter pin and the bigger diameter clamp or rear bracket holes. Conversely bigger spare clamps and rear brackets had to be used in case of size increase.

It is interesting to analyze the trend of the shift in sizes, as a statistical representation of the variation in support loads and sizes due to installation of gapped supports. The quantities of the different sizes had been estimated on the basis of the snubber sizes and design loads. The actual quantities obtained after analysis are given in the Table 3. It shows a size increase in 12 cases (10%) and a size decrease or deletion in 39 cases (34%). All but one of the size increase cases pertain to main steam lines, which are calculated for steam hammer loading (using time history analysis) and not only for seismic, as shown in Table 2.

Actual \ Estimated	1	2	3	4	5	6	TOTAL
1	13		5				18
2		2	6				8
3			16	4			20
4			1	14	5		20
5				6	16	6	28
6					5	0	5
Deleted	3		8	2			13
Rigid	1			1	1		3
TOTAL	17	2	36	27	27	6	115

TABLE 3
LIMIT STOP ESTIMATED SIZES VERSUS ACTUALS

INSTALLATION

All Limit Stops were adjusted during manufacturing to the pin to pin length specified from plant records and to the tension and compression gaps specified by the piping analysts. After measurement of the actual pin to pin length at the site, some adjustments were made by resetting the extension piece (51 cases) or by resetting the stops (12 cases) for the short models without extension piece.

As mentioned before, there were some changes to clamps or rear bracket or back up structure for 23 Limit Stops.

The installation was completed without any impact on plant outage schedule.

START UP TESTING

The installation of Limit Stops does not affect at all the thermal expansion displacements of the pipes. However, during the optimization process, there were 5 lines in which rigid supports had been either removed, or added in place of snubbers. For 3 of them, the impact on thermal displacements was totally negligible. For the other two, the thermal behaviour was verified during the plant heatup. For one line, the absence of interferences was verified by visual inspection. For the other line (main steam loop C) the thermal displacements were measured at key support locations, and found satisfactory.

CONCLUSIONS

The analysis conducted for the first 19 piping lines has demonstrated that 100 percent snubber elimination can be achieved, even in a modern plant such as Vandellos II which had already been optimized. It confirms the effectiveness of the Limit Stop technology and associated computer code GAPPIPE®, in eliminating snubbers. The success has been widely facilitated by the use of GAC for modal combinations.

The variations in support loads have lead to Limit Stop size increase in only 10 percent of cases, and to other support adjustement in less than 6 percent of cases.

At the time of writing this paper, the analyses for the next package of 47 lines are nearly completed and the elimination rate is also close to 100 percent.

ACKNOWLEDGEMENT

The authors want to thank J.L.Lopez Ansorena, lead engineer at Vandellos, for his dedication to this project, as well as the piping analysis team including M.Malfroy, S.Stoykof and J.L.Gheleyns (WESE), A.Tarin (UTE) and D.Gardiner (RLCA), who performed the analyses upon which this paper is based.

REFERENCES

[1] Cloud R.L, Leung J.S.M, Morgan Jr R.L, 1993, "Mc Guire snubber elimination program", ASME PVP conference Denver.
[2] Mertens P.G, 1994, "A new algebraic white noise modal combination rule – GAC(A)", Nuclear Engineering and Design 147 (p 299–309).
[3] NRC , 1976, Regulatory Guide 1.92 Rev.1 "Combining Modal Responses and Spatial Components in Seismic Response Analysis".
[4] NRC, 1995, letter to RLCA on Topical report [5] evaluation.
[5] RLCA/94/04–94/009, 1994, "A Topical Report on the Methodology, Verification and Applications of Computer Program, GAPPIPE".
[6] SVDB Nuklear Abteilung – Bericht Nr. A–9114–2, "Beurteilung der Kombination von Erdbeben Antworten fur das Projekt Requa II" (Verification by Swiss nuclear licensing authority of GAC method, applied at Beznau nuclear power plant).

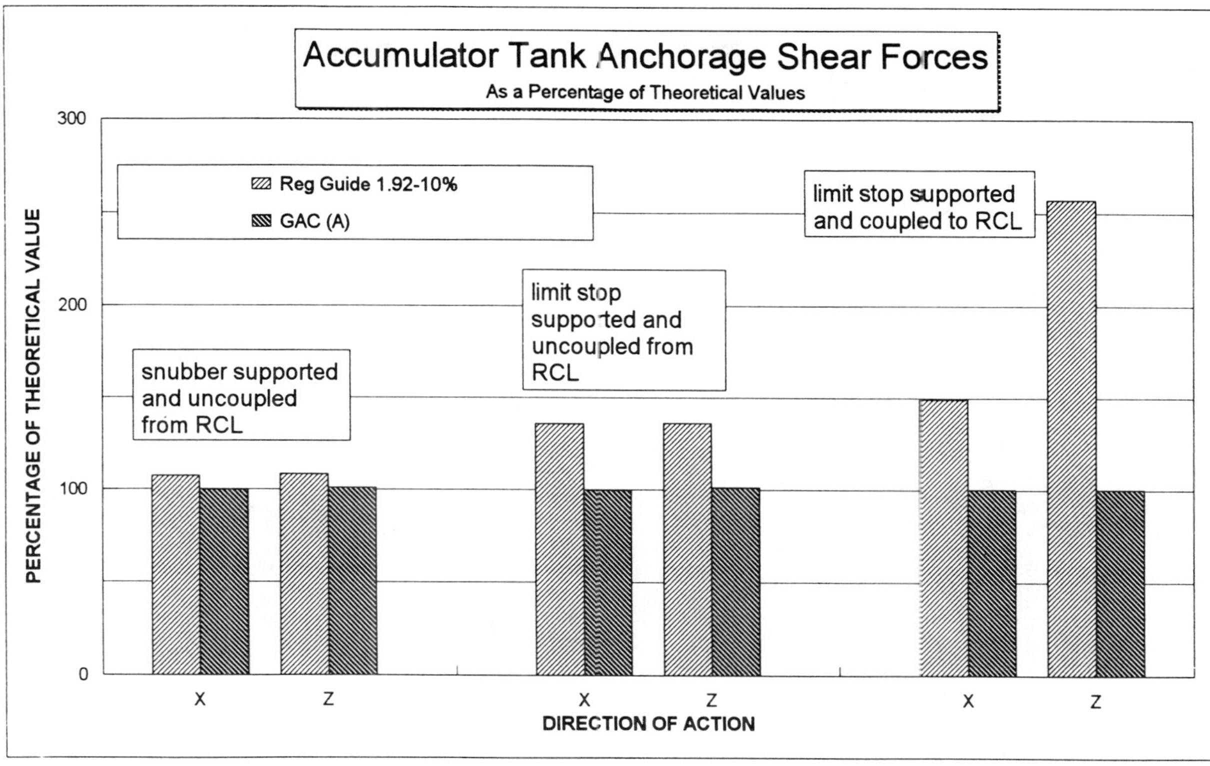

FIGURE 2 – COMPARISON OF THE REACTIONS AT THE ACCUMULATOR TANK BASE

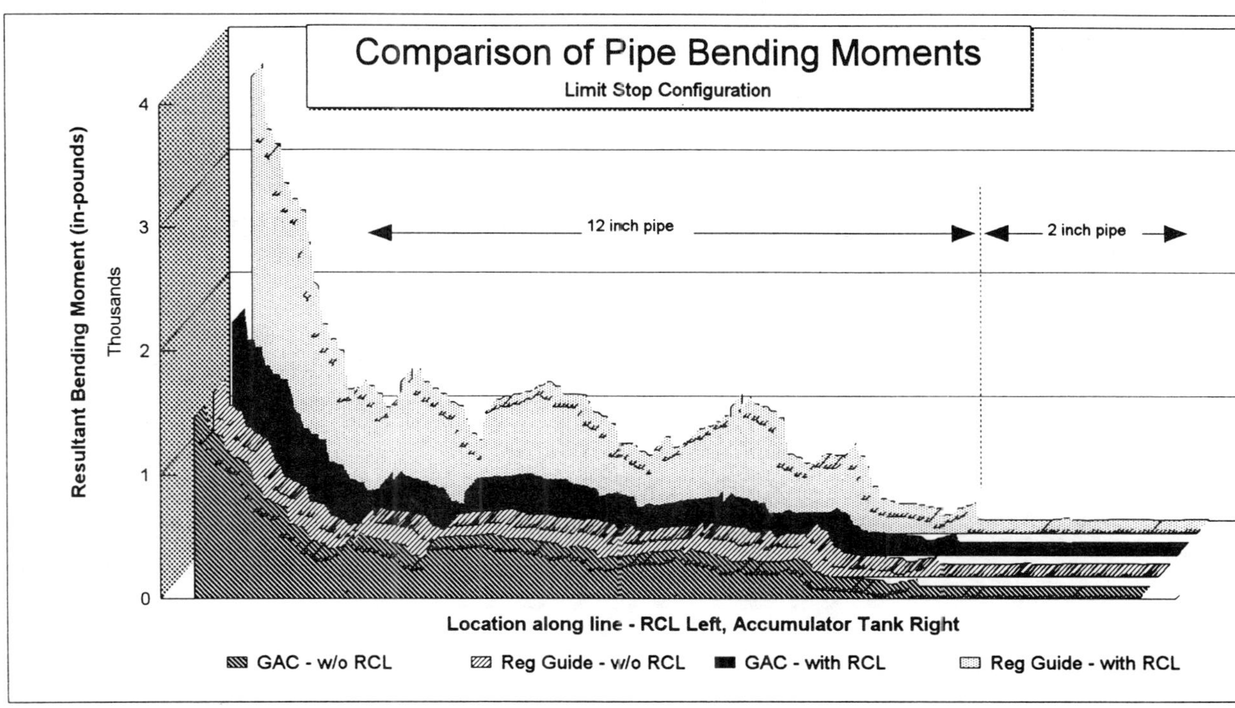

FIGURE 3 – COMPARISON OF PIPE BENDING MOMENTS BETWEEN RG 1.92 AND GAC

PVP-Vol. 340, Seismic Engineering
ASME 1996

SEISMIC MONITORING SYSTEM REPLACEMENT AT TEMELIN PLANT

R. Baltus
Westinghouse Energy Systems Europe
Brussels (Belgium)

S.S. Palusamy
Westinghouse Electric Corporation
Pittsburgh (PA)

ABSTRACT

The VVER–1000 plants under construction at Temelin (Czech Republic) were designed with an automatic reactor trip system triggered on seismic peak accelerations. Within the plant I&C upgrade, Westinghouse designed a digital Seismic Monitoring System to be integrated in an Artificial Intelligence based Diagnostic and Monitoring System. The system meets the requirements of the emerging standards prepared by the US NRC on the basis of EPRI studies, which recommend a detailed data evaluation and a pre–shutdown plant inspection before orderly shutdown, if required, rather than immediate emergency shutdown.

The paper presents the arguments about automatic trip, as discussed in an IAEA meeting attended by expert consultants from Japan, Russia, USA and Eastern and Western Europe. It describes the system installed at Temelin, including the plant specific criteria for OBE exceedance. Finally it presents the capabilities and limitations of the integration into an overall Diagnostic and Monitoring System.

INTRODUCTION

The requirements on seismic design of Nuclear Power Plants have evolved over time in relation with the perception of seismic induced risk of damage to the plant and to the population. In parallel with design criteria upgrade, the safety authorities of most countries have requested the installation of seismic sensors and monitoring systems in order to detect and measure seismic vibration amplitude in plant structures and components. Such systems are typically designed to meet one of two objectives:

a) To measure the seismic motion input and the plant structural response, in order to compare the actual demand of any specific event against the design basis of the plant.
b) To trigger as quickly as possible a reactor trip in order to limit the consequences of a potentially damaging earthquake.

If most countries agree on the first objective, the second one is very controversial and subject to international debates, especially for existing nuclear power plants in Eastern Europe countries.

The twin VVER 1000 units currently under construction at Temelin were initially designed with an automatic trip system similar to the SIAZ installed in several plants of Russian design. This system is being replaced with a digital Seismic Monitoring System (SMS) to be integrated in an Artificial Intelligence based Diagnostic and Monitoring System. The new system is aiming at providing support to the plant operator during and after a seismic event, instead of triggering an automatic trip of the reactor.

DEBATE ON AUTOMATIC SEISMIC TRIP SYSTEMS (ASTS)

In April 1995, one the authors participated to an international consultant meeting at IAEA on the advisability to install ASTS in Nuclear Power Plants [4] . This meeting was attended by experts from Japan, Russia, United States, Eastern Europe and Western Europe. The main conclusion was that ASTS is most favored when the risk of high level earthquakes is high in comparison to the seismic capacity of the plant. This is why ASTS is so common in Japan and in Eastern countries, and rare in Western Europe and US, where only a couple of californian plants are equipped with ASTS.

All plant operators and regulators are dreaming about a system which would shut the plant down before the earthquake hits it. However the plant shutdown does not mean only tripping the reactor, which takes only few seconds, but also starting up the emergency power supply system and ideally also cooldown systems, which takes several minutes or even hours. Unfortunately such a predictive system does not exist and in the best case, the seismic switches can only trip the reactor few seconds before the strong shaking part of the earthquake. Therefore the benefit of tripping quickly the reactor needs to be balanced against the risk of actuating pumps, valves and the associated relays during the strong shaking. Beside these considerations on safety, one must also consider other aspects such as the economical risk of tripping the reactor on spurious signals and on low level earthquakes unable to make any damage to the plant, as well as the loss of nuclear power supply to vital facilities of surrounding cities in case of medium size earthquakes.

The seismic switches which are currently used for ASTS are triggering when a preset level of peak acceleration is reached. Statistical studies performed by EPRI and other organizations have clearly demonstrated that the peak acceleration is not well correlated with the structural damage observed in industrial facilities subjected to real earthquakes. A new parameter named Cumulative Absolute Velocity (CAV) has been proven to exhibit a better correlation to the observed damage [1]. CAV is defined as the time integral of the absolute acceleration over the duration of the earthquake record. It is sensitive to the amplitude and to the duration of the acceleration time history, and it provides a better measure for the damaging energy injected into the structures. Therefore the emerging standards prepared by the US NRC [7] have redefined the plant operating limit as a combination of a CAV limit and the exceedance of OBE response spectra within a frequency band significant for structural response.

OBE Exceedance Criteria for Temelin. The seismicity at Temelin site is relatively low. Using IAEA seismic levels definition SL1 (equivalent to OBE) and SL2 (SSE), the SL2 has been set to an intensity 7 on MSK–64 scale with Peak Ground Acceleration of 0.1 g's. The SL1 is set at 0.05 g's. The complete criteria for OBE exceedance has been defined as follows:

a) CAV exceedance limit of 0.16 g*sec
   AND
b) OBE spectrum exceedance between 2 and 10 Hertz, with a minimum of 0.15 g.

These criteria are inspired from US NRC draft Regulatory Guide DG1017 [7] and permit to discriminate the low level, non–damaging events which would have caused non necessary trips with an ASTS. In parallel with the monitoring data evaluation, DG017 requests the operator to perform a plant inspection walkdown in order to detect any significant damage within 4 hours after the earthquake. The corresponding short term action plan logic recommended by EPRI [2] is shown on Figure 1. If damage is found or OBE exceeded, the plant must be shut down within the next 4 hours, i.e. maximum 8 hours after the earthquake. A pre–shutdown inspection is recommended, in order to facilitate an orderly shutdown generating less severe transients than an emergency shutdown.

The system installed at Temelin is comprised of two modules. The first one, named Seismic Monitoring System, performs data acquisition and OBE exceedance evaluation. The second one, named Post Earthquake Plant Diagnostic System (PEPDIS) is providing on line guidance to the operator within the Artificial Intelligence based Diagnostics and Monitoring system of Temelin (TMDS). These modules are described in the next sections.

SEISMIC MONITORING SYSTEM (SMS) FOR TEMELIN

Recent standards from US NRC [6] and IAEA [3] require the installation of sensors in the free field and at different elevations in seismic category I structures. As both units of Temelin have an identical design, a single SMS is required for the site. The block diagram is shown on Figure 2.

The main building of VVER 1000 units is a very compact structure constructed on a single square foundation (Figure 3). The lower part of the building houses the auxiliary systems and the control room. It is covered by a thick slab which supports the cylindrical containment and the peripheral building. The whole structure is fairly rigid and two sensors are judged sufficient to capture the building movements, one on the foundation and one at the elevation of the operating deck.

The free field control point is equipped with redundant sensors and separate data acquisition channels. This setup assures a very high reliability and a continuous availability, even during the automatic selftest of each sensor performed every day at request of the plant designer. The free field sensors are protected by a small shelter located outside of any vibration interferences. The outside temperature specified for the site may be as low as –40°C. Therefore the preference has been given to a robust sensor design based on a geophone, rather than the more delicate Force Balance accelerometer. These sensors have been tested at the Polythechnicum of Zurich (ETHZ) and proven to work efficiently many years under severe environmental conditions in the mountain stations of the Swiss National Strong Motion Network. A special software has been developed to trigger automatically the specified daily selftest of each channel. It performs a complete selfdiagnostics of the sensors and data acquisition units, and performs and automatic tracking of the main electronic calibration parameters. This trending permits to anticipate on any malfunction and to perform predictive maintenance by remote software adjustments, instead of periodic maintenance of the sensor hardware.

As shown on Figure 2, the SMS central computer is providing two alarms to each control room, the first one when an event has triggered the data recording, and the second one if OBE criteria has been exceeded. In parallel, more complete and detailed information is provided to the diagnostics and monitoring system network for distribution to the operator, engineering and maintenance groups.

INTEGRATION INTO TEMELIN PLANT MONITORING AND DIAGNOSTICS SYSTEM (TMDS)

TMDS is the customized version of the Artificial Intelligence based D&M system ALLY^{TM} [5]. It integrates a series of different modules, like fatigue monitoring, valve monitoring, pumps and reactor internals vibration monitoring, leak detection, loose part detection and SMS. It is connected to the plant I&C data highway and permits to perform very sophisticated diagnostics of different systems and components. The resulting information and alarm messages are directly transmitted to the different plant personnel groups in charge of the plant operation, engineering and predictive maintenance.

As explained before, the operators must perform an inspection walkdown within four hours after the earthquake in order to search for any significant damage to the plant [7]. As time is short, a series of key components have been selected as potential damage indicators [2]. This list contains components with low seismic capacity, like flat bottom tanks or switchyard ceramic isolators, rather than safety related components which are normally designed and qualified for high seismic levels. Certain damage indicator components are equipped with on–line monitoring sensors, which provide continuous information to TMDS. Therefore a special module has been developed to perform Post Earthquake Plant Diagnostics (PEPDIS) aiming at detecting perturbations and abnormal rates of change of the selected parameters. PEPDIS analyzes the variations with respect to the stable condition existing at the onset of the earthquake, and focuses the operator's attention on the most affected damage indicators.

It should be noted that the sensor inputs to PEPDIS are often displayed in parallel on main control room board. However the fast detection and ranking of abnormal changes, in conjunction with alarms from other D&M modules, will save a precious time to the operator for identifying damaged systems and components. A contrario, in case of low level, non–damaging earthquake, PEPDIS will provide quickly the evidence that no perturbation nor damage has occurred to the selected equipment, and thus it will support the justification for continued operation.

CONCLUSIONS

The SMS system installed at Temelin follows the new trend of US standards, which specify a drastic reduction of the seismic instrumentation hardware, combined with a faster data processing and more elaborated evaluation of the OBE exceedance criteria.

The relatively low seismicity of the site justifies the elimination of the automatic trip system.

SMS design permits to perform automatic selftest and selfdiagnostics suitable for predictive maintenance of the sensors and data acquisition units.

The integration within an upper level Diagnostics and Monitoring system presents several advantages, including sharing information from different modules between different functional groups of people, accessing plant I&C data and providing on–line guidance for the operator in support of post–earthquake actions and decisions.

REFERENCES

1. EPRI report NP–5930, "A Criterion for Determining Exceedance of the Operating Basis Earthquake", July 1988.
2. EPRI report NB–6695, "Guidelines for NPP Response to an Earthquake," December 1989.
3. IAEA Guide 50–SG–D15, "Seismic Design and Qualification of Nuclear Power Plants," 1992.
4. IAEA report, "Consultant Meeting on the Advisability of an Automatic Seismic Trip System (ASTS) in Nuclear Power Plants", Vienna, April 1995.
5. Kozlosky T.A, Lowenfeld S, Bauman D.A and Baltus R, "Advanced Plant Information System Using Intelligent Monitoring and Diagnostics and the ALLY™ Plant Monitoring and Diagnostics System", SMIRT12–PCS13, August 1993.
6. NRC DG 1016, "Nuclear Power Plant Instrumentation for Earthquakes", second proposed revision to NRC RG 1.12, November 1992 (slightly revised in February 1995 as DG 1033).
7. NRC DG 1017, "Pre–Earthquake Planning and Immediate NPP Operator Post–Earthquake Actions", November 1992 (slightly revised in February 1995 as DG 1034).

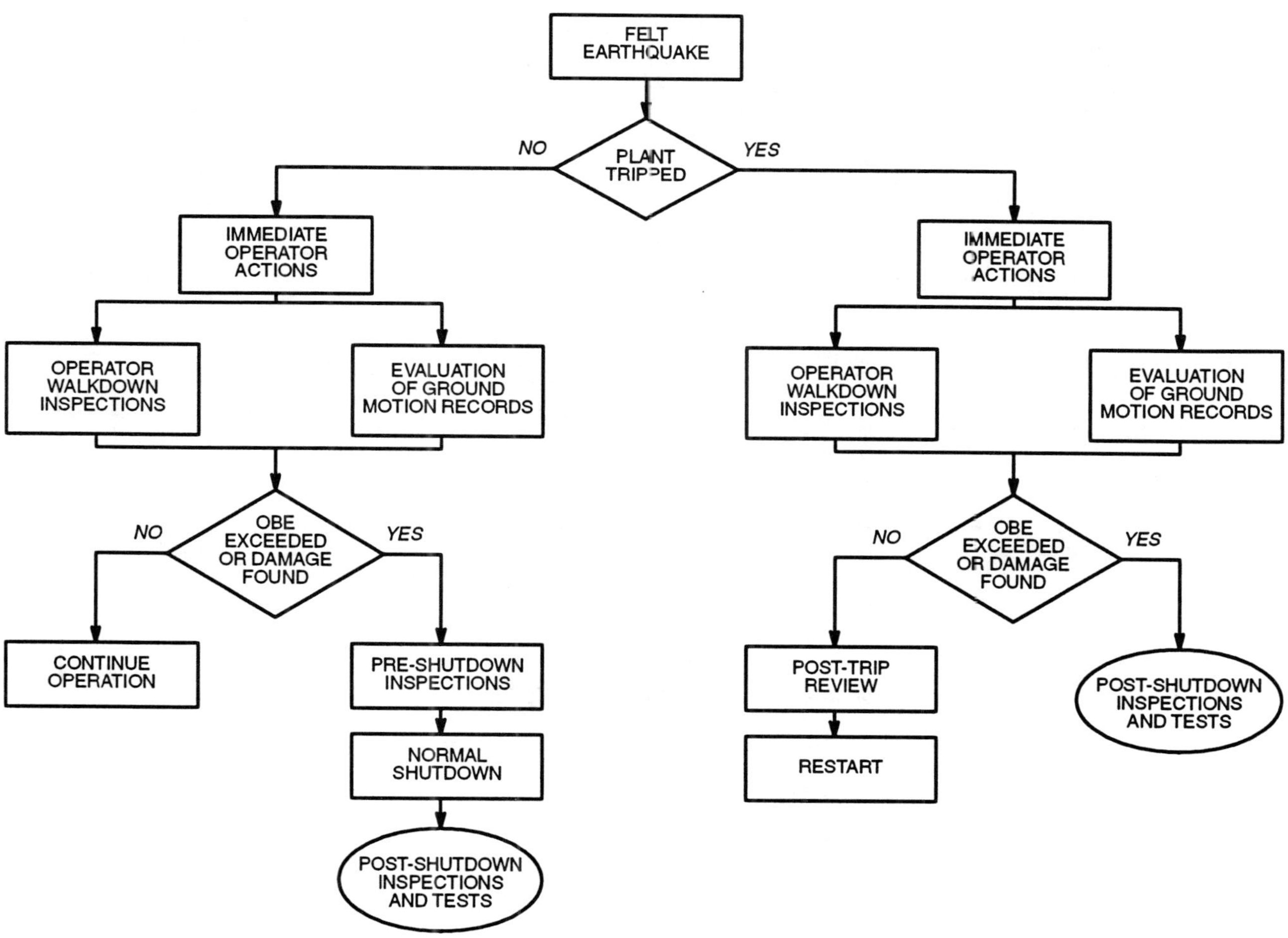

FIGURE 1 – FLOW DIAGRAM OF POST-EARTHQUAKE ACTIONS (FROM [2])

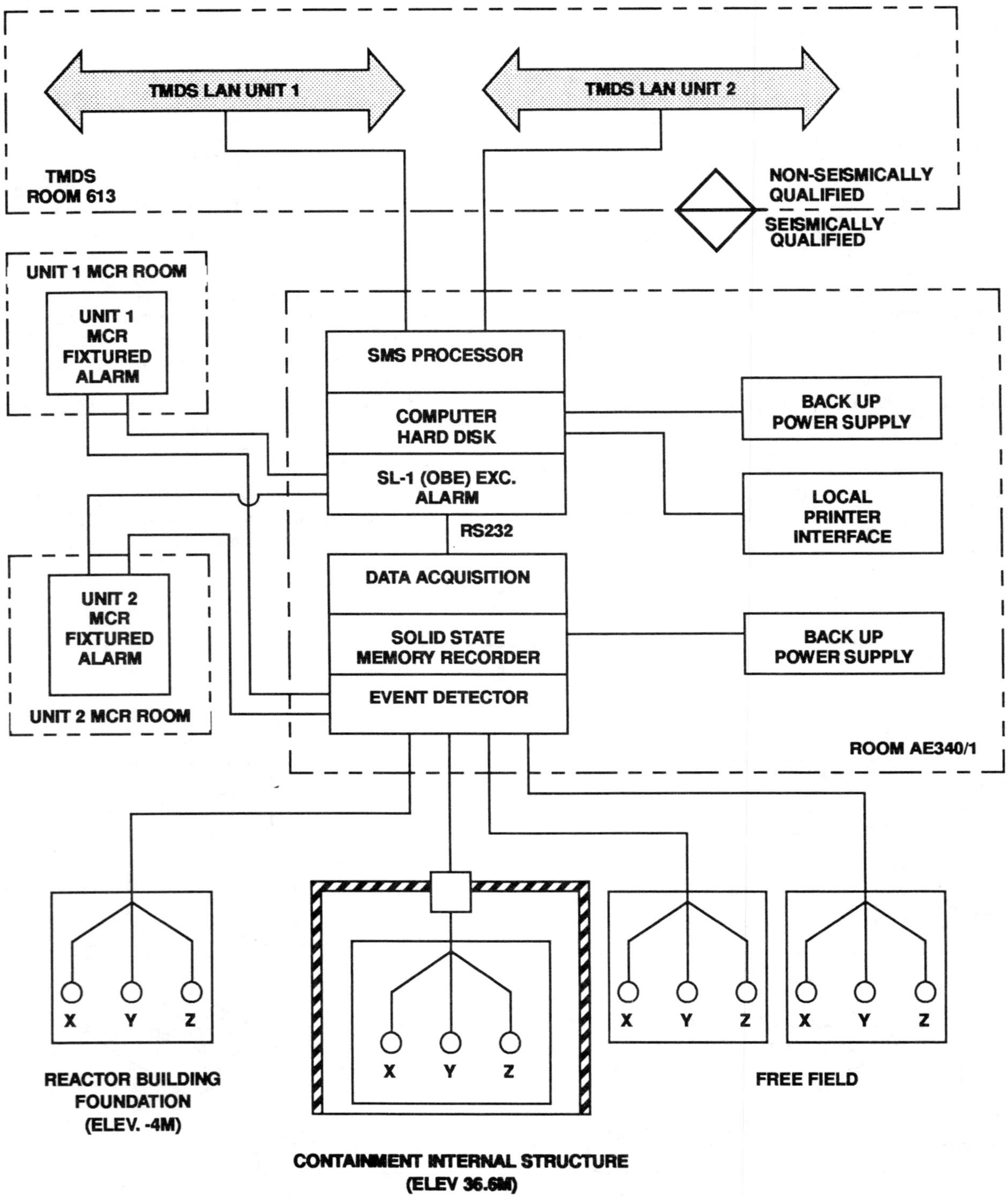

FIGURE 2 – SMS BLOCK DIAGRAM

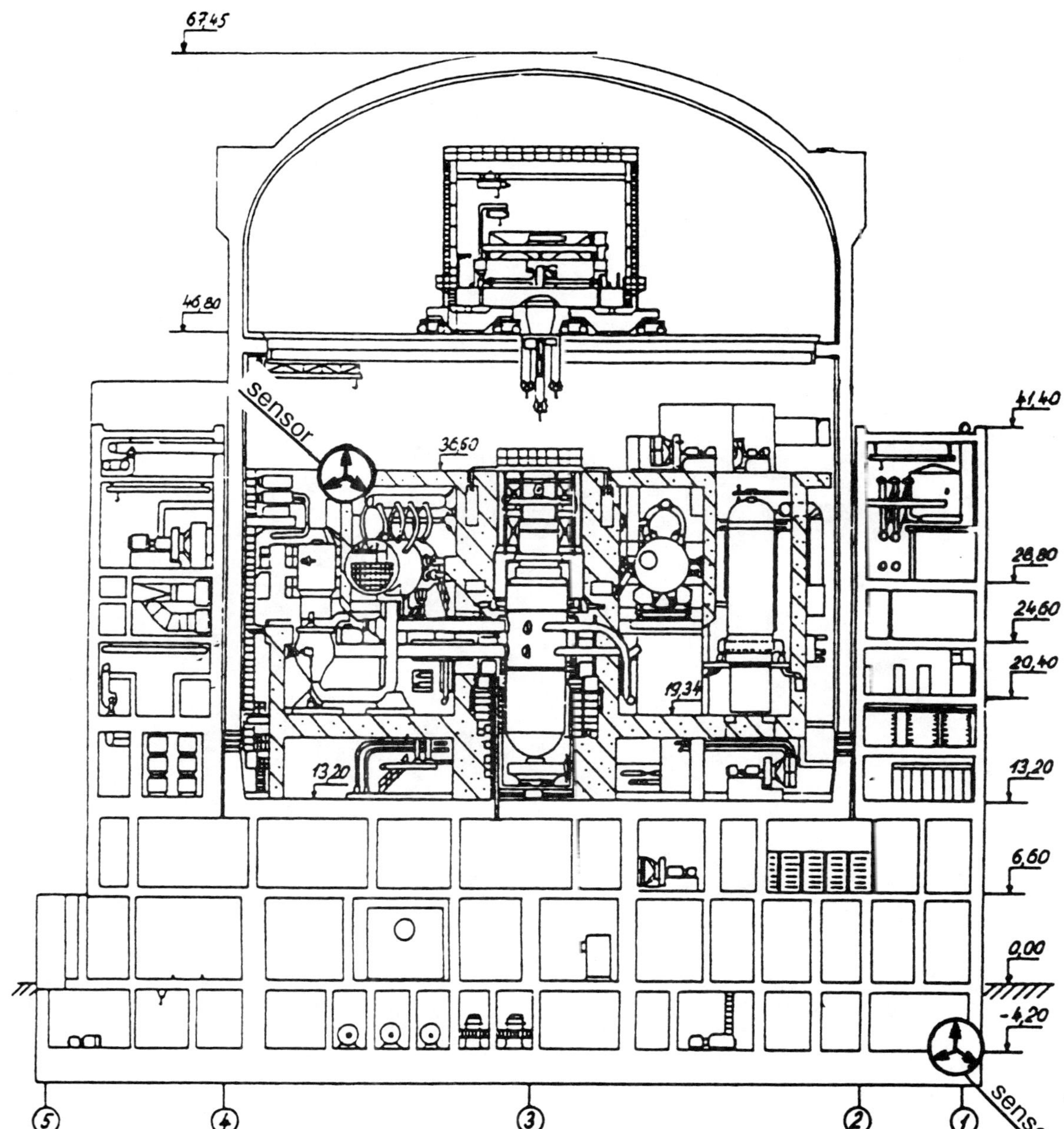

FIGURE 3 – REACTOR BUILDING CROSS SECTION WITH SENSOR LOCATIONS

PVP-Vol. 340, Seismic Engineering
ASME 1996

A STRATEGY FOR IMPLEMENTATION OF EXPERIENCE BASED SEISMIC EQUIPMENT QUALIFICATION IN IEEE AND ASME INDUSTRY STANDARDS

Timothy M. Adams
Stevenson and Associates
Cleveland, Ohio

ABSTRACT

In the past 20+ years, extensive data on the performance of mechanical and electric equipment during actual strong motion earthquakes and seismic qualification tests has been accumulated. Recognizing that an experience based approach provides a technically sound and cost effective method for the seismic qualification of some or certain equipment, the IEEE Nuclear Power Engineering Committee and the ASME Committee on Qualification of Mechanical Equipment established a Special Working Group to investigate the incorporation of experienced based methods into the industry consensus codes and standards currently used in the seismic qualification of Seismic Category Nuclear Power Plant equipment. This paper presents the strategy (course of action) which was developed by the Special Working Group for meeting this objective of incorporation of experience based seismic qualification of equipment into the industry consensus codes and standards used in the design and seismic qualification of seismic category nuclear power plant equipment. This strategy was recommended to both chartering organizations, the IEEE Nuclear Power Engineering Committee and the ASME Committee on Qualification of Mechanical Equipment for their consideration and implementation. The status of the review and implementation of the Special Working Group's recommended strategy by the sponsoring organization is also discussed.

INTRODUCTION

In the past 20+ years, extensive data on the performance of mechanical and electric equipment during actual strong motion earthquakes and seismic qualification tests has been accumulated. Intense review of this data has led to the identification of specific equipment attributes and inclusion rules which if met, will ensure, at a high confidence level, that the equipment will successfully operate after and in some cases during strong motion earthquakes. Recognizing that an experience based approach provides a technically sound and cost effective method for the seismic qualification of some or certain equipment, the IEEE Nuclear Power Engineering Committee (NPEC) and the ASME Committee on Qualification of Mechanical Equipment (QME) established this Special Working Group to investigate the incorporation of experienced based methods into the industry consensus codes and standards currently used in the seismic qualification of Seismic Category Nuclear Power Plant equipment. The members of this Special Working Group are given in Table 1.

This paper presents the strategy (course of action) which has recommended by the Special Working Group (SWG) for the incorporation of experience based seismic qualification of equipment into the industry consensus codes and standards used in the design and seismic qualification of seismic category nuclear power plant equipment. This recommended strategy was made jointly to both chartering organizations.

BACKGROUND

Component (excluding distribution systems) seismic qualification as it relates to nuclear power plants and facilities can be broken down into two primary areas: electrical and mechanical components. The qualification criteria for mechanical components can also be broken down into two categories: leak tight structural integrity and operability. The leak tight structural integrity for most pressure retaining mechanical components since at least 1971 has required some level of analytical evaluation either by meeting explicit ASME Boiler and Pressure Vessel Code (BPVC) equations or meeting layout and support requirements which can be demonstrated to meet ASME BPVC design equation requirements. The design of low or zero pressure retaining mechanical components (such as fans, air handling units, chillers, atmosphere storage tanks, etc.) is covered by other industry standards such as ASHRAE, SMACNA, API, etc.

For pressure retaining components most State Boiler Codes require that such components be designed to ASME BPVC. For safety related nuclear power plant components the applicable section of the code is Section III.[1] State laws or codes may also require that non-pressure retaining components conform to specific industry consensus standards. The state requirements are independent of and in addition to any nuclear plant licensing requirements put forth by the USNRC.

The operability seismic qualification of mechanical components used in nuclear power plants is currently covered by the ASME Committee on Qualification of Mechanical Equipment (QME) who have developed operability qualification standards for pumps and valves.[2] A non mandatory Appendix A of Section QR of the QME Standard, which was published by ASME in

June 1994, provides the requirements for the use of experience data for component seismic qualification. It is again emphasized that acceptance by the USNRC of the QME standards for operability qualification does not preclude or supersede the requirement of the State Boiler Codes to demonstrate conformance to the ASME BPVC Section III (for safety related pressure retaining components). It should be noted that historically, demonstration of operability has implied conformance to leak tight structural integrity but this concept has never been generically demonstrated nor has it been formally incorporated into industry consensus codes and standards.

Electrical components unlike mechanical components have historically not been required (for seismic qualification) to meet any explicit pressure retaining requirements. They have been qualified by demonstrating operability of performance requirements only in accordance with IEEE (primarily IEEE-344) requirements. IEEE-344-87[3] Section 9 provides requirements for the use of experienced based seismic qualification of components.

The most extensive application involving the use of experience based seismic data was the Seismic Qualification Utility Group (SQUG) Generic Implementation Procedure (GIP)[4] which provides a detailed technical approach, generic procedures, and documentation requirements for use in the verification of the seismic adequacy (operability) of mechanical and electrical safe shutdown equipment in operating nuclear power plants. The SQUG GIP methodology was developed to address unresolved safety issue A-46, "Verification of Seismic Adequacy of Equipment in Older Operating Nuclear Plants." Its application is for the verification of the seismic adequacy of existing equipment in older Nuclear Power Plants and not the design of new equipment.

Initiatives to apply seismic experience to the operability evaluation of new equipment and replacement equipment have been developed by the Advanced Reactor Corporation - Advanced Light Water Reactor Group and EPRI-PSE (STERI Guidelines)[5], respectively.

IMPLEMENTATION STRATEGY

Previous and current application methodologies for experienced based seismic qualification of equipment divides electrical and mechanical equipment into distinct and unique generic classes. The Special Working Group determined this generic class based approach was an appropriate approach for the application of experience data and proposed that the generic component classification for experienced based seismic qualification criteria be implemented in IEEE-344 (Section 9) for electrical equipment and Appendix A of Section QR of the QME Standard for mechanical equipment. The detailed application criteria for experienced based seismic qualification would then be placed in the IEEE and QME component specific standards. Furthermore, an approach is proposed to permit the implementation of the experience data method to show conformance to the seismic design criteria of mechanical component design codes, especially the ASME BPVC. This would not exempt the subject mechanical components from meeting all other design requirements of the applicable standard.

Strategy for Electrical Equipment

IEEE-344 is the generic standard which establishes the requirements for the seismic qualification of Class 1E electrical components. This standard provides for equipment qualification by the use of analysis, testing and the application of experience data. The other IEEE component specific standards in general use IEEE-344 (either by inclusion or reference) to identify component specific seismic qualification requirements. This general format of the IEEE Standards is as shown in the Figure 1. IEEE-344-1987 incorporated a Section 9.0 which provides guidance on the use of experience data for the seismic qualification of electrical components. This experience based approach focusses on the use of component similarity to previously qualified components.

It is suggested that IEEE-344 be enhanced to provide more concise guidance for utilization of experienced based qualification for new and/or existing electrical equipment. This may result in the development of new sections in the standard that address a generic classed based approach for the application of experience based qualification. This guidance will focus on the establishment of seismic capacities for generic equipment classifications and it is suggested that the component specific IEEE standards be modified to provide detailed experienced based seismic qualification criteria as it applies to a specific component. This is envisioned to include component specific criteria, inclusion rules, and caveats for the experienced based seismic qualification of the specific component type. These component specific specifications will use as input the generic classifications, inclusion rules, and criteria provided in IEEE-344.

Strategy for Mechanical Equipment

Incorporation of experience based methods for the qualification of mechanical equipment will require a more complex approach because 1) the development of a generic standard (similar to IEEE-344) for the "Seismic Qualification of Mechanical Components" is a recent occurrence and 2) design of various types of mechanical equipment is covered by a much larger range of industry consensus standards which span many different industry organizations (ASME, TEMA, API, SMACNA, etc.).

The recently issued QME standard on the qualification of pumps and valves now includes a non-mandatory appendix, Appendix A of Section QR, which provides a seismic operability qualification methodology (including experienced based qualification) for selected equipment (primarily pumps and valves). The overall format of the QME Standard is as shown in Figure 2. Component specific seismic qualification (for pumps and valves) is also provided in sections of the body of QV or QP. Hence, Appendix A of Section QR (which is non-mandatory) in many cases might not be used. Currently Section QV or QP provides very little guidance on experienced based seismic qualification. There appears to be no immediate plans in place to generate sections similar to QV and QP (applicable to pumps and valves) for other mechanical components.

As previously discussed in the QME Standards are only applicable to the operability qualification of mechanical components. The structural integrity and/or the leak tight/structural integrity design of these components falls under the jurisdiction of other industry consensus codes and standards. For safety related nuclear pressure retaining components, State Boiler Codes (which will not change) require that the leak tight/structural integrity be insured by conforming to the design and fabrication requirements of the ASME BPVC - Section III. The use of experience data to demonstrate conformance to leak tight/structural integrity requirements (for seismic loadings during an earthquake event) will require modifications to the ASME BPVC Section III. In addition similar changes will most likely be required in the other industry standards which govern the design of other mechanical components. (This will not exempt the component from meeting all other design criteria and requirements of the applicable code or standard. It would simply provide an alternative to use experienced based seismic qualification for the design of the component to earthquake loads.)

Considering these factors, the SWG suggests the following approach for incorporation of enhanced experienced based qualification methodology for mechanical equipment.

(1) Modify Section 7100 of Appendix A Section QR to provide basic guidance for the utilization of experienced based seismic qualification for new and/or existing equipment. This basic guidance will be generic in nature and will focus on providing generic equipment classification (as it relates to the applicability of experience data), identification and characterization of equipment categories, and incorporation into Section 7100 some of the generic information which is currently contained in Attachment C of Appendix A of Section QR. Also, editorial changes necessary to remove specific pump and valve qualification references and make Appendix A applicable to the qualification of all types of mechanical equipment should be made.

(2) Attachment C of Appendix A of Section QR should be broken down into subattachments. Each subattachment would provide the detailed experienced based qualification rules for a specific class of equipment (pumps, valves, chillers, etc.).

(3) Make Appendix A of Section QR a mandatory appendix. This is necessary to insure that the SWG's goal of standardization of experienced based seismic qualification is achieved.

(4) Section QV and QP should be modified to more clearly acknowledge the use of experienced based qualification of equipment and to reference the applicable sections of Appendix A of Section QR as the method which should be used for experienced based qualification.

(5) The SWG suggests the development of an ASME BPVC Section III Code Case which will permit the use of the applicable sections of Appendix A of Section QR of the QME standard to demonstrate leak tight/structural integrity for earthquake loadings. This code case should be submitted to the Subgroup on Design of the ASME BPVC Section III for consideration for incorporation into the Section III.

(6) The SWG also suggests development of an ASME BPVC Section VIII code case which will permit the use of the applicable sections of Appendix A of Section QR of the QME standard to demonstrate leak tight/structural integrity for earthquake loadings. This code case should be submitted to the Subgroup on Design of the ASME BPVC Section VIII for consideration for incorporation into Section VIII.

Implementation Strategy Summary

Upon completion of this effort, Appendix A of QME Section QR, the IEEE Standards, and the ASME BPVC will have a comprehensive set of experience based seismic qualification criteria for mechanical and electrical equipment. While for other mechanical and electrical components several additional industry codes, standards, and daughter standards may still require modification, the majority of mechanical components and all electrical components will have an acceptable experience based seismic qualification method. Therefore, standards could then be incorporated, by reference, in component design or purchase specifications as an acceptable method for seismic qualification of the component. This approach provides a vehicle for its near term use by the utility industry.

Logistical Strategy

Figure 3 presents the IEEE organization structure as

it relates to the maintenance of the affected IEEE Standards. Figure 4 presents the QME organizational structure as it relates to the development and maintenance of the QME Standard. Given these organizational structures and the proposed implementation strategy, the following logistical strategy is proposed:

(1) This Special Working Group would provide recommendations for the enhancement of IEEE-344 to the IEEE Working Group 2.5. This Special Working Group will also provide recommendations for the enhancement of QME Appendix A of Section QR (excluding Attachment C) and to Sections QV and QP to the QME committee.

(2) IEEE Changes Implementation: Working Group 2.5 would then prepare a revision request, known as a PAR, for revision of IEEE 344 and begin the process for revision of IEEE-344. Next, it is suggested that the SWG initiate work to develop recommendations for the detailed implementation of experienced based seismic qualification in the component specific standards.

(3) ASME-QME Change Implementation: It is first suggested that the now dissolved QME Working Group on Appendix A be reactivated and elevated to a Subcommittee entitled, "Subcommittee on Dynamic Qualification Techniques" and a Working Group be formed under this Subcommittee entitled, "Working Group on Experience Based Seismic Qualification." The proposed Appendix A (of Section QR, excluding Attachment C) and QP, QV changes would be referred to this new subcommittee for review and referral to the main QME committee for incorporation to the QME Standards.

    Next, it is suggested that the newly formed QME Working Group begin to develop the changes to Attachment C of Appendix A of Section QR for the detailed implementation of experience based equipment seismic qualification for the various classes of mechanical equipment.

(4) ASME-BPVC Implementation A task group of this Special Working Group (to be selected later) will prepare the necessary ASME BPVC code cases for submittal to the appropriate ASME-BPVC Code Committees.

LONG TERM GOALS

In implementation of experience based seismic qualification techniques, the two major issues discussed in both the Appendix A of Section QR and IEEE-344 documents are (1) the demonstration of component similarity to components which have experienced earthquake loading and (2) determination that the seismic demand experienced by the existing component is greater than or equal to the component seismic design basis demand.

To date, the focus of experience based seismic qualification has been the operability evaluation of existing equipment, not the design and installation of new equipment. Experience based qualification must be adapted to address the fact that in initial design you don't have a component to actually look at and you must maintain control over the final installation of the item.

The initial goal of this SWG is to facilitate revisions of IEEE-344, the QME Appendix A of Section QR and the component specific standards for equipment seismic qualification. The SWG believes the long term goal of both the IEEE and ASME should be to provide a definitive set of seismic design rules (component seismic "design by rule") to be incorporated into the appropriate design basis codes and standards which are based on test results and actual earthquake experience. A significant benefit to this goal is that it establishes definitive controls on the use of earthquake experience data, thereby reducing the "judgement calls" which are required in many of the currently applied screening criteria. It appropriately integrates industry concerns related to component manufacture by its inclusion into the applicable component design and fabrication standards. It also reduces the questions and concerns posed by similarity and demand considerations by providing specific seismic design rule criteria and methodology to address these issues.

This long term goal would include incorporation of this seismic "design by rule" approach to the applicable IEEE Standards, ASME BPVC sections and other appropriate industry consensus codes and standards. This goal will establish equipment - class specific seismic design rules, in the applicable design codes, which will permit seismic resistant capability to be designed into mechanical and electrical equipment. Further, this seismic "design by rule" will significantly reduce cost of seismic design and qualification of equipment. It is important to note that this long term goal addresses only the seismic (earthquake) design criteria. All other design requirements of the subject code or standard would still have to be met.

DISCLAIMER

The author was Chairman of the Strategy Task Group of the SWG which developed the SWG Implementation Strategy. This paper presents a summary of the Implementation Strategy affirmatively voted by the SWG and sent forward to the chartering committees. It does not necessarily reflect the opinions and views of all the individual working group members.

REFERENCES

1. American Society of Mechanical Engineers, Boiler and Pressure Vessel Code, Section III, Division 1, "Design of Nuclear Components", 1995 Edition.

2. American Society of Mechanical Engineers, ASME QME-1, 1994, "Qualification of Active mechanical Equipment Used in Nuclear Power Plants", June 30, 1994.

3. The Institute of Electrical and Electronics Engineers, ANSI/IEEE-1987, "Recommended Practices for Seismic Qualification of Class 1E Equipment for Nuclear Power Generator Stations".

4. Seismic Qualification Utility Group, "Generic Implementation Procedure for Seismic Verification of Nuclear Power Plant Equipment", Revision 2, February 14, 1992.

5. Electric Power Research Institute, EPRI-NP-7484s, "Guidelines for the Seismic Technical Evaluation of Replacement Items for Nuclear Power Plants", February 1993 (EPRI Proprietary).

Table 1 SWG MEMBERS

Stephen Eder, Co-Chairperson
George Antaki, Co-Chairperson

Timothy Adams	Robert Kassawara
Mostafa Ahmed	Kamal Manoly
Christopher Allen	Joseph Media
Jeffery Boucher	Kelly Merz
Howard Chung	William Metevia
Dennis Connelly	R.B. Miller
Robert Enis	Enver Odar
Jim Gleason	Joh Richards
Mike Griffen	William Schmidt
Brent Gutierrez	Satyan Sharma
Medhad Hassaballa	Z.T. Shi
Paul Ibanez	Dan Smith
John Jenkins	Paul Smith
	James Thomas

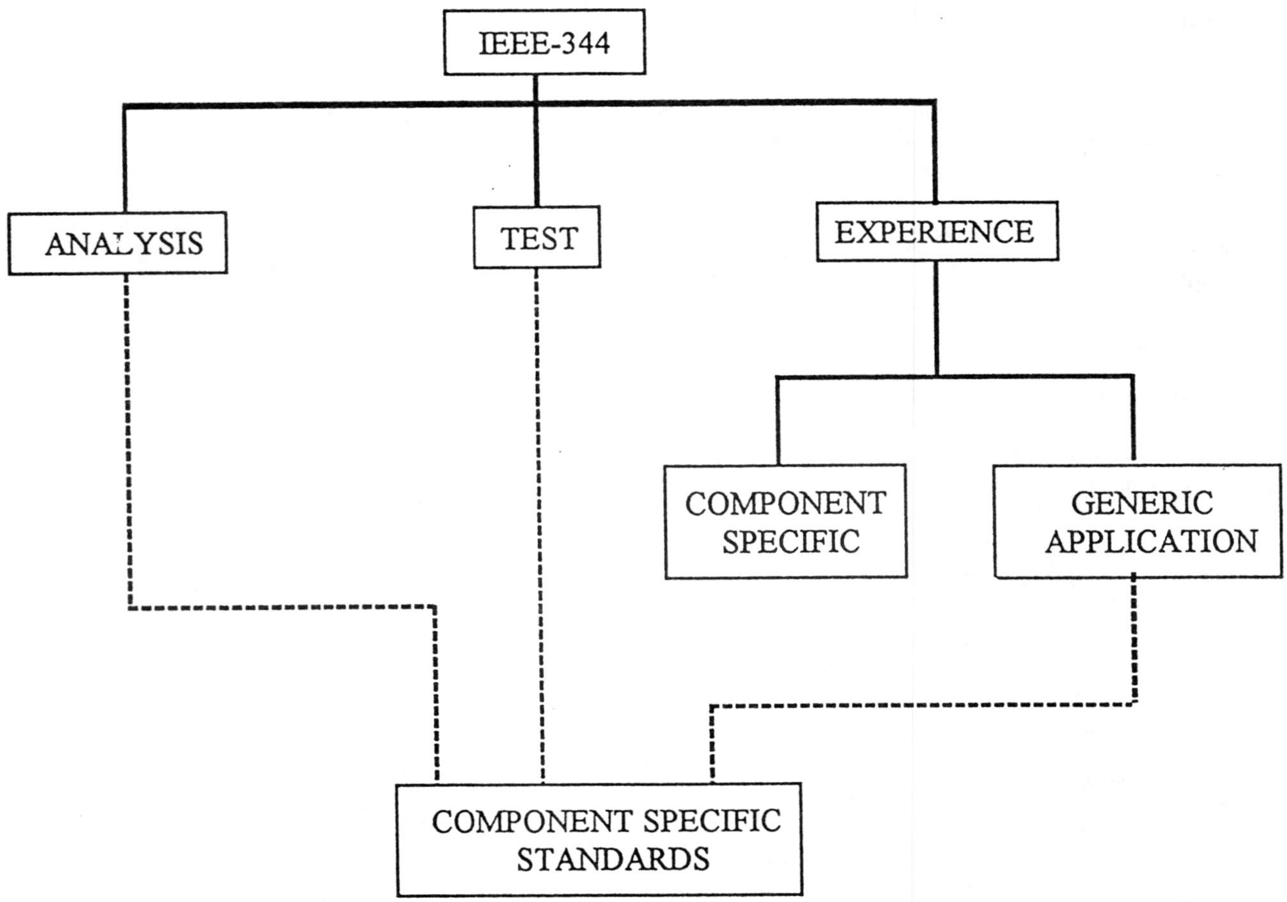

FIGURE 1 - OVERALL IEEE STANDARD FORMATS

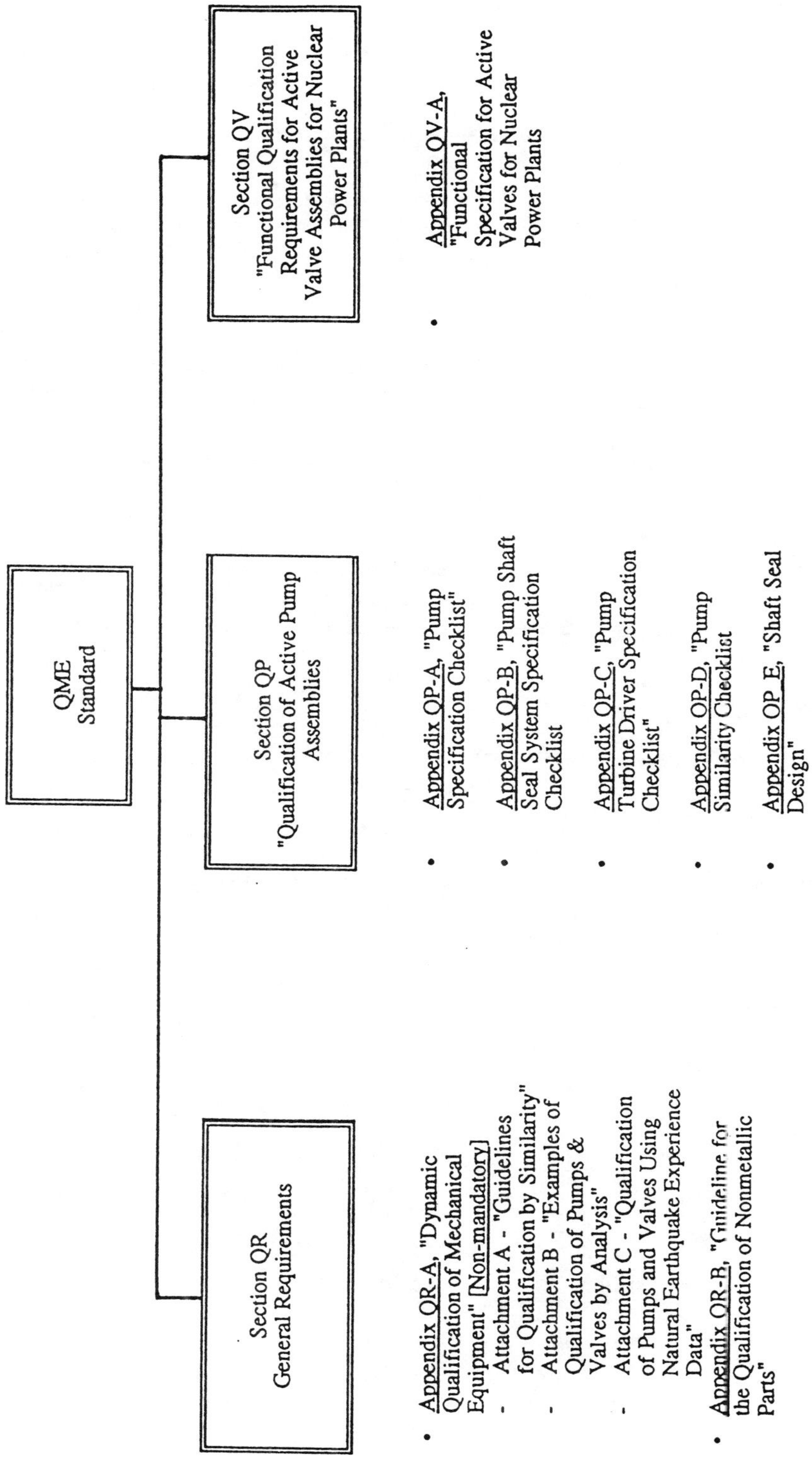

FIGURE 2 - QME STANDARD ORGANIZATION & FORMAT

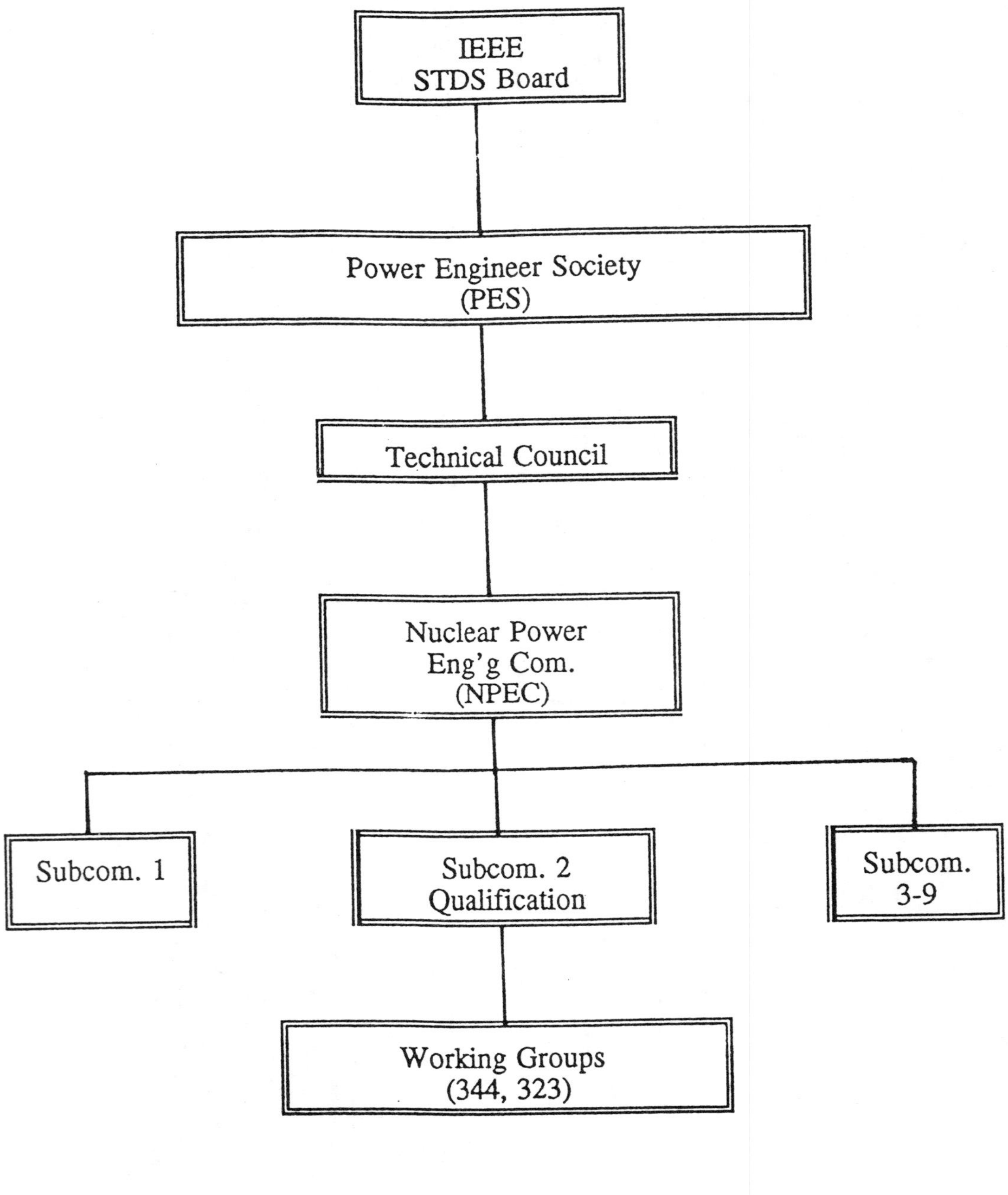

FIGURE 3 - IEEE ORGANIZATIONAL STRUCTURE

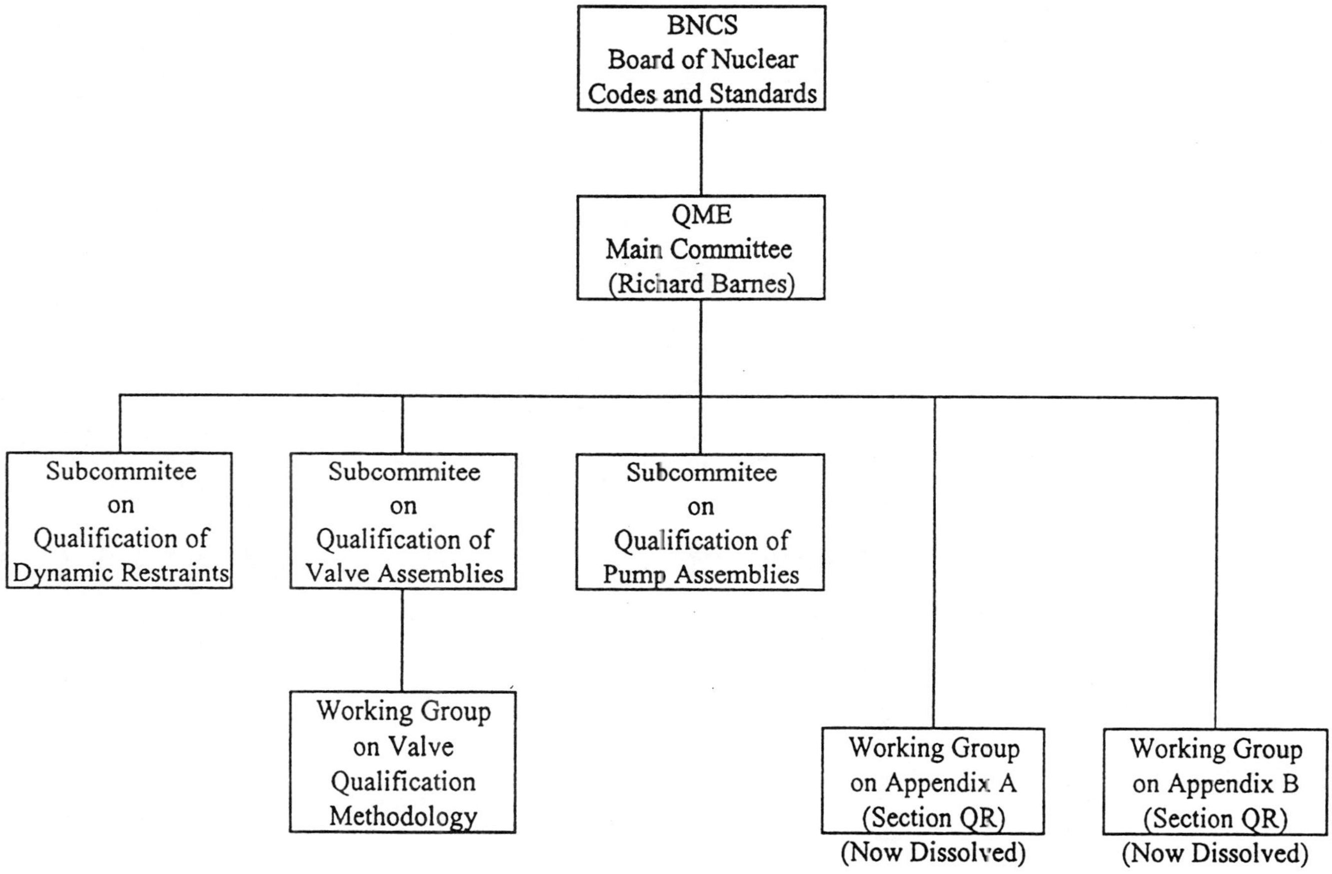

FIGURE 4 - QME ORGANIZATION

PVP-Vol. 340, Seismic Engineering
ASME 1996

A Case Study for Assessment of Consistency between ASME Section III and XI Rules

Yong Zhao and Gery Wilkowski
Battelle-Columbus
Columbus, Ohio

ABSTRACT

A case study was conducted to evaluate how the current ASME Boiler and Pressure Vessel (B&PV) Code Section XI nuclear piping flaw evaluation rules are affected by the design stress changes in increasing the allowable stress with the B&PV Code Section III piping design rules. A piping model was developed for this purpose from a realistic nuclear piping referred to as BM3 that has been used widely for developing or evaluating the nuclear piping design rules. The piping system was modified so that all the conditions required by the 1995 Section III rules are met. Seismic excitations and internal pressure were increased to reach the maximum allowable stresses under the Section III rules including the effects of pressure, deadweight, seismic inertia, seismic anchor motion, and thermal expansion. The stresses thus calculated elastically using a finite element analysis were then combined using the Section XI rules to evaluate whether the maximum stress ratios for flaw acceptance are exceeded that are allowed by Section XI IWB-3514 for the workmanship standard flaw sizes corresponding to current nondestructive examination capability at butt weld areas. Quantitative results were provided for discussions.

INTRODUCTION

New ASME B&PV Code (1995) Section III rules for the design of nuclear Class 1, 2, and 3 piping are included in the current code. The 1995 rules include several initiatives for attempting to solve issues about conservative seismic design margins relative to supporting structures or buildings. These rules are based on limited vibration tests of uncracked piping and piping fittings (EPRI, 1994; Slagis, 1993) and relevant elastic analyses, without considering the effects of flaws that are usually evaluated using Section XI rules of the code. The 1995 rules assume fatigue and ratcheting failure modes, rather than a structural collapse with reversing dynamic loads. The averaged through-wall ratcheting strain is limited to 5 percent for a 10-cycle event without leading to the fatigue-ratcheting failure. The 1995 rules greatly increase piping allowable primary stresses to 4.5 times the allowable design stress intensity, S_m, applicable to the Level D service loads of a safe shutdown earthquake (SSE) under the following conditions (Chen, et al, 1995):

1. The elastic response spectral dynamic approach is used to account for the seismic inertial stresses. The input response spectra should be 5% damped and ±15% peak-broadened;
2. The ratio of the dominant spectrum frequency to the lowest piping frequency is greater than 0.5;
3. The ratios of the piping outside diameter to wall thickness should not be greater than 50;
4. The B_2-based deadweight stress should be limited to $0.5S_m$;
5. The C_2-based stress ranges due to the SSE anchor motion effects should be limited to $6.0S_m$;
6. The averaged axial stresses caused by the SSE anchor motion should be limited to $1.0S_m$.

The above conditions can be satisfied easily in practical seismic piping design. However, the 1995 Section III rules are not accepted by the NRC due to some uncertainties and other related issues (NRC Letter, 1995).

The Section XI pipe flaw evaluation rules and criteria are based on elastic-plastic fracture mechanics and limit-load analyses using elastically-calculated piping stresses on the net-section of uncracked piping from Section III. In the Section XI pipe flaw rules, the allowable flow stress is defined as the maximum stress that an uncracked pipe could tolerate. The flow stress is $2.4S_m$ for ferritic pipe and $3.0S_m$ for austenitic pipe. The primary stresses plus the secondary stresses divided by a safety factor are used to determine the relevant loads for pipe flaw evaluations. The primary stresses include the bending stresses due to seismic inertial and dead-weight loads and the longitudinal stress due to pressure. The secondary stresses include thermal expansion and/or seismic anchor motion-induced stresses. The safety factor is equal to 1.39 for the SSE and 2.77 for the operating basis earthquake (OBE) controlled design. A Z-factor is introduced to account for elastic-plastic fracture considerations, due to the effects of pipe size and material toughness in Table H-6310-1 for the code specified material property S_m equal to 18.1 ksi and in Table H-6310-2 for other S_m values of materials. As a consequence, the allowable stresses changed in Section III may be significantly high enough that the Section XI rules can not be used to justify continued operation with any flaws above the workmanship flaw sizes in Section XI Article IWB-3514. This raises an issue of consistency between the Sections III and XI rules.

A number of references are available for evaluating the Section III rule changes from an uncracked pipe point of view (Tagart and Ranganath, 1992; Slagis, 1993; Barnes, Tagart, Branch and Landars, 1994; Adams, Branch and Landers, 1994; Terao, 1995; Chen, W. P., Jaquay, K. R., Chokshi, N. C. and Terao, 1995). In contrast, few studies were conducted for assessment of the impact on the Section XI flaw evaluation rules and acceptance criteria due to the Section III stress design changes (Yoon, 1994; Krishnawamy, Ghadiali and Wilkowski, 1995).

In this paper efforts were made to calculate the piping stress ratios using the Section XI rules and the elastically calculated primary and secondary stresses based on the 1995 Section III rules. Comparisons were made between the calculated stress ratios and the maximum allowable stress ratios in compliance with the Section XI acceptance flaw criteria. The purpose for this case study is to assess the consistency between the Section III stress design changes with the Section XI rules. To this end, a realistic nuclear piping model was developed specially for satisfying the 1995 Section III rules. A low frequency content SSE excitation was introduced.

PIPING SYSTEM MODEL

As shown in Figure 1, the BM3 water service piping system is 160-foot long and includes 3-inch, 4-inch, and 8-inch diameter Schedule 40 pipes and ten long radius elbows, one tee, one reducer, and various in- and out-of-plane bent structures. The piping is made of ASTM A-106 Gr. B carbon steel with the material properties listed in Table 1. The piping system is anchored at three ends of the 3-inch, 4-inch, and 8-inch diameter pipes. The original BM3 piping system was modified by changing supports between the anchors. The supports were designed on the basis of a support spacing concept using the NCIG-14 method (Stevenson and Associates and EQE Engineering, 1990) such that the piping system will behave more flexibly to accommodate an added thermal loading. It is understood that the NCIG-14 method was intended for small bore piping only, however, the basic concept has been applied literally to piping sizes up to 24-inch diameter (Adams, Branch and Landers, 1994). All five vertical supports thus defined were assumed to be rigid spring hangers. The seven horizontal restraints were specified as snubbers for seismic loading resistance purposes. Operating pressure and temperature were assumed to be 200 psi and 300 F.

The seismic excitation spectrum, as shown in Figure 2, was defined (Zhao, Stevenson and Gasparini, 1992) from a typical floor response spectrum (Chuang, et al, 1984) by converting to 5% damping after considering ASME B&PV Code Case N-411 (1995) as well as a broadening factor of ±15%, and then normalizing to 0.15g period ground acceleration (PGA) for an OBE earthquake assuming a floor peak acceleration two times real ground acceleration. The ratio of peak SSE spectral acceleration to peak OBE spectral acceleration was taken as 1.5. The floor response spectrum has a peak frequency range of 2.19 to 2.90 Hz that is expected low for discussing the spectral frequency content issue related to developing the 1995 Section III rules. The response spectra input in the two horizontal X and Z directions were assumed to be identical, and the response spectrum input in the vertical Y direction was assumed to be two-thirds of that input in the horizontal directions. Differential seismic anchor movements (SAM) were assumed at the ends of the 4-inch and 8-inch diameter pipes with 0.5-inch displacement for the OBE earthquake and 0.75-inch displacement for the SSE earthquake in the horizontal directions. No vertical SAM effects were taken into account, since the supporting structure was assumed to behave relatively stiffer in the vertical direction than the horizontal directions.

STRESS DESIGN USING SECTION III 1995 RULES

The Section III rules require separate elastic analyses of the piping system under seismic inertial, SAM, deadweight, and thermal expansion loadings corresponding to Level A, B, C and D service conditions. The resultant moments due to these loadings should be then combined in the most conservative manner.

The general purpose finite element code ANSYS version 4.4 (1990) was employed for the analyses. The piping model developed (Figure 1) consists of elastic pipe and elbow elements with a total of 67 nodes. The mass of the pipe and water contained is lumped at the element nodal points.

A response-spectrum dynamic analysis was performed for the seismic inertial effects. The natural frequency for the first five modes of the piping system was found to be 2.84, 4.35, 4.70, 4.86, and 6.68 Hz. The natural frequency for the 29th mode is 32.96 Hz. The modal combination of responses of pipe moments was done first for each of the three seismic inertial loadings using: (1) the square root sum of square (SRSS) and absolute sum approaches (absolute sum here is only for modes whose frequencies are within 10 percent of frequency of the previous mode grouped according to NRC Regulatory Guide 1.92) when modal frequencies are less than 33Hz; (2) algebraic sum for the modes whose frequencies are greater than 33Hz; (3) absolute sum of the results from (1) and (2). Then resultant responses of moments in the three directions were combined using the spatial SRSS method. The modal superposition was carried out from the first mode through the 74th mode in the X-direction, the 87th mode in the vertical Y-direction, and the 83rd mode in the Z-direction for a participation mass fraction of above 92 percent. As a practical matter, when the modal cumulative mass reaches about 80 percent of the total modal mass of the piping system, any increase in the number of modes to account for more mass will not change the dynamic analysis results by more than 10 percent, hence the requirements of the NRC Standard Review Plan Section 3.9.2 are typically met here.

The finite-element-analysis (FEA)-calculated moments at typical nodal points of the piping under SSE seismic inertial, SSE SAM, thermal, and dead-weight loadings were obtained by the ANSYS code. These moments were combined separately using the most conservative methods in each direction and then using the SRSS rule to get the resultant moments required for the stress calculations in the Code equations including Equations (9), (11), and (13), etc. in the Subsections NB/NC/ND-3600 for the failure modes of plastic collapse, fatigue, and ratcheting. The Code provides defined stress criteria for Equations (9) and (13). For the alternating stress intensity S_a criterion in the Code Equation (11), ten full cycles of earthquake excitations were assumed. Using the Code S-N fatigue design curve and considering a margin of 2 for fatigue failure and a factor of 20 for the corresponding cyclic loading, the value of S_a was obtained as 310 ksi for carbon steel with an ultimate tensile strength of 60 ksi at 300 F.

Both OBE-governing design and SSE-governing design analyses were performed at a number of nodal points to find the lowest safety margins at a particular nodal point associated with each Code equation. For the OBE-governing design, the lowest safety margin of 2.96 was obtained using the Code Equation (13) at Node 37, an elbow node of the 3-inch diameter pipe. The Code Equation (9) gives the lowest safety margin of 3.74 at Node 27 which is also an elbow node in the 3-inch diameter pipe, while the Code Equation (11) provides the lowest safety margin of 5.11 at straight pipe Node 37 of the 3-inch diameter line. When the OBE excitations are scaled up to 4.0 or 4.2 times current excitation values, the calculated stresses using the Code Equations (13) or (9) will reach the corresponding maximum allowable stress, respectively.

For the SSE-governing design, the lowest safety margin of 6.5 with Code Equation (9) was determined at straight pipe Node 27 of the 3-inch diameter line with the 1995 stress criterion of $4.5S_m$ for the primary membrane plus primary bending stress consideration. It is noted that the ratio of the dominant load-driving frequency to the lowest piping system natural frequency is in the range of 0.77 to 1.02, which is obviously greater than the prescribe limit of 0.5. When the SSE excitations are scaled up to 7.0 times current excitation values, the calculated stress in the Code Equation (9) will be equal to the maximum allowable stress of $4.5S_m$. All six conditions for application of this 1995 stress criterion, as listed previously in the introduction of this paper, are met herein for the normal SSE and scaled-up SSE loadings.

Table 2 lists the FEA-calculated primary stresses at selected points using the 1995 ASME Section III Code Equation (9) v.s. the new criterion of $4.5S_m$, and the FEA-calculated SAM bending plus averaged axial stresses v.s. the corresponding new criterion of $7.0S_m$ under Level D service condition. The inertial and SAM SSE loadings are scaled-up to 6.9 times the normal SSE specified earlier. The pressure is increased to 1,692 psi to reach the limit of $1.0S_m$. From the results shown in Table 2 it can be seen that the FEA-calculated primary stress at Node 27, an elbow end of the 3-inch line, exactly equals $4.5S_m$ and all other primary stresses are below $4.5S_m$. All the FEA-calculated SAM stresses do not exceed $7.0S_m$.

From the analysis results described above a conclusion may be made that the specific uncracked piping system was designed well in compliance with the 1995 Section III stress design rules and acceptance criteria.

For a further analysis of the cracked pipe using the Section XI rules, a maximum design pressure of 1,692 psi and the maximum seismic loading of 6.9 time the current SSE will be used to calculated piping stresses for the SSE-governing design or Level D service loading. The OBE excitation ratio value will drop to 2.8 for the maximum design stress using the Code Equation (13) for the OBE governing design or Level B service loading. This is for demonstration purpose for the Level B loading, since regulatory changes in process may essentially eliminate the OBE from design considerations.

APPLICATION OF SECTION XI RULES

Article H-6000 of Section XI is applicable to circumferential flaw evaluation of ferritic piping not less than 4.0 inches in diameter. This criterion is primarily used for determining the allowable stresses for a giving crack size. In this paper, the Code-recommended analytical approach in Article H-6000 was used to evaluate the 3", 4" and 8" lines using the Section XI flaw evaluation rules and acceptance criteria. Table 2 shows the FEA calculated primary bending stress due to the maximum seismic inertial load (scaled-up SSE and OBE levels), primary longitudinal stress due to the maximum pressure of 1,692 psi, and secondary bending stress due to thermal expansion without and with the maximum seismic SAM bending stress under the scaled-up SSE and OBE loadings. The resultant stresses corresponding to each loading were calculated using the most conservative combination approach.

For the Section XI flaw evaluation analysis, Article H-6320 of Section XI gives the equation for calculation of the allowable bending stress, S_c, as

$$S_c = [1/(SF)](P_b'/Z - P_e) - P_m[1 - 1/Z/(SF)] \quad (1)$$

where

P_m = primary membrane stress from pressure,
P_b' = limit bending stress from limit-load analysis defined in Article H-5320,
P_e = thermal expansion stresses (secondary stresses),
Z = a correction factor to account for elastic-plastic fracture behavior, and
SF = safety factor.

Rearranging Equation (1) and normalizing it with S_m gives

$$Z\{S_c + P_e/(SF) + P_m[1 - 1/Z/(SF)]\}/S_m = P_b'/(SF)/S_m \quad (2)$$

It is noted that Equation (2) is for the maximum loading of a cracked pipe. When the allowable primary bending stress S_c is replaced with actually calculated primary bending stress from dead-weight and inertial loads, the calculated stress ratio from the FEA analysis (SR_{FEA}) may then be defined as

$$SR_{FEA} = Z\{S_c + P_e/(SF) + P_m[1 - 1/Z/(SF)]\}/S_m \quad (3a)$$

Since it was somewhat unclear in Section XI Appendix H where the SAM bending stresses were to be included, Equation (3a) excludes them, whereas Equation (3b) includes them in the thermal expansion stress since they are generally considered to be secondary stresses. It should be noted that the effects of the averaged SAM axial stresses are small and not considered herein.

$$SR_{FEA+SAM} = Z\{S_c + (P_e+SAM)/(SF) + P_m[1 - 1/Z/(SF)]\}/S_m \quad (3b)$$

The Section XI stress ratio, SR_{XI}, can be defined from Equation (2) as

$$SR_{XI} = P_b'/(SF)/S_m \quad (4)$$

The limit bending stress, P_b', is defined in Article H-5320 as

$$P_b' = 2\sigma_f[2\sin\beta - (a/t)\sin\theta]/\pi \quad (5)$$

$$\beta = [\pi - (a/t)\theta - (P_m/\sigma_f)\pi]/2 \quad (6)$$

where,

σ_f = flow stress taken as 2.4 S_m,
a = flaw depth,
t = nominal wall thickness,
β = angle of pipe bending neutral axis as defined in Figure H-4221-1 of Section XI,
θ = angle of circumferential flaw as defined in Figure H-4221-1 of Section XI.

The maximum limit bending stress, $P_b'_{max}$, occurs as the flaw size goes to zero, where Equations (5) and (6) in combination of a = 0, and θ = 0 gives

$$P_b'_{max} = (4/\pi)(2.4S_m)(0.79) = 2.414S_m \quad \text{(for Level D, } P_m = 1.0S_m) \quad (7a)$$

$$P_b'_{max} = (4/\pi)(2.4S_m)(0.95) = 2.903S_m \quad \text{(for Level B, } P_m = 0.5S_m) \quad (7b)$$

Substituting Equation (7) into Equation (4) gives the maximum Section XI stress ratio, $SR_{XI\text{-}max}$, for cases of no crack or flaw as follows.

$$SR_{XI\text{-}max} = 1.74 \quad \text{(for Level D, SF = 1.39)} \qquad (8a)$$

$$SR_{XI\text{-}max} = 1.05 \quad \text{(for Level B, SF = 2.77)} \qquad (8b)$$

The Z-factors, accounting for elastic-plastic fracture conditions for ferritic pipe, are defined in Table H-6310-2 of Section XI when S_m is not equal to 18.1 ksi. The Z-factors are a function of the pipe diameter and the material where the flaw is located. In this case, we considered the flaws as being in an E7018 weld and hence the Z-factor equation for Material Category 1 in Table H-6310-2 was used. The Z-factors were computed as 1.110 for the 3-inch diameter pipe[1], 1.128 for the 4-inch diameter pipe, and 1.243 for the 8-inch diameter pipe. The value of S_m was 20 ksi for the A106 Grade B piping at 300 F.

The Section XI FEA-calculated stress ratios using Equations (3a) and (3b) and the FEA-calculated stresses for Levels B and D conditions were computed and listed in Table 3. The stress ratios are based on the maximum allowable loadings by the 1995 ASME Section III rules. The FEA-calculated stress ratios is a function of the pipe diameter (from the Z-factor) and the safety factor (SF).

The Section XI Ferritic Pipe Stress Ratios for flaws that just slightly exceed the Section XI Article IWB-3514 workmanship limits ($SR_{XI\text{-}max}$) are given in Table 3 for the pipe system at OBE and SSE loadings, respectively.

The key comparison of this effort was the determination of the $SR_{FEA}/(SR_{XI\text{-}max})$ ratios, i.e., the ratio of Equation (3a) or Equation (3b) to Equation (4). Table 3 shows the $SR_{FEA}/(SR_{XI\text{-}max})$ and $SR_{FEA+SAM}/(SR_{XI\text{-}max})$ values at selected piping points for the Level B and Level D conditions. If this ratio is greater than 1.0, then it means that that location has higher stresses than could be acceptable for flaws greater than the Section XI Article IWB-3514 workmanship flaw evaluation criteria, i.e., the Section XI Article IWB-3650 and Appendix H criterion could not be used. Hence, the pipe would have to be repaired or replaced if flaws exceeding the Section XI Article IWB-3514 flaw limits occurred.

DISCUSSION

For the Level D service condition from Tables 2 and 3 and Figures (including results at all piping points) 3 and 4, it can been seen that the stress ratios of the FEA-calculated primary stress over $4.5S_m$ using 1995 Section III Equation (9) does not exceed 1.0 at any nodal point of the whole piping system and reach 1.0 only at Node 27 of the elbow node of the 3-inch pipe. Higher stress ratios occur in the 3-inch pipe than the 4-inch and 8-inch pipes. This means the primary inertial plus dead-weight stresses more significantly affect the 3-inch pipe than the 4-inch and 8-inch pipes. However, when using the criterion of the C_2-based SAM bending stress less than $6.0S_m$ or $7.0S_m$ and considering the averaged axial SAM tension not greater than $1.0S_m$, the 3-inch pipe appears much stronger than the 4-inch and 8-inch pipes because the SAM-induced displacement loadings were input at the anchors of the 4-inch and 8-inch pipes. The results show that the 1995 Section III stress design requirements are met. The SAM bending stress is only 10 percent of the inertial stress at the most critical primary stress location of nodal point 27 (elbow nodal point) of the 3-inch pipe. The largest SAM bending stress is 0.44 times the corresponding inertial stress at Node 38 (elbow nodal point) of the 3-inch pipe, 3.70 times at Node 48 (elbow nodal point) of the 4-inch pipe, and 4.49 times at Node 60 (elbow nodal point) of the 8-inch pipe.

In evaluating the Section XI rules, Tables 2 and 3, and Figures 3 and 4 demonstrate that the 4-inch and 8-inch pipes are acceptable for the elastically calculated stresses consistent with the 1995 Section III rules, but the Section XI stress ratios are much larger than the Section III stress ratios at most of nodal points of all of the pipes for both cases of primary and secondary stress considerations. This means that the 1995 Section III rules significantly impact the ability to apply the Section XI rules. The comparisons of the Section XI stress ratios, either excluding or including the effects of secondary SAM bending stresses, indicate that SAM bending stresses are the key factors for evaluation of the 4-inch and 8-inch pipes that have higher SAM bending stresses relative to other stresses. The SAM bending stresses play an important role in determining the Section XI flaw acceptance rules and criteria. It is known that the SAM stress depends on both the piping supports and supporting structure for a given seismic excitation event, even though the 1995 Section III rules provide a condition to limit the SAM bending stress to $6.0S_m$ and the averaged SAM axial stress to $1.0S_m$.

For the 3-inch pipe, since the effects of the SAM bending stresses are small, no large differences exist between the Section XI stress ratios with and without consideration of the SAM effects. The Section XI FEA-calculated stress ratios exceed the maximum stress ratio criterion for flaw acceptance by up to 106% at twelve nodal points of both straight and elbow nodes. The twelve points represent 31 percent of the total length of the 3-inch pipe as defined in the FEA modeling.

From this case study a conclusion may be reached that piping designed in compliance with the 1995 Section III rules would have sufficiently high enough stresses that the Section XI pipe flaw evaluation rules and acceptance criteria for the Level D service condition or SSE dominant design could not be used.

For the Level B service condition, Table 3 shows that when the SAM stress is not included, the Section XI FEA-calculated stress ratio slightly exceeds the maximum stress ratios for flaw acceptance at only two locations of the 3-inch pipe, one at Node 29 and the other at Node 34. Both are straight pipe nodes. These results show that the 1995 Section III rule for the Level B service condition may be consistent with the Section XI rules.

For comparison purposes, the piping system was designed with the old (before 1994 Winter edition, say 1989 edition) Section III rule of primary stress not greater than $3.0S_m$. Table 4, similar to Table 2, lists the FEA-calculated pipe stresses at selected points using the 1989 Section III Equation (9) v.s. the old criterion of $3.0S_m$, and the FEA-calculated C_2-based SAM bending stresses v.s. an assumed criterion of $3.0S_m$ for the SAM-caused ratcheting failure mode from the thermal ratcheting failure mode as defined in Section III Equation (12). When the excitation was scaled up to 4.6 times current SSE value and the pressure was increased to 1,692 psi to reach to $1.0S_m$, the FEA-calculated primary stress at Node 27 (elbow nodal point) of the 3-inch pipe exactly equals to $3.0S_m$ and all other primary stresses are less than $3.0S_m$. Only the FEA-calculated SAM bending stresses at Nodes 60 and 61(elbow nodal points) of the 8-inch pipe exceed the limit of $3.0S_m$, since the SAM bending stresses at Nodes 60 and 61 are four times larger than the inertial

---

[1] Appendix H of Section XI is actually limited to pipe larger than 4-inches in nominal diameter. This was an arbitrary cut-off, hence calculations were made for the 3-inch pipe for illustrative purposes.

stresses due to larger torsional moments existing. Figures (including results for all piping points) 5 and 6 also show these results for the 1989 Section III rules.

Table 5, similar to Table 3, presents the Section XI FEA-calculated stress ratios based on the 1989 Section III Level B and D stress criteria. It is noted that only at Node 29 (straight pipe nodal point) of the 3-inch pipe the FEA-calculated stress ratios excluding and including SAM bending effects exceeds the Section XI maximum stress ratio criterion for flaw acceptance. This can be seen obviously from Figures 5 and 6. The reason for the higher Section XI FEA-calculated stress ratio occurring at Node 29 may be due to the large torsional inertial moment, and hence the largest B_2-based inertial bending stress. The same B_2 value of 1.0 for straight pipe was used in the Section XI stress calculations for all the nodal points of the piping loop assuming that the straight pipe side is of interest at elbow local areas. In the Section III stress calculations, the B_2 values of 2.57, 2.82, and 3.52 were computed for the 3-inch, 4-inch, and 8-inch long radius butt welding elbows in accordance with Article NB-3683.7 of Section III. This makes the B_2-based inertial stress at Node 27 (an elbow node) larger than that at Node 29 (a straight pipe node with B_2 equal to 1.0), even though the inertial moment at Node 27 is smaller than that at Node 29.

It is noted from Figures 3 through 6 that the Section III FEA-calculated stress ratios at some points are several times larger than those at other points for the same pipe size. The reason is that these higher stress points represent elbow stresses while the lower stress points represent straight pipe stresses. The straight pipe has B_2 (for inertial bending stress) and C_2 (for SAM bending stress) values equal to 1.0, but the elbow has higher values of B_2 listed above and C_2 of 3.85, 4.23, and 5.28 for the 3-inch, 4-inch, and 8-inch long radius butt welding elbows in accordance with Article NB-3683.7 of Section III.

Tables 4 and 5, and Figures 5 and 6 demonstrate that the old (before 1994 Winter edition) Section III rules are more consistent with the Section XI rule.

CONCLUSION

The specific case study results presented in this paper illustrates the inconsistency between the 1995 Section III piping stress design rules and the Section XI piping flaw evaluation rules and acceptance criteria for the SSE-dominant design or Level D service condition. These results essentially show that the 1995 Section III piping stress design rules result in sufficiently high enough stresses, such that if a flaw larger than the workmanship standards in Section XI IWB-3514 occurs, then the Section XI pipe flaw evaluation rules could not be used for justification of continued operation. Hence, the only options would be to repair or replace the pipe.

It should also be noted, that this particular case study had relatively low SAM stresses in the 3-inch pipe that dominates the stress design of the whole piping loop due to the relatively smaller SAM displacements (0.75 inches in the horizontal direction) input at the anchors of the 4-inch and 8-inch pipes. A piping system with SAM stresses approaching the maximum allowable by Section III would have even further difficulty in meeting the Section XI pipe flaw evaluation criteria. To raise the issue but without violation of the application conditions for the Section III design changes, numerous piping examples similar to this case could be developed to evaluate consistency between the Section III and XI rules, since a number of choices are available for defining the piping layout, supports, supporting structures, construction (including welding and installation), and analysis approaches used. There are many uncertainties existing behind these variables. Only sophisticated and comprehensive study could definitely address the inconsistency issue.

REFERENCE

Adams, T. M., Branch, E. B., and Landers, D. F. (1994), "Comparative Analysis Study for Piping Seismic Design Criteria: Parts I and II," ASME PVP-Vol. 275-1, pp. 3-31.

ANSYS (1990) Version 4.4, User Manual.

ASME (1995) Boiler and Pressure Vessel Code.

Barnes, R. W., Tagart, S. W., Branch, E. B., and Landars, D. F. (1994), "Proposed Revision to Section III Seismic Piping Rules," ASME PVP-Vol. 275-1, pp.67-71.

Chen, W. P., Jaquay, K. R., Chokshi, N. C. and Terao, D. (1995), "An Assessment of Seismic Margin in Nuclear Plant Piping," SMiRT13 Vol.III, pp.507-512.

Chuang, T. Y., et al (1984), "Impact of Changes in Damping and Spectrum Peak Broadening on the Seismic Response of Piping Systems," NUREG/CR-3526.

EPRI (1994), "Piping and Fitting Dynamic Reliability Program," EPRI TR-102792-Vol. 1 through Vol. 5.

Krishnaswamy, P., Ghadiali, N. And Wilkowski, G. (1995), "An Assessment of Proposed Changes in ASME Section III Allowable Stresses on the Critical Flaw Size of Piping," SMiRT13, Vol.II, 161-167.

NRC (1995), "ASME Code Revisions to the Design Rules for Piping Systems," Letter (and attached technical documents) to Mr. G. M. Eisenberg, Director of Nuclear Codes and Standards, ASME, from Mr. Brian W. Sheron, Director, Division of Engineering, Office of Nuclear Reactor Regulation, May 24.

Stevenson and Associates and EQE Engineering (1990), "Procedure for Seismic Evaluation and Design of Small Bore Piping (NCIG-14)," EPRI Report NP-6628, April.

Tagart, S. W. and Ranganath, S. (1992), "Proposed Dynamic Stress Criteria for Piping," ASME PVP-Vol. 237-1, pp.133-137.

Terao, D. (1995), "A Regulatory Perspective on Appropriate Seismic Loading Stress Criteria for Advanced Light Water Reactor Piping Systems," Journal Of Pressure Vessel Technology, Transactions of the ASME, Vol. 117, May, pp. 177-181.

Zhao, y., Stevenson, J. D. And Gasparini, D. A. (1992), "Failure Probability of Piping Systems in Nuclear Plant Facilities Subjected to Earthquake Loads," ASME PVP-Vol. 237-1, pp. 177-182.

Yoon, K. K. (1994), "A Review of Flaw Size Factor between Sections III and XI of ASME Boiler and Pressure Vessel Code," ASME PVP-Vol. 280, pp 41-43.

Table 1 Mechanical Properties of A106 Grade B Carbon Steel

Property	70 F	400 F
E	2.95×10^7 psi	2.77×10^7 psi
S_y	3.5×10^4 psi	3.1×10^4 psi
S_u	6.0×10^4 psi	6.0×10^4 psi
S_m	2.0×10^4 psi	2.0×10^4 psi
Thermal Expansion Coefficient	5.42×10^{-6} in/in.F	6.61×10^{-6} in/in.F
Poison's Ratio	0.3	

Table 2 1995 ASME Section III Stress Ratios of FEA-Calculated Equation (9) Primary Stress v.s. $4.5S_m$ and FEA-Calculated SAM Bending Plus Averaged SAM Axial Tension Stress v.s. $7S_m$ under Level D Condition (with Scaled-up SSE Loading, Deadweight, and Maximum Pressure of 1,692 psi) --- at Typical Nodal Points

Node #	Do inches	Equation (9) / 4.5Sm	SAM (bending + tension) / 7.0Sm		Node #	Do inches	Equation (9) / 4.5Sm	SAM (bending + tension) / 7.0Sm
8	3.5	0.207	0.010		42	4.5	0.312	0.577
13	3.5	0.551	0.026		47	4.5	0.178	0.506
20	3.5	0.222	0.006		54	4.5	0.194	0.000
26	3.5	0.424	0.023		60	8.625	0.181	0.775
27	3.5	**0.998**	0.098		61	8.625	0.219	**0.833**
28	3.5	0.846	0.103		62	8.625	0.072	0.130
33	3.5	0.462	0.012		67	8.625	0.087	0.000

Table 3 1995 ASME Section XI FEA-Calculated Stress Ratios v.s. Section XI Maximum Stress Ratios under Level B and D Conditions --- at Typical Nodal Points

Node #	Do inches	SR_{FEA} (SSE)	$SR_{FEA+SAM}$ (SSE)	SR_{FEA} / SR_{XI-max} (SSE)	$SR_{FEA+SAM}$ / SR_{XI-max} (SSE)	SR_{FEA} (OBE)	$SR_{FEA+SAM}$ (OBE)	SR_{FEA} / SR_{XI-max} (OBE)	$SR_{FEA+SAM}$ / SR_{XI-max} (OBE)
26	3.5	1.89	2.01	**1.09**	**1.16**	0.75	0.89	0.71	0.85
27	3.5	2.08	2.22	**1.20**	**1.28**	0.80	0.95	0.77	0.90
28	3.5	1.78	1.93	**1.03**	**1.11**	0.72	0.87	0.69	0.83
29	3.5	3.41	3.57	**1.97**	**2.06**	1.15	1.29	**1.10**	**1.23**
30	3.5	1.67	1.91	0.96	**1.10**	0.70	0.86	0.67	0.82
31	3.5	1.73	2.00	**1.00**	**1.15**	0.72	0.88	0.69	0.84
33	3.5	2.07	2.13	**1.19**	**1.23**	0.88	1.02	0.84	0.97
34	3.5	2.34	2.44	**1.35**	**1.40**	0.95	1.09	0.91	**1.04**
35	3.5	2.02	2.22	**1.16**	**1.28**	0.84	0.99	0.80	0.95
37	3.5	1.56	1.96	0.90	**1.13**	0.71	0.89	0.68	0.85
38	3.5	1.54	1.95	0.89	**1.12**	0.71	0.89	0.68	0.85
39	3.5	1.99	2.40	**1.15**	**1.38**	0.85	1.03	0.81	0.98
42	4.5	0.78	1.55	0.45	0.89	0.52	0.77	0.49	0.73
50	4.5	0.59	1.15	0.34	0.66	0.47	0.68	0.44	0.65
58	8.625	0.92	1.46	0.53	0.84	0.71	0.99	0.68	0.95
61	8.625	0.66	1.64	0.38	0.94	0.62	0.96	0.59	0.91

Table 4 1989 ASME Section III Stress Ratios of FEA-Calculated Equation (9) Primary Stress v.s. $3.0S_m$ and FEA-Calculated SAM Bending Stress v.s. $3.0S_m$ under Level D Condition (with Scaled-up 4.6 Times SSE Loading, Deadweight, and Maximum Pressure of 1,692 psi) --- at Typical Nodal Points

Node #	Do inches	Equation (9) / 3.0Sm	SAM (bending) / 3.0Sm		Node #	Do inches	Equation (9) / 3.0Sm	SAM (bending) / 3.0Sm
8	3.5	0.220	0.016		42	4.5	0.324	0.896
13	3.5	0.562	0.040		47	4.5	0.184	0.786
20	3.5	0.271	0.009		54	4.5	0.204	0.000
26	3.5	0.464	0.033		60	8.625	0.182	1.203
27	3.5	1.006	0.150		61	8.625	0.228	1.293
28	3.5	0.855	0.158		62	8.625	0.082	0.199
33	3.5	0.514	0.018		67	8.625	0.102	0.000

Table 5 ASME Section XI FEA-Calculated Stress Ratios v.s. Section XI Maximum Stress Ratios under Level B and D Conditions of 1989 Section III Rules -- at Typical Nodal Points

Node #	Do inches	SR_{FEA} (SSE)	$SR_{FEA+SAM}$ (SSE)	SR_{FEA} / SR_{XI-max} (SSE)	$SR_{FEA+SAM}$ / SR_{XI-max} (SSE)	SR_{FEA} (OBE)	$SR_{FEA+SAM}$ (OBE)	SR_{FEA} / SR_{XI-max} (OBE)	$SR_{FEA+SAM}$ / SR_{XI-max} (OBE)
27	3.5	1.45	1.54	0.83	0.89	0.80	0.95	0.77	0.90
28	3.5	1.25	1.35	0.72	0.78	0.72	0.87	0.69	0.83
29	3.5	2.32	2.43	**1.34**	**1.40**	1.15	1.29	**1.10**	**1.23**
30	3.5	1.18	1.33	0.68	0.77	0.70	0.86	0.67	0.82
31	3.5	1.22	1.40	0.70	0.80	0.72	0.88	0.69	0.84
32	3.5	1.01	1.11	0.58	0.64	0.68	0.82	0.65	0.78
33	3.5	1.47	1.51	0.85	0.87	0.88	1.02	0.84	0.97
34	3.5	1.65	1.72	0.95	0.99	0.95	1.09	0.91	**1.04**
35	3.5	1.43	1.56	0.82	0.90	0.84	0.99	0.80	0.95
37	3.5	1.12	1.39	0.64	0.80	0.71	0.89	0.68	0.85
38	3.5	1.11	1.38	0.64	0.80	0.71	0.89	0.68	0.85
39	3.5	1.42	1.69	0.82	0.97	0.85	1.03	0.81	0.98
42	4.5	0.60	1.12	0.348	0.643	0.52	0.77	0.49	0.73
50	4.5	0.47	0.84	0.272	0.484	0.47	0.68	0.44	0.65
58	8.625	0.73	1.10	0.423	0.631	0.71	0.99	0.68	0.95
61	8.625	0.55	1.20	0.316	0.693	0.62	0.96	0.59	0.91

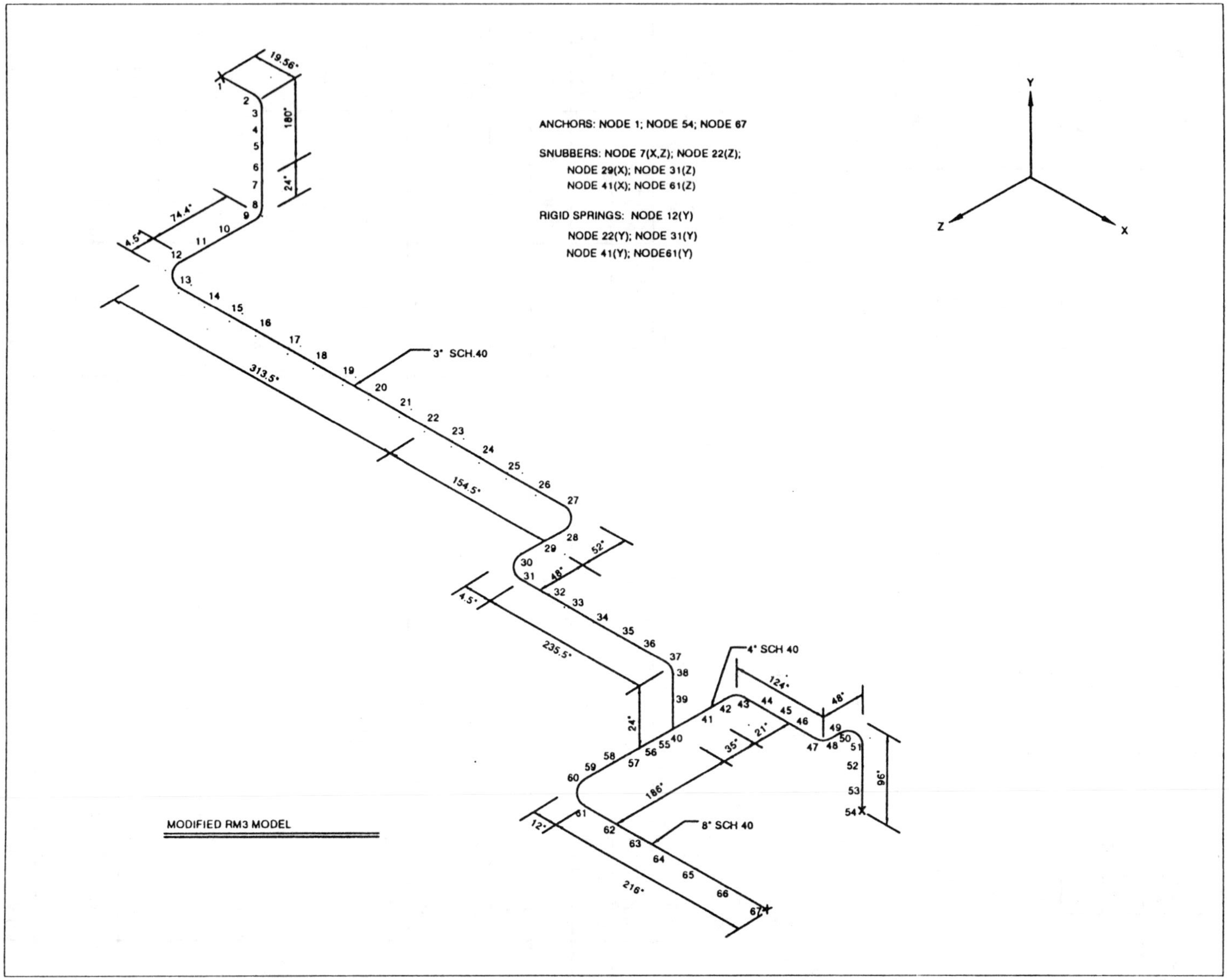

Figure 1 Modified BM3 Nuclear Piping System Model

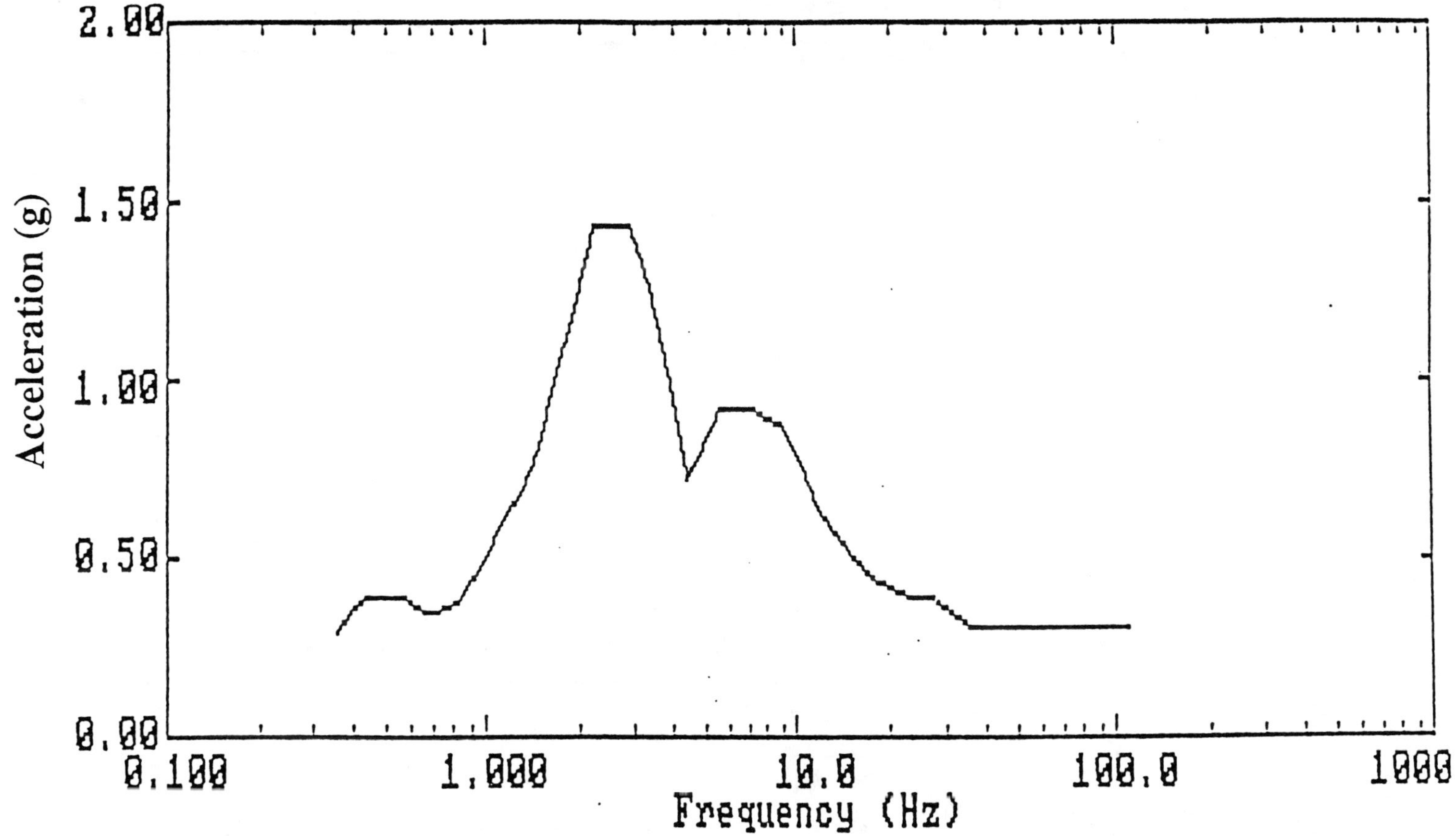

Figure 2 OBE Horizontal Floor Response Spectrum

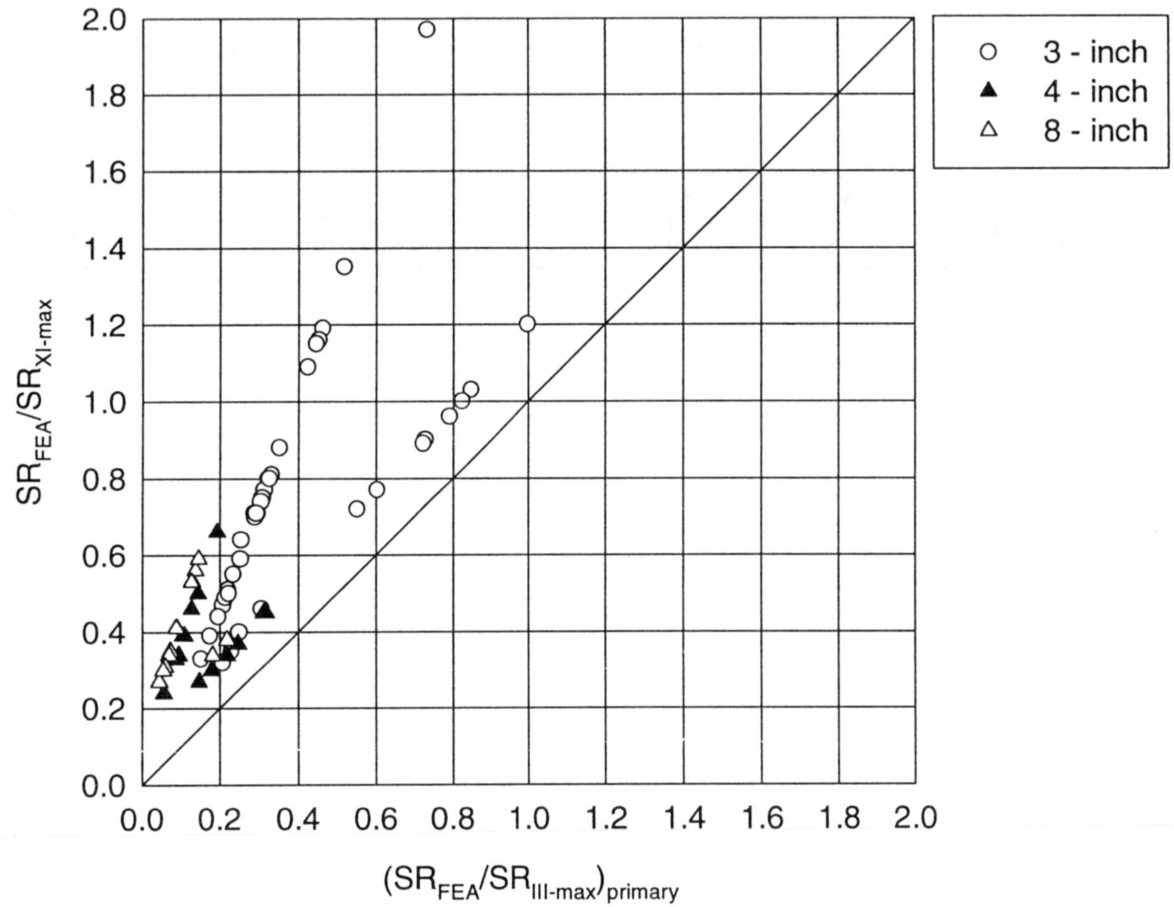

Figure 3. Comparisons of Stress Ratios at Level D Service Condition Using 1995 Section III Equation (9) Primary Stress Rules and Section XI Rules Excluding Secondary SAM Bending Stresses

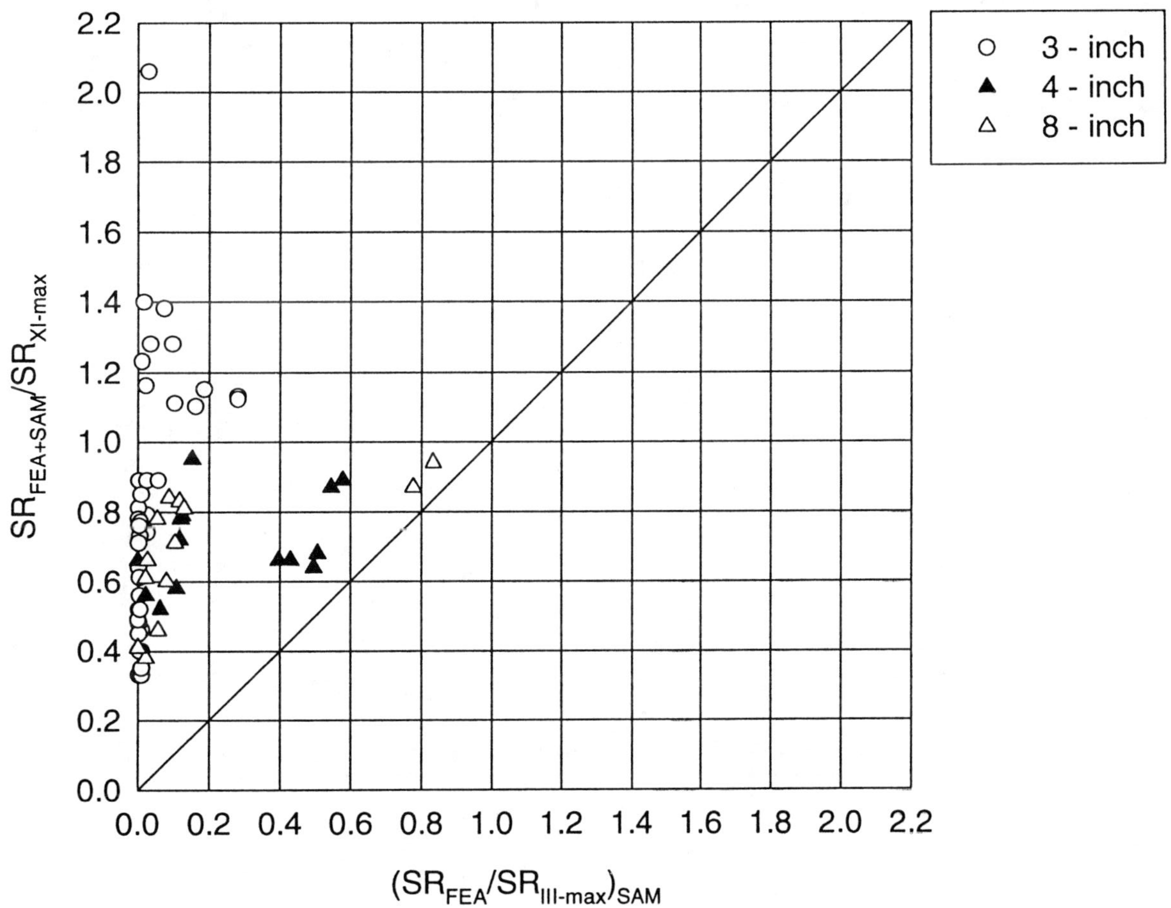

Figure 4. Comparisons of Stress Ratios at Level D Service Condition Using 1995 Section III Equation (9) Primary Stress Rules and Section XI Rules Including Secondary SAM Bending Stresses

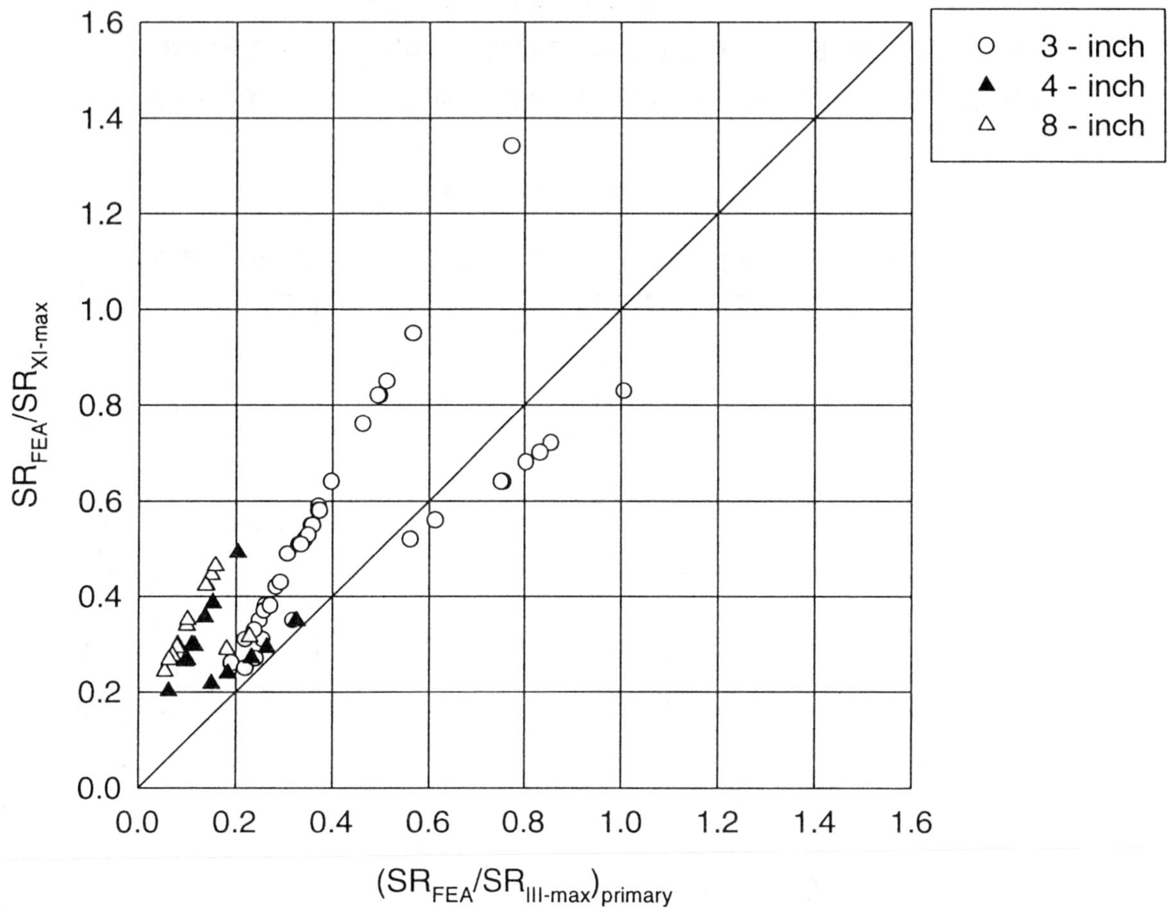

Figure 5. Comparisons of Stress Ratios at Level D Service Condition Using 1989 Section III Equation (9) Primary Stress Rules and Section XI Rules Excluding Secondary SAM Bending Stresses

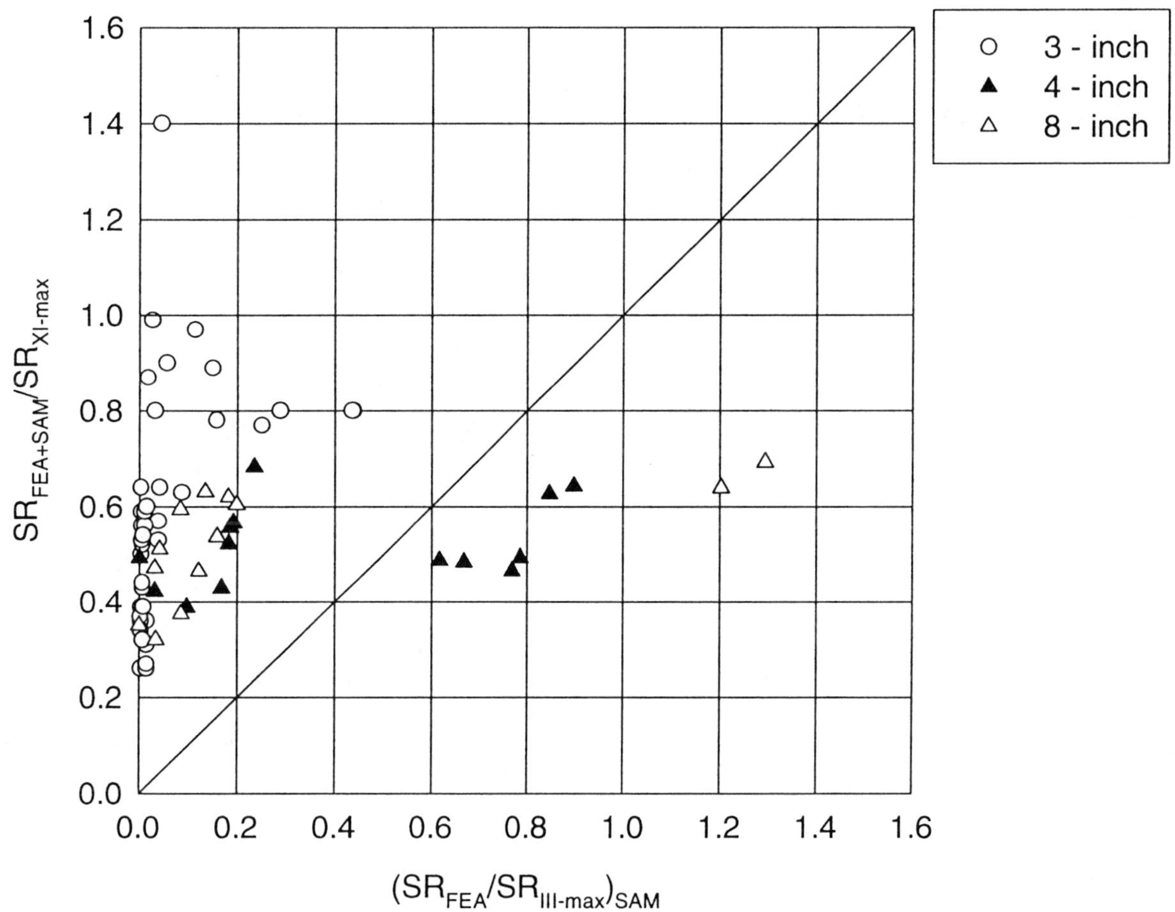

Figure 6. Comparisons of Stress Ratios at Level D Service Condition Using 1989 Section III Equation (9) Primary Stress Rules and Section XI Rules Including Secondary SAM Bending Stresses

PVP-Vol. 340, Seismic Engineering
ASME 1996

FAILURES IN PIPING MANUFACTURED TO NON-NUCLEAR STANDARDS

Richard W. Barnes
Member, ASME
ANRIC Enterprises
Etobicoke, ON, Canada

G. Douglas Cooper
Member, ASME
G.D. Cooper Consultants Inc.
Oakville, ON, Canada

Donald F. Landers
Member, ASME
Teledyne Brown Engineering
Woburn, MA, USA

ABSTRACT

Most non-nuclear process piping systems in Canada and the United States are constructed to the requirements of the piping codes of the American Society of Mechanical Engineers (ASME B31.1 and B31.3). Section III of the ASME Boiler and Pressure Vessel Code, has additional requirements for piping that are expected to provide further assurance of pressure boundary integrity.

This project[1] reviewed piping failures in the non-nuclear industry and attempted to determine whether the impact of the additional requirements of Section III were of benefit in preventing failure of the pressure boundary. The analysis of the database assembled for this study, indicated that the material requirements of Section III would have had an impact. It was found that in the area of design, all of the piping codes have very similar requirements and these requirements do not cover the typical failures that were identified and examined in this set of failure points.

INTRODUCTION

Most non-nuclear process piping systems in Canada and the United States are constructed to the requirements of the piping codes of the American Society of Mechanical Engineers (ASME B31.1 (ref. 1) and B31.3 (ref. 2)). The three piping codes of Section III (refs. 3, 4 & 5) of the ASME Boiler and Pressure Vessel Code, have additional requirements for piping that are expected to provide further assurance of pressure boundary integrity.

This project reviewed piping failures in the non-nuclear industry and attempted to determine whether the impact of the additional requirements of Section III were of benefit in preventing failure of the pressure boundary. The approach taken in this study was to determine the cause of failure of non-nuclear piping subjected to service similar to that experienced by piping in CANDU nuclear power plants.

Through a process of interviews and literature search, 186 failures were identified and assembled into a reference database. Many of the records were incomplete; therefore, the reference database was trimmed to include a subset of 65 failure points supported by complete data. This subset formed the basis for this study. The results from studies of other databases assembled for similar purposes were reviewed and compared with the conclusions reached in this study. These reviews tended to confirm the conclusions of the study.

BACKGROUND

There were eight tasks identified for this project. The first task consisted of gathering data on piping failures from any available non-nuclear source that had service conditions similar to those of a CANDU reactor. This limited the study to the collection and examination of failure information associated with carbon steel piping systems filled with water (steam), which operate at

[1]This project was sponsored by the Atomic Energy Control Board of Canada under its Regulatory Research and Support Program: AECB Project No: 2.275.1. The AECB is the government organization in Canada that is responsible for the regulation of nuclear power.

temperatures up to 600 ^{0}F and at pressures up to 1600 psi. Tasks two and three consisted of identifying the root causes of failure associated with each of the pipe failures, and then grouping the pipe failures according to root cause.

The next three tasks were: to analyse Section III requirements to determine whether or not application of these requirements would have prevented the failure; to summarize the results of the analysis and to identify the requirements of the Section III Code that would have prevented the failure from occurring; and finally, to develop the probability of failure for each one of the five piping codes. This last task could not be done. The data collected were not specific enough to enable that type of evaluation to be made. The difficulty in collecting detailed data is a common experience in this type of study. The remaining two tasks were the preparation of a report (ref. 6) and presentation of the findings of this work in a seminar to the sponsoring body for this project.

This study is an important first step in answering frequently asked questions about the value of the existing Section III requirements which exceed those of ASME B31 piping codes for power piping and other industrial piping.

DATA COLLECTION

Sources of Piping Failure Information

Piping failures occur at all industrial process facilities, ranging from fossil powered generating stations, to refining facilities, to chemical processing facilities. Much of the failure information in these industries is never seen in published literature. Probably the major reason for this paucity of information is the pressure on the plant operators to maintain production. The effort required to assemble the failure information in a form that can be presented in a public forum with credibility, is for most companies beyond the budget and perceived need to publish this data. They are more interested in spending the available money on prevention of failures than on publication of the causes of failures. The motivation for maintaining a degree of confidentiality is also intensified by the possibility of lawsuits and/or the need to keep sensitive operating information from competitors.

For large accidents, information is usually unavailable because of the existence of lawsuits and the potential for new lawsuits. As a result of the investigation done for this project, 186 failures were identified and assembled into a reference database that forms the basis for this report (ref. 7). For the purposes of this study a failure was taken to mean a breach of the pressure boundary except for a few cases that were found and repaired before a breach occurred.

About 100 of the 186 failure records assembled in the database may be called "raw data". The term "raw data" denotes failure information obtained directly from personnel at the plant site, either in discussion with them and/or from reference to documentation held at the plant. This term also includes failures reported in internal documents held in company libraries. Contributors included personnel from refineries, chemical plants, plastics manufacturing plants, thermal power plants, etc.

Information on another 60 failures was retrieved from failure reports in the files of the Inspection and Enforcement Branch, at the Technical Standards Division of the Ministry of Consumer and Commercial Relations of Ontario. The Boilers and Pressure Vessels Act of Ontario (ref. 8) requires plant owners to report to the Chief Inspector of the Province within 24 hours, any property damage and/or personal injury as a result of a break in the pressure boundary.

The remaining failures identified in the database have been gleaned from published literature such as papers from the annual ASME Pressure Vessel and Piping Conferences.

Other Sources of Data

Other sources of information were identified and yielded data on failures. In most instances, the information could not be independently verified or evaluated. This prevented the data from being accepted as valid "raw data". Data in this category is useful as a supplementary source to the primary data and is a valuable tool for checking and comparing with the conclusions reached from the primary data. Sources for this data type included general references provided by refineries, a database from the National Board of Boiler and Pressure Vessel Inspectors - Columbus, Ohio, a report from the Pressure Vessel Research Committee (PVRC) - Piping Review Committee, and limited databases from two inspection companies (see refs. 9-15 incl.).

Acceptability of Data

In the initial development of the failure database, it was agreed with the AECB staff, that for a data point to be considered valid, information was required for each of the variables shown in Table 1. The information for a large number of the 186 failure records was incomplete. Within the time constraints and budget of this study, it was not possible to generate the effort required to collect the missing information. In some cases it may not have been possible to get the complete information at all, since archives have been lost or destroyed. Meeting the above requirement resulted in a reduction of the collected records in the original database from 186 failure records to 65 Failure records.

Although many of the failure records in the reference database are incomplete, they are still real failures and can be of value in the development of valid conclusions, providing the missing information is not relevant to the conclusion.

It is important to recognize that the failure points assembled are not statistically random and therefore, they are not necessarily representative of the piping failures in industry. However, the analysis and comparison with the Section III requirements provide a good understanding of the type of protection received from the requirements of Section III, and whether or not Section III would have assisted in the prevention of the failures.

THE ASME CODE:
Its Approach to Pressure boundary Integrity of Piping

The ASME B&PV Code has developed a systematic approach to the application of requirements that if met, provide assurance of pressure boundary integrity of piping and other pressure boundary components. Firstly, the basic premise behind all of the ASME pressure boundary codes is "..no pressure boundary failure.." There is also a similarity of approach between Section III and other "construction" codes, viz., Section I (ref. 16) and Section VIII (ref. 17), of the Boiler and Pressure Vessel Code.

Section III has tightened the control of the design and manufacture of the pressure boundary items by defining specific responsibilities for the plant owner, the designer, the manufacturer and the installer of pressure boundary components and parts. Further, Section III defines additional requirements to be met in the areas of the design specification, the design, control of material, its manufacture and repair, the manufacture of the pressure boundary items, their examination and testing, and implementation of quality assurance programs combined with a classification approach.

The classification approach is based on the concept of providing greater assurance of pressure boundary integrity for components which have potential for a greater impact on the safety of the public. This classification approach is divided into three classes, Class 1, Class 2 and Class 3. There are specific requirements for each class, which are defined in the ASME B&PV Code , Section III, by Subsection NB (Class 1), Subsection NC (Class 2) and Subsection ND (Class 3), respectively. Each of these subsections can be regarded as a separate code.

Generally, Class 1 is applied to the systems associated with cooling the fuel, and these systems are more likely to experience high levels of radioactivity. Class 2 is mainly used for containment piping, and Class 3 is applied to systems expected to experience lower levels of radioactivity. Within each subsection in Section III, the construction of a component is divided into seven areas: material, design, fabrication, welding, examination, testing and overpressure protection, and each subsection contains requirements to be met for each of these areas.

A similar approach, although not exactly the same, is used for the ASME B31.1 and B31.3 piping codes. Section III has added two more areas: strict control on the content and responsibility for the design specification and design output documents, and definition of quality assurance requirements , which are applied to the control of materials and the construction of components.

Within the framework of these areas, Section III limits the type of materials that can be used in the construction of piping in the different classes; controls welding through limitation on material, procedures and personnel qualification; controls the joining of materials and products by practices that have been proven to be reliable and effective; places restriction and requirements on the examination methods used to determine the integrity and acceptability of base materials and joints, as well as, the repair of materials. Further, Section III provides specific requirements to protect piping from

- explosive rupture because strength capacity was exceeded,
- failure due to thermal fatigue,
- plastic collapse of the piping as a result of design loads and low cycle dynamic loads (including seismic), and
- brittle fracture through control of material properties.

Section III also provides a general guidance list of potential loads/root causes of failure.

ROOT CAUSES IDENTIFIED

The number of failures associated with the root causes identified are presented in Table 2 below. The terms used in the description of these root causes are generic and often include causes of failure related to the generic term, but often more complex than the generic term.

The terms, Manufacturing Defect and Mechanical, are significantly different from the other root causes because the other descriptions are explicitly definitive in their meaning. These two terms are collective by nature, and have failure records within their collection that are totally unrelated. For example under manufacturing defect, there is a machining tolerance defect and a forging defect. Both are obviously associated with manufacturing but unrelated, except that both may have been caused by human error.

IMPACT OF SECTION III REQUIREMENTS

Methodology

Outlined below is the methodology used to determine whether the application of Section III requirements would have prevented these failures from occurring. The root causes of failure were divided into two separate groups. The first group included the root causes which were clearly defined, such as corrosion/erosion, vibration fatigue, etc., and in which each failure was similar to the other failures collected under that root cause description. The second group, as discussed previously, consisted of two root causes which were collective in nature, and resulted in the collection of failure records that were associated with the title description. For this second group of root causes, each failure had to be considered individually for coverage by the Code.

The process used to analyse each of these groups was to identify all relevant references in the five codes, B31.1, B31.3, Subsection NB, Subsection NC, and Subsection ND that applied to that particular root cause/failure. From this review, the extent of the coverage of these root causes by each code could be determined and compared. The results are summarized in the discussion below and compared in Tables 3 and 4.

Root Cause of Failure and ASME Code Requirements

Each root cause identified in Table 2 of this paper is evaluated against the requirements specified in the five ASME piping codes B31.1, B31.3 and Section III, Subsections NB, NC, ND. As far as could be determined, all the piping systems associated with the failures were constructed to ASME piping codes B31.1 or B31.3.

In all five piping codes, there are two types of requirements provided. The first type are general advisory requirements, for which the responsibility has been placed directly on the designer to understand and provide a design capable of withstanding the loadings and conditions that would result in pipe failure from the root cause. The second type of requirements are specific requirements that the design must meet, and in doing so, if the loads and conditions have been specified correctly, the root cause is negated.

Root Causes Vs. Code Requirements

Set out below in Table 3 is a summary of the coverage of the root causes by the various ASME Codes, B31.1, B31.3, Section III, Subsection NB, NC and ND. The Mechanical and Manufacturing Defects are covered by Table 4 where each individual failure is examined.

One point immediately evident from Table 3 is the close relationship between the various pressure boundary codes that cover piping. Of the 65 failures included, the choice of any one of the recent codes would have made little or no difference for about 75% of these failures. The only categories listed in Table 3, for which the nuclear/non-nuclear code coverage is significantly different , are manufacturing defects, freezing and abrasion.

The rules are similar, if not the same, in many areas of the five codes. This should not be a surprise, since the basic premise behind the ASME code development is, "No pressure boundary failure". Therefore, when one code committee finds it necessary to make a specific change, the other code committees will usually examine that action and follow along with similar changes, if applicable.

For most of the root causes of failure there is only general guidance given in the codes. It became obvious during the data collection process, that when a code does not contain specific requirements for recurring failures, often a company produces its own design and fabrication procedures. For example, in an attempt to reduce the number of failures experienced with small nipples, which often fail as a result of vibration fatigue, companies often specify installation procedures to reduce and if possible, elminate vibration.

There is another example of this type of proactive action. Recently, the petroleum industry has become aware of the potentially large economic and safety impact that can result from the use of the wrong material. With the advent of relatively cheap and accurate material analysers, the plant inspection departments have begun to analyse the chemical content of each piece of material used in their plants and replace the piece when it does not correspond to the design specification. This is not required by any of the codes.

All the mechanical failures would not have been impacted by the Section III requirements, because the root causes are all related to an operating plant and therefore, are not covered by a construction code.

More significantly, the three codes in Section III seem to be more effective than the other two codes, in the area of manufacturing defects. The control of material, the requirements for volumetric and surface examination of welds associated with the manufacture of fittings and pipe, the rules for repair of defects in material, together with control of the manufacturing process, provide the capability for detection of the defects that resulted in failures under this category. It is not possible to say that they would have prevented the occurrence 100% of the time but the potential for finding the flaws was there.

Failures due to Manufacturing Defects

It is clear from Table 4 that the increased requirements of Section III in the control, examination, and repair of defects, and the control of manufacture and welding provides the potential to discover the defects that lead to the pipe failures. In the case of Item 108 (lap opened), the failure was discovered during the hydro test. All five codes require a hydrostatic test of about the same magnitude, although the NB test pressure is lower. However, there is no guarantee that another lap defect would be discovered by a hydrostatic test. If the forging had been slightly stronger, it might have withstood the hydrostatic test, and the flaw might not have been detected. The volumetric examination requirements for materials in Subsection NB, might have picked up this lap defect. This requirement is unique to Subsection NB. Subsections NC & ND have surface examination requirements for forgings under certain circumstances and therefore, the lap might have slipped detection.

A somewhat similar comment could be made for Item 279 as has been made above for Item 108. The crack in the elbow of Item 279 was discovered before it was placed in the system, but it did pass the manufacturing process. In this actual example the hydro test would have certainly have caused it to leak, but a crack that did not go through the whole wall might have withstood the hydro test.

For Item 243 (wrong materials), general guidance is provided in B31.3 about the use of halides with stainless steels. This advice is not present in the other four codes. Since this system was constructed to B31.3, this would have been a design error since the code had warned against it. The design requirements of Section III would not have prevented this, (except that Subsection NB does not allow bellows). On the other hand, if this had been a supplier's mistake, the material controls of Section III should have detected the mistake.

In the case of Item 229 (incorrect hole drilled), all the codes would have required it to be corrected. The defects of Items 212 (lack of overlap pass) and 213 (lack of overlap pass), might have been discovered by meeting the surface examination requirements for fillet welds in Subsection NB and NC but not ND. Subsection ND only requires surface examination of fillet welds for diameters greater than two inches.

In Table 5, the types of failure resulting from manufacturing defects are summarized. In the sample of 65 failures, one of the failures due to the manufacturing defects resulted in an explosive type release. The remaining defects were either leaks (and one of them was a large leak), or failure without fluid loss.

Why do the Failures Still Occur?

With all the effort, research and development that goes into the development of codes of practice, failures still occur. Systems and components that are designed to meet the code requirements still experience failure due to these root causes. Several factors could give rise to this problem. The first is the ability to correctly predict the loadings and conditions in the design specification. Next the designer has to understand the detailed nature of the impact of these conditions on the design and what must be done in the design to handle/compensate for the impact. This can be a significant problem at times even for the most experienced designer. A third factor may be the increased tendency of designers to rely on analysis instead of the application of sound design practice and layout practices based on past experience. The failures identified in this subset cannot be covered successfully by analytical techniques.

Components and systems are also significantly impacted by maintenance and operation. When these activities result in the system operating outside the original design envelope, the design itself can be severely compromised. The complexity of the ageing mechanisms associated with many of the root causes also prevents development of satisfactory criteria, The design solution is left up to the judgment, knowledge and experience of the designer to provide a satisfactory answer.

In this sample of 65 points, the areas in which Section III has the potential to make an impact, is in the control, use and preparation of material and the more stringent requirements for manufacture that are part of the Section III requirements. While the potential for exclusion or detection of the defects in this sample is increased very significantly, there is still not a 100% assurance that all of the defects would have been found before operation. The size of the equipment, the method of manufacture and the sensitivity of the measurement associated with these variables are all factors that leave room for some uncertainty to exist.

REVIEW OF THE LARGER DATABASE

A review of the larger database of 186 points shows some interesting correlations. Corrosion (31%) and fatigue (23%) failure processes account for 54% of the total failures. Of these categories, 45% of the total corrosion failures are due to corrosion/erosion and 33% are due to external corrosion. For fatigue, failure due to high cycle vibration make up 75% of the total fatigue failures.

Approximately 84% of all failures occurred for pipe diameters twelve inches and less, 70% of the failures occurred for pipes six inches in diameter or less and 36% of all the failures occurred for pipes two inches in diameter or less. Smaller pipes tend to be more prone to failure. Of course, account must be taken of the fact that the total length of smaller pipe is greater; hence, a larger number of failures would be expected in the smaller pipe.

Of the variables that influence the failures experienced by pipes of different diameters, one major impact is the type of service for which the pipe is used. A review of the database shows that process lines and header type supply lines are generally two inches in diameter or greater. The smaller lines are often used as drain lines, sample lines, nipples for instruments, thermowell connections, recirculation lines at the discharge of pumps, and low flow type systems that provide a more utilitarian type of service.

The corrosion/erosion failures in this sample appear to be concentrated heavily in the smaller diameter set of six inches and less. This is best explained by the type of service for which this size of piping is used in this sample.

Approximately 55% of the fatigue-vibration failures occurred in lines one inch diameter and less, and the fatigue-vibration root cause of failure accounts for 45% of all the failures in that diameter range. With increasing diameter size and therefore increasing stiffness, vibration failure is no longer dominant. There is also an indication that the cause of the fatigue failure changes with the larger pipe diameters, from a vibration loading to a thermal or a low cycle loading, which once again can also be shown to be strongly related to the service experienced by these pipes.

The analysis showed that 68% of the failures were leaks, cracks, pinhole leaks and a category called "other". This category includes situations when the pressure boundary did not break because the thinning of a wall was found during a maintenance procedure, allowing correction of the problem before there was an actual pressure boundary failure. The remaining 32% of the failures are categorized as ruptures which indicates a more violent, often explosive type release.

A pinhole leak is almost exclusively related to corrosion. Rupture is often associated with water hammer, overpressure, overload, manufacturing defect, mechanical damage and even some fatigue failures. This type of failure associated with fatigue or corrosion/erosion has to be a combination of fatigue or corrosion/erosion with a dynamic loading of some type, such as a water hammer pulse.

Finally, this discussion points to another advantage of Section III, which is associated with the materials allowed in Section III. Ductile material usually results in a leak through the wall before a rupture occurs. The materials available for Section III piping are very ductile and operate at relatively low pressure and temperature. The materials allowed by the other codes can be of high strength with substantial reduction in ductility. They can operate at high pressure and temperatures and therefore, much smaller flaws could lead to explosive type failures.

COMPARISON WITH OTHER DATABASES

In the literature search stage of this project, there were several studies of pipe failures identified that had databases related to this project. However, these studies were not acceptable as original databases because they could not be verified independently. These studies were briefly reviewed and a comparison of the conclusions reached by them was made with the conclusions arrived at from analysis of the reference database assembled for this study.

Discussion

These studies tended to confirm the existence of the same root causes of failure as identified by the present AECB project. Corrosion/erosion and vibration fatigue are the more common failures and have been occurring in pipes for a long time. The 1964 GE Report (ref. 10) found that not enough attention was paid to corrosion/erosion during the design phase. All the studies suffer from the same problem of being unable to obtain complete data. As a consequence, it forced them also to be more qualitative in their conclusions. There is also similarity between studies in the relationship between pipe diameter, root cause and service experienced by the pipe.

SUMMARY AND CONCLUSIONS

Summary

1. There were 186 failure points assembled into a reference database. Since only failure points for which complete information was available could be used in the evaluation, the database was trimmed to 65 points. The root causes of failure for this subset were identified and then analysed according to the following methodology:

- Based on the root causes identified in the trimmed database, it was found that the process of evaluation required the database to be divided into two separate groups. The first group included the root causes which were clearly defined such as corrosion/erosion, vibration fatigue, etc. The second group consisted of the two root causes that were collective in nature, namely, Mechanical and Manufacturing Defect. In this second group each individual failure of the group was analysed for coverage by the Code.

- The process used was, identify for each root cause in Group 1 and each failure in Group 2, all relevant references in the five codes, B31.1 (ref. 1), B31.3 (ref. 2), Subsection NB (ref. 3),

Subsection NC (ref. 4), and Subsection ND (ref. 5), that applied to that particular root cause or failure. From that review, the extent of the coverage by each code of these root causes/failures, was determined and compared.

- From this analysis, the impact of the requirements of Section III was evaluated to determine whether the use of Section III would have prevented these failures from occurring.

2. Once the analysis of the subset had been completed, the reference database was analysed to gain an understanding from the larger database of the relationships between the root causes of failure and the variables of pipe size, service for which the pipe was used, location of failure and wall thickness.

3. Other databases that had been assembled previously in other studies or separately by other people for their own purposes were also reviewed and compared to the database assembled for the study. They confirmed the conclusions reached.

Conclusions

The conclusions reached in this study are:

1. For this sample of 65 failures, about 75% of the root causes of failure are addressed in all five codes either through specific rules or by general reference. The design requirements in one code are usually quite similar to the others. Therefore, the application of Section III design requirements rather than the requirements of the Code of Record used, would not have been a significant factor in preventing failure.

2. Failures in piping do occur and will continue to occur because of the occurrence of situations that were not covered by the design specification, ageing processes and human errors which cause the system to be operated outside the design envelope.

3. Section III requirements have merit because of the control, examination and repair of materials, and the requirements for welding, manufacturing and installation. Section III also limits the material use to piping specifications that specify high ductility. This will tend to mitigate the type of failures that occur towards leaks rather than explosive type ruptures.

4. The review of the full database collected for the study, indicated that there is a significant correlation between the root cause of the failure and service for which the piping is used. For this sample, the correlation also translated to pipe sizes, probably because for the industries from which this database was collected, certain pipe sizes are associated with the processes and services that result in these failure types. This finding was also confirmed by the other databases referenced in this study, which would seem to indicate that certain processes and

   services across a broad range of industries use similar pipe sizes for these services.

5. The large majority of failures occurred in smaller pipes (NPS 6 and less). The root causes of corrosion/erosion and vibration fatigue were mainly responsible for the pipe failures in these sizes. The ASME codes handle these root causes of failure only by general advice and warning.

The conclusions reached in this study are based on a limited set of data. The application of the findings of this study must take this limitation into consideration. Nevertheless, the other databases reviewed, confirmed the root causes of failure identified in the reference database assembled for the study both in type and relative frequency, giving credence to the conclusions reached.

In conclusion, Section III does have an advantage in the detection of material and manufacturing defects over the other piping codes and therefore, will make a difference. It must once again be kept in mind, that these findings are the result of a very small sample of pipe failures, and the conclusions reached might be different if another set of failures were studied.

REFERENCES:

1. ASME B31.1, Power Piping, 1992 Edition including Addenda A92 (ASME B1.1a - 1992).

2. ASME B31.3 Chemical Plant and Petroleum Refinery Piping, 1990 Edition up to and including the 1992 Addenda (ASME B31.3c - 1992).

3. ASME Boiler and Pressure Vessel Code, Section III, Subsection NB, 1992 Edition with Addenda up to and including A93.

4. ASME Boiler and Pressure Vessel Code, Section III, Subsection NC, 1992 Edition with Addenda up to and including A93.

5. ASME Boiler and Pressure Vessel Code, Section III, Subsection ND, 1992 Edition with Addenda up to and including A93.

6. Barnes, R.W. and Cooper, G.D.; Failure Rates in Piping Manufactured to Different Standards; a report prepared for the Atomic Energy Control Board, Ottawa, Canada under

its Regulatory Research and Support Program, AECB Project No. 2.275.1; November, 1995

7. Reference Database of Piping Failures, compiled by G.D. Cooper Consultants Inc., 1994. (Atomic Energy Control Board Reference Library - Category: Restricted Report)

8. Boilers and Pressure Vessels Act of Ontario, Revised Statutes of Ontario, 1980, Chapter 46 as amended by 1983, Chapter 33.

9. National Board of Boiler and Pressure Vessel Inspectors, Incident Reports, 1971-1991.

10. W.S. Gibbons and B.D. Hackney, Survey of Piping Failures for the Reactor Primary Coolant Pipe Rupture Study, US AEC Report GEAP-4574, General Electric Company, May 1964.

11. S.L. Basin and E.T. Burns, Characteristics of Pipe System Failures in Light Water Reactors, EPRI NP-438, August 1977, prepared by Science Applications, Inc. for Electric Power Research Institute.

12. T.A.W. Geyer, N.W. Burst, L.J. Bellamy, and J.A. Astley, Causes of Pipe-work Failures: Implications for Assessing the Effects of Plant Management of Failure Rates, for presentation at A.I.Ch.E. National Meeting, Orlando, Florida, March 18 - 21, 1990, Copyright, Unpublished.

13. Piping Failures Reported during the Period of 1991 to 1993, Factory Mutual Engineering, Facsimile Transmission to R. W. Barnes from Arthur Spencer, April 21, 1994.

14. J.F. Copeland, Environmental Effects on Components: Commentary for ASME Section III, EPRI NP-5775, April 1988, prepared by Structural Integrity Associates, Inc. for Electric Power Research Institute

15. Welding Research Council Bulletin, Mechanisms and Corrective Actions, WRC 382, June 1993.

16. ASME Boiler and Pressure Vessel Code, Section I, Power Boilers, 1992 Edition with Addenda up to and including A93.

17. ASME Boiler and Pressure Vessel Code, Section VIII, Division I, Pressure Vessels, 1992 Edition with Addenda up to and including A93.

Table 1 - Information Fields for Record of Failure

General Information	Process Information	Failure Details
Case Number	Pipe Diameter	Date of Failure
Location of Files	Pipe Thickness	Location of Failure
Background Reference	Material	Phase of Life Cycle
Company/Plant	Pressure	Nature of Failure
	Temperature	Cause of Failure
	System	
	Fluid	
	Construction Code	

Table 2 - Root Causes Identified

No.	Basic Root Cause	No. of Failures
1	Abrasion	1
2	Corrosion	22
3	Fatigue	16
4	Mechanical	2
5	Manufacturing Defect	12
6	Water Hammer	1
7	Overpressure	1
8	Erosion	3
9	Freezing	5
10	Overheating	1
11	Overload	1
Total		65

Table 3 - Coverage of Root Causes by ASME Codes

No.	Root Causes	B31.1	B31.3	Class 1	Class 2	Class 3
1	Abrasion	No	No	Yes	Yes	Yes
2	Corrosion					
	- Internal Corrosion	Yes	Yes	Yes	Yes	Yes
	- External Corrosion	Yes	Yes	Yes	Yes	Yes
	- Corrosion/Erosion	Yes	Yes	Yes	Yes	Yes
	- Galvanic Corrosion	Yes	Yes	Yes	Yes	Yes
	- Stress Corrosion	Yes	Yes	Yes	Yes	Yes
3	Fatigue					
	- Dynamic Load (not seismic)	Yes (S)	Yes (S)	Yes (S)	Yes (S)	Yes (S)
	- Thermal Loads	Yes (S)	Yes (S)	Yes (S)	Yes (S)	Yes (S)
	- Vibration High Cycle	Yes	Yes	Yes	Yes	Yes
4	Mechanical	Table 4	Table 4	Table 4	Table 4	Table 4
5	Manufacturing Defect	Table 4	Table 4	Table 4	Table 4	Table 4
6	Water hammer	Yes	Yes	Yes	Yes	Yes
7	Overpressure	Yes (S)	Yes (S)	Yes (S)	Yes (S)	Yes (S)
8	Erosion	Yes	Yes	Yes	Yes	Yes
9	Freezing	Yes	Yes	No	No	No
10	Overheating	Yes (S)	Yes (S)	Yes (S)	Yes (S)	Yes (S)
11	Overload	Yes (S)	Yes (S)	Yes (S)	Yes (S)	Yes (S)

Note: "Yes" means it is covered by the Code and "(S)" means specific requirements

Table 4 - Coverage of Mechanical & Manufacturing Defect Root Causes

No.	Root Causes	B31.1	B31.3	Class 1	Class 2	Class 3
	Mechanical					
242	- Stepping on pipe	No	Yes	No	No	No
231	- High velocity impact	No	No	No	No	No
	Manufacturing Defect					
215	- Poor machine tolerances	No	No	Yes	Yes	Yes
279	- Crack through wall of elbow	No/Yes	No/Yes	Yes	No/Yes	No/Yes
230	- Flange cracked due to overheating	No	No	Yes	Yes	Yes
241	- Pipe Scarf weld with inclusions	No	No	Yes	Yes	No
226	- Forging defect	No	No	Yes	No	No
243	- Wrong use of materials; bellows split	No	Yes	NA	No	No
108	- Lap opened up in hydro test	No/Yes	No/Yes	Yes	Yes/No	Yes/No
219	- Improper formation of ERW joint	No	No	Yes	Yes	No
232	- Fitting fails; weld process defective	No	No	Yes	Yes	No
229	- Hole drilled by accident	Yes	Yes	Yes	Yes	Yes
212	- Lack of overlap of passes on fillet weld	No	No	Yes	Yes	No
213	- Lack of overlap of passes on fillet weld	No	No	Yes	Yes	No

Note: "Yes" means there are Code requirements that might have resulted in its detection and when combined, "Yes/No" means that detection was not an absolute certainty. When reversed, "No/Yes", it means the were no code requirements, but it might have still been detected. NA means bellows are not allowed for Class 1 piping.

Table 5 - Types of Failures from Manufacturing Defects

Type of Failure	No.	Item Nos.
Leak	3	215, 279, 226
Explosive Release	1	241
Pinhole Leak	3	212, 213, 219
Large Leak	1	243
None	4	108, 229, 232, 230

PVP-Vol. 340, Seismic Engineering
ASME 1996

PRIMARY STRUCTURE
-- AN IMPORTANT CONCEPT TO DISTINGUISH PRIMARY STRESSES

LU Ming-Wan
Dept. of Eng. Mechanics, Tsinghau University, Beijing 100084, P.R. China
Member of Design Subcommittee of the China National Standards Committee on Pressure Vessels

LI Jian-Guo
China National Standards Committee on Pressure Vessels,
Beijing 100088, P.R. China, P.O. Box 8012

ABSTRACT

A new Chinese design code JB 4732-95, "Steel Pressure Vessels -- Design by Analysis", was issued at March 1995. In this paper we will discuss some concepts and ideas, which formed in the introduction and application of the design by analysis approach in China. They are multi-possibility of stress decomposition, self-limiting stress, classification of constraints, and finally an important concept to distinguish primary stress -- primary structures.

INTRODUCTION

"Design by Analysis" approach, initiated by the ASME Boiler and Pressure Vessel Code, Section III and VIII, Div. 2, is an advanced and widely used design method in the world. Since 1982, the China Pressure Vessels Committee, after July 1984 renamed as the China National Standards Committee on Pressure Vessels (CNSCPV), is devoted to introduce and popularize this advanced approach. A great deal of seminars and courses on design by analysis were organized by CNSCPV. A new Chinese design code JB 4732-95, "Steel Pressure Vessels -- Design by Analysis", was issued at March 1995. In this paper, we will discuss some concepts and ideas, which formed in the introduction and application of the design by analysis approach in China.

Design by analysis approach is a design method based on elastic stress analysis and plastic failure criteria. In the approach stresses are classified as general primary membrane stress, local primary membrane stress, primary bending stress, secondary stress and peak stress. A significant problem in engineering applications of the ASME Code III and VIII-2 is how to decompose a total stress field obtained by the finite element analysis or the experimental stress analysis into above mentioned different stress categories. A lot of papers related to this problem were published, e.g., [2]-[17] and more than 20 papers published in the Chinese journals.

A typical and widely used method is based on the concept of equivalent linearization of elastically calculated stress field of considered structure along several checklines specified by user. Some well-known finite element analysis software, for example ANSYS, MARC, have developed special postprocessor corresponding to the ASME Code. The equivalent linearization method was initiated by Kroenke et al. for the 2-D problems [2-4]. Three approaches for 3-D conditions, stress-at-a-point; stress-along-a-line; and stress-on-a-plane, were compared and discussed in a series of papers written by Hollinger and Hechmer [5-9]. A summary of three dimensional stress classification was given in [9]. In this paper following problems related to the equivalent linearization method will be discussed:

(1) How should membrane and bending stresses obtained by the linearization procedure be classified into primary or secondary stress categories?

(2) Is it possible to get a reasonable stress decomposition only based on a total stress field of the considered original structure?

An alternative was of stress classification, so called reduced modulus method, was originally proposed by Dhalla to estimate the elastic follow-up effect of piping [11]. In this method, elastically calculated stress is partitioned into primary and secondary components by comparing the simulated inelastic response of a structure with ideal models of primary and secondary behavior. An iterative elastic finite element analysis procedure by systematically reducing the elastic modulus of highly loaded elements is needed to simulate the stress redistribution. Roche [12] pointed that the elastically calculated stress is characterized by its intensity Q and its spring effect factor r. When r is small, the stress is secondary; when it is high, it is primary. Some improved methods were also proposed, e.g., the lower bound theorem approach (Marriott [13]), the GLOSS R-Node method (Seshadri & Fernando [14]), the elastic compensation method (Mackenzie & Boyle [15, 16]), and the simplified elastic ASP method (Dhalla [17]).

In this paper, a new method based on the concept of "primary structures' is proposed. This concept is exactly based on the original definition of primary stress. Some typical examples are demonstrated. Some unreasonable stress decomposition schemes are discussed. Preliminary to introduce this concept, we discuss firstly ideas of the multi-possibility of stress decomposition and of the classification of constraints.

MULTI-POSSIBILITY OF STRESS DECOMPOSITION

Stress classification is a key problem of the design by analysis approach. A recommended classification procedure consists of two steps:

Step 1: According to their function and property, stresses are classified as primary stress and self-limiting stress.

Primary stress (P) is a stress which is necessary to satisfy the equilibrium conditions with the external imposed loading. It may also be called load-controlled stress [18] or equilibrium-controlled stress [16]. The basic characteristic of a primary stress is that it is not self-limiting.

Self-limiting stress, including secondary stress and peak stress (Q+F), is a stress which is necessary to satisfy the continuity conditions in the structure or with the external constraint. It may also be called deformation-controlled stress [16,18] or continuity-controlled stress. The basic characteristic of a self-limiting stress is that local yielding and minor distortion can satisfy the continuity conditions which cause the stress to occur and failure form one application of the stress is not to be expected. This property is call "self-limiting".

Step 2: According to their distribution and effective range, primary stress is classified as general primary membrane stress (Pm), local primary membrane stress)PL), primary bending stress (Pb); self-limiting stress is classified as secondary stress (Q) and peak stress (F).

The basic characteristic of a peak stress is that it is localized and self-limiting, therefore it does not cause any noticeable distortion and is objectionable only a possible source of a fatigue crack or a brittle fracture. Stress at the action point of external concentrated loading is not a peak stress, although it is highly localized. Because it is necessary to equilibrate the external imposed loading, therefor is not self-limiting. It should be classified as primary bearing stress and evaluated by ASME III, NB-3227.1.

Following discussion will focus attention on Step 1 of the stress classification procedure.

The most significant problem of stress classification is how to distinguish primary stress. Mistaking primary stress for secondary stress may result in reduced safety margin. If the primary stress is found, self-limiting stress can be directly obtained by subtracting it from the total stress. But this is a more difficult problem, we receive frequently user's inquiry about it.

Answer comes directly from the definition of primary stress [1]:

"Primary stress is any normal stress or a shear stress developed by an imposed loading which is necessary to satisfy the laws of equilibrium of external and internal forces and moments".

This definition means that in terms of the limit analysis theory of plasticity the primary stress field is a statically admissible stress field which has to satisfy the equilibrium conditions with external loading, but need not to regard continuity conditions.

It is well known in plasticity that there exists a lot of statically admissible stress fields for a given structure under specified external loading. Therefore, there also exist a lot of reasonable primary stress fields, so long as they satisfy the equilibrium conditions. The first task of designer is to find reasonable primary stress fields in considered structure, and then to select the best one from them. This is an optimization problem. The optimization criterion is defined by designer on the basis of cost consideration, material supply, fabrication condition, etc..

Some examples to explain the multi-possibility of stress decomposition and a final answer of the significant problem are given later.

CLASSIFICATION OF CONSTRAINTS

According to their function, constraints are classified as essential constraint and redundant constraint.

Essential constraint is a constraint which results in necessary reactions to equilibrate the external imposed loading. If it is removed, the structure becomes unable to bear the loadings. Examples of essential constrain are the saddle support of horizontal vessels, the skirt support of tower etc.. Stresses caused by the reaction of an essential constraint include certainly a primary stress component.

Non-essential constraint is all the constraints except essential ones. In general, non-essential constraints are of benefit for whole structure to bear the external loadings, but some ones may result in higher local stress in the neighborhood of them. For this reason, non-essential constraints can be further classified as favorable constraint and unfavorable constraint.

Favorable constraint is such a constraint, if it is removed, the maximum primary stresses (Pm, PL, or PL + Pb) of reduced structure comparing with original structure will increase, although the reduced one is still able to bear the loadings. In order to get a more economical and light structure, designer should keep the favorable constraint together with essential constraints, and classify the stresses caused by such constraint as primary stresses.

A typical example of favorable constraint was given by Lu and Li [10]. We consider a clamped circular plate under a uniform load, see Figure 1(a). If we remove the circumferential bending constraint, a simply supported circular plate in Figure 1(b) is obtained. On the basis of the elastic plates theory, radial bending moments at the center A and boundary B for a simply supported circular plate are:

$$M_A^s = \frac{3 + \mu}{16} qa^2; \quad M_B^3 = 0 \qquad (1)$$

for a clamped circular plate are:

$$M_A^c = \frac{1 + \mu}{16} qa^2; \; M_A^c = \frac{1}{8} qa^2 \qquad (2)$$

where q is load per unit are, a is radius of plate, and u is the Poisson ratio.

Comparing Eq. (1) and (2), it is clear that the maximum bending moment M_A^s of simply supported plate is greater than one of the clamped plate M_B^c in absolute value. Therefore, the circumferential bending constraint is a helpful constraint, it should be kept. The bending stress caused by M_B^c should be classified as primary stresses rather than secondary stress.

Unfavorable constraint is such a constraint, if it is removed, the maximum primary stresses (Pm, PL, or PL + Pb) of reduced structure comparing with original structure will decrease, and the structure could bear higher loadings. Designer may classify stresses caused by unfavorable constraint as secondary stress, then should remove such constraint when the primary stress field is calculated.

If designer can not classify a constraint, an effective approach is to try removing it and observing its influence. Pay attention to that general speaking the maximum primary stress of reduced primary structure and one of original structure occur at different point in structure.

PRIMARY STRUCTURES

Now we discuss the main problem of this paper - how to find the reasonable primary stress fields in considered structure, and to select the best one from them. At first, we introduce an important concept -- primary structures.

Primary structure is a reduced structure which is obtained by removing the unfavorable non-essential constraints of considered original structure and is able to bear the external imposed loadings. If no unfavorable constraint is removed, the original structure is also a primary structure.

Stress field in a primary structure satisfy the equilibrium conditions with external imposed loadings, therefore it is a reasonable primary stress field. Once a primary structure is found, the problem to find reasonable primary stress field is immediately solved by finite element analysis of the primary structure. Stress field caused by removed unfavorable non-essential constraints is a self-limiting stress (Q + F).

Generally speaking, deformation of a primary structure does not satisfy the continuity conditions. Following ideas related to primary structure should be pointed:

(1) If there exists favorable non-essential constraint, it is sagacious to keep it in primary structure.

(2) The only way to get primary structure is to remove constraint from a considered structure. It is not permitted to attach any favorable non-essential constraint which does not exist in considered structure. If not, the primary stress will be lower-estimated.

(3) All the stresses in primary structure should be evaluated by limits of primary stress intensity. If designer want to specify some higher local stress in primary structure as secondary stress, constraint causing the secondary stress has to be removed and a new reduced primary structure is obtained. Stresses in the new primary structure but not in original one are primary stresses corresponding to the specified secondary stress.

As we know that any stress which is not necessary for equilibrium with external loadings is not classified as primary stress. In calculating of primary stress field from external loadings, all stresses except for primary stress should be omitted. If the constraint causing secondary stress is not removed, secondary stress existing in the calculated structure will help the primary stress to bear external loadings, thus maximum primary stress will be lower-estimated. This is the reason why we introduce the important concept of primary structure.

An usually used approach of stress decomposition is only based on a total stress field of considered original structure. The total stress is decomposed in to membrane stress, linearized bending stress, and nonlinearly distributed stress by using equivalent linearization along a checkline specified by designer. Then designer by experience specify them as primary, secondary or peak stress respectively. As mentioned above that if there exists secondary stress components, maximum primary stress obtained by such way is lower-estimated.

Let us return to the example in Figure 1. We can take the original clamped circular plate as a primary structure and classify the maximum bending stress of M_B^c at the boundary as primary stress. In the case secondary stress is equal to zero. An alternative reasonable decomposition is to classify the bending stress of M_B^c as secondary stress and to take the simply supported circular plate as a primary structure. In the case the maximum primary stress is the bending stress of M_A^s at the center of simply supported plate. It is obvious that primary stress will be lower-estimated, if we, as usually made, classify the bending stress of M_B^c at the center of clamped plate as maximum primary stress and the bending stress of M_B^c as secondary stress. Above discussion comes to the conclusion that the first stress decomposition scheme is reasonable and optimal, the second one is reasonable but conservative, the third one is un-conservative and unavailable.

The second example is a cylindrical vessel with circular plate head, see Figure 2(a). We discuss three reasonable stress decomposition schemes:

(1) Take the original structure in Figure 2(a) as a primary structure. All the stresses in plate and vessel should be evaluated by limits of primary stress intensity.

(2) Classify the bending stress at connection B as secondary stress. Remove the rotation constraint and radial displacement constraint at B, and take the primary structure as a cylindrical vessel in membrane stress state together with a simply supported circular plate, see Figure 2(b).

(3) Classify the bending stress at connection B as secondary stress. Remove only the rotation constraint at B. The primary structure is shown in Figure 2(c). In this case, a lateral shear force occurs at B.

In the usually adopted scheme (2), maximum primary stress in vessel is the lowest, but in circular plate is the highest. The connection of vessel and plate at B is a helpful constraint for the plate, but it is harmful for vessel. If designer, in consideration of material supply or fabrication condition, want to thin off the thickness of plate, scheme (1) may be selected. Scheme (3) is an intermediate one. The radial displacement constraint at B is helpful for plate, and slightly harmful for vessel. If any one of the three schemes can satisfy all limits of stress intensity specified in ASME Code III and VIII-2, the design is surely acceptable.

Two un-conservative and unavailable scheme are as follows:

(1) Classify the bending stress at B as secondary stress, and take the primary structure as a cylindrical vessel in membrane stress state together with a clamped circular plate. In this case, an originally not existed favorable constraint, clamped boundary condition, is imposed on the circular plate.

(2) Consider only the original structure. Classify the bending stress at B as secondary stress, and take the bending stress at center A as the maximum primary stress of circular plate. In this case, the primary stress of plate is lower-estimated. In case of this vessel, the bending moment at B allowed by limit of secondary stress intensity even greater than the plastic hinge moment of the vessel. It is obviously an originally not existed helpful constraint for the circular plate.

The third example is a connection of cylindrical nozzle and conical vessel under internal pressure, see Figure 3. If we classify the bending stress at connection as secondary stress, two reasonable primary structures are shown in Figure 3(a) and Figure 3(b).

(1) Assume the cylindrical nozzle is in membrane stress state. According to the reaction principle, the connection force T1 acting on the conical vessel has to be vertically upward as in Figure 3(a). Decompose it in the directions of the radius of parallel circle and of the meridian of conical vessel, we get Q1 and N1. Force N1 is corresponding to a membrane solution of conical vessel. Force Q1 results in a local primary membranae stress in the vessel.

(2) Assume the conical vessel is in membrane stress state. Now the connection force T2 acting on cylindrical nozzle has to be in the directions of the meridian of conical vessel and downward. Decompose it in horizontal and vertical directions, we get Q2 and N2. Force N2 is corresponding to a membrane solution of cylindrical nozzle. Force Q2 results in a local primary membranae stress in the nozzle.

Many reasonable primary structures can be obtained, if we assume the direction of connection force T is between the T1 and T2. In these cases, there exists local primary membrane stress in both conical vessel and cylindrical nozzle.

If designer wants to thin off the thickness of cylindrical nozzle, scheme (1) may be selected. Otherwise scheme (2) can be used to thin off the thickness of conical vessel.

It should be pointed out that both membranae states of cylindrical nozzle and conical vessel together can not constitute a reasonable primary structure. Because the membranae force T1 of cylindrical nozzle and T2 of conical vessel can not satisfy the reaction principle, i.e., the equilibrium condition. In other words, the local primary membranae stress at connection B is surely necessary to equilibrate the external loading.

CONCLUSION

In this paper, we introduce and explain the following important concepts and ideas:

(1) Primary stress field is a statically admissible stress field which has to satisfy the equilibrium conditions with the external imposed loading, but need not to regard the continuity conditions.

There exist a lot of reasonable primary stress fields. How to find the best primary stress field is an optimization problem. The optimization criterion is defined by designer on the basis of cost consideration, material supply, fabrication condition, etc..

(2) Constraints can be classified as essential constraint, favorable non-essential constraint and unfavorable non-essential constraint. Designer should keep the favorable non-essential constraint together with the essential constraint, and classify the stresses caused by such constraint as primary stresses. Stresses caused by unfavorable non-essential constraint my be classified as secondary stresses, and then such constraint should be removed when the primary stress field is calculated.

(3) Primary structure is a reduced structure which is obtained by removing the unfavorable non-essential constraints of considered original structure and is able

to bear the external imposed loadings. All the stresses in primary structure should be classified as primary stress.

Constraint causing secondary stress has to be removed and a new reduced primary structure is obtained. Stresses in the new primary structure but not in original one are primary stresses corresponding to the specified secondary stress. If there exists secondary stress components, the maximum primary stress which is obtained by usually decomposition approach based only on a total stress field of original structure will be lower-estimated.

REFERENCES

[1] ASME Boiler and Pressure Vessel Code, Section III and VIII, Div. 2, 1989, The American Society of Mechanical Engineers, New York, NY.

[2] Kroenke, W. C., Classification of Finite Element Stresses According to ASME Section III Stress Categories, Pressure Vessels and Piping, Analysis and Computers, ASME, June 1974.

[3] Kroenke, W. C., Addicott, G. W., and Hinton, B. M., Interpretation of Finite Element Stresses According to ASME Section III, Paper 75-PVP-63, ASME.

[4] Kroenke, W. C., et al., Component Evaluation Using the Finite Element Method, Pressure Vessel and Piping Technology - 1985 - A Decade of Progress, ASME.

[5] Hollinger, G. L. and Hechmer, J. L., Three-dimensional Stress Criteria - A Weak Link in Vessel Design and Analysis, PVP-Vol. 109, pp. 9-14, ASME, July 1986.

[6] Hechmer, J. L. and Hollinger, G. L., Three-dimensional Stress Criteria - Application of Code Rules, PVP-Vol. 120, ASME, July 1987.

[7] Hechmer, J. L. and Hollinger, G. L., Code Evaluation of 3D Stress on a Plane, PVP-Vol. 161, ASME, July 1989.

[8] Hechmer, J. L. and Hollinger, G. L., The ASME Code and 3D Stress Evaluation, Journal of Pressure Vessel Technology, Vol. 113, pp. 481-487, 1991.

[9] Hollinger, G. L., Summary of Three Dimensional Stress Classification, 7.ICPVT, Vol. 1, 1992.

[10] Lu, M. W. and Li, J. G., Chinese PV Code and the ASME Code - Remarks on Design by Analysis and Role of Stress Categories, PVP-Vol. 109, pp.33-37, ASME, July 1986.

[11] Dhalla, A. K., Verification of an Elastic Procedure to Estimate Follow-up, PVP-Vol. 86, pp. 81-86, ASME, 1984.

[12] Roche, R. L., Modes of Failure - Primary and Secondary Stresses, PVP-Vol. 120, ASME, July 1987.

[13] Marriott, D. L., Evaluation of Deformation or Load Control of Stresses Under Inelastic Conditions Using Finite Element Stress Analysis, PVP-Vol. 136, pp. 3-9, ASME, 1988.

[14] Seshadri, R. and Fernando, C.P.D., Limit Loads of Mechanical Components and Structures Using the GLOSS R-Node Method, PVP-Vol. 210-2, pp. 125-134, ASME, 1991.

[15] Mackenzie, D. and Boyle, J. T., Assessment of Classification Procedures for Finite Element Stresses, 7.1 CPVT, Vol. 1, 1992.

[16] Mackenzie, D. and Boyle, J. T., A Computational Procedure for Calculating Primary Stress for the ASME B&PV Code, Journal of Pressure Vessel Technology, Vol. 116, pp. 339-344, 1994.

[17] Dhalla, A. K., Stress classification for Elevated Temperature Service, Journal of Pressure Vessel Technology, Vol. 113, pp. 488-496, 1991.

[18] ASME Boiler and Pressure Vessel Code, Case N-47-28, Class 1 Components in Elevated Temperature Service, Section III, Div. 1, 1988, The American Society of Mechanical Engineers, New York, NY.

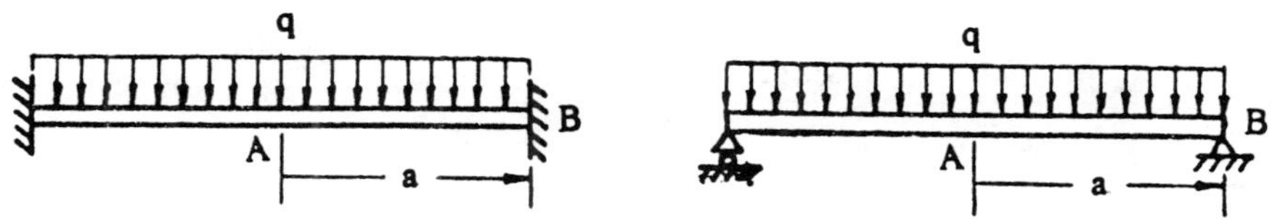

(a) Clamped circular plate (b) Simply supported circular plate

Fig.1 Helpful Constraint

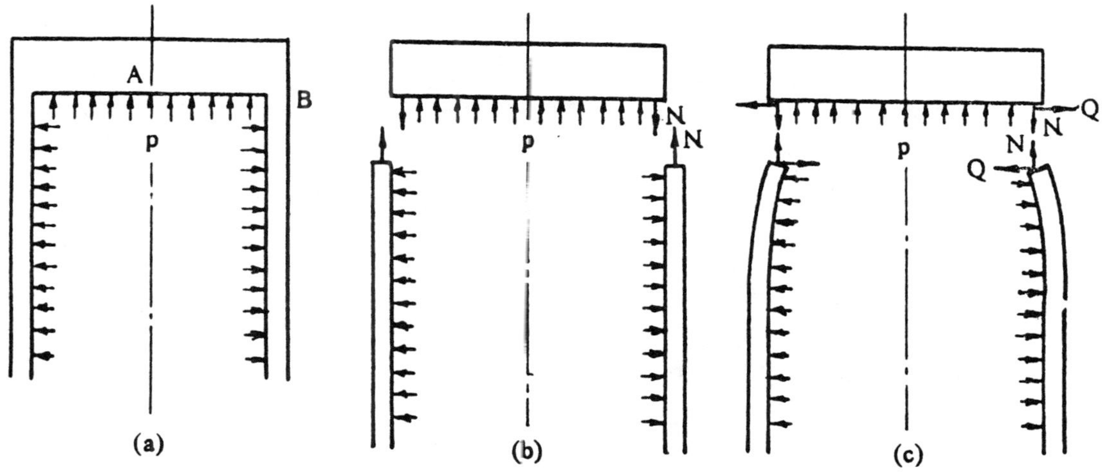

Fig.2 Primary Structure of Cylindrical Vessel with Circular Plate Head

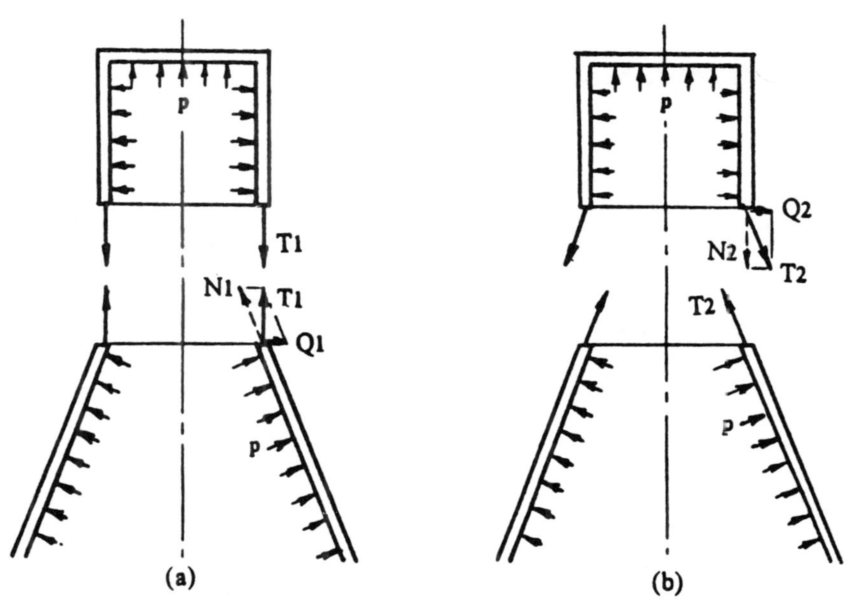

Fig.3 Primary Structure of Connection of Cylindrical Nozzle and Conical Vessel

AUTHOR INDEX

PVP-Vol. 340
Seismic Engineering — 1996

Book Number: H01068